CURRENT SOURCES & VOLTAGE REFERENCES

CURRENT SOURCES & VOLTAGE REFERENCES

Linden T. Harrison

ELSEVIER

AMSTERDAM • BOSTON • HEIDELBERG • LONDON
NEW YORK • OXFORD • PARIS • SAN DIEGO
SAN FRANCISCO • SINGAPORE • SYDNEY • TOKYO
Newnes is an imprint of Elsevier

Newnes

Newnes is an imprint of Elsevier
30 Corporate Drive, Suite 400, Burlington, MA 01803, USA
Linacre House, Jordan Hill, Oxford OX2 8DP, UK

Recognizing the importance of preserving what has been written, Elsevier prints its books on acid-free paper whenever possible.

Library of Congress Cataloging-in-Publication Data
Application Submitted.

British Library Cataloguing-in-Publication Data
A catalogue record for this book is available from the British Library.

ISBN: 0-7506-7752-X

For information on all Newnes publications
visit our Web site at www.books.elsevier.com

Transferred to Digital Printing 2009

DEDICATION

This book is dedicated to the memory of these eight outstanding pioneers, now passed away, but whose contributions to the semiconductor industry have been immeasurable:

Robert J. Widlar
Legendary analog IC designer and engineer. Created the first monolithic op amps, bandgap voltage references, current sources, and linear regulators. In 1981 co-founded Linear Technology Corp. with friends Robert Dobkin, George Erdi, and Robert Swanson.

David Talbert
Widlar's processing specialist at both Fairchild Semiconductor, then later at National Semiconductor. He transformed Widlar's revolutionary designs into real silicon devices, and created the first "super beta" transistors that are used in bipolar analog ICs.

Dr. Robert Noyce
Physicist, and co-inventor of the IC (planar design) in 1964. Created several types of alloy transistors. Co-founded Fairchild Semiconductor, Intel, Sematech, and the Semiconductor Industry Association. Previously worked with Nobel Laureate, Dr. Shockley. Oversaw the development of Intel's earliest microprocessor and memory products.

Dr. Jean Hoerni
Swiss-born physicist, mathematician, and semiconductor researcher. Invented the all-important Planar™ process (1959), and created some of the first silicon BJTs and JFETs. Previously worked with Nobel Laureate, Dr. Shockley. Co-founder of Fairchild Semiconductor, Teledyne Amelco, Union Carbide Electronics, and Intersil.

Dr. Willis Adcock
Chemist, and semiconductor researcher. In 1954 he created the industry's first silicon-based semiconductor material at Texas Instruments Inc., who then introduced the world's first silicon transistors. Dr. Adcock later recruited Jack Kilby, who went on to co-invent the IC (mesa design), and receive the Nobel Prize for physics in 2000.

Dr. Karl Lark-Horovitz
Austrian-born chemist, physicist, and materials researcher. Led the Physics Dept. at Purdue University for several decades. It was his work with germanium during WWII, that enabled making the most durable rectifiers available, for military radar systems.

Dr. Russell Ohl
Chemist, materials researcher, and HF radio pioneer at AT&T Bell Labs. Discovered the PN junction, and created the first silicon diodes in 1940. Invented the modern solar cell.

Dr. Julius E. Lilienfeld
German-born physicist, researcher, and US Patent-holder (1930), who laid the foundations for the voltage-controlled FET. Also invented the electrolytic capacitor.

"Tall oaks from little acorns grow"
David Everett (1769 - 1813)

Contents

List of Figures and Photos

List of Tables

Acknowledgements

The author would like to acknowledge the following individuals for their help and advice in supplying information for this book:

Robert Dobkin:
Co-founder and CTO—**Linear Technology Corporation**
Inventor of the buried-zener voltage reference
Inventor of the monolithic temperature sensor
Co-inventor of the first commercial bandgap voltage reference
Creator of the first high-speed monolithic op amp
Creator of the first adjustable linear voltage regulator

Professor Thomas H. Lee:
Stanford Microwave ICs Laboratory
Department of Electrical Engineering
Stanford University

Robert A. Pease—Staff Scientist, and renowned columnist and guru
Naomi Mitchell—Senior Manager—Marketing Communications
National Semiconductor Corporation

Lisa Wong—Marketing Manager—Voltage References
Scott Wayne—Technical Communications Manager
Analog Devices, Inc.

Joe Neubauer—Technical Business Manager SP & C
John Van Zand—Publicity Manager
Maxim Integrated Products, Inc.

Carlos Laber—VP, Engineering
John M. Caruso—VP, Engineering
Andy Jenkins —Director of Marketing, Precision Analog Products
Jules Farago—Product Marketing Manager
Intersil Corporation

Robert Chao—President and Chief Engineer
John Skurla—Director of Marketing
Advanced Linear Devices, Inc.

Dr. Michael Wyatt
Honeywell Fellow and researcher
Space Systems Divn.
Honeywell Corporation

Don Howland—Director of Marketing
Linear Integrated Systems, Inc.

Dr. Mike Chudobiak
Avtech Electrosystems Co.

David L. Anderson
Snr. Applications Engineer
Caddock Electronics, Inc.

Vernon Bluhm
Former transistor researcher
(Syracuse, NY)
General Electric Company

In addition, the author would like to thank the following individuals for their great help and guidance with this book:

Harry Helms—my Editor at Elsevier Science & Technology
Don Snodgrass—who created the book's original master CD
George Morrison—Senior Project Manager, Elsevier Science & Technology
Tara Isaacs—Senior Marketing Manager, Elsevier Science & Technology
Lori Koch—Promotions Coordinator, Elsevier Science & Technology
Mona Buehler—Pre-Production, Elsevier Science & Technology

Very special thanks go to:

Alan Rose, Tim Donar, and Ginjer Clarke at Multiscience Press, Inc., New York for transforming my original Mac manuscript into this fine book.

Thank you all so much !

Linden Harrison
lindenh248@aol.com

Chapter 1

A Short History of References

1.1 Introduction

Current sources and voltage references both depend on inherent characteristics of the transistor, either the bipolar junction transistor (BJT) or the field-effect transistor (FET), in order to operate properly. It seems only fitting then, in beginning this book, that we first take a brief trip back in time, and trace a history of how and when some of these products originated.

Actually, current sources and voltage references both predate the integrated circuit by several decades. Current sources were originally created by simply using resistors or by using resistors together with vacuum tubes. However, their combination did not provide much accuracy or stability. In the 1940s and '50s, it was common to see vacuum tube voltmeters (VTVMs) and other instrumentation based on the tube. Before being used within integrated circuits, or designed as commercial IC products, voltage references took the form of bulky and expensive laboratory standards. These included the *Weston cell*, the Clark cell, as well as some types of batteries. Those types of laboratory standards were used for decades. The best-known standard cell was the Weston cell, which produced a constant voltage output of 1.019 volts and was virtually independent of temperature change. The Weston cell utilized an H-shaped glass container and was powered by unique chemical actions. Great care was needed not to tip or knock over the cell or to load the output with any appreciable current, because it would temporarily cease operation. Recovery time could take days or even weeks. Both types of cells produced stable voltages, though, accurate to a few parts per million (ppm) or better. Modern Weston cells require being held in a temperature-controlled bath. Mercury cells, originally developed during World War II, were also used as voltage references, mostly because they were small, cheap, and immune to the physical problems of the fragile glass standard cells. Mercury cells provided an output of 1.35 volts at several milliamps (mA) and for more than 1,000 hours, but had much lower accuracy and a higher temperature coefficient. All of these types have been mostly superseded by semiconductor voltage references.

In modern times, the first milestone reached for a semiconductor-based reference was the introduction of the *zener diode*, which serves as a voltage reference. This was created in the late 1950s by its inventor Dr. Clarence Zener, a researcher at Westinghouse Electric Company. Operation of the zener diode relies on some unique characteristics of a reverse-biased P-N junction and is usually operated so that it is in parallel (shunt) with the load. It can be produced by one of several techniques, the most common being a planar epitaxial method. Each zener diode's particular specifications are mostly created by small differences in processing. Because it is a small single component, it is still a popular workhorse today in many industrial and commercial designs, because of its reasonable accuracy, ease of use, and low cost. The downside of the zener was that it was temperature-sensitive, drifted, was noisy, and above all required more than 6 volts to function. Low-voltage zeners were deemed to be too unpredictable to use in any precision circuitry at the time.

As electronic instrumentation gradually became more sophisticated in the early 1960s, some instrument designs called for an internal calibrated reference, which would enable the instrument to calibrate itself. At first this took the form of using mercury cell batteries, but then gradually, small discrete transistor-diode-zener circuits were used, often housed in sealed plug-in modules. Along the way, engineers discovered that the normal positive temperature coefficient of the zener could be compensated by the negative temperature coefficient of one or more series-connected rectifier diodes. They found that by burning in, characterizing, and selecting zeners with a nominal value of 5.6 volts, and adding a rectifier diode (having a forward voltage of approximately 0.7 volts) in series, it would create a *temperature-compensated (TC) zener* reference. However, it had to be run at a particular current of 7.5 mA, for which it was specified. Burn-in time could be hundreds of hours or more. The longer the burn-in time and the more characterization, the higher the cost. At the time, these specially created TC zeners were the most accurate devices available. Attempts at using zeners of different voltages did not produce as low a tempco as this TC zener combination. While the temperature coefficient of the TC zener was very low (i.e., 1N827, 0.001%/°C), the combined voltage was typically around 6.3 volts, and so required a power supply of at least 7 volts or higher to function properly. That was a serious limitation for any low-voltage designs.

The year 1958 was memorable for many reasons. This was the year that National Semiconductor was originally founded, that *photolithography* was invented at Bell Labs, that Dr. Jean Hoerni started developing his *planar* transistor prototypes at Fairchild Semiconductor, and that a small semiconductor manufacturer called Crystalonics Inc. was founded in Cambridge, Massachusetts, by two former Raytheon engineers, and soon began making transistors. A year or two later they began to make JFETs, which at the time had only been made at Fairchild Semiconductor's development lab in California (again by Dr. Jean Hoerni). Crystalonics was first in the industry to put N-channel JFETs into production and pioneered many of the original JFET designs and applications, including *current regulator diodes* (CRDs), as shown in Figure 1.1B.

The CRD is a small single component, made by adding (or diffusing) a resistor internally between a JFET's gate and source terminals. This establishes a fixed voltage between the gate and source, so that a constant current flows between the drain and the source, irrespective of changes in supply voltage. For many engineers, the CRD is

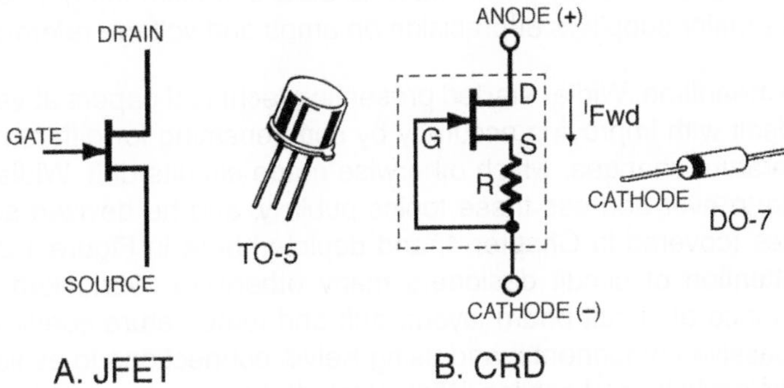

Figure 1.1. Crysalonics and Siliconix pioneered some of the industry's earliest JFETs in the early 1960s, as well as making current regulator diodes (CRDs), as shown here.

considered to be equivalent to the zener diode, except that it provides a constant current rather than a constant voltage.

1.2 The first JFETs and op amps

Another major contributor to both JFET and CRD technology was Siliconix, which was founded in 1962 by several former Texas Instruments' engineers, and a former Bell Labs researcher. During the first few years of its existence, Siliconix created and developed JFETs, in which it became the undisputed market leader. Siliconix also developed several product families of CRDs, as well as later pioneering the power MOSFET.

Another important milestone was reached in 1964, when the first commercial monolithic op amp (Fairchild Semiconductor's μA702) was introduced. It had been designed by the legendary Robert J. Widlar. (This analog breakthrough was partly a result of Widlar's use of his precision current source references, which he used for biasing the various amplifier stages, and five years later would lead to the creation of the monolithic voltage reference.) The μA702 was followed about a year later by the μA709, which Fairchild introduced in November 1965. The μA709 became an industry standard, although it required a resistor and two capacitors for compensation, and cost around $50 apiece. The other side of the story was that it was hard to compensate and harder still to manufacture, because it had a poor yield. Fairchild could barely keep up with demand for this op amp, though, and had back orders for several years after it was introduced. Two people at Fairchild Semiconductor were teamed with the 26-year-old Widlar: David Talbert, his processing specialist who actually made Widlar's designs in silicon and who invented the *super-beta* transistors that the designs often required, and Jack Gifford, the 24-year-old product manager. He was one of the few people who understood Widlar, amplifiers, feedback and control theory, and whom Widlar got along with. Gifford helped make these early op amps such a huge success, introduced the *DIL package* to the industry, and laid the foundations for today's analog

IC marketplace. (Today Jack Gifford is CEO of Maxim Integrated Products, one of the world's major suppliers of precision op amps and voltage references.)

In the meantime, Widlar started presenting technical papers at various industry forums that dealt with improving accuracy by compensating for differences in beta, V_{BE}, and temperature changes, which otherwise made circuits drift. Widlar was one of the first people to ever address these topics publicly, and he devised several unique current sources (covered in Chapter 4, and depicted here in Figure 1.2). He also brought to the attention of circuit designers many otherwise overlooked factors, such as the importance of circuit board layout, drift and temperature coefficient, the use of precision passive components, and using Kelvin connections to avoid voltage drops in PC boards, sockets, and cabling. Widlar included current sources in all his IC designs as a means of biasing and compensation.

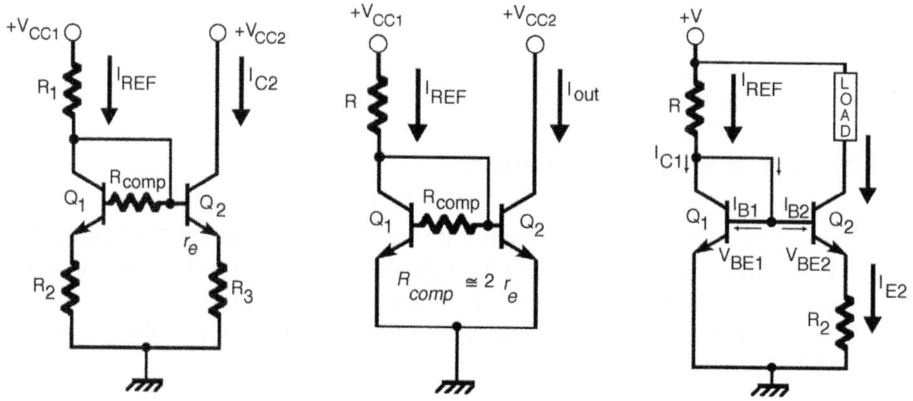

Figure 1.2. Some examples of Widlar's current sources. These were designed to compensate for differences in the betas and V_{BE}'s of the transistors.

In 1968 another Fairchild designer, Dave Fullagar (originally from the United Kingdom, with a degree from Cambridge University), while at Fairchild's Mountain View facility, designed the legendary bipolar op amp—the μA741. It was reliable, and better still, it required *no* compensation. Above all, it was much easier to manufacture and had good yields. In the early 1970s, designers began using it to buffer the TC zener diode, to provide a more stable reference voltage, as shown in Figure 1.3A. Equally important, the op amp also provided an easy way for circuit designers to create precision current sources with low values. It was found that by combining a zener diode with precision resistors and an op amp, one could create a stable current source, as shown in Figure 1.3B. More about this topic is covered in Chapter 11.

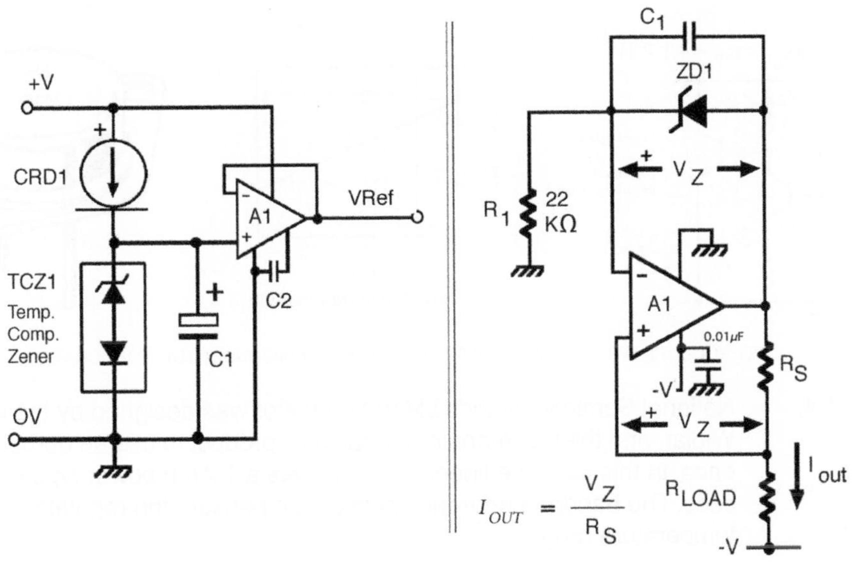

A. A CRD + TC zener + op amp voltage reference.

B. A zener + op amp current source.

Figure 1.3. Precision reference circuits from the mid-1970s.

1.3 The first bandgaps

Another milestone was reached in 1969 when Bob Widlar, then at National Semiconductor, became the first analog IC designer to create an integrated voltage reference (based on the bandgap principle he had conceived), as part of a regulator IC design, the LM109, shown in Figure 1.4. This was the first high-power monolithic linear regulator. It proved that it was possible to build such a device, which was temperature-compensated by a precision reference, on *one* monolithic chip, despite significant changes in chip temperature. Like all of Widlar's (bipolar) designs, it included current sources in the circuitry in order to establish correct bias levels. Yet another milestone was reached in 1971, when Widlar, together with his friend and fellow designer Bob Dobkin, co-designed the industry's first commercially available monolithic *bandgap shunt reference*—National Semiconductor's LM113.

With the advent of the first monolithic A/D and D/A converters in the early 1970s, much development focused on creating monolithic voltage references, as well as on improving overall precision. Besides National Semiconductor, other early innovators included Texas Instruments, Fairchild, RCA, Analog Devices, and Intersil. Some of the early pioneering work was shared on an industry-wide basis in the form of published technical papers. Technical articles also appeared in the electronics industry's leading

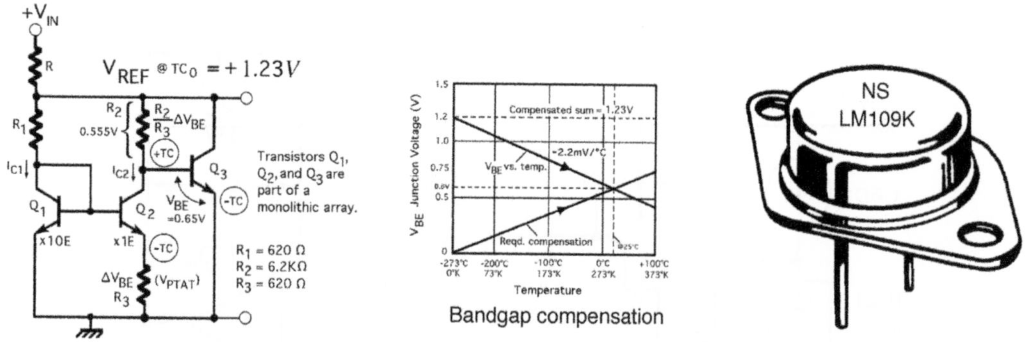

Figure 1.4. National Semiconductor's LM109 regulator was designed by the legendary Bob Widlar, and the first monolithic analog IC product to use an on-board voltage reference. In this case the linear regulator was a 1-Amp power type in a steel TO-3 case. The bandgap reference helped compensate the regulator over its operating temperature range.

magazines, as well as in manufacturers' own in-house application notes. Up until this point, these first-generation voltage reference ICs were marketed and referred to as though they were (improved) zener diodes, which were the established industry standard at the time.

At about this time, a process called *zener zapping* was introduced, which would impact all future precision monolithic analog circuitry, including today's voltage references. It was designed by George Erdi at Precision Monolithics, while he was designing the first precision op amp, the OP07. It coupled his op amp design with a technique he developed for adjusting the op amp's input offset voltage (V_{OS}) to a very low level, by using a computer-controlled laser to short a string of parallel-connected resistors and zeners. This trimming technique effectively shorts the zener to obtain a precise voltage level. Erdi's ultra-stable OP07 op amp brought with it such a major improvement in precision that it became the industry standard and is still a steady production item to this day. George Erdi later left Precision Monolithics and in 1981 co-founded Linear Technology Inc. The technique he created has since been modified and refined and applied to many other products besides op amps, but it is still used throughout the analog semiconductor industry. Many of today's voltage references are laser-trimmed at the wafer-sort level, then trimmed again after the die has been mounted into its package. Precision trimming is applied to certain internal monolithic thin-film resistors, which helps enhance certain features of the device, such as its initial accuracy, drift, and tempco curve-correction circuitry.

1.4 The buried-zener debuts

In the mid-1970s, another milestone was reached when National Semiconductor introduced the LM199, the first *buried-zener* monolithic voltage reference, shown in Figure 1.5. This legendary product, designed by Robert Dobkin, offered a 6.95-volt reference, with a 0.3-ppm/°C temperature drift and a noise spec of about $7\mu V_{rms}$ (10Hz to 10KHz). This was better than anything at the time and still better than most devices are even today. Uniquely, the device included an on-chip substrate heater for stabilizing the chip's temperature, which helped provide it with great accuracy and

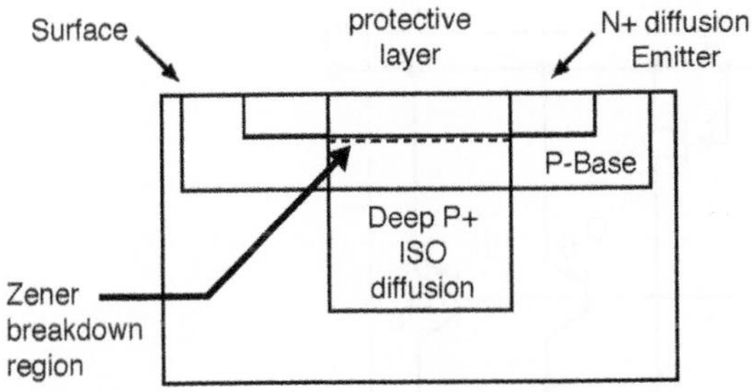

Figure 1.5. Robert Dobkin's novel buried-zener structure was revolutionary, because it provided ultra-low noise, a 0.3ppm/°C tempco, and greatly improved stability.

the ultra-low tempco. Another important milestone was reached in the late 1970s, when Paul Brokaw at Analog Devices created the first series (three-terminal) precision bandgap reference (the AD580), based on what is referred to today as the *Brokaw cell.* This product was destined to become one of the most successful voltage reference ICs ever introduced and ushered in a new level of precision for voltage reference ICs. Up until then, voltage references were either 6.9-volt or 10-volt (buried-zener) types, and the remainder were 1.2-volt shunt bandgap devices. The introduction of a 5-volt voltage reference that had a separate input and output terminal like a linear regulator, as well as not requiring any input resistor, was another remarkable achievement.

In the late 1970s, National Semiconductor addressed the needs of instrumentation designers by introducing the LM134 family of current source ICs. These were dedicated three-terminal adjustable bipolar devices and had been designed by another outstanding National designer, Carl Nelson. For the first time, a *monolithic current source* IC was available, and it quickly became a favorite design-in of instrumentation and other designers of precision circuitry. The versatile LM134 provided a current source that was adjustable over a practical range of 1 μA to 5 mA and with an operating voltage range of between 1 to 40 volts. It offered excellent current regulation and the ability to create a true floating current source, so that it could be used as either current source or a current sink. Figure 1.6 shows a simplified view of the LM134's internal circuitry, which incorporates both NPN and PNP bipolar transistors, as well as some very-low-voltage P-channel JFETs, and an integrated capacitor. The JFETs are used for start-up biasing and enable the internal BJT current mirrors. Uniquely, the device also doubled as a linear temperature sensor, which made it even more attractive at the time. Carl Nelson later joined Linear Technology, once again working for his former boss, Bob Dobkin. The LM134 family is still in production today and available from both National Semiconductor and Linear Technology. More on this product family can be found in Chapter 10.

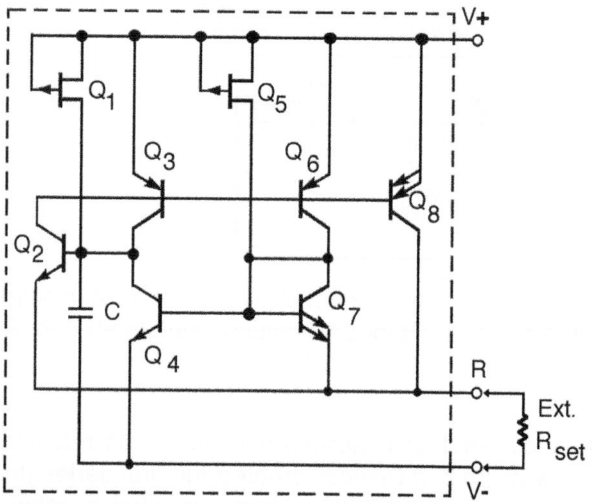

Reproduced with permission from National Semiconductor Corp.

Figure 1.6. LM134 monolithic adjustable current source. A simplified view of its internal circuitry.

1.5 Advancements in technology

Over the next few years, other analog semiconductor companies looked at getting into the voltage reference market and followed National Semiconductor and Analog Devices by introducing some outstanding products of their own. Most of them also incorporated the laser-trimming technique developed at Precision Monolithics, which enabled them to provide devices with enhanced precision. These included Precision Monolithics (now part of Analog Devices), with its 10-volt REF01 and 5-volt REF02 series bandgap products; Burr-Brown (now part of Texas Instruments), with its 10-volt buried-zener REF102; Linear Technology's LTZ1000 (an exceptional 7-volt buried-zener designed by Robert Dobkin) and its LT1021 (a buried-zener family designed by Carl Nelson); Maxim's innovative MAX676 family of bandgap products, with an on-board temperature sensor, a temperature-correction ROM, and output Kelvin connections that provided a tempco of 0.6-ppm/°C; and National Semiconductor again with its low-cost bandgap devices LM185 family (previously designed by Bob Dobkin).

While current source references used either the LM134, dual/quad matched BJTs, or JFET CRDs, in 1990 Burr-Brown introduced its remarkable REF200. This unique device contained two fixed 100-μA current sources, as well as a precision current mirror, all packaged in an eight-pin *surface-mount* SOIC package. It used a proprietary Burr-Brown dielectrically isolated (Difet®) process, which completely isolated the three circuits, making them independent of one another. Each of the 100-μA current sources used a precision bandgap cell to provide a near-zero tempco, while the current mirror used the reliable Full Wilson architecture. Each of the circuits was laser-trimmed at the wafer level to provide the highest accuracy. The REF-200 could be pin-strapped to provide currents of 50 μA, 100 μA, 200 μA, 300 μA, and 400 μA. Or by adding external circuitry, one could create virtually any current, either smaller or greater than these. (This is covered fully in Chapter 10.)

During the 1990s, the surface-mount package became widely used, so that most analog products, including the current source ICs mentioned, most voltage references, as well as discretes like transistors, JFETs, CRDs, and zener diodes were available in this form of package. Pioneers of this were National Semiconductor, Maxim, TI, Analog Devices, and others. Until then, references were available in the dual-in-line ceramic or plastic package. Today most references, including many mature products, are only available in surface-mount packages. The most commonly used surface-mount package for a voltage reference is the SOIC-8.

Another factor that started to emerge in the late 1980s and throughout the '90s was that as A/D and D/A converters evolved, some of those products included their own on-board reference. However, in every case, the on-board reference was a bandgap type, which is inherently limited to about 12-bit resolution, even though the converter may be capable of much higher resolution. It is fairly straightforward for the manufacturer to build the bandgap on the same die as the converter, but it is mostly a user convenience enabling a design to get off the ground faster. To support higher bit resolutions (e.g., 14-, 16-, 20-bit) and lower noise levels, an external precision series voltage reference is recommended. Of course, the chosen converter must have the option of using an external reference, and of having its own internal reference turned off.

1.6 Other topologies emerge

By the mid-1990s, the existing types of voltage reference available were shunt and series bandgaps and variations of the buried-zener. In the late '90s, Analog Devices developed and introduced the first generation of its XFET™ voltage reference, a topology (type) based partly on the characteristics of the JFET. It had been about two decades since a completely new voltage reference topology had been introduced. In fact, Analog Devices had introduced the series bandgap in the form of its AD580.

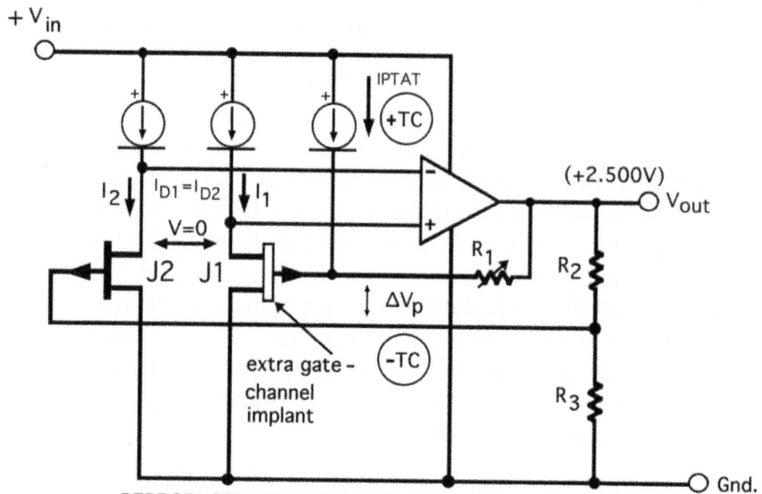

Figure 1.7. Analog Devices' XFET™ technology provides a low-cost, but very high-performance reference.

The XFET, shown in Figure 1.7, was purposely designed to get around many of the limitations of the bandgap and buried-zener types. These had noticeably begun to impact A/D and D/A converter systems, whose operating voltages were increasingly headed below 5 volts and whose increased resolutions depended on low-noise precision references. The main limitations of those references included operating voltage, quiescent current/power dissipation, noise, and nonlinear temperature coefficients. Even the best bandgaps and buried-zener products still have nonlinear tempcos (something that is inherent in their design), particularly at the extremes of their temperature ranges. (Laser trimming the reference's internal resistors helps compensate for that.) By contrast, the inexpensive XFET provided a low linear tempco and allowed for lower voltage operation, over a wider operating temperature range (automotive versus industrial and commercial) than the others. It also featured a lower noise level, a lower quiescent current, lower thermal hysteresis, and a very low long-term drift. Two further generations of the XFET have since followed, as the product evolves with the ever-changing marketplace.

Early in the 21st century, a new type of bandgap emerged and is referred to in this book as the *super-bandgap*. The main difference is that it is built with more advanced technology, including processing, which all results in a series bandgap product with many of the best features of the buried-zener. Super-bandgaps can be seen to have three key features: (1) excellent initial accuracy, (2) a very low temperature coefficient (tempco), and (3) an ultra-low noise level. Several manufacturers make the super-bandgap type of product. Some examples include National Semiconductor's LM4140 family, built with a proprietary CMOS process (±0.1% initial accuracy, 3 ppm/°C tempco, and less than 2.5-μV pk-pk noise voltage); Linear Technology's LT1461 (±0.04% initial accuracy, less than 3 ppm/°C tempco, and a typical 8-μV pk-pk noise voltage); Analog Devices' AD780 (±0.04% initial accuracy, less than 3 ppm/°C tempco, and a 4-μV pk-pk noise voltage); Maxim's BiCMOS MAX6126 family (±0.02% initial accuracy, 3 ppm/°C tempco, and a 1.3-μV pk-pk noise voltage), and also Maxim's MAX6325 family (±0.02% initial accuracy, 1 ppm/°C maximum tempco, and a 1.5-μV pk-pk noise voltage). The super-bandgap type is certainly a major step forward for this particular topology, and no doubt more of these products will be introduced soon.

More recently, in 2003, Xicor Corporation (now part of Intersil Corporation) announced a completely new type of series voltage reference, using its proprietary Floating Gate Analog (FGA™) technology. This new CMOS-based topology offers the lowest quiescent current of any voltage reference—less than 0.8 μA. It also has other characteristics that challenge some of those of the buried-zener, as well as having voltage options available of between 1.25 and 5 volts. It is a cousin of the EEPROM memory, which provides nonvolatile storage of digital data, but Xicor's FGA is a far more complex component, storing analog voltage levels for more than 10 years. The unique characteristics of the FGA translates into an exceptional voltage reference, which can be used in up to 24-bit systems. Other analog products based on this exciting technology are future possibilities.

In summary, the monolithic current source and the monolithic voltage reference can both be traced back to the designs of Bob Widlar, who first implemented them. For many current source designs at the circuit board level, it is often necessary to create one's own, unless the desired current level is within the range of the few dedicated IC products available. Several chapters in this book will be helpful in that regard and deal with creating good current sources using different types of transistors (e.g., BJTs, JFETS, MOSFETs) or combinations of them. One area that appears to have great potential for the development of small precision current sources is with the new EPAD® and ETRIM™ families of *matched* MOS transistors from Advanced Linear Devices, Inc. These are both exciting new technologies that can be directly applied to current sources.

As for the voltage reference IC, today there is a convergence in the marketplace as the best of each topology attempts to compete with some of the most exotic specifications first established by the buried-zener, a topology invented by Bob Dobkin, one of the world's greatest analog IC designers. Generally, the other topologies are becoming more accurate, irrespective of whether they are shunt or series types. Over the past decade, the constant need to support higher-resolution A/D and D/A converters in a myriad of digitally based products has fueled the growth and improvement of the precision monolithic voltage reference. The 1 ppm/°C tempco specification over a product's temperature range has become an industry-wide goal for series references. Besides a few buried-zener products, only one or two *super-bandgap* products and a 5-volt version of the FGA™ have yet broken through that challenging barrier. While the buried-zener has already reached beyond that barrier and peaked as a topology, further advances in the other topologies will come in time. Designing voltage references is not an easy matter, though, because characteristics are often interlinked. For example, as one reduces the device's current drain, its noise level usually increases or its operating voltage range is adversely affected. Improvements will happen, but as a combination of advancements in processing, design, packaging, and thermal management. During the past five years, the industry has witnessed the arrival of two completely new topologies (XFET™ and FGA™) and an existing one has been enhanced (super-bandgap). This is an exciting time in the evolution of electronic references and seeing how they will be applied in the products of tomorrow's world.

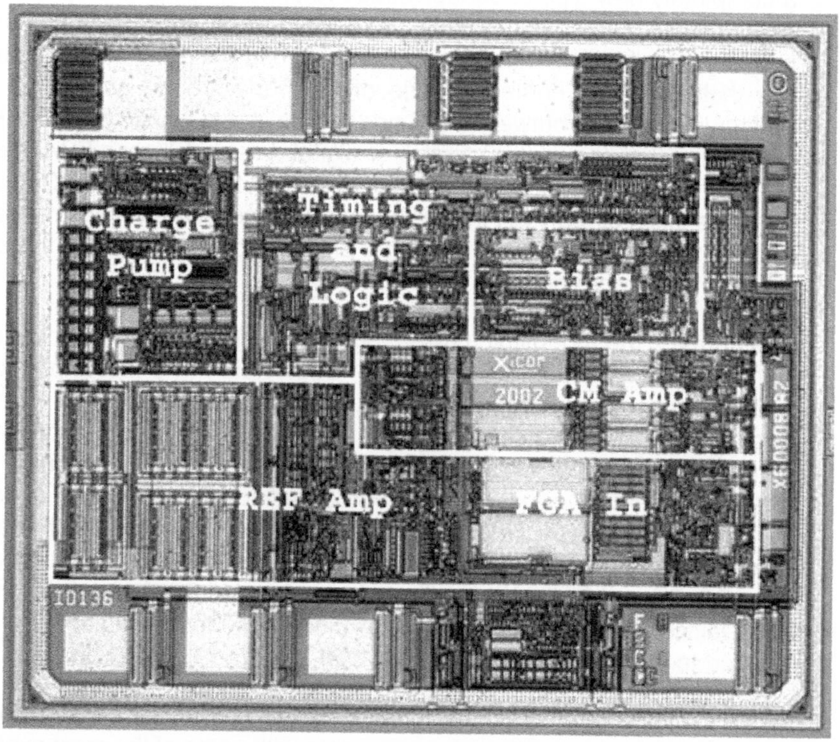

Photo 1.1. A photomicrograph of the Intersil x60008 die. This is the world's newest and most technically advanced voltage reference. (Photo courtesy of Intersil Corporation).

Chapter 2

An Introduction to Current Sources

2.1 An overview

Current sources are basic electronic building blocks that are used extensively in the architectures of analog ICs, as well as in OEM circuit board designs. In both cases, current sources are created by combining diodes, resistors, and transistors (BJTs or FETs). They can also be created at the circuit board level by using discretes, matched pairs, transistor arrays, or by combining op amps with precision voltage references. The various techniques for doing so will be reviewed in Part 1 of this book. Although a few dedicated monolithic current source ICs are commercially available, it is often necessary to create one's own circuit to match the particular needs of the application.

Although most forms of today's instrumentation use either voltage or current references, the former are far more available. As a result, designers frequently use voltage references together with precision resistors, so that a stable reference voltage is converted into a precise current. Applications for current sources range from biasing and stabilization to reference and linearizing. For example, in the design of an op amp, the IC designer will use current sources to create active loads for the amplifier stages and to establish precise bias levels. By providing a constant current, this forces amplifier stages to stay at the Q-points within their active linear regions (see Figure 2.1). In a circuit board design, a current source may be used for linearly charging a capacitor with a constant current, as in a precision timing circuit or in a peak detector. In a medical instrument application, a sensor and a low-noise, front-end amplifier could be biased using precision current sources to assist in recovering very-low-level signals.

The advantages of using these building blocks is their inherent constant current outputs, which are mostly independent of changes in supply voltage, temperature, load resistance, or load voltage. One could liken the current source to that of a precision current regulator. These advantages, when compared with using a simple fixed resistor load, include the following:

- Greater precision
- Better repeatability
- Improved temperature stability

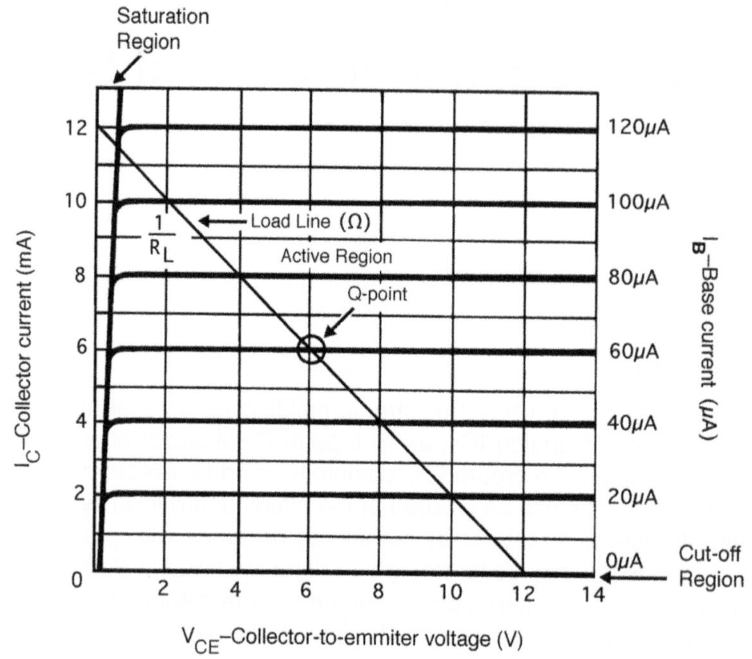

Figure 2.1. *A typical transistor amplifier's characteristics showing the Q-point, in the center of its active region.*

- Lower long-term drift
- Higher output impedance
- Increased bandwidth
- Larger signal range

Current sources are not a new innovation. They predate the integrated circuit by at least a couple of decades. Before their implementation in integrated circuits, they were used in vacuum tube–based circuits (triodes and pentodes). When both NPN and PNP silicon transistors became readily available during the 1960s, analog designers were able to build current sources that connected to either a positive or negative supply rail. Then when the silicon bipolar IC became a practical reality in the late 1960s, the current source had already become an integral part of the internal architecture, for the purposes of biasing and stability. Much development focused on creating current sources for various types of monolithic analog ICs between the mid-60s and the mid-80s, although it still continues today, but at a slower pace. Early innovators included Philbrick, Texas Instruments, Fairchild Semiconductor, National Semiconductor, GE,

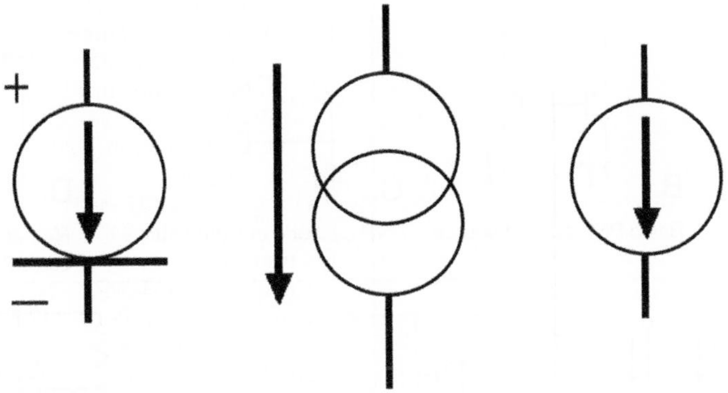

Figure 2.2. *Common current source symbols.*

Analog Devices, RCA, and Motorola Semiconductors, among many others. Some of the early pioneering work was shared on an industry-wide basis (a small fledgling industry at the time), in the form of published technical papers such as the IEEE's *Journal of Solid State Circuits.* Technical articles also appeared in some of the electronics industry's leading-edge magazines of that era, including *Electronics, Electronic Design,* and *EDN,* as well as in manufacturers' own in-house application notes. Over the past decade, however, the subject of current sources has not otherwise been a much-publicized topic.

Although most engineers have seen a current source's symbol depicted in an analog IC's circuit schematic (see Figure 2.2), in most cases even if its circuit is shown, it may only be a simplified version because the actual circuit is proprietary and may be patented. Few IC manufacturers, if any, give out specifics about how their circuits are built and biased, other than on a need-to-know basis and how it applies to the end user. However, the current source is an essential element in most analog circuits' architecture and operation, irrespective of the IC's function (e.g., op amp, A/D converter, voltage regulator, RF receiver, video amplifier, D/A converter).

While loosely referred to as current sources, there are actually three specific types. Typically, a current *source* (usually comprising P-type devices) connects between the positive supply rail (+V) and the load, while a current *sink* (usually comprising N-type devices) connects between the load and a negative (-V) or ground (0V) potential. A current *mirror* (also known as a current *reflector*) can connect to either rail and usually provides multiple current sources/sinks that either mirror (1:1 match) or are arranged in preset current ratios (e.g., 1:2, 1:4, 1:8). This is sometimes shown in a circuit schematic as a multicollector or multiemitter structure, commonly created by analog IC designers. Figure 2.3 shows some examples of current sources, current sinks, and current mirrors, created in different technologies. We will focus on the specifics of these different types in subsequent chapters.

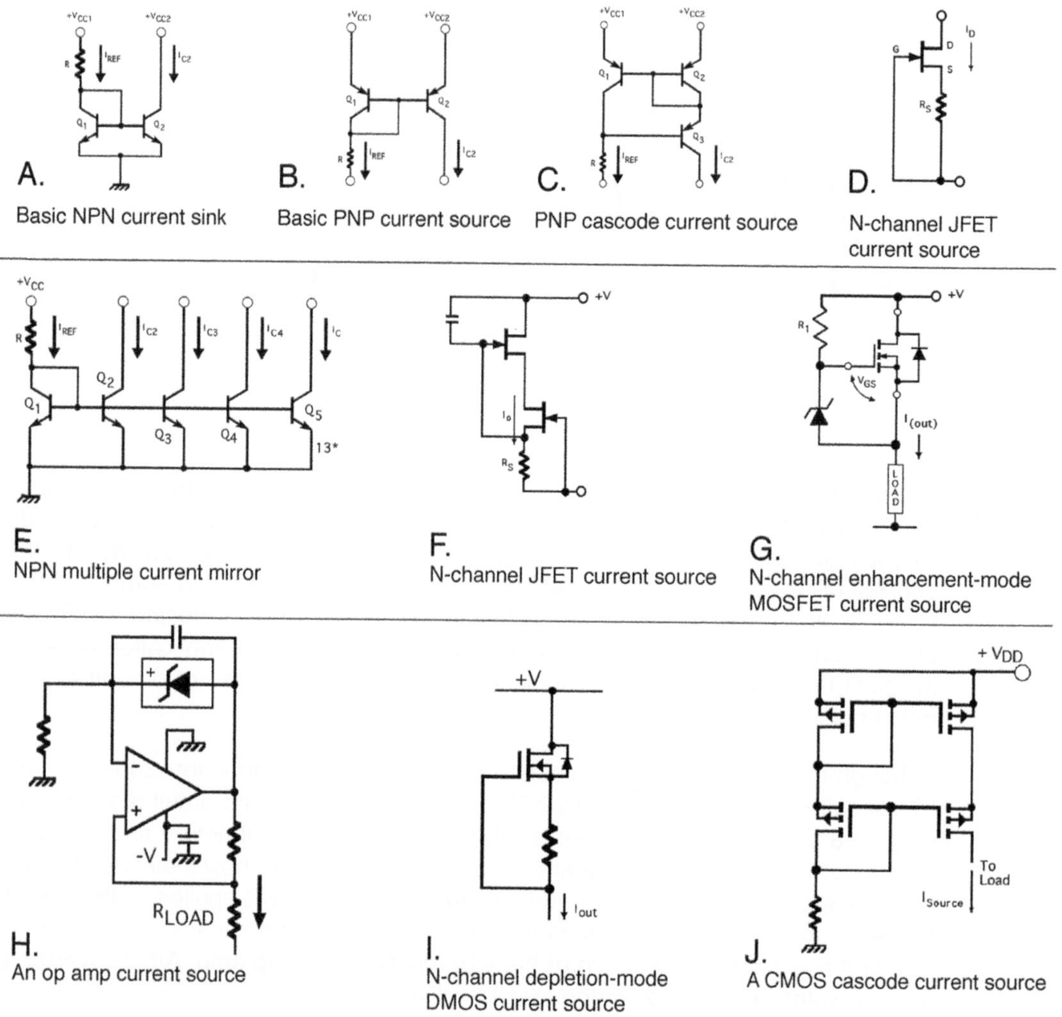

Figure 2.3. *Examples of different types of current sources.*

Over the years, current sources have been applied in subsequently newer technologies and processes, such as high-speed bipolar, complementary-bipolar, bipolar-FET, analog metal-gate CMOS, and more recently in very-low-voltage processes. The examples in Figure 2.4 show part of the internal circuitry for two typical op amp input stages (one is a bipolar-FET device in Figure 2.4A and the other is a CMOS device in Figure 2.4B). As you can see, current sources are employed throughout.

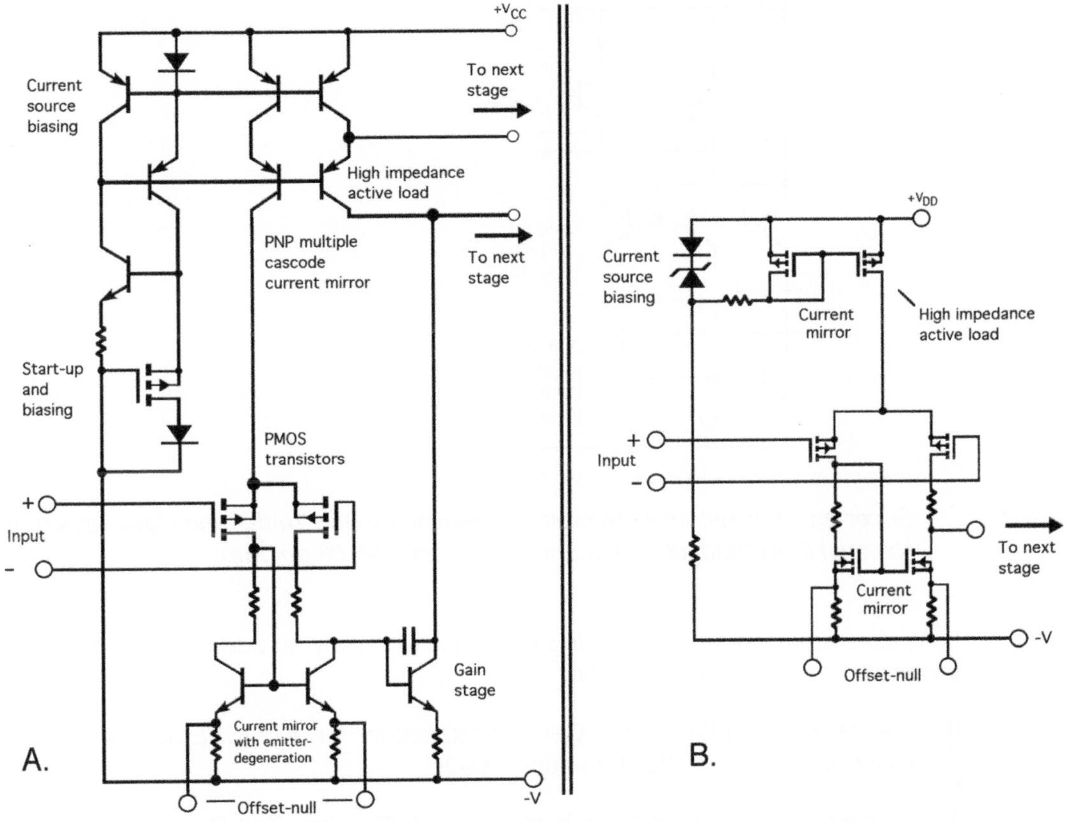

Figure 2.4. *These examples show how current sources, sinks, and mirrors are used as active loads, and for biasing in the front-end amplifiers of typical bipolar-FET (left) and CMOS (right)voltage-feedback op amps.*

Two major advances with the op amp came in those early days from National Semiconductor. The first was Bob Widlar's implementation of what are referred to as *super-gain* transistors. These were specially processed, very-low-voltage NPN transistors, which had very high h_{FE}s (typically of around 5000, at a very low collector current of about 1 μA). For the first time, it uniquely enabled a monolithic bipolar op amp to have ultra-low-input bias currents similar to those of a JFET, but over the full military temperature range. They were a concept devised and implemented by Widlar for some of his designs and created at the wafer level by his processing specialist, Dave Talbert. Super-gain transistors were first used in the LM108 precision op amp, which National introduced in 1969. In a typical op amp schematic, they are depicted as a regular NPN transistor, but with a hollow base junction, as shown in Figure 2.5. Today super-gain

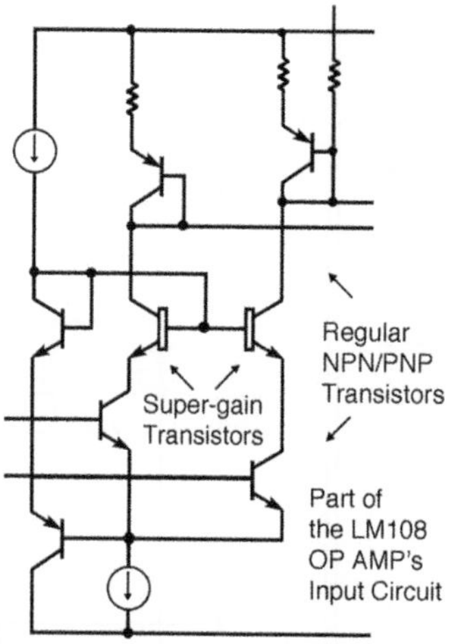

Figure 2.5. *Super-gain transistors were orginally devised by Bob Widlar and Dave Talbert at National Semiconductor, and first used in the LM108 op amp.*

transistors are used in many analog IC products, usually as part of a gain stage, but sometimes as a current source too.

(Reference: R. J. Widlar, "Super-Gain Transistors for the IC," *IEEE Journal of Solid-State Circuits*, Vol. SC-4, No. 4, August 1969.)

The second major advance came in the early 1970s when Rod Russell (another brilliant young analog IC designer at National Semiconductor at the time) invented the *bipolar-FET* op amp. National subsequently introduced these to the market in the mid-70s. Russell's design ushered in a whole new kind of op amp that offered very low noise, ultra-low-input bias currents, at an economical price. By the early 1980s, various manufacturers, including Fairchild Semiconductor, Texas Instruments, Analog Devices, National Semiconductor, and others, were researching ways that would help improve speed and bandwidths, because the transistor-based technology of that era had serious limitations. At that time, PNPs had much lower fTs than their NPN counterparts, and this bottleneck limited the overall bandwidth of the typical bipolar op amp to just a few MHz. Up until then, high-speed op amps relied on all-NPN transistors, which in turn limited the input and output signal swings, not to mention the operating supply range. It was also possible to create high-speed op amps, but with costly hybrids. Analog Devices eventually introduced a working process that became known as *complementary-bipolar* (C-B). This process used ion-implantation, and for the first time was able to include high-speed PNP transistors with fTs of more than 500 MHz, alongside high-speed NPN devices, in the actual signal path. The C-B process was a major advancement in the overall evolution of the op amp, because it provided symmetrical operation internally, lower power, with higher output currents, and effectively improved the bandwidth by at least an order of magnitude. Current sources are an integral part in any C-B op amp too.

Other technology leaders, like Texas Instruments with its Excalibur™ process and National Semiconductor with its Vertically Integrated PNP (VIP®) process, also contributed enormously to the advancement of this important high-speed op amp technology. Figure 2.6 shows a simplified example of National's original VIP low-power, C-B architecture, the first generation of which they introduced in the late 1980s, in the form of three op amps (LM6361, LM6364, and LM6365). National had developed the proprietary VIP process that fabricated lateral PNP transistors on-chip, having almost equal characteristics (in terms of speed and gain) to their NPN counterparts. Figure 2.6 shows the LM6361 op amp, built using the VIP process, which includes some NPN current sources, as well as two lateral PNPs, which boosted performance.

The current sources used in older-generation ±15-volt bipolar-FET op amps are not the same as those used in today's single-supply rail-to-rail input/output op amps. To bear this out, let's look at some examples. Figure 2.7 shows a simplified example of a typical C-B op amp, with rail-to-rail input. In this circuit you will notice that there are two differential pairs at the input—one an NPN pair and the other a PNP pair, and remarkably with both operating in parallel. Each differential pair has a current source in series with its emitters, for biasing. The voltage gain of the input stage is deliberately kept low to

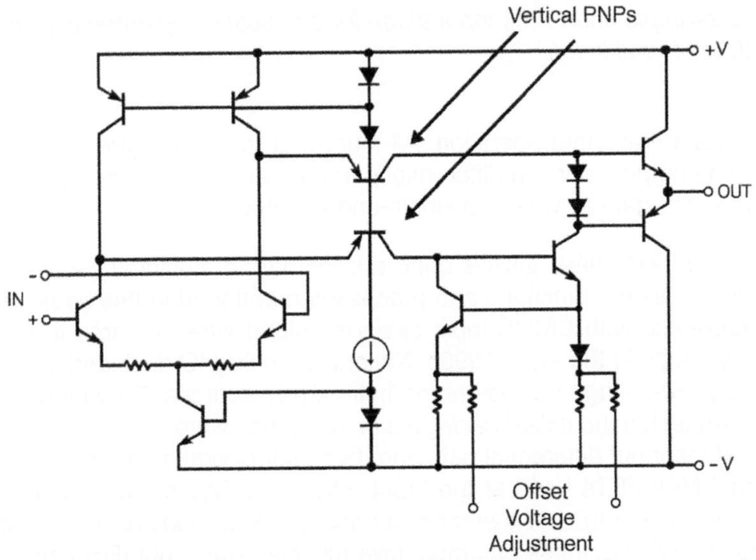

Figure 2.6. *A simplified schematic of National Semiconductor's LM6361 Fast VIP® op amp. Note the vertical PNP transistors and the NPN current sources..*

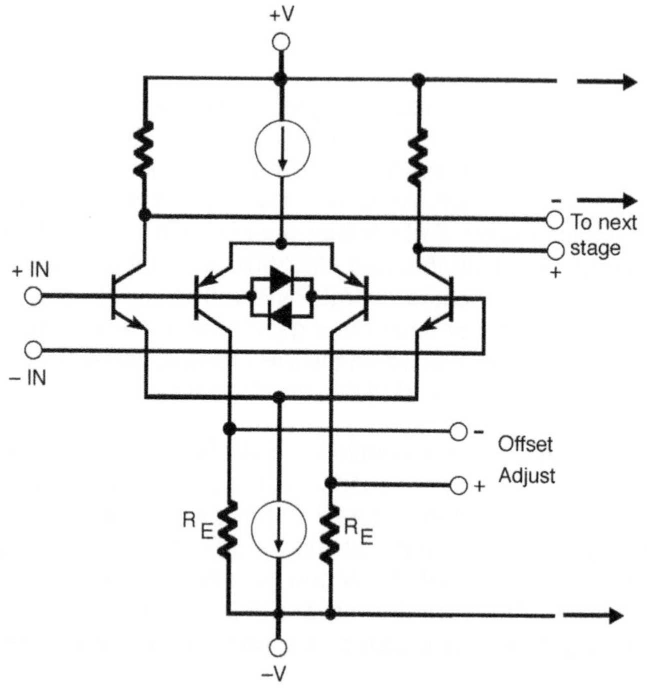

Figure 2.7. *This example shows the input stage for a typical complementary-bipolar op amp, with rail-to-rail input.*

facilitate rail-to-rail input operation. All four input transistors are protected by the two diodes. The outputs of each differential pair are connected to the op amp's next stage, which provides gain as well as a single-ended output.

Innovative designs have always been a trademark of National Semiconductor, which range across all the functions and processes mentioned in this book. The company's long experience with CMOS logic designs helped when it came to designing 3-volt CMOS op amps in the early 1990s. National's analog IC designers had to first create some very-low-voltage transistors for their current sources. They found that by using a little-known technique called *back-gate biasing*, they were able to control the threshold voltage of the input differential pair, and thus their operating mode. Remarkably, these P-channel MOSFETs used at the input (shown in Figure 2.8) operate as *depletion-mode* devices near the positive rail and *enhancement-mode* devices near the negative rail, and in so doing, these op amps have full rail-to-rail input (RRI) and rail-to-rail output (RRO) capability. The current sources operate to within 200 mV of either supply rail and play a major role in the performance of these dependable op amps. Even more astonishing is the fact that the signal level can even briefly exceed the supply

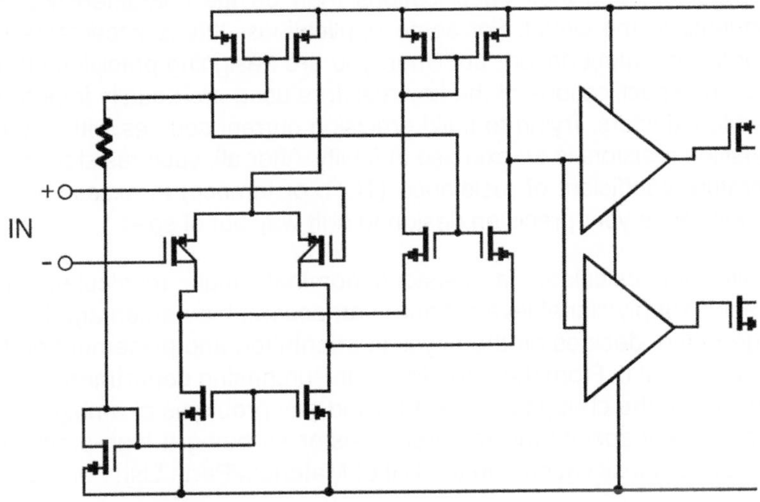

Figure 2.8. *National Semiconductor's LMC6482/4 CMOS op amp can handle rail-to-rail input signals due to its depletion-mode P-channel MOSFETs and specially designed low-voltage current mirrors.*

rails (which can range from between 3 and 15 volts), without phase inversion or catastrophic failure.

As you can see, current sources play a vital role in the operation and overall performance of all of today's analog semiconductors, no matter the type or the process by which they are made. The challenge for the analog IC designer becomes increasingly more difficult as desired voltage operation gets ever closer to the 1-volt level (as with digital designs), along with a miserly power consumption, good precision, and increased bandwidth—characteristics that are always at odds with one another.

2.2 Precision resistors, networks, and trimmers

Creating accurate and reliable current source designs depends on using a precision resistor to convert a reference voltage into a precise and stable reference current. The precision resistor(s) used in this type of application need to be very-high-quality, low-drift (i.e., low ppm/°C) types that match the application. The choice of components should be practical and within the limits of the design budget. Assuming that the semiconductors have minimal drift, the circuit's overall stability and temperature drift will be largely governed by the resistor's drift characteristics. This applies from the Q&A stage at the factory, where power is first applied to your design, throughout the entire life of the end product, where its real working environment is involved.

It is worth taking time here to review what kinds of resistors are available for any current source design project. In many cases, the design engineer focuses more on the semiconductors and their specifications than on the characteristics of the passive components in the circuit. For some applications, this is acceptable (e.g., the load resistor for an output driver), but when you are designing precision current or voltage circuits, the specifications of the key resistors used are equally important as those of the semiconductors. Trying to build precision current sources with cheap 5% carbon-composition resistors is an exercise in futility. After all, such resistors can often have a temperature coefficient of resistance (TCR, or tempco) in excess of ±4000 ppm/°C, which will cause your precision design to drift way out of spec.

Normally, one calculates a resistor's nominal value, recalculates for increased/decreased supply/signal levels, figures an acceptable percentage tolerance and the wattage rating, decides on the physical orientation and placement on the PC board, and that's about it. From the parts list your Purchasing department then finds a product, usually at the cheapest price. To avoid any problems of ambiguity, it is essential that you clearly specify the required resistor as being a high-precision type, along with its basic specifications on the Bill of Materials/Parts List. For example:

1 x Caddock TN-130

10 KΩ precision film resistor

±0.05% tolerance

20 ppm/°C tempco

0.3 watt

With many of the circuits reviewed in this book, using anything but a true precision resistor could be a fatal mistake for your design. Also, it is important to remember that precision resistors can often cost as much or more than many of the semiconductors used in one's design. In fact, the higher precision they are, the more expensive they can be; unfortunately, there is no shortcut to true precision.

Various types of fixed resistors are available to the designer, but for precision circuits the list quickly shortens to these types, in the following order:

- Metal-foil (best)
- Wirewound
- Thin-film
- Metal-film
- Thick-film (lowest cost)

It might be helpful here to briefly define some of a precision resistor's more important characteristics, which are important when creating precision circuits and include the following:

- *TCR.* This is the resistor's temperature coefficient of resistance, which relates to the change in resistance for a given change in temperature. It is expressed in parts per million per degree Celsius (ppm/°C). A percentage-to-ppm converter is shown in Table 2.1. For ultra-high-precision metal-foil resistors, this value can be as low as 1 ppm. In a current source design, the resistor sets the desired current value, which should change as little as possible over the operating temperature range.

ppm	%
100K	10
10K	1
1K	0.1
100	0.01
10	0.001
1	0.0001
0.1	0.00001
0.01	0.000001

Table 2.1 Percentage (%) to parts-per-million (ppm) converter.

- *Tolerance/accuracy.* The tolerance refers to the resistor's measured deviation in resistance from its nominal value and is usually expressed as a percentage (i.e., 1 KΩ ±5%). In this example, the actual measured resistance will occur somewhere between 950 Ω and 1,050 Ω. For very-high-precision resistors, because this deviation from the nominal value is so small, it is often referred to as accuracy rather than tolerance, and as a result is expressed in parts per million (ppm). So, for a high-precision 1-KΩ resistor with a tolerance of ±0.01%, this means that the actual measured resistance will occur somewhere between 999.9 Ω to 1,000.1 Ω, and relates to an accuracy of 100 ppm (see Table 2.1). Accuracy is a major factor in most current source applications, because the resistor's value is used to precisely set the desired current value.
- *Tracking TCR.* This refers to two or more similar resistors usually sharing the same package and substrate (such as in a resistor network/array) and how their resistance changes relative to one another, in response to changes in temperature (i.e., how they track). It is expressed in parts per million per degree Celsius (± ppm/°C). Good precision resistor networks

have very low tracking TCR values, as shown in Table 2.2. This is important in op amp–based circuits where the network sets the gain. Ideally, the gain will not change over the operating temperature range.

- *Long-term stability.* This is a measure of how well the resistor retains its normal TCR characteristics over long periods of time and is usually specified by the manufacturer in ppm/year. This is very important for all precision designs, particularly current sources.
- *Wattage.* This relates to the nominal amount of power that the resistor can safely dissipate. As current flows through the resistor, heat is produced. The material that the resistor is composed of, along with its physical size, lead length, and ambient temperature, all combine to determine the wattage rating. This rating can typically range from about 100 mW to more than 10 W. For many low current source applications, the current-setting resistor will probably have a rating of less than 0.5 W. To avoid self-heating effects, it is advisable to use a higher wattage resistor than needed (i.e., use a 1/2-watt resistor instead of, say, a 1/8-watt device).

Table 2.2 gives a brief summary of these different types and lists some of the key U.S. manufacturers. With the exception of wirewound or cylindrical types, the other types are mostly manufactured with techniques similar to those of semiconductors, because they are created mostly in a planar fashion, with the resistive element bonded to a flat ceramic or other neutral substrate. As a result, their construction and packaging lends them to be more readily surface-mountable, which most wirewound types are not.

Metal-foil resistors (a.k.a. *bulk metal-foil*) offer the ultimate in precision, along with excellent long-term stability, low noise, a tight tolerance, and with ultra-low TCRs (tempcos), down to approximately 1 ppm/°C. They provide normal resistance values up to approximately 250 KΩ. Due to their typically small, low-inductance construction, they have also become a popular choice for surface-mount applications. This type of resistor is made by bonding a 2.5-μm strip of resistive foil, made of a specially formulated nickel-chromium (NiCr) alloy, onto a flat square ceramic substrate. The foil is then photo-etched with a serpentine-like pattern and laser-trimmed. Because of differences in their respective coefficients of expansion, the metal foil, substrate, and adhesive are all carefully engineered to have minimum temperature sensitivity. Unlike *wirewound* resistors, foil types can be used even at RF frequencies, but they do not have high wattage ratings or much overload handling capability. Of all the types mentioned here, they are clearly the most expensive, but they do represent the ultimate in precision. There are several U.S. manufacturers of precision bulk metal-foil resistors, as shown in Table 2.2, the largest of which is Vishay Intertechnology Inc., which introduced the technology some years ago. Precision multiturn, bulk-metal trimmers can also be made using this technology and have equally outstanding specifications. They too are available from Vishay.

Typical Characteristics:	Metal-foil	Wirewound	Thin-Film	Thick-Film	Metal-film
Classification	Ultra-high precision	Ultra-high precision	Very-high precision	Semi-precision	Semi-precision / High precision
Maximum Range (Ω)	250KΩ	1MΩ	25MΩ	2GΩ	10MΩ
Accuracy (%)	0.01 to 0.001%	0.01 to 0.005%	0.01 to 2%	2 to 5%	0.1 to 5%
TCR (±ppm/°C)	0.5 to 10 ±ppm/°C	1 to 20 ±ppm/°C	10 to 100 ±ppm/°C	25 to 300 ±ppm/°C	15 to 100 ±ppm/°C
Tracking TCR (±ppm/°C)	0.5 to 3 ±ppm/°C	Not applicable	1 to 5 ±ppm/°C	2 to 50 ±ppm/°C	Not applicable
Long-term stability (±ppm/year)	5 ±ppm/year	20 ±ppm/year	50 to 100 ±ppm/year	30 to 1000 ±ppm/year	1000 ±ppm/year
Cost	High	Medium	Medium	Low	Low
Disadvantages:	Cost	Low to medium frequency; size; inductive effects	Often fragile; low power	Poor TCR, and long-term stability	Differences between mfrs.; long-term stability
Principal U.S. manufacturers:	* Imperial Astronics * Isotek * Wilbrecht Elect. * Vishay	* Isotek * Ohmite * Precision Resistor Co. * Prime Tech./ General Resistance * Process Instr./ Julie Research * Vishay	* AVX Corp. * Caddock * Intl. Mfg. Services * IRC * Kamaya Corp. * Mini-Systems * Thin Film Tech * Vishay	* AVX Corp. * Intl. Mfg. Services * IRC * Kamaya Corp. * Mini-Systems * Ohmite * Vishay	* Brel Intl. Corp. * IRC * Kamaya Corp. * Thin Film Tech * Vishay

NOTES:
1. Contact manufacturers for more information.
2. See Contact Info Section in Appendix.
3. Percentage to ppm conversion:
   % = ppm x 0.0001
   Thus 0.01% = 100-ppm
4. Parameters vary between manufacturers.

Table 2.2 Comparison of various types of precision resistors.

Wirewound resistors are the oldest, most mature technology of the group, having been manufactured in similar fashion for decades. Precision wirewounds offer very high precision along with ultra-low tempcos, also down to approximately 1 ppm/°C. Like metal-foil resistors, they are intended for designs requiring absolute accuracy. They offer great stability, low noise, and a high wattage rating. They are usually made by loosely winding (bifilar) a special NiCr alloy wire element onto a cylindrical insulating core, which is then either encased later in a molded housing or is hermetically sealed. Because of their very high wattage ratings, they usually have excellent overload handling capability, but because of their inherent inductor-like construction, they also have inductive effects that preclude them from high-frequency applications. Wirewounds provide resistance values below 1 Ω up to approximately 1 MΩ, with typical long-term stability of less than 20 ppm/year.

Compared to the other types, wirewounds are also a natural choice for power circuits. They often provide a lower cost advantage above 50 KΩ. Although wirewounds, because of their size and shape, do not naturally lend themselves to being used in precision (matched) networks like foil and *film* types do, their same technology is used to create precision multiturn trimmers. There are several U.S. manufacturers of precision wirewound resistors, the largest of which is again Vishay Intertechnology Inc. As a general rule of thumb, the lower a precision resistor's percentage tolerance, its absolute TCR (tempco), and its long-term stability, it will usually provide the lowest drift and best overall performance. In any design project, be prepared to pay for this level of precision.

Film resistors represent a lower level of precision and at a lower cost than either metal-foil or wirewound types. Their respective technologies have been around for more than two decades, with the early implementation of thick-film and thin-film being mostly used in custom hybrid microcircuits. Of these types, thin-film resistors offer higher performance, with tempcos of between 10 to 100 ppm/°C and accuracies down to 0.01%. By contrast, thick-film resistors offer somewhat lower performance and are classed as semiprecision types with tempcos of between 25 to 300 ppm/°C and accuracies from approximately 1% to 5%. Metal-film resistors are also classed as semiprecision types even though their 0.5% to 5% accuracy and less than 100 ppm/°C TCR is better than thick-film. However, their long-term stability can be as much as 1000 ppm/year, and overall performance varies significantly between manufacturers. Metal-film resistors typically have an ohmic range of up to about 10 MΩ. This contrasts with thin-film resistors, which typically have an ohmic range of up to around 25 MΩ, while thick-film devices are offered in a much higher resistance range of more than 1 GΩ.

Thick-film resistors are generally made from proprietary formulations of metal-oxide mixed with a binder, then screenprinted onto a ceramic substrate and fired in a kiln at high temperature. Thin-film resistors are made in an entirely different manner, usually by sputtering NiCr or TaN (tantalum nitride) onto a ceramic substrate, either flat or cylindrical. Metal-film resistors are made with a partial combination of these two techniques. Of the three types, metal-film and thick-film provide the lowest cost, while thin-film clearly provides the best overall performance. Various U.S. companies manufacture film resistors, as shown in Table 2.2, three of the most prominent being Caddock Electronics, IRC, and again Vishay. If resistors are used in pairs, such as in gain setting an op amp or in some other kind of network, then close tracking between the individual resistors will be important. Precision thin-film resistor arrays, which have excellent

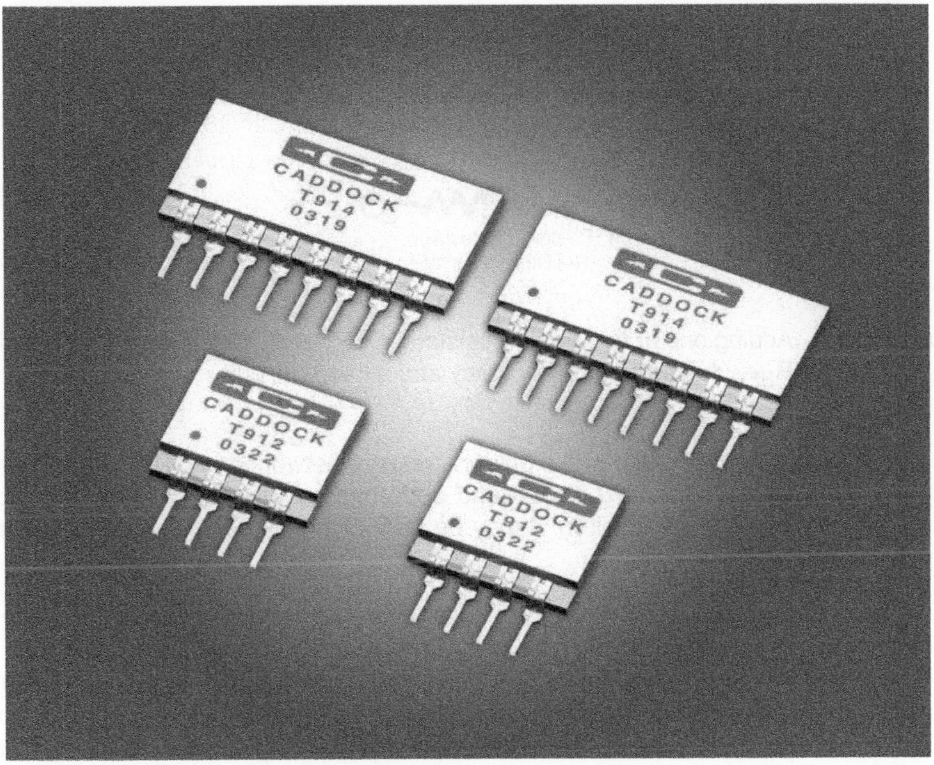

Photo 2.1. Precision thin-film resistor arrays from Caddock Electronics, Inc.

tracking and low-drift characteristics, will usually be the best solution. You should contact some precision resistor network manufacturers for their product information.

In some cases, one may have to trim a voltage across a fixed current-setting resistor, thereby adjusting it to an exact current level, which means there are two courses of action. One is to use one or two additional precision resistors in series with the main resistor, in order to achieve the desired current/voltage level (see Figure 2.9). The downside of this is that it requires various different resistor values in order to achieve the desired value, and this may not be too practical in a volume production situation. It does, however, maintain the accuracy and overall tempco of the precision resistors involved, assuming that they are mounted horizontally and in proximity to one another. The second option is using a potentiometer (i.e., a multiturn trimmer), but including a trimmer to adjust for a precise voltage/current (which is easy to do) will usually degrade the performance of the fixed resistor, particularly its low TCR drift. Because the trimmer and the fixed resistor are likely to be made of different materials, from two different manufacturers, not to mention in two physically separate packages, there will be little or no matching characteristics between them, as there would be if both resis-

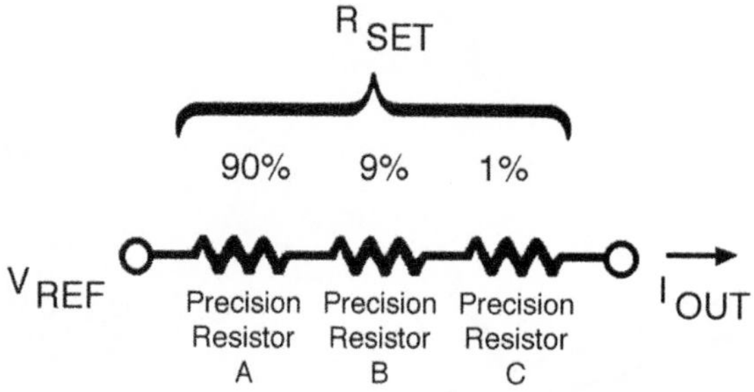

Figure 2.9. Adding one or two similar resistors in series with the main current-setting resistor, R_{SET}, maintains high accuracy and a low TCR drift.

tors were part of the same physical resistor network/array. This will ultimately degrade the circuit's overall TCR drift and performance. In most cases where high accuracy and low drift are desired, it may be best to simply use the fixed precision resistor(s) and not use a trimmer.

If this is not possible, then the only option left is to attempt to minimize the effects of the trimmer as much as possible. This can be done by using a small value for the trimmer to give the desired adjustment range (i.e., less than 10% of the value of the fixed resistor), and preferably by making the critical adjustment value occur approximately in the center of the trimmer's range. In this scenario it would be more advisable to use a fixed precision resistor with a higher tolerance (0.1% instead of say 0.01%, or 1% instead of say 0.1%), because there is no sense in paying for a more expensive part if its precision characteristics are undermined by those of the trimmer. Remember that a typical cermet trimmer will usually have a tolerance of ±10% and a tempco of more than ±100 ppm, which will virtually negate the benefits of using the precision resistor (i.e., 0.01% with a 20 ppm/°C tempco). The combination of these two may well equate to tolerance of ±10%, with an overall tempco of ±120 ppm.

So a better, but more expensive solution is found by using either of the following possibilities: (1) a precision wirewound trimmer with a low ±5% tolerance and a tempco of ±50 ppm or (2) a metal-foil trimmer that also has a ±5% tolerance but with a lower tempco of ±10 ppm. Unfortunately, such products are made by very few U.S. companies, although Bourns (precision wirewound trimmers) and Vishay (precision metal-foil trimmers) do have some excellent products. The combined tolerances and tempcos of the fixed resistor and the trimmer would have to be measured over the operating temperature range, in order to determine the overall stability and drift characteristics of the circuit. However, this may be acceptable for many designs.

2.3 Essential development equipment

2.3.1 Bench power supply unit

An essential requirement for developing any precision current source, sink, or mirror would be the use of a high-quality, variable, calibrated, low-noise bench power supply unit with either an analog or digital readout. The bench power supply acts as a stable reference for your current source design. In this manner, it is easy to set the supply level(s) at a particular nominal setting and measure the circuit under test. With such a unit it is an easy matter to vary the supply level(s) up and down, and measure any change in the current source output, according to one's test criteria.

At the circuit board level, it is also a good idea to power the precision current source circuit from its own on-board voltage regulator. For very low current requirements (less than 5 mA), one could substitute a low-noise voltage reference IC instead. For low to medium current requirements, one could use an inexpensive low-noise, linear voltage regulator IC, such as a National Semiconductor LM2936 (as shown in Figure 2.10). A minimum of a 10-μF electrolytic capacitor, in parallel with a 0.01-μF ceramic disc capacitor, may be needed to bypass the regulator's output to the current source. The 1N4001 switching diode is included simply to protect the regulator from being reverse-biased. The LED indicator shows when the circuit is on.

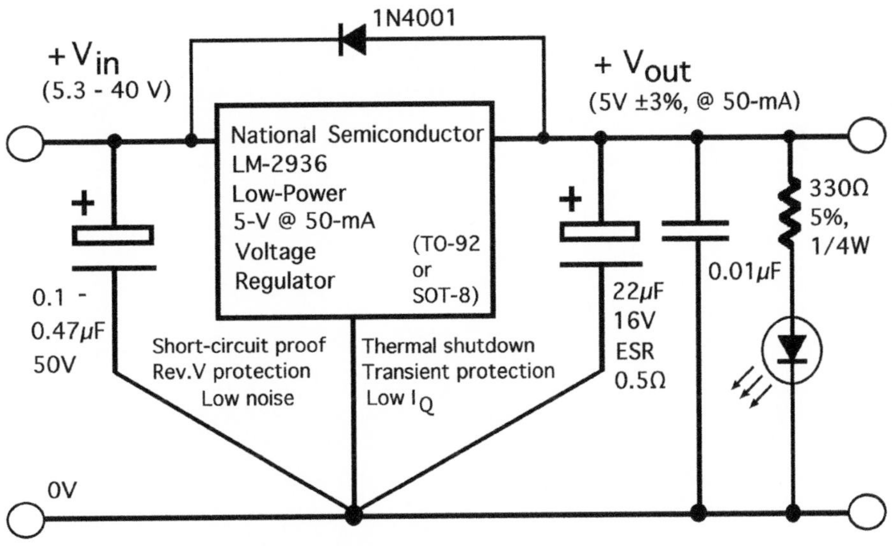

Figure 2.10. A simple low-noise regulated power supply.

2.3.2 A thermostatically controlled oven

Another essential piece of lab equipment needed for developing precision current sources is a high-quality, thermostatically controlled oven. This is needed to determine just how well your current design performs over any particular temperature range. Much of today's test equipment can be controlled from your lab computer, networked, or even controlled remotely over the Internet. A good source of possible vendors for lab ovens and environmental chambers can be found in *Test & Measurement World* magazine. If purchasing such equipment is not an option, then it may be leased from various equipment rental companies nationwide.

2.3.3 A calibrated, precision ammeter

This instrument is required for accurately measuring and monitoring the current level in your design. It should be a high-quality calibrated instrument, which can be connected to a chart recorder for a permanent printout or connected via RS-232 or IEEE-488 to your lab computer. As mentioned previously, this type of equipment may be easily located in *Test & Measurement World* magazine and either purchased or leased.

In summary, the current source is an important building block in many analog designs. It is this book's intention to show how current sources can be created using various techniques by the OEM circuit designer.

Chapter 3

The P-N Junction

Before looking at the transistor in Chapter 3, it will be helpful to first briefly review the diode, and thereby some of the basic characteristics of the P-N junction, because diode action is an essential ingredient in making current sources and reference voltages. In a circuit's design, this can be induced by using either the forward voltage (V_{FWD}) of a dedicated switching diode or a transistor's base-emitter junction voltage (V_{BE}), or even a JFET's channel diode voltage. In each case the P-N junction is used, either as a stable predictable reference or for compensation purposes.

Early semiconductor diode types were silicon and germanium point contact devices, developed by Dr. Russell Ohl and others at Bell Labs, as well as higher voltage types developed by Dr. Karl Lark-Horovitz and his group at Purdue University. These were developed primarily for the U.S. military and were used in radio, radar, sonar, and other equipment in World War II. Today's silicon junction diodes followed the introduction of the silicon transistor in the early 1960s. Some of the earliest silicon diode manufacturers included Texas Instruments, Fairchild Semiconductor, Transitron, Hughes Semiconductor, and General Electric.

3.1 Characteristics of the P-N junction

The diode is a two-terminal, single-junction semiconductor device that passes electrical current in only one direction. It is commonly used for rectification, switching, and clamping, as well as in oscillator, varactor, and reference circuits. Small-signal diodes are typically used in relatively small current circuits (less than 100 mA), whereas *rectifiers* (which imply power devices) are generally used for switching 1 Amp and higher.

Most modern-day semiconductor devices, and all of those covered in this book, are made of *Silicon* (Si), which is in Group IV of the Periodic Table of Elements. Silicon is a nonmetallic element making up around 25% of the Earth's crust. It is one of the most common elements on Earth, the most common source being sand. When raw silicon is purified, it can produce 99.99% pure silicon. Silicon's atomic structure (Figure 3.1) has a nucleus at the center, composed of positively charged protons and electrically neutral neutrons. Around the nucleus are several orbits of negatively charged electrons; the outermost orbit is referred to as the *valence orbit*. Electrons in this orbit are referred to as *valence electrons*. Electrons have energy because they have mass and are moving within their respective orbits. Those in the valence orbit have more energy than electrons closer to the nucleus. When an electron becomes separated from its parent atom, it takes a negative charge with it and gains even more energy. These are referred to as *free electrons* and are free to wander and join other atoms. This loss of an electron imbalances the normally neutral atom and gives it a net positive charge of one hole, thus making the atom a *positive ion.* (Conversely, an atom with a surplus of electrons is referred to as a negative ion.) Pure silicon is *quadravalent*, meaning that its valence orbit contains four valence atoms.

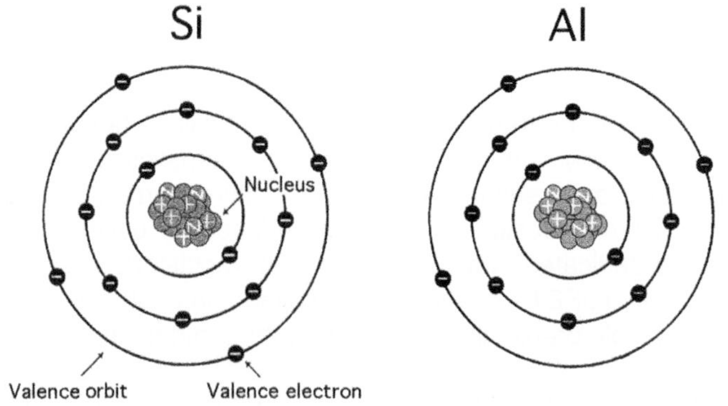

Figure 3.1. *The Bohr model of a silicon atom on the left and an aluminum atom on the right. Combining trace amounts of aluminum to the silicon makes P-type semiconductors.*

When tiny amounts of particular impurities are added, this creates semiconductor material. When forming *N-type* semiconductor material, elements in Group V of the Periodic Table of Elements (see Figure 3.2) are used, such as arsenic (As), antimony (Sb), or phosphorus (P). These are all pentavalent (five atoms) and are used as *donors.* When a pentavalent atom replaces a quadravalent atom in the crystal lattice, because it contains an excess electron in its valence band, it is able to donate one free electron. When forming *P-type* semiconductor material, elements in Group III of the Periodic Table of Elements are used, such as aluminum (Al), indium (In), boron (B), or gallium (Ga). These are all trivalent (three atoms) and are used as *acceptors.* When a trivalent atom replaces a quadravalent atom in silicon's crystal lattice, because the trivalent atom lacks an electron in its valence band, it is able to accept a free electron from an adjacent quadravalent atom, thus producing a hole.

The donor electron is free to cross the *bandgap* and move into the conduction band, if given a small amount of additional energy, either thermal or an electric field is applied. The bandgap is one of the key properties of any semiconductor and is a measure of the energy difference between the valence and conduction bands. This energy gap lies between the top of the valence band and the bottom of the conduction band, as

	Group 3	Group 4	Group 5
SYMBOL:	Al	C	As
NAME:	Aluminum	Carbon	Arsenic
ATOMIC #:	13	6	33
ATOMIC WT:	26.9815	12.01115	74.9216
TYPE:	metal	non-metal	non-metal
SYMBOL:	B	Ge	Bi
NAME:	Boron	Germanium	Bismuth
ATOMIC #:	5	32	83
ATOMIC WT:	10.811	72.59	208.98
TYPE:	non-metal	metal	metal
SYMBOL:	Ga	Si	P
NAME:	Gallium	Silicon	Phosphorus
ATOMIC #:	31	14	15
ATOMIC WT:	69.72	28.0855	30.9738
TYPE:	metal	non-metal	non-metal
SYMBOL:	In	Sn	Sb
NAME:	Indium	Tin	Antimony
ATOMIC #:	49	50	51
ATOMIC WT:	114.82	118.71	121.75
TYPE:	metal	metal	metal

Figure 3.2. *Part of the Periodic Table of Elements.*

shown in Figure 3.3. The valence band is the lower band that is usually full of electrons, whereas the conduction band is the upper band that holds the electrons that have been excited to a higher state. Electrons are free to move back and forth between these two bands. The energy required to create a free electron is directly related to the number of valence electrons an atom has. Generally speaking, the fewer the number of valence electrons, the smaller the amount of energy needed to release an electron. The *electron-volt* (eV) is the standard unit of measurement for an electron's energy (1 eV is defined as the energy acquired by an electron moving through a potential of 1 volt in a vacuum; i.e., 1 eV = 1.6×10^{-19} joule).

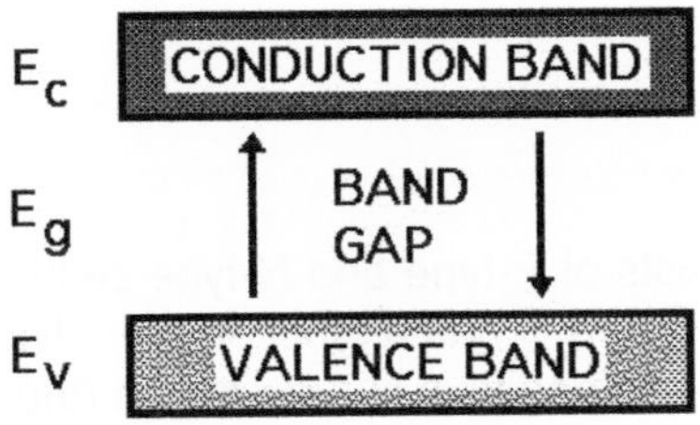

Figure 3.3. *The bandgap.*

In its pure form (intrinsic) silicon is similar in many respects to an insulator, but an insulator has a much wider bandgap, typically greater than 2.5 eV wide, which makes electrical conduction virtually impossible. At the other end of the scale, a conductor has many available and overlapping energy levels, which results in electrons moving through the bandgap very easily. As a result, conductors have a bandgap of 0 eV. Semiconductors, which lie between conductors and insulators, typically have a bandgap value of less than 2.5 eV. Each semiconductor material has a different bandgap value (e.g., Silicon's is 1.15 eV, while Germanium's is 0.67 eV). Gallium arsenide (GaAs) has a bandgap value of 1.43 eV, while gallium phosphide (GaP, used for making green LEDs) is higher at 2.26 eV. One can see from these eV values that the energy needed to transfer an electron from the valence band to the conducting band for a silicon device is nearly double that of a germanium device. This means that at room temperature, germanium is a better conductor than silicon, but it is also more sensitive to heat and has a much higher leakage current.

This information may be somewhat academic for the average electronic engineer who generally works with silicon devices, but it is more important for the IC designer or the electrooptics physicist who may work with different semiconductor materials. It is visibly apparent to everyone when one considers light-emitting diodes (LEDs) and various types of laser diodes, all of whose emitted color depends on their particular material's bandgap. It is also important in the creation and use of optical detectors and solar cells, which rely on a certain range of light wavelengths to function most efficiently. For many voltage references that use the bandgap effect, this is also important, as we will see in Part 2 of this book.

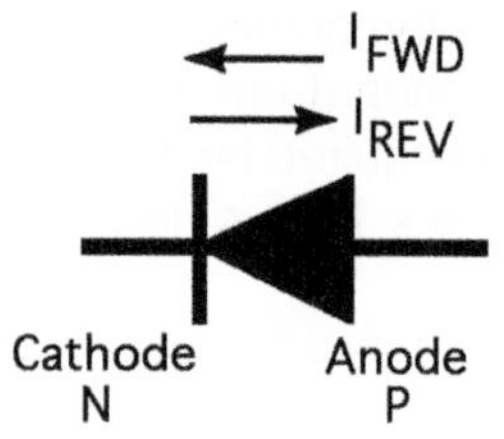

Figure 3.4. *The diode symbol.*

The silicon junction diode consists of P-type and N-type semiconductor material having a continuous crystalline structure. It has a junction that divides the two regions. Its symbol is shown in Figure 3.4. The negative N-type region (the *cathode*) contains free (donor) electrons, while the positive P-type area (the *anode*) contains free (acceptor) holes. Electrons and holes are the electric *charge carriers* that make conduction possible. After the diode is created, some of the electrons cross over the junction to fill some of the holes in the P-type material. As a result, a *depletion region* forms on both sides of the junction, which contains no free carriers. A negative electric potential, called the *barrier potential*, is created internally on the P-side, which repels further electrons from crossing over.

The diode can be biased into conduction simply by setting the bias conditions across the junction. This is achieved by applying a positive voltage to the anode, while the cathode is made more negative. This will create a condition known as *forward bias*, which turns the diode on. The externally applied forward voltage (V_F) eliminates the effect of the depletion region and causes the diode to conduct, as shown in Figure 3.5.

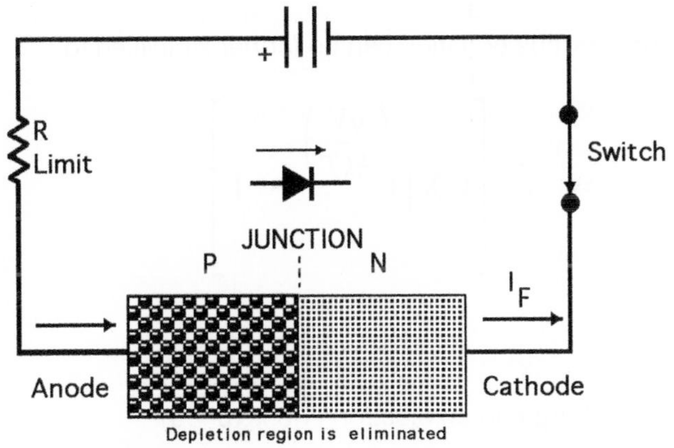

Figure 3.5. *Forward-biasing the diode.*

Several currents are involved in both forward conduction and under reverse-bias conditions, but the two main ones are the *diffusion* current (I_{diff}) and the *drift* current (I_{drift}). The diffusion current comprises both electrons and holes, and refers to the movement of *majority carriers* (electrons) in the conduction band that travel from the N-side to the P-side, and holes that travel from the P-side to the N-side. This concentration of majority carriers is both voltage and temperature dependent. The drift current also comprises both electrons and holes, but refers to the movement of *minority carriers* across the junction. The direction of their movement is opposite that of the majority carriers and is independent of both voltage and temperature. The total current (I_{total}) through the P-N junction may be expressed simply as:

$$I_{total} = \left(I_{drift\,\mathrm{e}} + I_{drift\,\mathrm{h}}\right)\left[\exp^{\left(\frac{eV}{hkT}\right)} - 1\right] \qquad \text{(Eq. 3.1)}$$

The quantity $I_{drift\ e} + I_{drift\ h}$ refers to the electron and hole drift current and to the theoretical *reverse saturation current* (I_S) of the system. This current, referred to in many textbooks, is based on an old model and does *not* take into account other minor currents in the crystal's depletion and surface regions. It is a theoretical quantum mechanics value anyway, given as approximately 18 femto-amps (1.87×10^{-14} A), and is probably not too relevant to practical circuit design, including current sources or voltage references. One should also keep in mind that because the reverse current never actually saturates, the term I_S is somewhat misleading. Instead, the diode/P-N junction's *reverse current* (I_R) is of more practical value and is usually shown on device data sheets. Reverse current is usually specified at both 25°C and either 125°C or 150°C.

However, the previous equation can be further simplified to:

$$I_{total} = I_S\left[\exp^{\left(\frac{qV}{hkT}\right)} - 1\right] \qquad \text{(Eq. 3.2)}$$

where:

q is the charge on an electron (1.6×10^{-19} J)

V is voltage (volts)

h is the ideality factor, which varies per diode and is caused by such physical phenomenon as the surface effect, tunneling, recombination, etc.

K is Boltzmann's constant (1.38×10^{-23} J/°K)

T is the temperature (°Kelvin; 298°K = 25°C)

Because the junction has an electric field, it is expressed as a voltage, which is equivalent to an internal battery. A silicon P-N junction will have an equivalent internal battery field of at least 0.62 volt (germanium's is lower, at about 0.3 volt). This means that the external voltage applied should be greater than the diode's internal battery-field voltage, so that the diode is fully turned on at the desired current. Usually a current-limiting resistor is necessary to prevent the diode from being destroyed, because of excess forward current flow ($I_{f\ max}$). This resistor drops the voltage between the positive voltage source and the diode's anode. From an electrical viewpoint, its resistance value (in Ω), its tolerance (±%), temperature coefficient (TC in ppm), and power rating (W) should be considered in most applications (see Figure 3.6). From a practical mechanical viewpoint, its size, ruggedness, suitability, position, and orientation matter

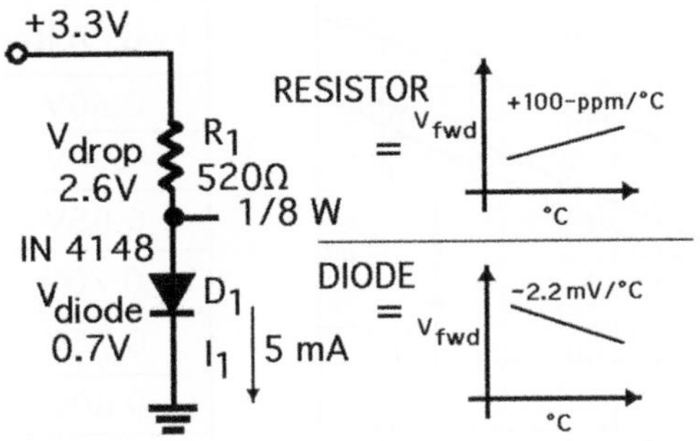

Figure 3.6. *A simple bias scheme for the diode, also showing how the diode's negative TC compensates for the resistor's positive TC.*

the most. Also to be considered are its proximity to other components (possible thermal interchange) and any cooling factors. As temperature increases, the resistance and voltage drop across the diode decreases and the diode conducts better. The diode's inherent action helps compensate for the resistor's typically positive temperature coefficient. This is *important*, because the same action is used to compensate transistors that are used in current sources and voltage references.

When the diode is forward-biased, it takes only a small forward voltage to cause a large current flow. For example, with a forward-bias of 0.70 volt, a IN4148 silicon switching diode at 25°C room temperature will have a typical forward current of 5 mA. Increasing the forward-bias to 0.92 volt boosts the forward current by 20 times, to 100 mA. The diode's resistance, forward voltage, and forward current are temperature dependent, as seen in Figure 3.7. At a forward current of 5 mA, the forward voltage is 0.8 volt at –40°C, while at +65°C the forward voltage drops to 0.66 volt. This confirms that the forward voltage has a *negative temperature coefficient* with increasing temperature, of between –1.4 to –2.2 mV/°C (usually calculated for a worst case of –2.2 mV/°C).

If the DC voltages are now reversed, so that a positive voltage is applied to the cathode (negative, N-material) and a negative voltage is applied to the anode (positive, P-material), then a state of *reverse-bias* will exist (see Figure 3.8). Under these conditions the depletion region will widen and electrical charge movement will be at a minimum. The diode is effectively off, and only a small leakage current, called the reverse current (I_R), will exist. A diode's reverse current is a practical, measurable amount, composed of several minor currents resulting from surface leakage effects, recombination, diffusion, generation, and various other nonideal factors. Reverse current also depends on doping levels and the junction's physical area. The reverse current includes various other currents, including the reverse saturation current (I_S), mentioned previously.

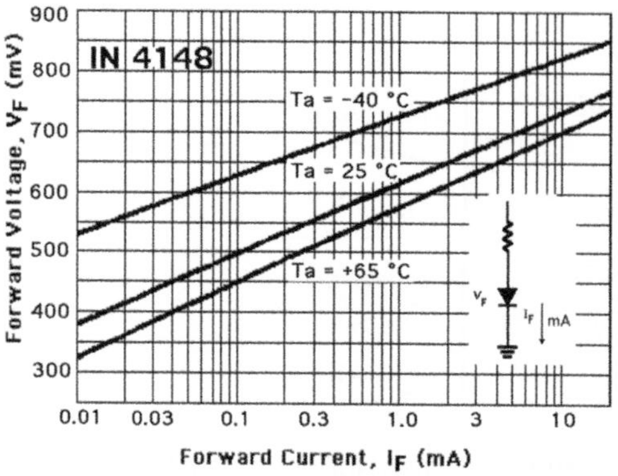

Fwd. Voltage	Fwd. Current
0.50V	0.1 mA
0.60V	0.8 mA
0.65V	2.0 mA
0.70V	5.0 mA
0.75V	15.0 mA
0.80V	30.0 mA

@ 25° C

Data and curves courtesy of Fairchild Semiconductor Corp.

Figure 3.7. *Forward voltage versus forward current versus ambient temperature.*

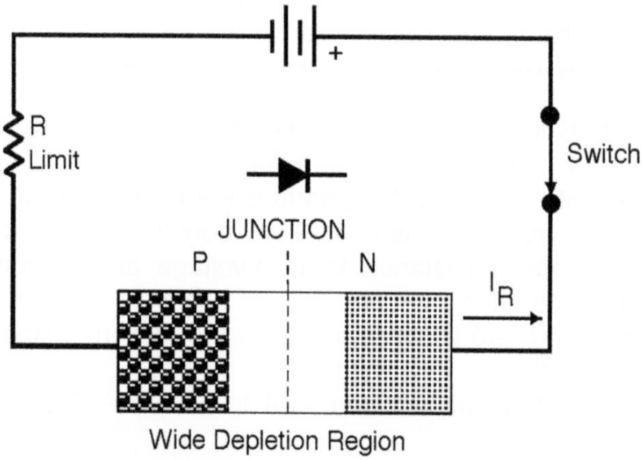

Figure 3.8. *Reverse-biasing the diode.*

These unique forward and reverse characteristics give the silicon junction diode a nonlinear, exponential curve, as shown in Figure 3.9. It is important to remember that the leakage current (I_R) increases with both a rise in temperature *and* the magnitude of the applied reverse voltage. When reverse-biased at say 80 volts, the IN4148 will have a typical I_R leakage current at room temperature of about 50 nA (5×10^{-8} A), as shown in Figure 3.10A. That's equivalent to an open circuit resistance of approximately 1.6 GΩ. That looks impressive if your application's operating temperature range can be controlled, or it does not change much, but one should remember that

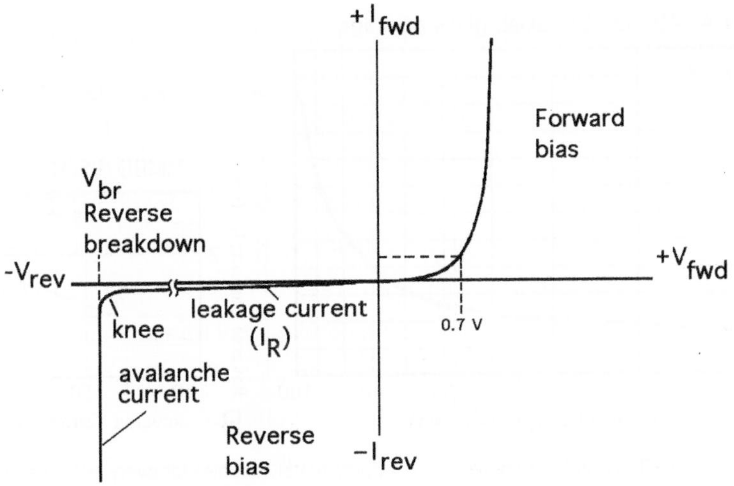

Figure 3.9. *A typical silicon diode's characteristic nonlinear V/I curve.*

with silicon diodes, this reverse leakage current *doubles* for every 10°C rise in temperature. (This was of much greater importance with germanium devices, which had reverse leakage currents about 1,000 times greater than silicon.) As an example, if the temperature is increased to 150°C and a reverse-bias of 20 volts is applied, the IN4148 diode will have a typical leakage current of less than 50 μA (5×10^{-5} A). This would give an open circuit resistance of only 400 KΩ, which is not very impressive, except that one's circuit would probably have cooked long before it reached 150°C. As you can see in Figure 3.10B, a specially made, low-leakage diode does a much better job, at an even higher reverse voltage. For some applications, using general-purpose switching diodes and a controlled ambient temperature, a certain level of leakage current can be largely ignored; for others it is very important and cannot be ignored. This is where you should probably use a low-leakage device (shown here as a Fairchild Semiconductor MMBD-1501, in an SOT-23 surface-mount package).

If reverse leakage is important in your design, then consider using dedicated low-leakage diodes such as a Fairchild Semiconductor FDH300, FDH3595, or similar. These diodes' specs include a maximum reverse current (I_{Rmax}) of 1 nA at 125 volts at 25°C (or 3 μA max at 150°C); a forward voltage (V_F) of 0.75 volt at 5 mA (at 25°C); a maximum DC forward current (I_{Fmax}) of 500 mA; and a maximum working reverse voltage (WIV) of 125 volts. They are available in an axial DO-35 package. Another low-leakage diode from Fairchild, with similar specs (but in an SOT-23 SMD package), is the MMBD1501. The specs include a maximum reverse current (I_{Rmax}) of 5 μA at V_r 180 volts at 150°C max; a forward voltage (V_F) of 0.72 volt at 1 mA (at 25°C); and a breakdown rating of 200 volts.

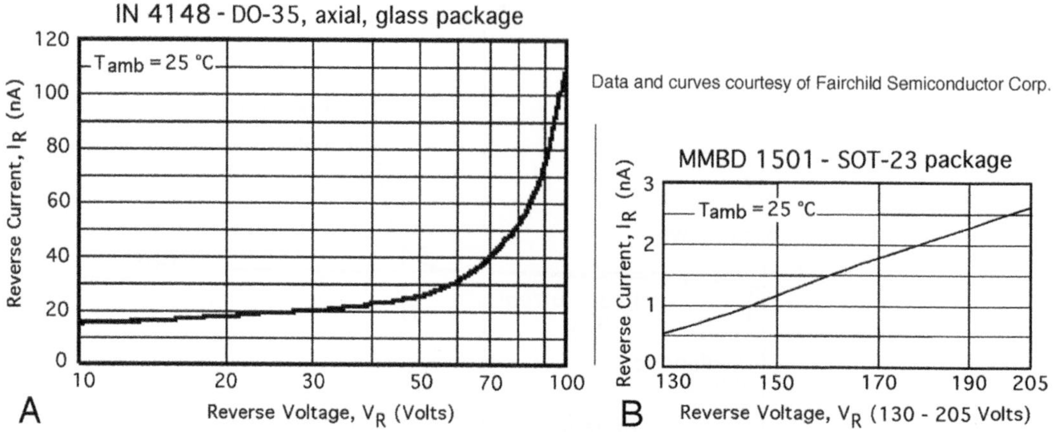

Data and curves courtesy of Fairchild Semiconductor Corp.

IMPORTANT NOTE: A diode's reverse current approximately doubles for every 10° C rise in temperature.

Figure 3.10. Reverse current versus reverse voltage for a IN 4148 general purpose switching diode, versus a low-leakage, surface-mount MMBD 1501 type. Notice that the low-leakage diode has a much lower reverse current at an even higher reverse voltage.

If the reverse voltage continues to increase, it will reach a point called the *reverse breakdown voltage point* (V_{BR}) (see Figure 3.9), where the current across the junction increases rapidly. This point, where the reverse current changes from a very low value to heavy conductance, is known as the *knee* and is well characterized for most P-N junctions. At this point, minority carriers in the semiconductor material gain enough energy to collide with valence electrons and dislodge them from the crystal lattice, causing an avalanche of carriers. A current-limiting resistor is necessary to prevent the diode from being destroyed by excess current flow. There is nothing inherently destructive about operating in this region, so long as current limiting is always employed.

When reverse voltages greater than V_{BR} are applied, the voltage drop across the junction remains constant at the value of the breakdown voltage, over a wide range of currents. At very low voltages (less than 5 volts), the mechanism by which this works is known as the *Zener effect*, while beyond that (more than 6 volts), the *avalanche effect* is responsible. In both cases a critical electric field in the junction region is required. This creates the mechanism by which zener, avalanche diodes, and silicon transient voltage suppressors function. (This topic is looked at more fully in Part 2 of this book.) Such diodes are specially doped and made of different resistive materials than regular switching or low-leakage type diodes. An alternative to using a low-leakage silicon diode is to use a JFET as a diode, such as the popular N-channel 2N4117A or a specially made JFET diode such as the JPAD 5. These devices have leakage currents of less than 10 pico-amps (1×10^{-11} A), although their forward currents are also less.

One should pay careful attention to the diode's specifications, particularly the maximum forward current (I_{Fmax}), the maximum working inverse (reverse) voltage (WIV), the maximum reverse current (I_{Rmax}), and the forward voltage (V_F). If an AC voltage is applied to the diode, there will be a current through it for every half cycle when the voltage goes positive. This process of converting AC to DC is known as *rectification* and has traditionally been the largest application for diodes from the earliest days.

Knowing the operating point (Q-point) of a general-purpose diode in a circuit will help one determine other useful data about it, such as its *forward resistance* (R_{fwd}). The forward resistance of a diode (the series or differential resistance) refers to its small-signal resistance, when the diode is operating in the forward direction at the Q-point. It is determined by the forward current and voltage, or the small-signal AC values. To determine the forward resistance, we simply apply Ohm's law, so that the forward resistance at 5 mA is:

$$Rfwd = \frac{Vfwd}{Ifwd} = \frac{0.7V}{0.005A} = 140\Omega \qquad \text{(Eq. 3.3)}$$

Whereas at 100 mA, the forward resistance becomes:

$$Rfwd = \frac{0.92V}{0.100A} = 9.2\Omega \qquad \text{(Eq. 3.4)}$$

The classic diode equation is shown as follows, which describes the forward current:

$$I_F = I_S\left(\exp\frac{V}{\frac{KT}{q}} - 1\right) \qquad \text{(Eq. 3.5)}$$

where:

I_S the theoretical reverse saturation current (= 1.87 × 10^{-14} A)

q is the charge on an electron (1.6 × 10^{-19} J)

V is the applied voltage (volts)

K is Boltzmann's constant (1.38 × 10^{-23} J/°K)

T is the temperature (°Kelvin; 298°K = 25°C)

KT/q, the *thermal voltage* equals 0.025875 volt (approx. 26 mV), at room temperature (25°C)

A more modern version of this equation is:

$$I_F = I_R\left(\exp\frac{qV}{KT}\right) \qquad \text{(Eq.3.6)}$$

Notice that the reverse current (I_R) replaces the reverse saturation current (I_S).

It is interesting to note that a forward-biased diode and a forward-biased junction transistor share similar characterstics. These include the diode's forward voltage (V_{fwd}) and the transistor's base-emitter voltage (V_{BE}); the diode's forward current (I_{fwd}) and the transistor's emitter current (I_e); the diode's reverse current (I_{rev}) and the transistor's I_{ebo}. The thermal voltage KT/q (approximately 26 mV) and the V_{fwd} and V_{BE}'s negative temperature coefficient are also similar (see Figure 3.11). We will see just how closely they apply in the next chapter, when we look at using a transistor to create diode action.

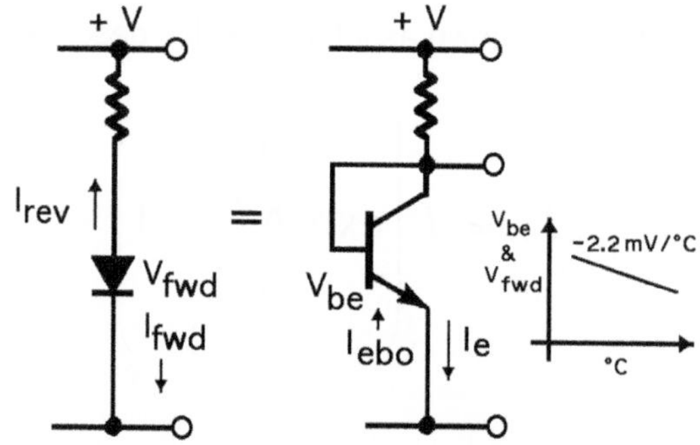

NOTE: Both V_{fwd} and V_{be} have a negative temperature coefficient of approx -2.2mV/°C

Figure 3.11. Comparison of a diode's I_f and V_{fwd} characteristics with similar characteristics of a bipolar junction transistor.

Silicon junction diodes can be made by any one of several standard techniques, which include diffusion, growing, alloying, or meltback, but the most common means of manufacture is by the diffusion process, together with the planar epitaxial technique. An important factor in creating a rugged diode is the *passivation* of its surfaces. This is achieved by forming the small planar junction under the silicon dioxide layer (an insulator), which helps reduce capacitance. It also helps create a more stable, reliable, and repeatable device, which has a low reverse leakage current. The epitaxial structure helps reduce the bulk resistance, which in turn provides higher conductance and low reverse recovery time. A diode's particular specifications can be mostly created by processing (e.g., low-leakage or high-voltage, fast-recovery or fast-switching). Major U.S. manufacturers of small-signal diodes include the following:

- Calogic Inc.
- Central Semiconductor
- Diodes Inc.
- Fairchild Semiconductor
- General Semiconductor/Vishay
- Microsemi
- ON Semiconductor
- Philips Semiconductor
- Zetex Inc.

Diodes come in many different shapes and sizes, which include small plastic or glass DO-7 packages, tiny surface-mount packages, or even large metal-stud packages, when used as power rectifiers. A diode's cathode is usually marked with a "+" symbol, a ring, or a band on the outer package. They are available as single or dual devices, even as a dual-in-line (DIL), or surface-mount arrays. By utilizing different properties of the diode, this has led to the creation of many special types, some of which include the following:

- *Tunnel diode.* An ultra-fast switch (less than 25 pico-seconds)
- *Varactor diode.* Exploits a diode's capacitance for RF tuning
- *Zener diode.* Uses a diode's reverse-voltage characteristics
- *LED.* Light-emitting type using different semiconductor materials
- *PIN diode.* An ultra-fast switch
- *Photo diode.* A high-speed optical detector

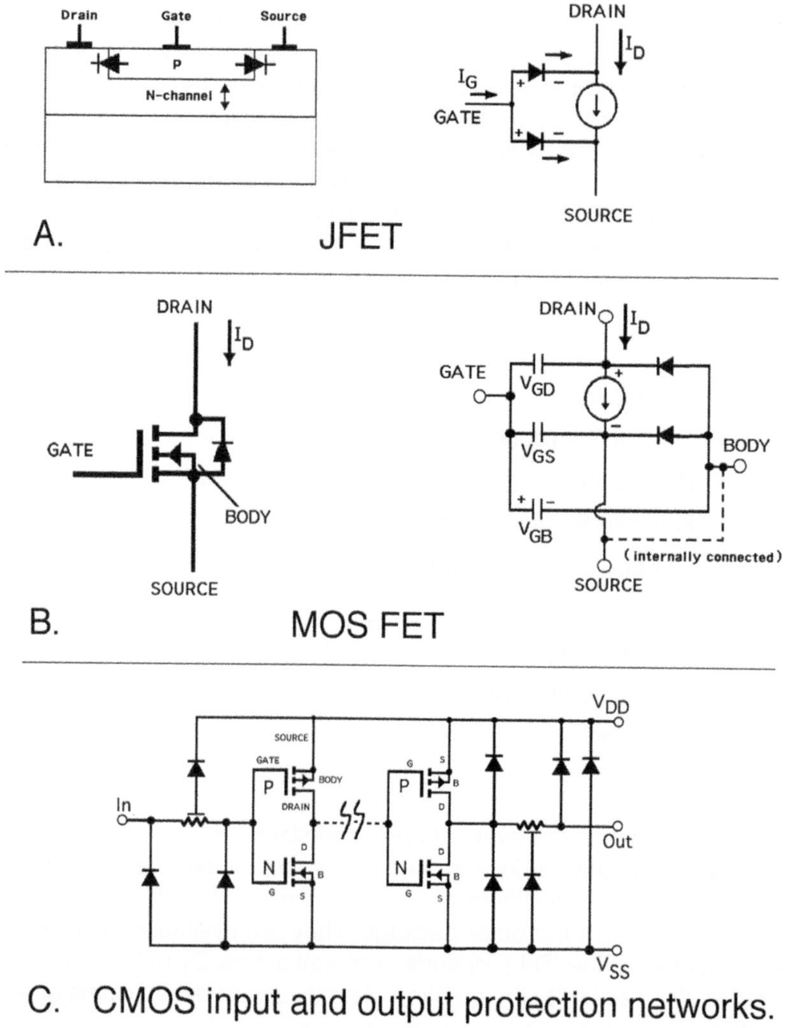

Figure 3.12. *JFET, MOS FET, and CMOS input and output protection networks.*

The most common type is the planar epitaxial switching diode.

In the manufacture of certain discretes and integrated circuits, the creation of neighboring P- and N-regions inherently forms diodes. Depending on the circuit design, this may be beneficial or undesirable. The IC designer has to be vigilant that the intended design does not create an unwanted P-N junction, because in some circumstances it may be necessary to include some form of isolation. For some products the formation of a diode may be beneficial or necessary. For example, with the JFET, its operation depends on either forward-biasing or reverse-biasing its channel-diodes (see Figure 3.12A). Another example is in the making of MOS transistors, where a reverse-biased parasitic diode is created between the drain and the source/body regions (see Figure 3.12B). This diode has a capacitance associated with it (C_{DS}), which plays a part in the MOSFET's output-capacitance and frequency response. However, the diode beneficially acts as a protective clamp when the FET is switching inductive loads. With CMOS circuits, their ESD-sensitive inputs and outputs are deliberately protected by reverse-biased diodes (see Figure 3.12C).

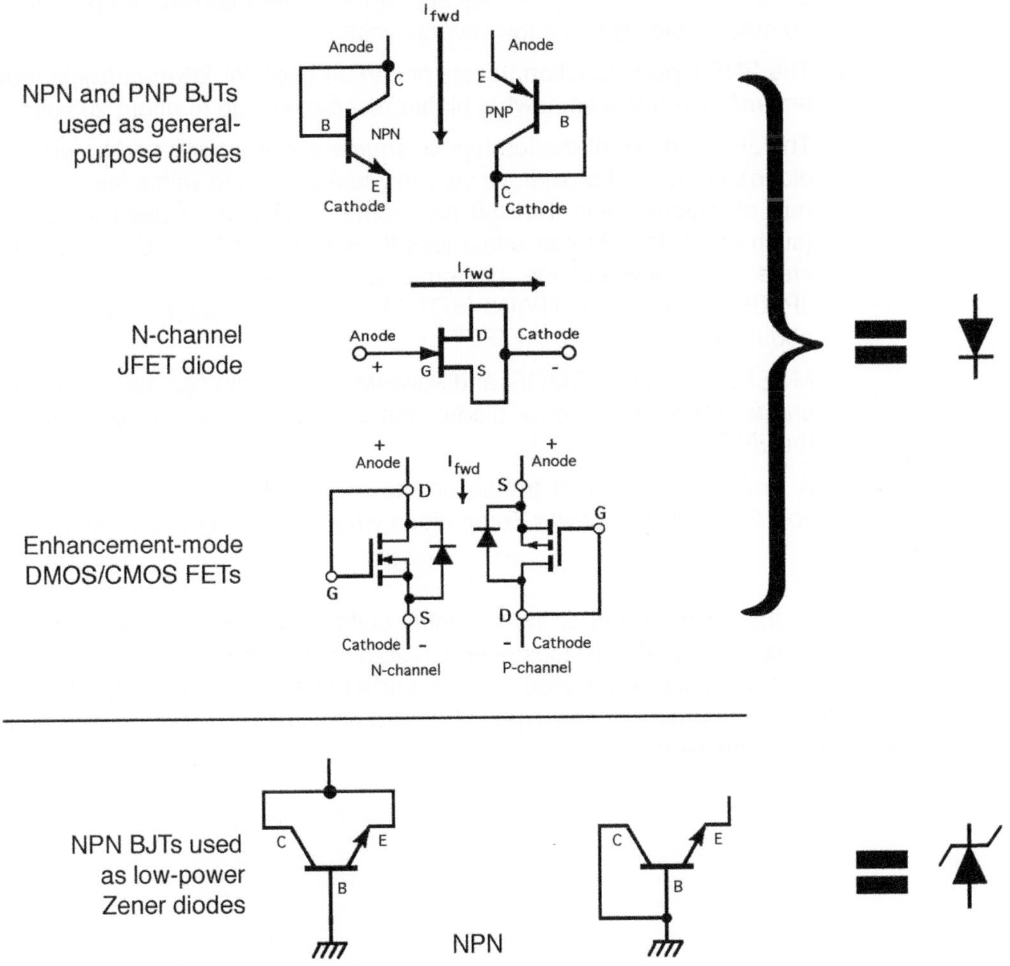

Figure 3.13. *Creating diodes from transistors.*

It is often necessary to create diodes in the architectures of integrated circuits, particularly in start-up, current-steering, current sources/sinks, and voltage reference circuits. Diodes may be easily created from transistors in the design of linear ICs, whether bipolar, bi-FET, or MOS. Figure 3.13 shows how the different types of transistors are used. These include the BJT, the JFET, and MOSFETs. There are several advantages for the IC designer in creating diodes from transistors (e.g., the tight control of characteristics and thereby matching, including thermal matching, and reduced cost), which include the following:

- The NPN bipolar junction transistor with its collector and base shorted together provides a general-purpose diode, with a forward voltage of about 0.7 volt. Although it may support a forward current of up to about 50 mA, it typically has a low reverse voltage.
- The PNP bipolar junction transistor can be used for lower currents (several mA) but has a somewhat higher reverse voltage of about 20 volts.
- The JFET shown at the top has its source and drain shorted to form the diode's cathode. It provides a very-low-leakage diode with a leakage current of typically less than 50 pA. Some JFETs when used as diodes (such as 2N4117A) can attain less than 10 pA at 25°C. Dedicated discrete JFET low-leakage pico-amp diodes, such as Vishay-Siliconix JPAD5 (TO-92) or SSTPAD5 (SOT-23), can attain less than 5 pA. (More about this in Chapter 6 on JFETs.)
- MOSFETs (DMOS, CMOS, and PowerMOS devices) can also be used to create simple yet effective diodes, but with higher leakage currents than the JFET.
- A reverse-biased NPN transistor can make a simple but practical 6.5-volt, 250-mW, low-power Zener diode when it is configured as shown.

In summary, the diode is one of the essential building blocks of modern electronics and one of the most useful devices ever invented. It has come a long way since Dr. Russell Ohl at Bell Labs discovered the P-N barrier in 1939. In the early years of the 21st century, development of new types of diodes continues at research labs and universities around the world.

Chapter 4

Using BJTs to Create Current Sources

Today's silicon planar, epitaxial, bipolar junction transistors are but distant cousins of the early transistors invented at Bell Labs. Transistors have been the subject of hundreds of millions of dollars worth of research over the years, yet today we can buy them for pennies. As with the early development of the Internet by the U.S. Defense Advanced Research Projects Agency (DARPA), so it was with the support of the U.S. Department of Defense, particularly by the Air Force, that fueled the early R&D work into both the transistor and the integrated circuit. Airplanes, rockets, and missiles have payload, weight, and reliability concerns, which solid-state electronics helped improve dramatically. The Cold War was a propelling factor in this advance for the U.S. military, paralleled by the race to the moon and into space by NASA.

It is interesting to note that the earliest transistors were germanium (Ge) point-contact and mesa PNP types, which suffered from severe leakage, low resistance, and temperature drift, among other things. One recurring problem was that impurities seemed to settle on the surface of the chip (surface states) or at its junctions, causing the intended design specifications to shift unpredictably. Another more baffling question was why germanium reportedly turned from N-type to P-type material when it was heated, then allowed to cool. Then there were the persistently poor yield problems, which plagued transistor manufacturers for years. In some cases, manufacturers would get as little as one working transistor per wafer—not a very efficient or profitable situation. Although the semiconductor industry has always been competitive, it has for the most part shared much of the basic technology within itself. With that in mind, research continued at a frantic pace as every manufacturer tried to find the solutions to the yield and reliability problems that plagued the industry.

It was not until the early 1950s that the first germanium transistor was commercially developed and more than a decade later that practical NPN silicon types became commercially available and affordable. In the ten years following the debut of the first point-contact germanium transistor at Bell Labs, about a dozen other types were introduced (each of which depended on the process used in its manufacture). These included the grown junction, the alloy junction, the micro-alloy diffused junction, the epitaxial base type, and many others—all germanium. By the mid-50s, Raytheon was the largest transistor maker, producing roughly half of the world's market needs. It's largest market was supplying transistors to makers of hearing aids.

Although both germanium and silicon display a negative temperature coefficient (-ve TC), in that their resistance decreases with increasing temperature, it was found that a germanium transistor's resistance was far lower than that of its silicon counterpart, by a ratio of about 1000:1. It also had a correspondingly higher leakage current than silicon. Germanium offered a lower forward voltage than silicon (Figure 4.1), but it had a sloppy reverse characteristic, resulting in a soft knee (see Figure 4.1). Because of silicon's improved temperature characteristics and operating range, designers gradually switched away from the troublesome germanium transistors and used silicon almost exclusively.

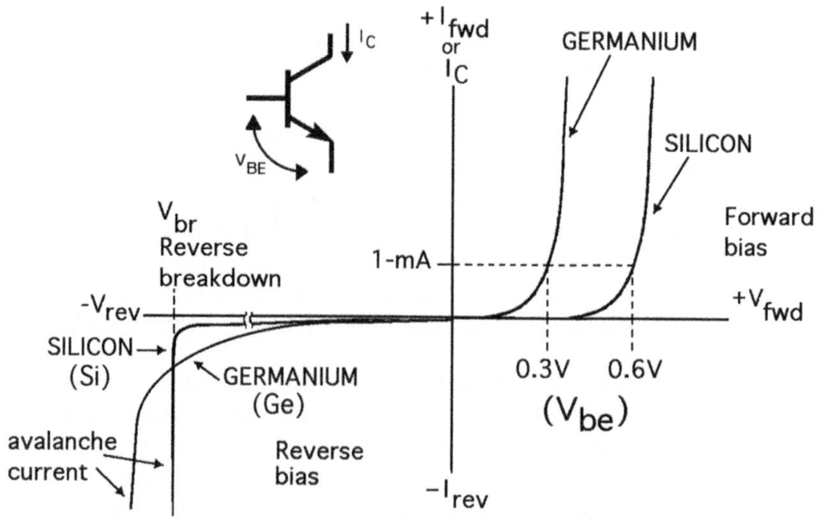

Figure 4.1. *Comparing Germanium's V-I curve with Silicon's. While Germanium has a lower Vf_{wd} and V_{be}, it also has a very poor reverse characteristic.*

So it followed that after 1954, when the first silicon transistors were introduced by Texas Instruments, most transistors and then integrated circuits used silicon, whether digital or analog, until the advent of gallium arsenide (GaAs) microwave transistors in the 1960s. The *epitaxial process* developed by Dr. Ian Ross's group at Bell Labs helped cure some of these manufacturing problems, while a little later (1959), Fairchild Semiconductor's Dr. Jean Hoerni's introduction of the *planar* silicon structure cured many more. Those two processes are the foundation of all of today's semiconductors. As other processes were developed over the course of time (such as Radiation Inc.'s *dielectric isolation* in the mid-60s, then Fairchild Semiconductor's isoplanar™ [isolated-planar] process in 1971), the yields went up, as did quality and device repeatability. Others contributed greatly, such as Transitron, National Semiconductor, Raytheon, Delco, Motorola Semiconductor, and General Electric, along with its subsidiary RCA, which created some of the best power transistors ever made. The development and manufacture of a low-cost, reliable silicon transistor paved the way for all of today's integrated circuits.

In the United States, the following semiconductor manufacturers make today's discrete, small-signal silicon BJTs:

- Central Semiconductor
- Crystalonics
- Diodes, Inc.
- Fairchild Semiconductor
- Linear Integrated Systems
- Microsemi
- ON Semiconductor
- Philips Semiconductors
- Vishay
- Zetex

It's a tribute to all of the companies mentioned here, as well as the thousands of researchers and BJT engineers from the 1960s and '70s, that many of the products they created and modeled are still available today, some 40 years later, and still finding new applications in today's wireless digital world.

4.1 Characteristics of the BJT

The silicon planar, epitaxial, bipolar junction transistor (BJT) is both remarkably simple and complex at the same time. It is a current-operated semiconductor device, usually having three terminals that are designated the *base*, the *collector*, and the *emitter*. It is referred to as *planar* because of its design, which creates all three terminal connections at the surface of the chip. It is *epitaxial* because the chip is created in precisely controlled layers (epitaxy). It is referred to as a *junction transistor* because its inherent design creates distinct junctions between the P and N materials of which it is made. Actually, it has two junctions, unlike a diode, which has just one. It is referred to as a *bipolar* device because both electrons and holes are involved in the current flow. The transistor is available in either polarity, referred to as either an NPN or a PNP type. This is shown in Figure 4.2, together with the relevant polarities, currents, and voltages.

NPN types have free electrons, which are negative carriers, whereas PNP types have free holes, which are positive carriers. One sometimes hears the expression *hole current*, which refers to PNPs, or likewise *electron current*, which refers to NPNs. Electrons have a higher mobility than holes, and as a result tend to move faster through

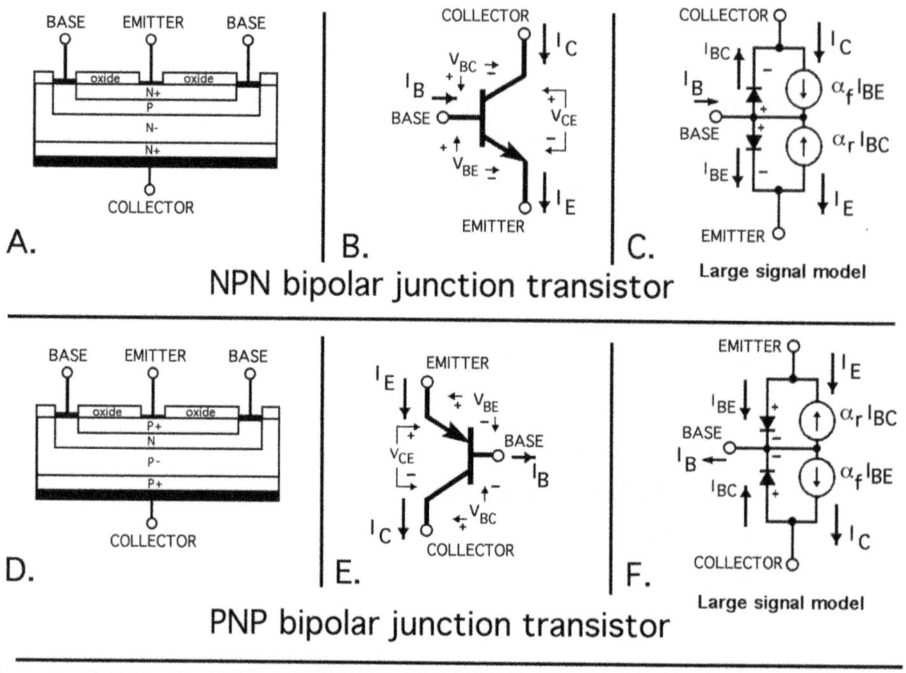

Figure 4.2. *Showing the structure of the BJT, as well as its various voltages, currents, and polarities. Notice the two inherent diodes that provide forward and reverse bias.*

the semiconductor crystal material. This higher speed capability translates into the fact that NPN types are more popular than PNPs. (This is true for not only BJTs but FETs too.) Advances over the past decade have closed the gap considerably between PNP/P-types and NPN/N-types, so that their speeds are much closer. BJTs are available as individual discretes and also as matched-pairs, duals, quads, and arrays, in standard metal, plastic, and surface-mount packages (Figure 4.3). In creating any kind of current source with BJTs, the best results will always be obtained when the devices used are matched-pairs or quads in a single package.

The BJT, being a three-lead device, is capable of three distinct modes of operation. These are the common base (CB), the common collector (CC), and the most familiar—the common emitter (CE). Table 4.1 summarizes some of the basic characteristics of the BJT's three operating modes. As you can see, the CE configuration provides a current gain, a voltage gain, and a power gain, which underscores its popularity. Small-signal transistors are generally categorized into several different types, namely general-purpose amplifiers, switching transistors, high-frequency RF (VHF/UHF) amplifiers, choppers, high-voltage, and low-noise. In addition, there are Darlington, medium-power, and power transistors. In theory, most of these can function as current sources,

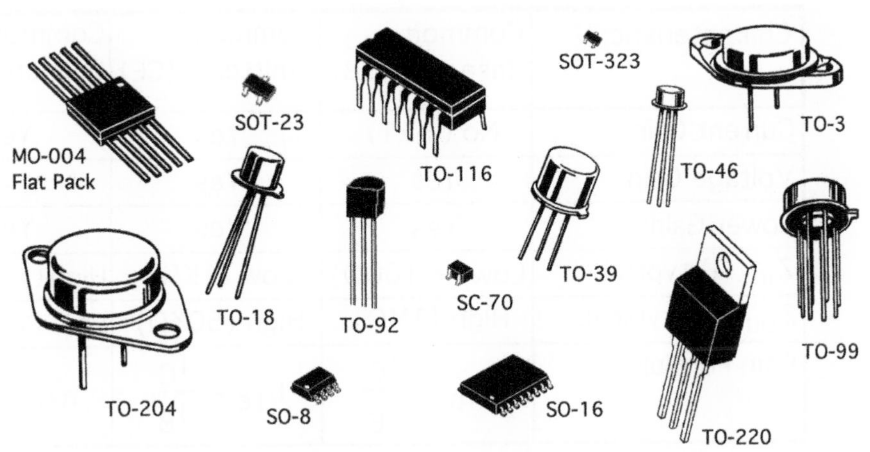

Figure 4.3. *Examples of popular transistor packages.*

with the exception of the Darlington type, which is generally considered to have poor bandwidth. Typically, general-purpose, low-noise, and high-frequency (small-signal) types are used in creating current sources.

In addition, the BJT has some unique subscript notations, which reference either the connection, current flow, or test condition used. In some situations one, two, or three subscript letters may be used. For example, I_C refers to the collector current, while C_{CB} refers to the capacitance between the collector and base. The term I_{CES} refers to the collector-to-emitter current, with the base shorted. (The last letter "S" refers to the condition of the remaining terminal, in this case the base, which is designated as being shorted to the common terminal, always the middle of the three subscript letters.) Thus, I_{CBO} refers to the collector-to-base leakage current, with the emitter terminal open (O) circuit and not shorted. The term $V_{CE(sat)}$ refers to the collector-to-emitter saturation voltage, measured at a particular collector current. The term h_{fe} refers to the forward (designated by "f") gain characteristic in CE mode (designated by "e"). A detailed study of the BJT is beyond the scope of this book, but some additional BJT equations and models are given in Appendix A. We will look here at some of the transistor's more relevant characteristics when viewed as potential current sources.

Characteristic:	Common Base (CB)	Common Emitter (CE)	Common Collector (CC)
Current Gain:	No (< x1)	Yes	Yes
Voltage Gain:	Yes	Yes	No (< x1)
Power Gain:	Yes	Yes	Yes
Z_{in} (Ω) typical:	Low (< 100Ω)	Low (1KΩ)	High (>250KΩ)
Z_{out} (Ω) typical:	High (1MΩ)	High (50KΩ)	Low (< 500Ω)
Gain Formula:	$h_{fb} = \frac{I_C}{I_E}$	$h_{fe} = \frac{I_C}{I_B}$	$h_{fc} = \frac{I_E}{I_B}$

Table 4.1 Transistor operating modes

In the structure of the BJT, the base area is made narrow and is lightly doped. The NPN transistor has a P-type base, whereas the PNP has an N-type base. In order to function properly, the two junctions must be correctly biased. With an NPN transistor, the collector-base junction must be reverse-biased as seen from the base terminal, while the base-emitter junction must be forward-biased. This means that its collector must be positive/more positive than its base, and its emitter more negative than its base. With a PNP transistor the opposite is true: its collector-base junction must be forward-biased, while the base-emitter junction must be reverse-biased. This means that its emitter must be positive/more positive than its base, and its collector must be more negative than its base (see Figure 4.2).

The bipolar transistor (both NPN and PNP) is a normally-off device until sufficient voltage (V_{BE}) and base current (I_B) are provided to turn it on, because it is a current-controlled device. The transistor can be biased into any one of three states: *cutoff* (off), *saturated* (full-on), or *active* (a predetermined state facilitated by biasing it halfway between saturation and cutoff) (Figure 4.4A). For example, in a 12-volt amplifier circuit, the transistor would normally be biased midway between the supply rails at 6 volts, to ensure a perfectly symmetrical AC signal swing (assuming that the input signal was also at the correct level) (Figure 4.4B).

The BJT relies on a small forward base current (I_{BE}) and a small forward voltage (V_{BE}) at its base, inducing and controlling a much larger forward current between the collector and the emitter. The base current (I_B) is very small in comparison with the collector and emitter currents. For small-signal transistors switching a few milliamps, the base current will likely be in microamps. The base current can be determined by:

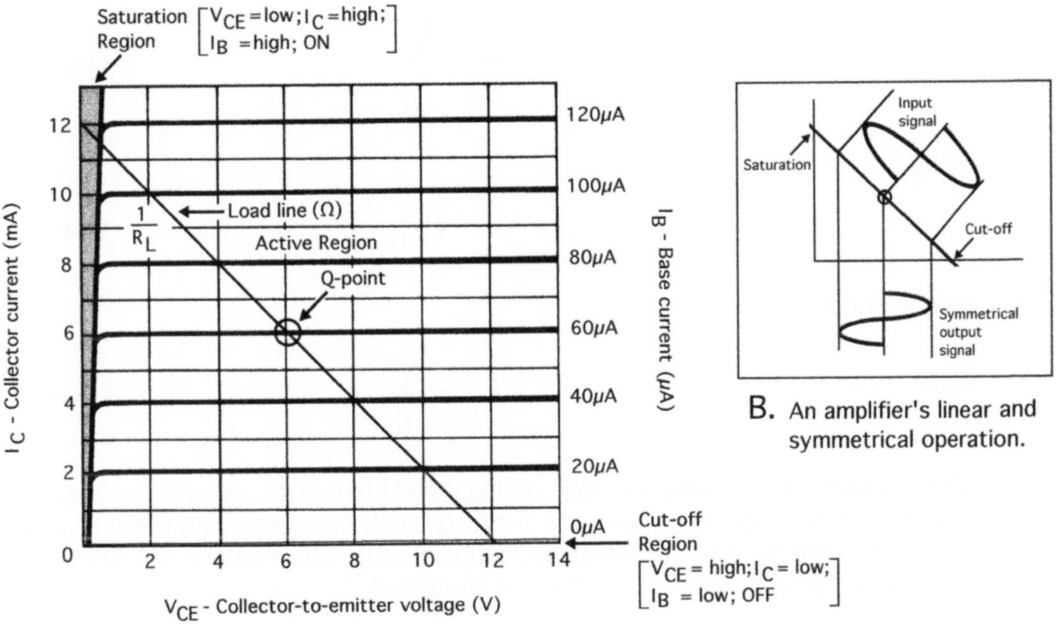

A. Typical transistor amplifier characteristics showing the load line and the three operating regions.

B. An amplifier's linear and symmetrical operation.

Figure 4.4. *Some BJT characteristics.*

$$I_B = I_E - I_C \qquad \text{(Eq.4.1)}$$

This current transfer relationship between base and collector is referred to as the *forward beta* or the *forward current gain* (aka β, βf, h_{FE}, or h_{fe}) in the CE configuration. To be meaningful, the gain should be specified at a particular collector-emitter voltage (V_{CE}), collector current (I_C), base current (I_B), and temperature. For example, the popular 2N3904 NPN general-purpose amplifier is specified at 25°C as having a minimum DC current gain of 70, when measured with a V_{CE} of 1 volt and a collector current of 1 mA. Its counterpart, the equally popular 2N3906 PNP general-purpose amplifier, is specified as having a minimum DC current gain of 80, under the same conditions. There is also a *reverse beta* (βr),which is sometimes referred to in technical material. However, for our purposes, gain is assumed to mean forward beta unless otherwise specified.

Under normal conditions, gain can be simply defined by the following:

$$\beta = \frac{I_C}{I_B} \qquad \text{(Eq.4.2)}$$

where,

$$I_C = \beta I_B \qquad \text{(Eq.4.3)}$$

and therefore:

$$I_B = \frac{I_C}{\beta} \qquad \text{(Eq.4.4)}$$

For example, let's assume we have an ideal NPN transistor, with no leakage. If we applied 10-μA current to its base and 1-mA current is induced to flow between its collector to emitter, then the transistor would theoretically have a beta of 100. From equation 4.2 we can calculate this as shown in equation 4.5:

$$\beta = \frac{1000\ \mu A}{10\ \mu A} = 100 \qquad \text{(Eq.4.5)}$$

A standard test circuit for measuring the BJT's gain is shown in Figure 4.5. Here two current meters are used, which allow one to see the corresponding collector current, for any particular applied base current.

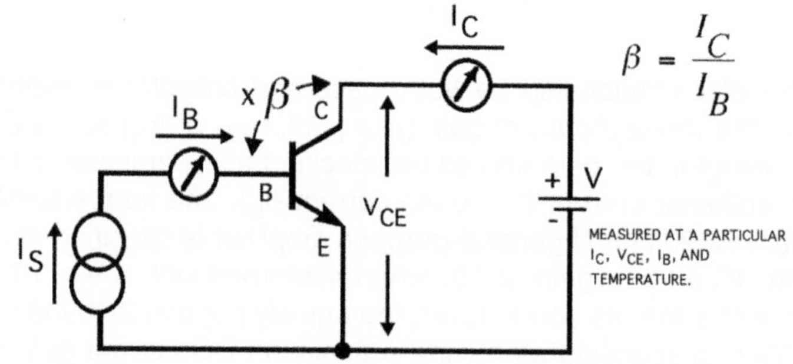

Figure 4.5. Current gain (h_{FE}) test circuit.

While in theory the base current alone directly controls the collector current via the transistor's beta, other factors can also contribute. For example, the base-to-emitter voltage affects gain, because it determines the base current. The typical base voltage for a small-signal transistor ranges from about 0.65 to 0.95 volt. Too low a base voltage will not permit enough base current to flow, which will in turn restrict the amount of collector current. Another factor that can affect gain and cause unforeseen problems is by operating the device too close to its maximum V_{CE} voltage rating—a situation that should definitely be avoided.

Temperature can also affect beta, as shown in Figure 4.6. The two graphs clearly show that for most transistors gain increases with an increase in temperature. It shows the performance of two popular metal can transistors—the NPN 2N2222A and the PNP 2N2907A—over the full military temperature range. (Both devices are available in various other packages, including plastic, DIP, and surface-mount duals and quads.) In the case of the 2N2222A, the graph shows that with a V_{CE} of 1 volt and a collector current of 15 mA, gain increases by about 90% at 175°C, from its normalized value at 25°C. Likewise, h_{FE} decreases by about 50% at –55°C. With the 2N2907, gain also increases with temperature, although not quite as much (75%) because it is PNP, and it too decreases by about 50% at –55°C. In either case, increasing V_{CE} to 10 volts produces similar results over the same temperature range. Note that for either device, gain decreases beyond 60 or 70 mA, irrespective of collector-emitter voltage or temperature. As you can see, gain is most affected by collector current and temperature.

Gain is therefore temperature sensitive, and so with a small-signal transistor operating at low to medium currents, its gain will have a small positive temperature coefficient of between +0.3% to approximately +0.7%/°C, partly because of leakage currents, which we will cover shortly. In the transistor's saturation region, this changes to a small negative tempco of about –0.25%/°C.

Gain is a variable parameter that is difficult to control in processing and manufacturing. As a result, manufacturers normally provide only a minimum gain specification. The particular process will yield a range of gains, where any one device's value may typically range from, say, 100 to 350. Usually the manufacturer will test devices according to gain or some other desirable parameter such as $V_{CE\ max}$, and sort them into separate bins, categories, and part numbers. Devices with higher gains, or with gains close together, may cost more. Small-signal transistors tend to have gains ranging from around 60 to several hundred, whereas power transistors tend to have much lower gains, typically less than 50.

Exceptions to this include Darlington transistors, which can have gains of several hundred to several thousand, and special super-gain transistors created by semiconductor makers in the manufacture of analog ICs. These are specially processed BJTs that give very high gains (typically several thousand), but at the expense of breakdown voltage (just a few volts). However, by combining these super-gain transistors with regular IC transistors in a cascode configuration, it allows an IC design (typically an op amp) to achieve both high gain and a normal level of breakdown voltage. Super-gain transistors were first used in the architecture of the LM108 op amp in 1969.

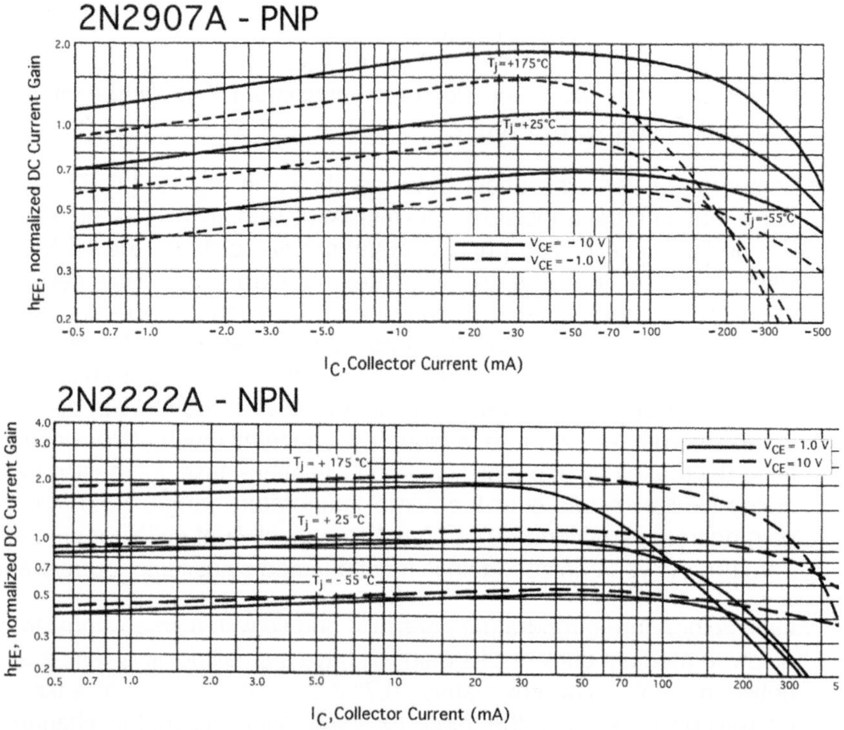

Figure 4.6. *Showing how temperature affects h_{FE}, and therefore the collector current for two popular general-purpose transistors, the PNP 2N2907A (upper graph) and the NPN 2N2222A (lower graph). Notice particularly how as temperature increases, that h_{FE} also increases, irrespective of whether the collector-emitter voltage is 1 volt or 10 volts.*

The transistor's beta is also interrelated with another mechanism called the *alpha* (α). Both of these characteristics are shown in Figure 4.7. Alpha is a figure of merit and a measure of the current gain between the emitter and collector, in the common-base configuration. This is not usually mentioned on a transistor's data sheet, as it is with beta or h_{FE}. It stands to reason that because few current carriers ever reach the collector from the emitter, the collector current will always be slightly less than the emitter current. Alpha usually ranges from 0.98 to 0.999 for today's silicon transistors. (Although beyond the scope of this book, alpha is composed of three other factors, namely the collector efficiency, the emitter efficiency, and the base transport factor.) Alpha is related to these and can easily be determined from:

$$\alpha = \frac{I_C}{I_E} \qquad \text{(Eq.4.6)}$$

also:

$$\alpha = \frac{\beta}{1 + \beta} \qquad \text{(Eq.4.7)}$$

The alpha is directly related to the beta by:

$$\beta = \frac{\alpha}{1 - \alpha} \qquad \text{(Eq.4.8)}$$

As an example, if alpha equals 0.99, then from equation 4.8:

$$\beta = \frac{0.99}{1 - 0.99} = \frac{0.99}{0.01} = 99 \qquad \text{(Eq.4.9)}$$

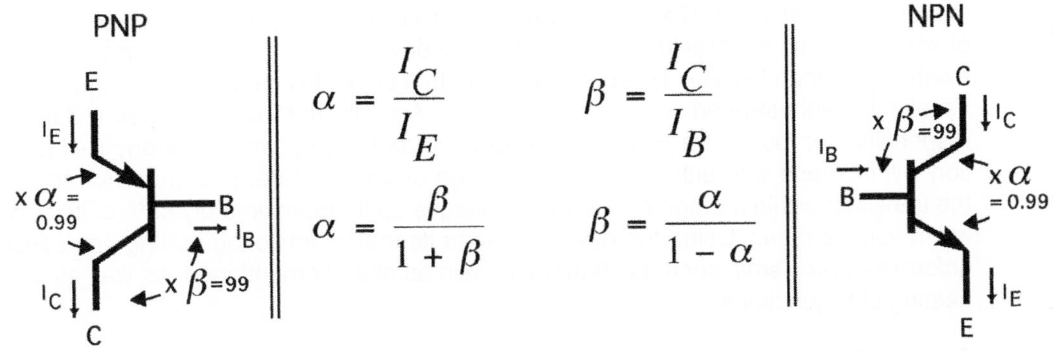

Figure 4.7. *Showing the mechanisms of alpha (α) and beta (β) in the bipolar junction transistor. Alpha relates to the collector current (I_C) versus emitter current (I_E), while beta relates to base current (I_B) versus collector current.*

As with beta, which has a forward and a reverse component, so too with alpha. The equations given so far all imply a forward alpha and a forward beta. The *reverse alpha*

(α_r) is typically half the value of the forward alpha's value, so that with the value given in equation 4.9, the reverse alpha's value would be 0.495. This may be described by:

$$\alpha_R = \frac{\beta_R}{1 + \beta_R} \quad \text{(Eq.4.10)}$$

If one needed to compute the reverse beta for a circuit, it is given as:

$$\beta_R = \frac{\alpha_R}{(1 - \alpha_R)} \quad \text{(Eq.4.11)}$$

So that using the same values as above would yield:

$$\beta_R = \frac{0.495}{(1 - 0.495)} = \frac{0.495}{0.505} = 0.98 \quad \text{(Eq.4.12)}$$

Reverse beta is often just considered as being 1.0 and calculated as such.

Another important set of considerations are the BJT's various breakdown voltage specifications. These characteristics include the all-important collector-emitter breakdown voltage $V_{(BR)CEO}$, the collector-base breakdown voltage $V_{(BR)CBO}$, the collector-emitter breakdown voltage $V_{(BR)CES}$, and others. Many of these characteristics are usually specified at the beginning of a transistor's data sheet in the "absolute maximum ratings" section. BJTs are manufactured according to a particular *process*, much of which dictates the breakdown voltage of the device. In normal operation the V_{CE} is much higher than the V_{BE}. However, from the data sheet it is clear that the $V_{(BR)CBO}$ is usually the highest-rated voltage characteristic. Actually, this collector-base voltage is responsible for much of a BJT's tiny leakage current (I_{CBO}), which like any P-N junction will increase with either increased voltage or temperature. It is important to use the transistor within a range of operating voltages up to approximately 65% of its maximum voltage rating. Using the device close to its maximum voltage rating can cause unforeseen problems, such as changes in gain as already mentioned, as well as self-heating of its junctions.

Figure 4.8 shows a test circuit for measuring some of the NPN transistor's voltage characteristics. For example, $V_{(BR)CEO}$ (shown in Figure 4.8 A) is measured at a specified collector current, with the BJT's base open. $V_{(BR)CBO}$ (shown in Figure 4.8B) will test the breakdown voltage of the collector-base junction. It is measured at a specified collector current, with the emitter terminal open-circuit. $V_{(BR)CES}$ (see Figure 4.8A) tests the breakdown voltage of the collector-emitter junction, with the base shorted to the emitter. $V_{(BR)CER}$ (see Figure 4.8A) tests the breakdown voltage of the collector-emitter junction, with the base connected to the emitter via a specified resistor.

$V_{(BR)CEX}$ (see Figure 4.8A) tests the breakdown voltage of the collector-emitter junction, measured with the base-emitter junction forward-biased at a particular voltage or current. $V_{(BR)EBO}$ (shown in Figure 4.8B) will test the breakdown voltage of the emitter-base junction, when measured at a specified emitter current (I_E), with the collector terminal open-circuit.

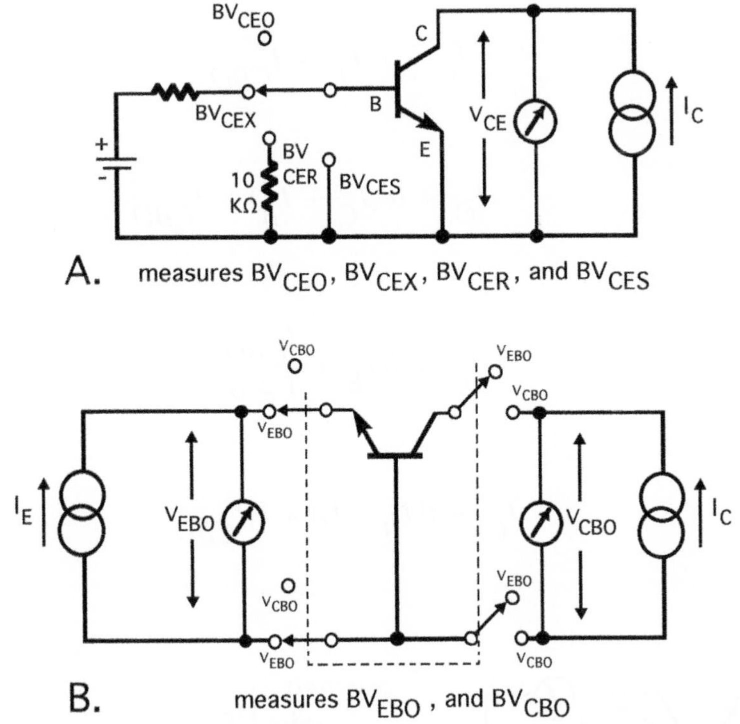

Figure 4.8. *An NPN transistor's various $V_{(BR)CE}$ breakdown voltage test circuits.*

Incidentally, the collector-base voltage V_{CBO} as well as the collector-emitter voltage V_{CEO} both have a small positive temperature coefficient of between +0.07 to +0.09%/°C.

No BJT is perfect: they all have some small degree of leakage currents, caused by minority carriers moving across the junctions. As mentioned previously, the collector-base junction is one of the main contributors to the BJT's leakage current. This particular current is known as the collector-to-base saturation current or the collector cutoff current (I_{CBO}), or as it is more commonly referred to, the I_{CO} leakage current. It is defined as the collector-base current with the junction reverse-biased and with the emitter open-circuit. It is depicted in Figure 4.9 and is inherently related to the collector current, beta, and alpha by the following:

$$I_C = \frac{\alpha}{1-\alpha} I_B - \frac{I_{CBO}}{1-\alpha} \quad \text{(Eq.4.13)}$$

$$I_C = \alpha I_E + I_{CBO} \quad \text{(Eq.4.14)}$$

$$I_C = \alpha (I_C + I_B) + I_{CBO} \quad \text{(Eq.4.15)}$$

$$I_C = \beta I_B + \frac{I_{CBO}}{1-\alpha} \quad \text{(Eq.4.16)}$$

$$I_C = \beta I_B + (\beta + 1) I_{CBO} \quad \text{(Eq.4.17)}$$

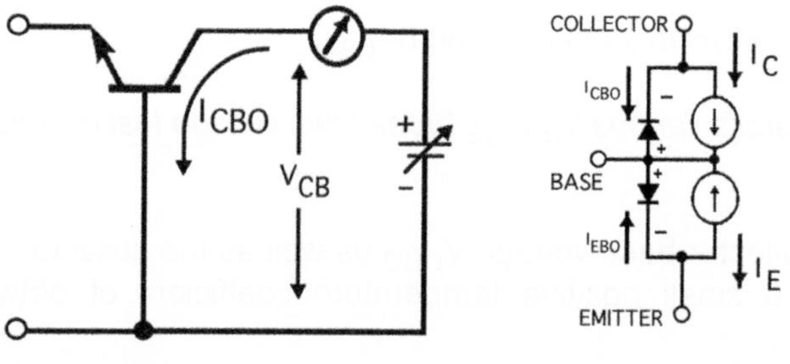

Figure 4.9. *I_{CO} -(I_{CBO}) the BJT's reverse collector-to-base leakage current.*

The I_{CO} leakage current comprises junction leakage currents, surface effects, recombination, and thermal effects. Just as with any P-N junction, leakage currents are caused by thermally generated electron-hole pairs (EHPs) and will increase exponentially with either increased voltage or temperature. I_{CO} has a temperature coefficient that doubles for each 11°C rise in temperature. For silicon devices, the I_{CO} leakage current is usually extremely small, as shown here (where α_f and α_r are the forward and reverse alphas, and I_{CS} is the theoretical collector junction saturation current, given as 1.87×10^{-14}):

$$I_{CBO} = \left(1 - \alpha_f \alpha_r\right) I_{CS} \qquad \text{(Eq.4.18)}$$

$$= 9.536 \times 10^{-15} A$$

That's approximately 10 femto-amps! Manufacturers often specify the I_{CO} leakage current on their data sheets, and a typical maximum value at room temperature is 50 nA at a collector-to-base reverse voltage of 25 volts. This applies to either NPN or PNP devices. Figure 4.10 is an example of such, which shows a graph of the I_{CO} leakage current for the popular 2N5551 (TO-92) and the MMBT5551(an SOT-23 surface-mount version). The 5551 is a high-voltage NPN general-purpose amplifier specified for operation at up to 160 volts. These devices are specified at room temperature, for a *maximum* I_{CO} value of 50 nA, at a collector-to-base reverse voltage of 120 volts. The graph shows an impressive 20 nA at 125°C.

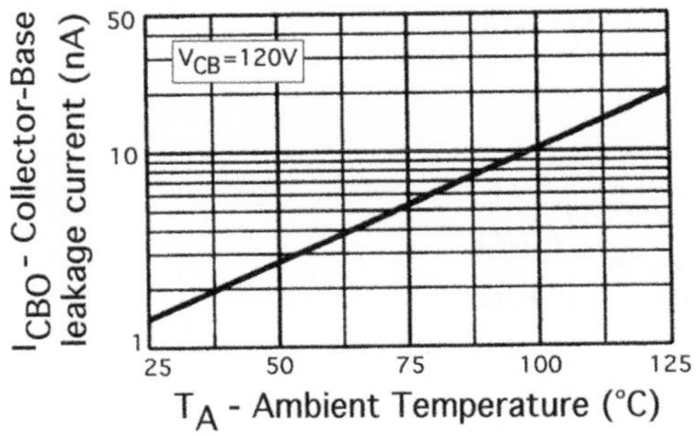

Figure 4.10. *I_{CEO} reverse leakage current versus temperature for 2N5551/MMBT5551.*

Another leakage current is the emitter-to-base leakage current (I_{EBO}), which is depicted in Figure 4.11. It is also referred to as the *emitter cutoff current* and is defined as the emitter-to-base current with the junction reverse-biased and with the collector open-circuit. I_{EBO} is often specified on a BJT's data sheet, at a typical maximum value at room temperature, of 100-nA leakage at an emitter-to-base reverse voltage of 5 volts for NPN devices. This is usually higher (1 μA) for PNP devices under similar bias conditions.

The largest leakage current for a BJT is the collector-emitter leakage current (I_{CEO}), which is the amplified leakage current of I_{CO}. This is because the I_{CO} leakage current (with the base open-circuit) will appear to the base terminal as though it were normal base current. As a result the transistor amplifies this base leakage current, which

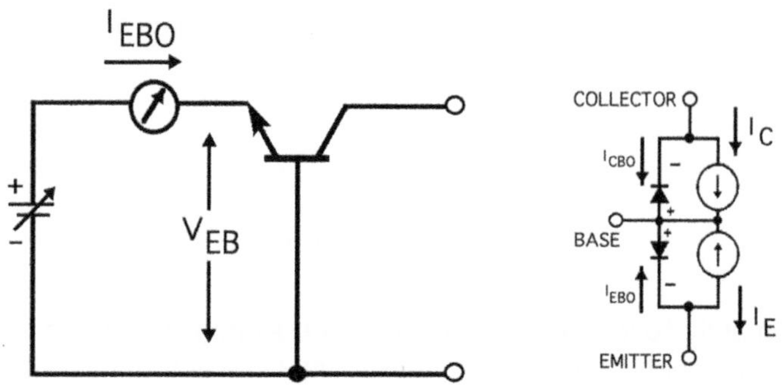

Figure 4.11. *I_{EBO}: the BJT's emitter-to-base reverse-biased leakage current.*

results in I_{CEO}. It is defined as the collector-emitter current with the junction reverse-biased and with the base open circuit. It is worth mentioning that this collector leakage current is also affected by changes in temperature, so that I_{CEO} increases in value four times for each 11°C rise in temperature, and is also gain-dependent. I_{CEO} is depicted as one of the test circuits in Figure 4.12, and can be expressed by the following:

$$I_{CEO} = (\beta + 1)\, I_{CBO} \qquad \text{(Eq.4.19)}$$

also:

$$I_{CEO} = \frac{I_{CBO}}{1 + \alpha} \qquad \text{(Eq.4.20)}$$

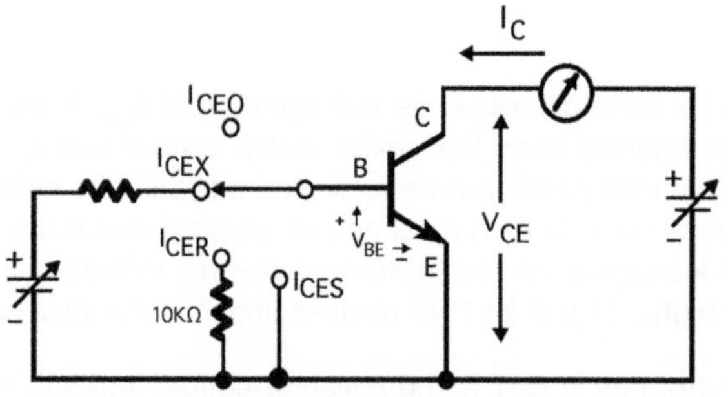

Figure 4.12. *This circuit measures I_{CEO}, I_{CEX}, I_{CER}, and I_{CES}.*

Some other I_{CE} test circuits are also shown in Figure 4.12, which are sometimes referred to in device data sheets. These are the I_{CEX}, the collector-to-emitter leakage current specified at a particular voltage, or current bias at the base. Also I_{CES}, the collector-to-emitter leakage current specified at a particular collector-to-emitter voltage, with the base shorted to the emitter. Another is I_{CER}, the collector-to-emitter leakage current specified at a particular collector-to-emitter voltage, and with the base connected to the emitter via (typically) a 10-KΩ resistor. This leakage current increases in value three times for each 11°C rise in temperature.

To conclude our review of currents, remember that the transistor's largest current is the emitter current (I_E), which can be described by the following equations:

$$I_E = I_C + I_B \quad \text{(Eq.4.21)}$$

$$I_E = (\beta + 1)\left(I_B + I_{CBO}\right) \quad \text{(Eq.4.22)}$$

In addition, the collector current (I_C), which was previously described in equations 4.13 to 4.17, is the next largest current.

As mentioned previously, in order to function, the BJT relies on a small forward current and a small forward voltage at its base, where the base controls the current between the collector and emitter. This small base-to-emitter voltage is known as the base-to-emitter saturation voltage $V_{BE(sat)}$, and is required to correctly forward-bias (NPN) or reverse-bias (PNP) the junction. The transistor is composed of two junctions: the forward-biased base-emitter junction (similar in many respects to the diode, which we looked at in a previous chapter) and the reverse-biased collector-base junction. A transistor's DC performance often depends on its optimum bias point (Q-point, shown in Figure 4.4), within the circuit, and it must be correctly biased with respect to the supply voltages. This is the active state mentioned previously. To turn it on, the voltage applied at the base must be at or above the transistor's inherent V_{BE}, which for silicon is typically 0.65 volt. The theoretical relationship between this base-to-emitter voltage and the collector current (I_C) can be expressed by:

$$V_{BE} = \frac{kT}{q}\ln\left(\frac{I_C}{I_S}\right) \quad \text{(Eq.4.23)}$$

where kT/q = 25.7 mV; I_C = the desired collector current; I_S = the theoretical reverse saturation current ≈ 1.87×10^{-14} Amps. At a collector current of 1 mA, the resulting V_{BE} of a single transistor can be determined as follows:

$$V_{BE} = 0.0257 \times \ln 5.347 \times 10^{10}$$
$$= 0.635V \qquad \text{(Eq.4.24)}$$

Table 4.2 shows various collector currents and related V_{BE}s based on this relationship expressed in equation 4.23.

V_{BE} (V)	I_C
0.457	1-μA
0.516	10-μA
0.558	50-μA
0.576	100-μA
0.617	500-μA
0.634	1-mA
0.676	5-mA
0.694	10-mA
0.735	50-mA

Table 4.2 The Relationship between V_{BE} versus I_C

$V_{BE(sat)}$ is usually specified on a transistor's data sheet at a particular base current and collector current and can be measured as shown in Figure 4.15 (which is the same circuit for testing $V_{CE(sat)}$). For example, at 25°C the popular 2N3904 NPN general-purpose transistor has a $V_{BE(sat)}$ specification of 0.65 volt minimum to a maximum of 0.85 volt, at a base current of 1 mA and a collector current of 10 mA. With a 5-mA base current and a collector current of 10 mA, the maximum may be as high as 0.95 volt. The popular 2N3906 PNP general-purpose transistor has the same specifications, although other small-signal PNP devices will have different specs. This $V_{BE(sat)}$ voltage directly affects gain, because it determines the base current. Too low a base voltage will not permit enough base current to flow, which will in turn restrict the amount of collector current. Too high a base-emitter voltage can destroy the transistor, although most have an absolute maximum rating of several volts. This specification ranges from about 3 volts for some devices up to about 6 volts for others; the 2N3904 is rated for up to 6 volts, while the 2N3906 is rated for up to 5 volts. (Darlingtons have a higher absolute maximum rating often reaching 10 volts, but they are not really suitable as potential current sources.) Typically, a resistor is inserted in series with the base terminal, to limit the current and voltage to an appropriate level.

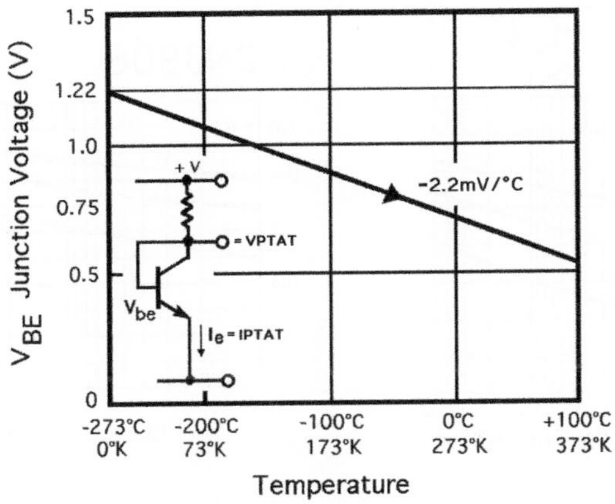

Figure 4.13. *Showing the variation of base-emitter voltage (V_{BE}), with the emitter current proportional to absolute temperature ($I_E \alpha T$) in degrees Kelvin (°K).*

When a silicon transistor's V_{BE} is correctly forward-biased, it can be used as a predictable voltage reference. Device physics suggests that BJTs have a V_{BE} yielding an approximate –2 mV/°C negative temperature coefficient. By extrapolating this temperature characteristic out to "absolute zero" (°K), with its emitter current considered proportional to absolute temperature (I_{PTAT}), it has been calculated that the V_{BE} will reach 1.205 volts at 0°K (–273°C). This V_{BE} can also be considered as proportional to absolute temperature (V_{PTAT}) and is shown in Figure 4.13. (It is one of the fundamental mechanisms by which the *bandgap cell* works and which results in a stable voltage reference. The typical bandgap voltage reference IC provides a compensating voltage, by purposely driving two transistors at different current densities and amplifying the resulting ΔV_{BE} difference, which has a positive TC. This is summed with the original negative TC V_{BE}, to provide a stable 1.205-volt reference voltage, but now with a zero TC. This is the basis for half of the monolithic voltage references on the market today. See Part 2 of this book for more on that topic.)

The V_{BE} voltage is usually specified as having a negative temperature coefficient of between –1.8% to –2.4mV/°C, depending on the process involved in the transistor's manufacture. As a rule of thumb, V_{BE} is usually calculated as having a tempco of –2.2 mV/°C. The two graphs in Figure 4.14 illustrate the general V_{BE} temperature sensitivity, using the 2N3904 and 2N3906 transistors as examples.

As current flows between the collector and emitter, a small voltage drop will occur between the collector and emitter. This is usually referred to as the collector-to-emitter saturation voltage or $V_{CE(sat)}$ and is measured in the common-emitter configuration at a particular collector current, and with a specified base current. This characteristic is shown on any BJT data sheet and usually specified at two different collector currents.

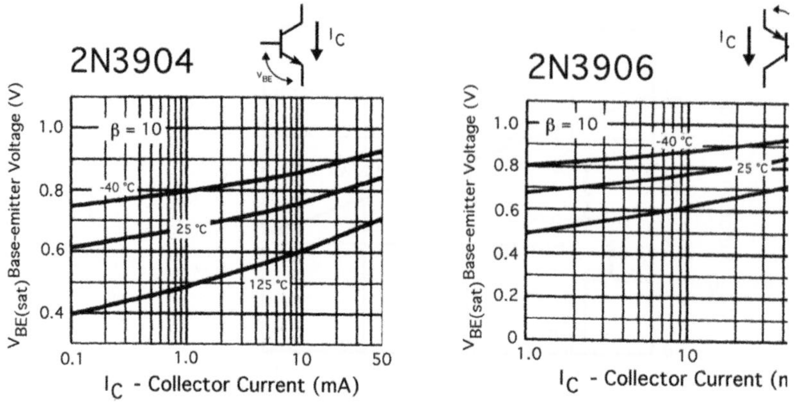

Figure 4.14. *$V_{BE(sat)}$: Base-emitter saturation voltage versus I_C: collector current for 2N3904 and 2N3906.*

It can be measured as shown in Figure 4.15. For most applications, a lower $V_{CE(sat)}$ is more desirable.

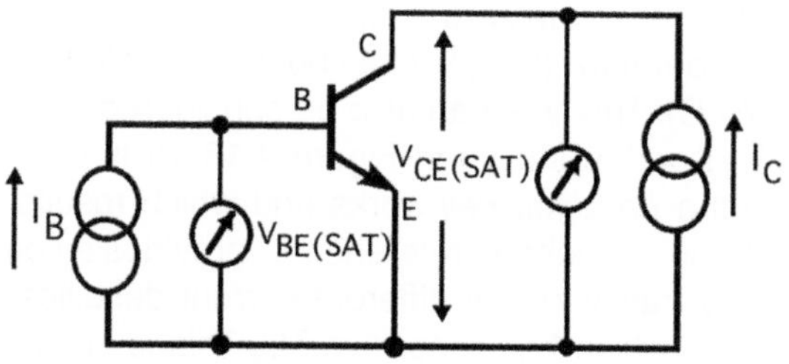

Figure 4.15. *An NPN transistor's $V_{CE(sat)}$ collector-to-emitter saturation voltage measurement circuit.*

$V_{CE(sat)}$ is a variable parameter that is difficult to control in processing and manufacturing. As a result, manufacturers normally provide only a maximum $V_{CE(sat)}$ specification. A particular process may yield a range of $V_{CE(sat)}$ values, where any one device's value may typically range from, say, 0.3 to 0.5 volt. Usually the manufacturer will test devices according to maximum $V_{CE(sat)}$ values and sort them into separate part numbers. Devices with lower values of $V_{CE(sat)}$ may cost more. Typically, small-signal general-purpose amplifiers and switches will have $V_{CE(sat)}$ values of between 0.2 and 0.35 volts. Because this collector-to-emitter voltage is temperature sensitive, it is usually specified (for small-signal transistors) as having a small positive temperature coefficient of between +0.22% to +0.4mV/°C. Higher-power devices have about half these values.

Another important characteristic of the BJT, which is uniquely related to V_{BE} and I_C, is its small-signal forward transconductance (aka g_m). G_m measures the effect of a change in collector current (I_C), for a specific change in V_{BE}, referenced to common-emitter mode. The input voltage and the output current are related by the transconductance, because the base voltage is *transferred* to the collector current and because g_m applies to *conductance*. In practical terms, transconductance is a measure of the BJT's ability to amplify and is a figure of merit. On a graph it appears as a steep or shallow slope and is usually measured in milli-Siemens (mS). G_m is usually measured as milliamps per volt (mA/V). G_m is seldom specified on data sheets, though, and should be either calculated or measured, thus:

$$I_C = g_m V_{BE} \qquad \text{(Eq.4.25)}$$

The transconductance value can be found by the following:

$$g_m = \frac{\Delta I_C}{\Delta V_{BE}} \qquad \text{(Eq.4.26)}$$

It can also be determined from:

$$g_m = \frac{1}{V_T} I_C \qquad \text{(Eq.4.27)}$$

where V_T is the BJT's thermal voltage (kT/q = 25.7 mV). It can also be expressed as:

$$g_m = \frac{I_C}{26} \qquad \text{(Eq.4.28)}$$

where I_C is calculated in milliamps.

These small voltages that occur in $V_{BE(sat)}$ and $V_{CE(sat)}$ can be viewed as internal bulk resistances within the transistor. Actually, as a rule of thumb, the small-signal BJT's base may be considered as having a resistance (R_B) of about 100 Ω; its collector (R_C) about 50 Ω; and its emitter (R_E) about 1 Ω. A reverse-biased junction such as the collector-base junction will typically have a resistance value of around 10 KΩ.

4.2 Using the BJT as a current source

The earliest current sources were built with vacuum tubes back in the 1940s and '50s. Then following the commercial introduction of the junction transistor, the earliest transistor-based current sources were germanium PNP types. As discussed in a previous chapter, because of germanium's unpredictability—particularly its instability with temperature—these current sources were not too practical. Eventually, when both NPN and PNP silicon transistors became available during the 1960s, analog designers could now build current sources that connected to either a positive or negative supply rail.

However, while better than their germanium counterparts in that they were more temperature stable, these current sources were neither accurate nor repeatable, because they were built using discrete transistors, diodes, and resistors (see Figure 4.16). When the silicon bipolar analog IC became a practical reality in the late 1960s, simple current sources began to be used in their architectures for purposes of biasing and stability. Their performances far outperformed their discrete predecessors, because all of the necessary components were integrated into the chip's design. As a result, much development focused on creating current sources for bipolar circuits between the mid-60s and the early '80s. At last, analog designers had found a reliable method of biasing their circuits and thereby improving the overall quality and accuracy of their products. As other technologies have since evolved, such as complimentary-bipolar (CB), analog CMOS, and gallium arsenide (GaAs) RF amplifiers, current sources were used in them also, with similarly good results.

After much research over two decades, the essential requirement found when using BJTs for creating precision current sources, sinks, and mirrors, was in using BJTs with tightly matched V_{BE}s, h_{FE}s, and used in a *monolithic* array. The most critical parameter is the ΔV_{BE}s, which can be viewed collectively like an op amp's offset voltage (V_{OS}), between a monolithic pair of transistors used as a current source (see Figure 4.17). To be clear, the ΔV_{BE} is the difference in voltage between the two V_{BE}s needed to equalize the two collector currents. If the transistors' V_{BE}s are the same and the collector voltages are the same, then the collector currents will also be the same. Actually, current source errors are exponentially proportional to the transistors' V_{BE}s, therefore it is *very important* that one begins with as closely matched devices as possible. Table 4.3 shows that even with a 1-mV ΔV_{BE} at a 100-μA collector current level, there will be a 3% current matching error.

Additionally, in a monolithic array, devices are inherently thermally matched (sharing the same package or substrate), otherwise the slightly different ground potentials, temperature differences/gradients, and overall long-term drift will quickly degrade a discrete circuit's performance. In fact, once a current source design has been built, one should measure its long-term TC V_{OS} stability (measured in μV/month). Ideally, one should aim for 1 μV/month or something close to that. Any accompanying resistors used should be tight tolerance (0.1% or better), low TC types (less than 50 ppm).

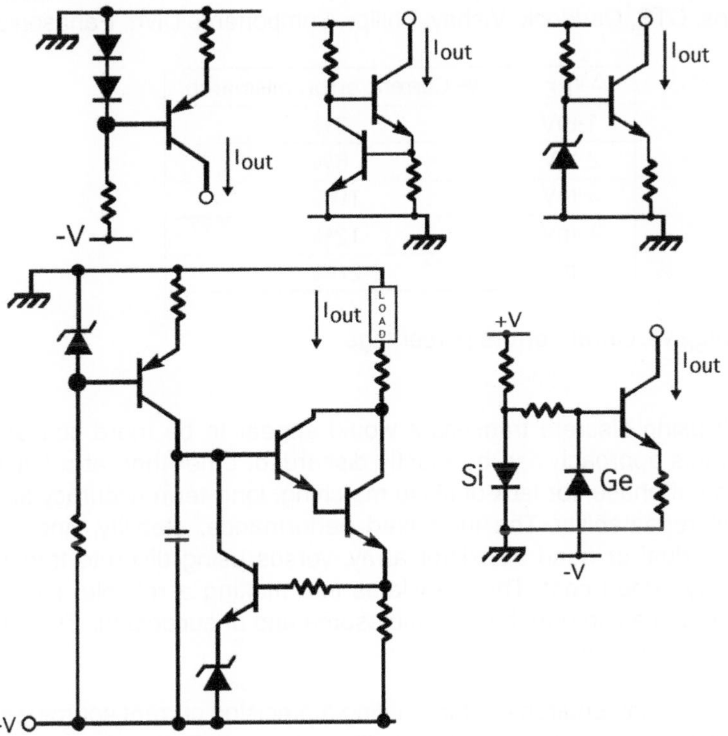

Figure 4.16. Early BJT current sources circa 1960s, and 1970s.

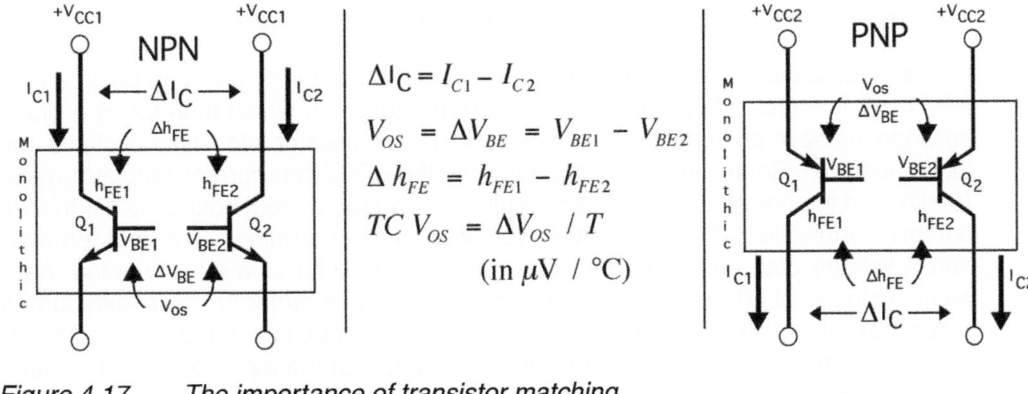

Figure 4.17. The importance of transistor matching.

Ideally, they are also thermally matched, as in a monolithic resistor network (as made by Bourns, CTS, Caddock, Vishay, Philips Components Divn, Panasonic, and others).

ΔV_{BE}	% Current error/mismatch
1-mV	3%
2-mV	6%
3-mV	10%
4-mV	12%
5-mV	20%

Table 4.3 ΔV_{BE} vs. current errors percentage

Although using discrete transistors would appear to be more convenient and economical, this approach can be mostly discarded, other than at a feasibility or initial development phase, for lack of close matching, long-term accuracy and drift, as well as circuit repeatability. The increased performance, stability, and low drift of the monolithic dual or quad transistor array, versus using discrete transistors, far outweighs any added cost. This translates into making a reliable, successful product compared to making one that is troublesome and unsuccessful. The same holds true today and will tomorrow.

Another essential requirement for creating a precision current source, sink, or mirror is running it from a stable, low-noise voltage source. (This subject was discussed in Chapter 2, which suggested using a low-power, low-noise linear regulator such as a National Semiconductor LM2936 or similar.) For low current requirements (i.e., less than 5 mA), a precision low-noise voltage reference may even be used. In some applications, a small electrolytic capacitor (10 μF), in parallel with a 0.01-μF ceramic disc capacitor, may be required to bypass the regulated output to the current source supply. A hash-free regulated supply is essential in all precision work.

Unlike other semiconductors, the BJT does not lend itself very easily to being a two-lead current source. One can tie its base to the collector, which then forms a diode between the base and emitter, or one can invert the transistor into a type of clamp, but that's about all. Some other devices, such as the JFET, the current regulator diode (CRD), and the power MOSFET, can result in a two-lead current source configuration, but any worthwhile BJT current source requires an input terminal, an output terminal, and a voltage bias point. In some configurations, four terminals are needed. As a result, virtually all BJT current sources are configured as *current mirrors* (aka current reflectors), where an input current is mirrored at the output, in either a matched 1:1 or other ratio. This may be either a fraction or a multiple of the input, or scaled in some other manner. (We will look at multiple current mirrors—those with more than one output—as well as scaling later in the chapter.)

Although various BJT current source designs originally appeared, few of them were stable or else used an excessive number of components, as shown previously in Figure 4.16. The simplest two-transistor current sources of modern times are shown in Figure 4.18A and B, which can be built using a pair of either NPN or PNP devices. Technically speaking, the NPN circuit is a *mirror-sink*, while the PNP circuit is a *mirror-source.* This is because I_{C2} mirrors the current in I_{REF}, and the I_{C2} current is either flowing into the circuit (NPN) or out of it (PNP). In either case, the pair of transistors are assumed to have identical characteristics. The PNP version of this circuit works identically, but with all polarities and currents reversed. Of the two types, the PNP version is more commonly used.

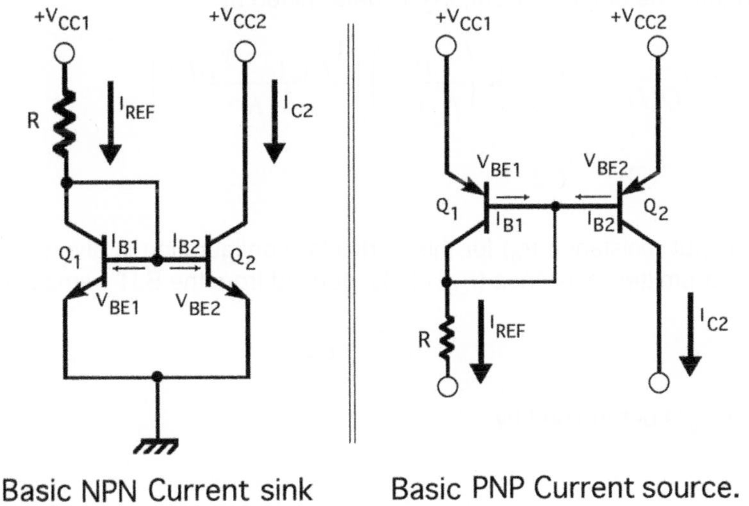

Figure 4.18. Basic BJT current mirrors.

In both current mirror circuits, Q_1's base and collector are shorted together, while the base and emitter act as a diode. The series resistor, R, provides a fixed voltage and current to Q_1's collector, as well as base current for Q_1 and Q_2. In the NPN circuit, because Q_1's collector and base are shorted, and its emitter is grounded, this clamps the V_{BE} at approximately 0.65 volt above ground. The PNP circuit has its V_{BE} at approximately 0.65 volt below VCC. Because both V_{BE}s are assumed here to be identical, so too will be their collector currents. If Q_1's V_{BE} is now impressed across transistor Q_2, then theoretically the reference current, I_{REF}, will be mirrored (duplicated) by Q_2.

Given:

$$I_{B1} = I_{B2};\ V_{BE1} = V_{BE2};\ \beta_{Q1} = \beta_{Q2} \qquad \text{(Eq.4.29)}$$

then to determine the reference current, I_{REF} in either NPN or PNP circuit:

$$I_{REF} = \left(\frac{V_{CC1} - V_{BE1}}{R}\right) \quad \text{(Eq.4.30)}$$

Also, because the reference current is composed of both Q_1's collector current and both base currents, then:

$$I_{REF} = I_{C1} + 2I_B \quad \text{(Eq.4.31)}$$

Therefore, the output current, I_{C2}, is determined by:

$$I_{OUT} = I_{C2} = \left(\frac{\beta}{\beta + 2}\right)\left(\frac{V_{CC1} - V_{BE1}}{R}\right) \quad \text{(Eq.4.32)}$$

$$\therefore I_{C1} \cong I_{C2} \quad \text{(Eq.4.33)}$$

The output resistance (r_o) for this particular configuration is given by the internal collector-to-emitter resistance (r_{ce}) of Q_2, derived from the BJT's small-signal model:

$$r_o = r_{ce2} \quad \text{(Eq.4.34)}$$

where r_{ce} is determined by:

$$r_{ce} = \frac{V_{CE}}{I_C} \quad \text{(Eq.4.35)}$$

As a working design example, the previous equations can be used to design two simple, practical 500-μA current sources in both NPN and PNP, with the following characteristics: Transistors: NMT-2222 (2N2222 type), a general-purpose NPN amplifier pair, in a SOT-6 surface-mount package. Also NMT-2907 (2N2907 type), a general-purpose PNP amplifier pair, in a SOT-6 surface-mount package. The +VCC equals 6 volts; V_{BE} equals 0.65 volt; and h_{FE} equals 50. Assuming that the base currents, gain, and base-to-emitter voltages are identical for both transistors, then:

$$V_{CC1} = V_{CC2} = 6V$$
$$V_{BE1} = V_{BE2} = 0.65V$$
$$\beta_{Q1} = \beta_{Q2} = 50$$
$$I_{B1} = I_{B2} = 10\mu A$$
$$I_{OUT} = 500\mu A \qquad \text{(Eq.4.36)}$$

The resulting circuits are shown in Figure 4.19.

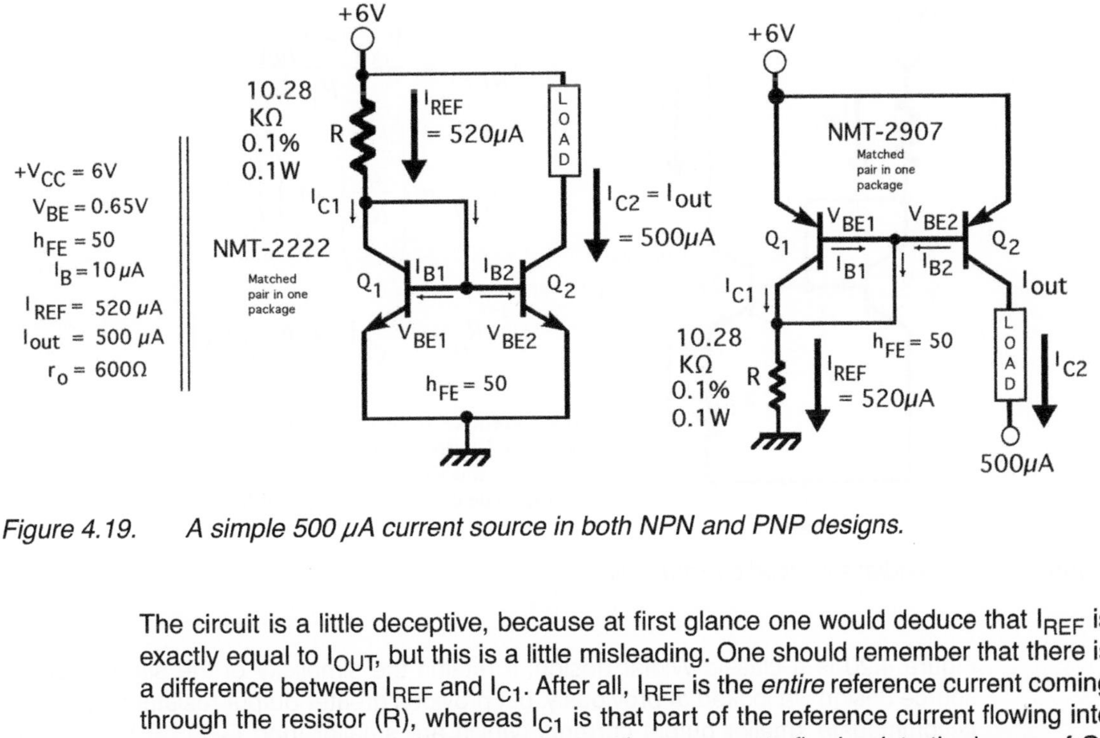

Figure 4.19. A simple 500 μA current source in both NPN and PNP designs.

The circuit is a little deceptive, because at first glance one would deduce that I_{REF} is exactly equal to I_{OUT}, but this is a little misleading. One should remember that there is a difference between I_{REF} and I_{C1}. After all, I_{REF} is the *entire* reference current coming through the resistor (R), whereas I_{C1} is that part of the reference current flowing into Q_1's collector. The rest makes up the two base currents flowing into the bases of Q_1 and Q_2. This was defined previously in equation 4.36. As a result, the reference current should *not* be set at 500 μA, but slightly higher at 520 μA (to accommodate the two times I_B). This will result in an output current of 500 μA, as shown in Figure 4.19. Although this current source is one of the simplest (three components), not to mention quite popular, its regulation is soft and its output resistance is likewise poor at less than 1 KΩ. Others we will review shortly do a much better job in both regards. Nonetheless, if you want a simple, three-component current source, this may be all you need for your application. Also, this current source is often used in conjunction with additional mirrors, which can either multiply or divide its output current (I_{C2}), or even invert it. We will look at this subject more, later in the chapter.

4.3 Widlar current sources

It is likely that the legendary Bob Widlar (who at 25 years old designed the first monolithic op amp—the μA702 at Fairchild Semiconductor in the mid-60s) ran into the same problem when creating the internal biasing for his op amps: he needed to design a better current source. In fact, Widlar designed several different BJT current sources, after all transistors were his pets, and he could make them do some amazing tricks. Widlar loved a challenge, and throughout his short but colorful career, he strove to create excellent products, so he created better current sources to fill his requirements.

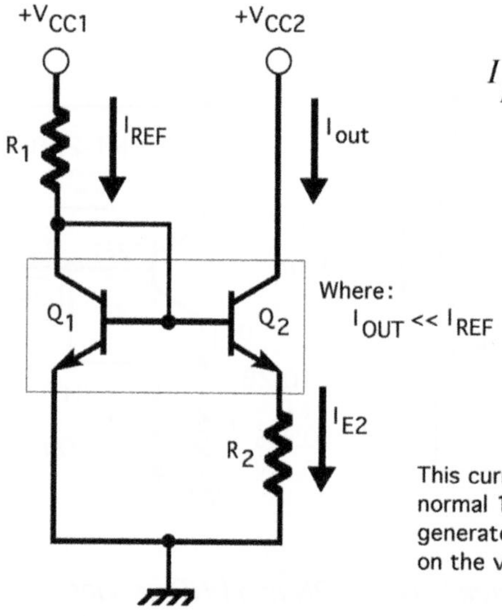

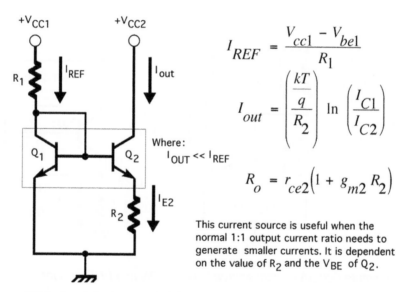

$$I_{REF} = \frac{V_{cc1} - V_{be1}}{R_1}$$

$$I_{out} = \left(\frac{\frac{kT}{q}}{R_2}\right) \ln \left(\frac{I_{C1}}{I_{C2}}\right)$$

$$R_o = r_{ce2}\left(1 + g_{m2} R_2\right)$$

This current source is useful when the normal 1:1 output current ratio needs to generate smaller currents. It is dependent on the value of R_2 and the V_{BE} of Q_2.

Figure 4.20. Widlar's original current sink.

Figure 4.20 is a classic Widlar design; it retains the simplicity of the simple current source design we looked at previously, but provides greater output resistance and can accommodate smaller output currents (which are smaller than I_{REF}). At first glance, the only visible difference between this circuit and the one in Figure 4.19 is the addition of one more resistor, R_2. It is almost certainly a circuit he designed for his internal op amp biasing, because it works best for very small currents (10–500 μA). It may have caused him to look further into the subject of the silicon bandgap (V_{go}), and proportional currents and voltages (I_{PTAT} and V_{PTAT}), and their dependence on temperature. Alternatively, it may have been a by-product of this same subject. It is remarkable to note that with the addition of two components, the circuit transforms from this basic current source into a simple bandgap voltage reference (see Figure 4.21). No doubt this entire subject caused Widlar and his process engineer, Dave Talbert, to look further into creating multiemitter transistor structures and ratioed emitters. Probably Widlar's greatest design (LM-109, the first 5-volt linear power regulator) was based on the bandgap cell, which he pioneered and which relied on ratioed emitters. The LM-109's control circuitry got its temperature compensation from the stable bandgap cell. Without it, the monolithic IC regulator would not have functioned properly at high currents and temperatures. Widlar was the first analog designer to do that.

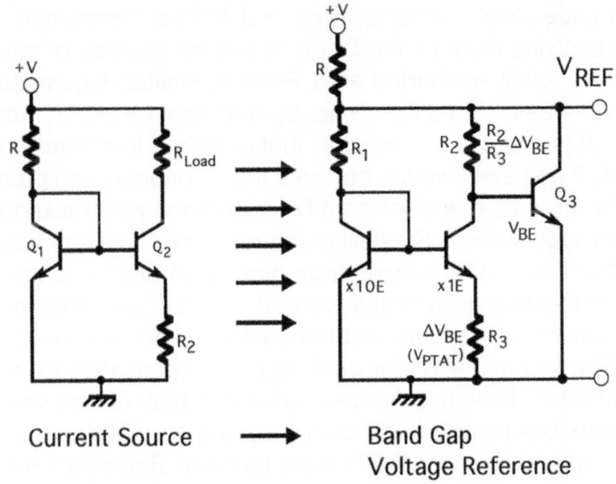

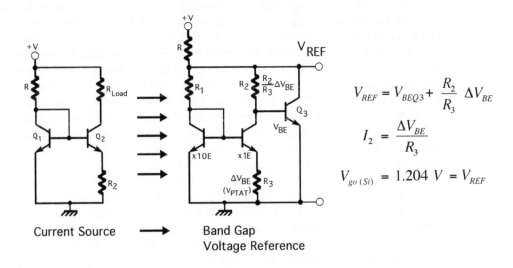

$$V_{REF} = V_{BEQ3} + \frac{R_2}{R_3}\,\Delta V_{BE}$$

$$I_2 = \frac{\Delta V_{BE}}{R_3}$$

$$V_{go\,(Si)} = 1.204\ V = V_{REF}$$

Figure 4.21. Widlar's Leap: Simple but ingenious, the transformation of his simple current-scaling current source into a simple bandgap voltage reference, by the addition of just two components. He did this in the mid-60s, as he created the world's first monolithic op amp, the Fairchild Semiconductor μA702. He was just 25 years old, and would become a legend.

(*Note*: For anyone wondering why I combined the subjects of current sources and voltage references into one book, Figure 4.21 should answer your question.)

Meanwhile, returning to Figure 4.20, by adding R_2 at Q_2's emitter (*emitter degeneration*), the output is no longer near-matched in a 1:1 current ratio, but is divided into a smaller current, which is also proportional to absolute temperature (I_{PTAT}). By changing R_2, you can produce a different output current ratio, such as 4:1, 10:1, and so on. In other words, it provides division of I_{REF} where I_{C2} (the output current) is always smaller than I_{REF} Whereas I_{REF} is calculated in the same manner as previously described (and in equation 4.30), both I_{OUT} and the output resistance (r_o) are determined differently, as shown in Figure 4.20. As a result, the output current depends on both the value of R_2 and of transistor Q_2's V_{BE}. Actually, this technique helps compensate for mismatches in V_{BE}. Typically, the voltage value across R_2 should be made quite small (between 25 to 100 mV). The output resistance (r_o) for this circuit is significantly higher (between 10 KΩ to 100 KΩ) than in the simple current sources of Figs. 4.18 and 4.19, depending on the voltages and currents involved. Although the circuit works well, it is not the best BJT current source (it has its limitations), nor the easiest to work with, but remember it was part of an op amp's internal design and the particular needs of that circuit.

Widlar did a lot of work with current sources and produced several novel circuits and features for his op amp and regulator designs. Some of these were designed to compensate for the little-understood characteristics and limited capabilities of the early BJTs, not to mention applying them for the first time into the creation of monolithic analog integrated circuits—a totally uncharted area. Besides emitter degeneration, another technique he created or was first to utilize in his designs was *beta-compensation.* Even to this day, gain (h_{FE}, β) is a variable parameter that is difficult to control in manufacturing. Back in the 1960s, it was even worse, because the technology and processes were still in their infancy. As a result, manufacturers both then and now have only ever provided a *minimum* gain specification. Small-signal transistors, whether discretes or in a monolithic array, tend to have gains ranging from around 30 to several hundred, at low currents. So that when creating even simple current sources, as we have seen so far, where the initial calculations all assume that the characteristics are identical, in reality this is *not* the case. The technique Widlar devised for compensating for the beta mismatch in a pair of transistors (whether the beta value was high or low) was called *beta compensation.* The compensation was done by inserting a small resistance (typically between 22 to 100 Ω) in series with a BJT's base terminal. Because beta is also voltage-sensitive, this technique also helped compensate for changes in the input voltage. Figure 4.22 illustrates this concept more fully. Another compensation technique he developed (but more for his bandgap voltage references, rather than current sources) was the use of current scaling by emitter ratioing (emitter scaling) (Figure 4.22D). It was a key factor in making his bandgap cell work, which in turn led for the first time to real temperature compensation.

Later, as Widlar refined these techniques, he combined them into other designs, which took the form shown in Figure 4.23. The result was improved output regulation, less sensitivity to beta and V_{BE} mismatches, and increased output impedence.

A generic current source (inventor unknown), which also provides good regulation, is shown in Figure 4.24. Although this current source does not have the output resistance of the Widlar design and uses three transistors instead of two, it does provide reasonable accuracy and is easy to use. The design also reduces the effects of base currents and is often used in analog ICs.

In this circuit, Q_3's base and emitter form a diode in parallel with Q_1's collector and base. The series resistor, R, supplies a fixed voltage and current to Q_1's collector, as

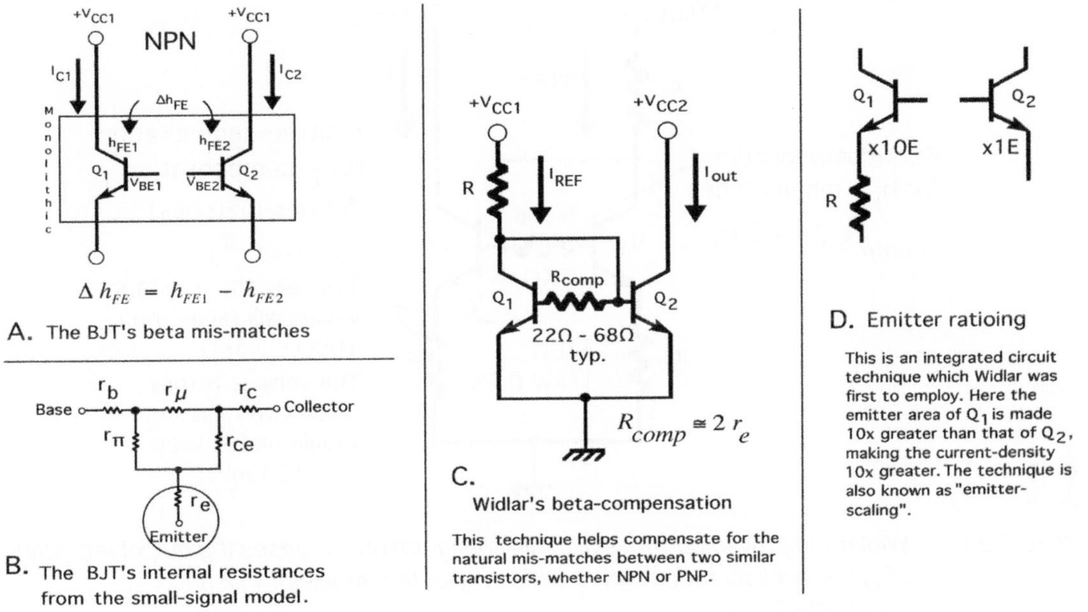

Figure 4.22. *Some of Widlar's compensation techniques used in his current sources and band-gap voltage references.*

well as base current to Q_3. Q_3's emitter and base (V_{BE}) clamps Q_1's collector at approximately 0.65 volt above its base. Q_3's emitter sits at approximately 0.65 volt above ground, because of the V_{BE} of both Q_1 and Q_2. Because both of their V_{BE}s are identical, so too will be their collector currents. As a result, the reference current, I_{REF}, will be duplicated by Q_2.

Given:

$$I_{B1} = I_{B2};\ V_{BE1} = V_{BE2};\ \beta_{Q1} = \beta_{Q2} \qquad \text{(Eq.4.37)}$$

The following determines the reference current, I_{REF}:

$$I_{REF} = \left(\frac{V_{CC1} - V_{BE3} - V_{BE1}}{R}\right) \qquad \text{(Eq.4.38)}$$

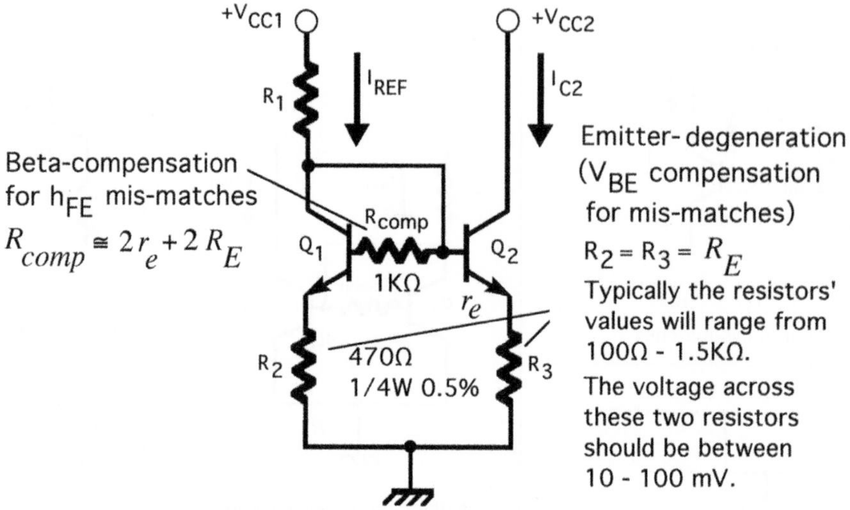

Figure 4.23. *Widlar's improved current sink. This configuration increases the output impedance (Z_{out}), as well as reducing the sensitivity of the transistor h_{FE} and V_{BE} mismatches.*

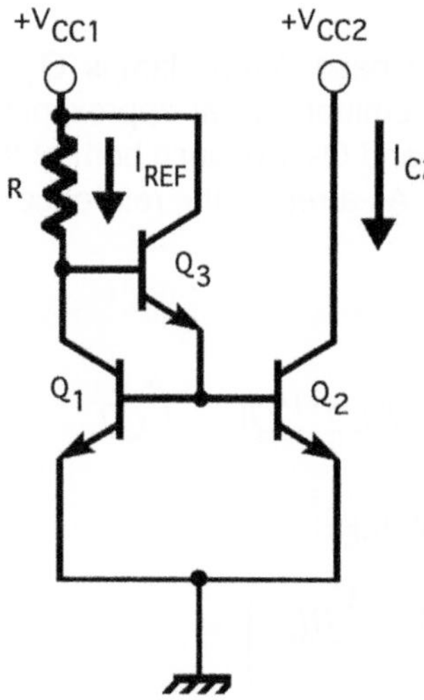

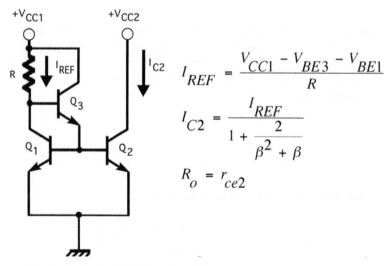

$$I_{REF} = \frac{V_{CC1} - V_{BE3} - V_{BE1}}{R}$$

$$I_{C2} = \frac{I_{REF}}{1 + \dfrac{2}{\beta^2 + \beta}}$$

$$R_o = r_{ce2}$$

Figure 4.24. *Basic three-transistor mirror-sink.*

Therefore, the output current, I_{C2}, is determined by:

$$I_{C1} \cong I_{C2} = I_{OUT} \quad \text{(Eq.4.39)}$$

$$I_{OUT} = \frac{I_{REF}}{\left[1 + \frac{2}{\left(\beta^2 + \beta\right)}\right]} \quad \text{(Eq.4.40)}$$

The output resistance (r_o) for this three-transistor current source is given by the internal collector-to-emitter resistance (r_{ce}), derived from the BJT's small-signal model:

$$r_o = r_{ce2} \quad \text{(Eq.4.41)}$$

where again r_{ce} is determined by:

$$r_{ce} = \frac{V_{CE}}{I_C} \quad \text{(Eq.4.42)}$$

As another working design example, the previous equations can be used to design a simple, practical NPN 250-μA current sink with the following characteristics: Transistors = MMPQ3904 (2N3904 type), a general-purpose NPN quad amplifier, in a SOIC-16 surface-mount package. The +VCC equals 9 volts; V_{BE} equals 0.65 volt; and h_{FE} equals 50. Also assuming that the base currents, gain, and base-to-emitter voltages are identical for all four transistors (we will only use three of them), then:

$$I_{REF} = \frac{(9 - 0.65 - 0.65)}{30.8K\Omega} = 250\mu A \quad \text{(Eq.4.43)}$$

The output current is:

$$I_{OUT} = \frac{250\mu\text{A}}{\left[1 + \frac{2}{\left(50^2 + 50\right)}\right]} = \frac{250\mu\text{A}}{1.000784} = 249.8\mu\text{A} \quad \text{(Eq.4.44)}$$

$$I_{C1} = I_{C2} = I_{OUT} = 250\mu\text{A} \quad \text{(Eq.4.45)}$$

If the transistor's beta changes by ±20% because of temperature changes, then we can calculate the difference as follows:

1) *With* $\beta = 40$ (-20%):

$$I_{OUT} = \frac{250\mu A}{\left[1 + \frac{2}{\left(40^2 + 40\right)}\right]} = \frac{250\mu A}{1.0012195} = 249.69\mu A$$

$$\therefore I_{OUT} = 249.69\mu A \qquad \text{(Eq.4.46)}$$

2) *With* $\beta = 60$ (+20%):

$$I_{OUT} = \frac{250\mu A}{\left[1 + \frac{2}{\left(60^2 + 60\right)}\right]} = \frac{250\mu A}{1.000546} = 249.86\mu A$$

$$\therefore I_{OUT} = 249.86\mu A \qquad \text{(Eq.4.47)}$$

The maximum difference is 0.31 microamps (or less than 0.2%), which is acceptable.

It is common to portray a circuit's characteristics by using the temperature coefficient (TC) as either a percentage or in parts per million. This is referred to as the circuit's *tempco.* For a current source, the tempco of the output current ($\alpha\, I_{O(TC)}$) is the ratio of change in output current, to the change in temperature (say, from 20°C to 85°C). This is usually expressed as a *percentage* of the output current (e.g., 0.025%) and is the average value for the total temperature change. You can determine the tempco from your data by the formula in equation 4.48:

$$\alpha I_{O\,(TC)} = \pm \left(\frac{I_o @ T_2 - I_o @ T_1}{I_o @ 25°C}\right)\left(\frac{100\%}{T_2 - T_1}\right) \qquad \text{Eq.4.48}$$

For example, we needed to check a current source design over a temperature range of 0°C to +85°C. After taking some current measurements in a controlled environment, we obtained the following results:

I_{out} at 25°C = 252 μA

I_{out} at 0°C = 248 μA

I_{out} at 85°C = 267 μA

Using the equation 4.48 would yield:

$$T_C = \left(\frac{267\mu A - 248\mu A}{252\mu A}\right)\left(\frac{100}{85}\right) = 0.089\% \qquad \text{Eq.4.49}$$

We can also express the tempco as a value in *parts per million per degree Celsius* (ppm/°C), by using the formula in equation 4.47:

$$T_C = \left(\frac{I_o @ T_2 - I_o @ T_1}{I_o @ 25°C}\right)\left(\frac{1 x 10^6}{T_2 - T_1}\right) \qquad \text{Eq.4.50}$$

In this example, the output current's tempco could also be expressed as 887 ppm/°C—only a mediocre current source design over the temperature range. This would suggest that some more work may be needed to bring its tempco down to a few hundred ppm or lower.

In summary, the lower the tempco, the less the current source will drift over the intended temperature range, and the more stable and reliable your circuit will be.

4.4 Wilson current mirrors

Of all BJT current sources, the most popular and widely used are the Wilson family, named after its inventor. The Wilson family can be configured as either a three- or a four-transistor cell, in either NPN or PNP versions. The three-transistor cells are referred to as *Basic Wilson* current mirrors, and the four-transistor versions are referred to as *Full Wilson* current mirrors. These current sources are configured as either mirror-sinks (NPNs) or mirror-sources (PNPs). They are all fast and have exceptionally low sensitivity to transistor gain mismatches. In addition, they provide high output resistance. We will look at the four types here, together with some practical examples.

The three-transistor current mirror cells are the NPN or PNP Basic Wilson circuits. Both versions are shown in Figure 4.25, along with the relevant equations. Technically speaking, the NPN circuit is a mirror-sink, while the PNP circuit is a mirror-source. The PNP version of this circuit (shown in Figure 4.25A) works identically, but with all polarities and currents reversed. Of the two types, the PNP version is more commonly used.

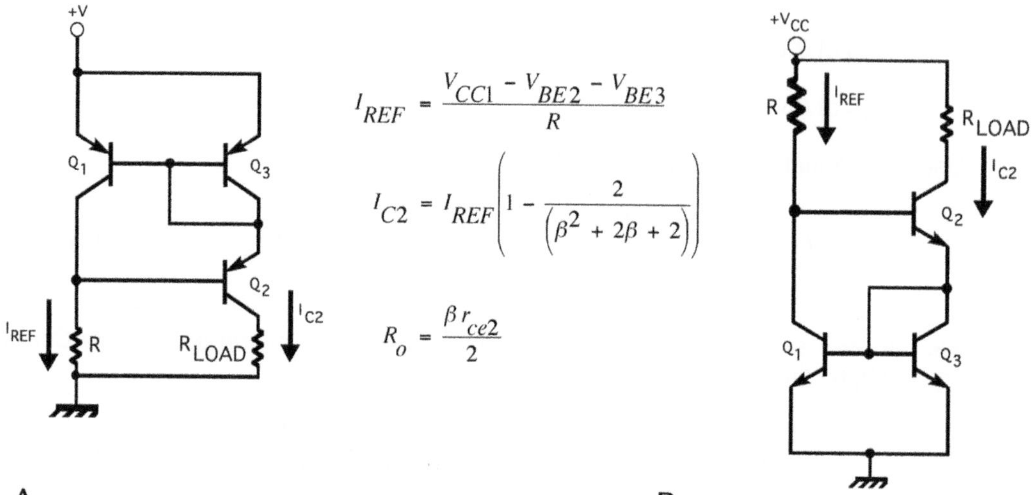

$$I_{REF} = \frac{V_{CC1} - V_{BE2} - V_{BE3}}{R}$$

$$I_{C2} = I_{REF}\left(1 - \frac{2}{\left(\beta^2 + 2\beta + 2\right)}\right)$$

$$R_o = \frac{\beta r_{ce2}}{2}$$

A.
The basic Wilson PNP current mirror.

B.
The basic Wilson NPN current mirror.

Figure 4.25. Basic Wilson current mirrors.

As shown in the diagram, a reference current (I_{REF}) flowing through R splits between Q_1 and a small current flowing through the base of Q_2. This causes Q_2 to supply current to Q_3, while a small base current now also flows through Q_1. Transistors Q_1 and Q_3 function as a current mirror, with Q_1 supplying just enough current to keep Q_2 on, with negative feedback from Q1 and Q_3 regulating Q_2's current. This action cancels the base current drawn and makes Q_2's current relatively independent of gain (meaning that it is beta-compensated, but differently from Widlar's). As a result of the negative feedback, the output resistance (r_o) is higher, and the regulation is much sharper than any of the BJT current sources we have looked at so far. Although Q_2 is not critical, both Q_1 and Q_3 should be matched for gain and V_{BE}. Ideally, one should use a quad array device for best thermal and electrical matching, which will provide the best performance and repeatability. (A dual and a discrete will work okay, but not as well.)

Because both Q_1 and Q_3's V_{BE}s are assumed here to be identical, so too will be their collector currents. When Q_1's V_{BE} is impressed across transistor Q_3, then the reference current, I_{REF}, will be mirrored by Q_2 and Q_3. Given:

$$I_{B1} = I_{B3};\ V_{BE1} = V_{BE3};\ \beta_{Q1} = \beta_{Q3} \qquad \text{(Eq.4.51)}$$

To determine the reference current, I_{REF} in either NPN or PNP circuit:

$$I_{REF} = \left(\frac{V_{CC} - V_{BE2} - V_{BE3}}{R}\right) \qquad \text{(Eq.4.52)}$$

The output current, I_{C2}, is determined by:

$$I_{OUT} = I_{REF}\left(1 - \left(\frac{2}{\beta^2 + 2\beta + 2}\right)\right) \qquad \text{(Eq.4.53)}$$

$$\therefore I_{C1} \cong I_{C2} \qquad \text{(Eq.4.54)}$$

The output resistance (r_o) for this configuration is given by the internal collector-to-emitter resistance (r_{ce}) of Q_2 (derived from the BJT's small-signal model), times the beta, then divided by two, thus:

$$r_o = \frac{\beta r_{ce2}}{2} \qquad \text{(Eq.4.55)}$$

where r_{ce} is determined by:

$$r_{ce} = \frac{V_{CE}}{I_C} \qquad \text{(Eq.4.56)}$$

As a working design example, the previous equations can be used to design two simple, practical 220-μA current sources in both NPN and PNP versions, with the following characteristics: Transistors: MMPQ-3904 (2N3904 type), a general-purpose NPN amplifier quad, in a SOIC-16 surface-mount package. Also MMPQ-3906 (2N3906 type), a general-purpose PNP amplifier quad, also in a SOIC-16 surface-mount package. The +VCC supply equals 3.5 volts; V_{BE} equals 0.65 volt; and h_{FE} equals 50.

Assuming that the gain (h_{FE}), base currents (I_B), and V_{BE}s are identical for these transistors as given in equation 4.51, then:

$$V_{CC1} = 3.5V \qquad \text{(Eq.4.57)}$$
$$V_{BE1} = V_{BE3} = 0.65V$$
$$\beta_{Q1} = \beta_{Q3} = 50$$
$$I_{B1} = I_{B3} = 10\mu A$$
$$I_{OUT} = 220\mu A$$

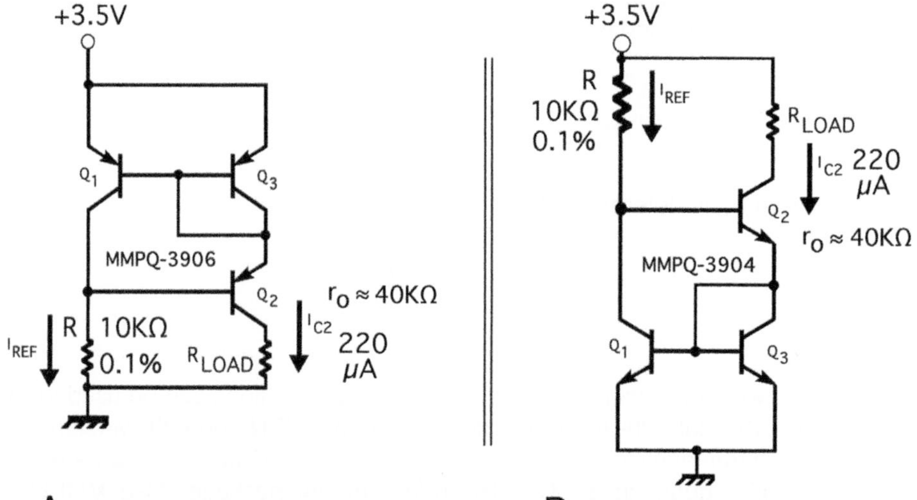

A. PNP current mirror. B. NPN current mirror.

Figure 4.26. Examples of low-voltage basic Wilson current mirrors.

Let's begin by determining the resistor setting value (R), so that:

$$R = \frac{3.5V - 0.65V - 0.65V}{220\mu A} = 10K\Omega \qquad \text{(Eq.4.58)}$$

The output current ($I_{C2} = I_{OUT}$) is given by:

$$I_{OUT} = 220\mu A\left(1 - \left(\frac{2}{50^2 + 2 \bullet 50 + 2}\right)\right) = 219.83\mu A \qquad \text{(Eq.4.59)}$$

The minimum output resistance (r_o) is approximately:

$$r_o = \frac{80000}{2} = 40K\Omega \qquad \text{(Eq.4.60)}$$

The resulting two circuits are shown in Figure 4.26. Although these current sources are the basic type of Wilson, they are very popular, with good regulation and output resistance. The Full Wilsons and cascodes, which we will review shortly, do an even better job. Nonetheless, if you want a simple, reliable four-component current sink or source, one of these may be all you need in your application. The best part is that they are both beta-compensated, and to illustrate this we can show that if the gain (h_{FE}) that we previously calculated at a nominal 50 changes by ±50%, the output current is still regulating really well. For a 50% reduction in gain, this would mean that the gain is now 25, thus:

$$I_{OUT} = 220\mu A\left(1 - \left(\frac{2}{25^2 + 2 \bullet 25 + 2}\right)\right) = 219.35\mu A \qquad \text{(Eq.4.61)}$$

For a 50% increase in gain, this would mean that the gain is now 75, thus:

$$I_{OUT} = 220\mu A\left(1 - \left(\frac{2}{75^2 + 2 \bullet 75 + 2}\right)\right) = 219.92\mu A \qquad \text{(Eq.4.62}$$

When you consider that the nominal output current with a gain of 50 is 219.83 μA (as shown in equation 4.59), this equates to a change in output of –0.22% for the 50% reduction in gain or a +0.041% for a 50% increase in gain. Even if the gain is increased to a phenomenal 150 (an increase of three times), the output is still at 219.98 μA, or an increase of only +0.068%. Even if the gain drops to 15, the change in output is still only –0.7%. That kind of regulation is impressive by any standard and a good example of beta-compensation (thanks to Mr. Widlar and Mr. Wilson). By the way, these circuits will probably work well down to about 2 volts or above 30 volts (but you don't want to get too close to any V_{BR} ratings or start to heat up the transistor junctions). These current sources are sometimes used in conjunction with additional mirrors, which can either multiply or divide their output current.

For much increased output resistance and equally good regulation, the Full Wilson current mirror is highly recommended. These circuits are based on a four-transistor cell, as shown in Figure 4.27, along with the relevant equations. Again, the NPN circuit functions as a mirror-sink, while the PNP circuit is a mirror-source. Of the two types, the PNP version is more commonly used, because it is used primarily for precision biasing and as a high-impedance active load. Adding emitter degeneration (equal emitter resistors, with values typically of between 100 Ω and 1.5 KΩ), as shown in Figure 4.27, compensates for any slight mismatches in the transistors' V_{BE}s. This helps improve overall accuracy and fully optimizes the current source for best overall performance.

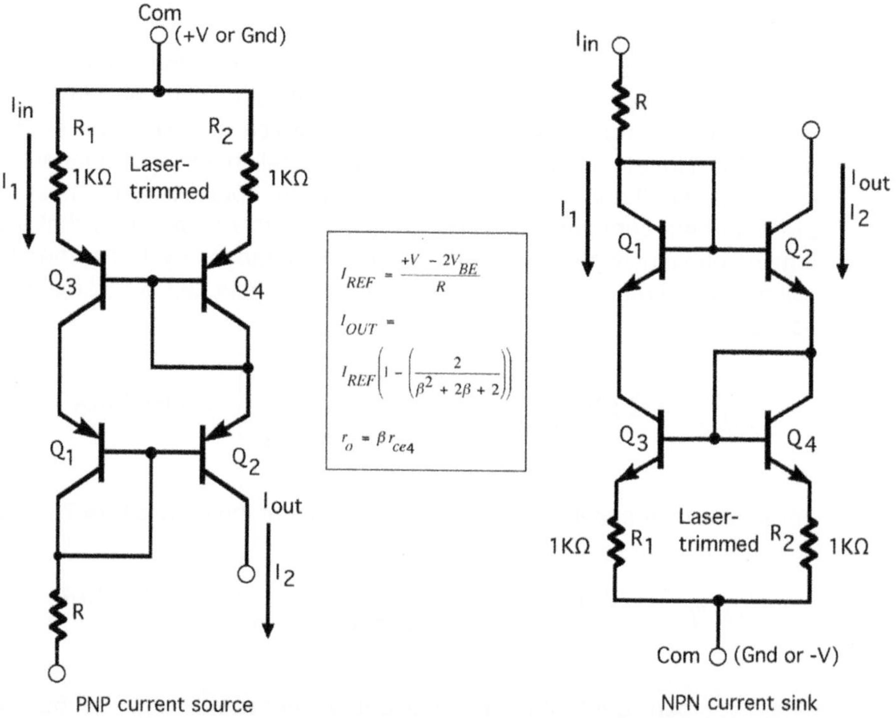

Figure 4.27. *The best current mirror built with the BJT is the Full Wilson with emitter-degeneration. It provides high accuracy, excellent regulation, and very high output impedance. Building the circuit with a quad matched array improves the circuit's temperature coefficient.*

While the calculations and equations are virtually the same as with the three-transistor cell, the noticeable difference is the equation for the output resistance, shown here:

$$r_o = \beta r_{ceQ4} \qquad \text{(Eq.4.63)}$$

where again r_{ce} is determined by:

$$r_{ce} = \frac{V_{CE}}{I_C} \quad \text{(Eq.4.64)}$$

Here the output resistance is "beta times" the collector-emitter resitance of the transistor Q_4 (in both circuits), which can provide a value from hundreds of KΩ to tens of MΩ, depending on the voltage, current output, beta, and r_{ce} values. It's worth noting that the Texas Instruments/Burr-Brown REF-200 contains a Full Wilson current mirror and two current sources in a choice of either an eight-pin DIP or an eight-pin SOIC surface-mount package. That current mirror is specified as having a minimum output impedance of 20 MΩ, and the emitter degeneration resistors are laser-trimmed for optimum accuracy. It would be hard to match those specs, building your own.

In summary, the Wilson current mirrors are probably the first choice in building any current mirrors with the BJT. They offer higher accuracy (better than 0.01%), as well as a high output impedance, and building them in a monolithic rather than discrete form will render improved temperature tracking and a low tempco.

4.5 Wyatt current source

It's perhaps a coincidence that the three American engineers who have developed far-reaching and incredibly useful current sources all have their surnames start with a W, but it's true. So far we have met two of them, Mr. Widlar and Mr. Wilson. Now we come to the third member of the trio, Michael Wyatt. The *Cascode Peaking Current Source*, depicted here in Figure 4.28, was developed at Honeywell's Space Systems Division (Clearwater, Florida) by Michael Wyatt during development of a proprietary monolithic RF IC project in the late 1990s. Dr. Wyatt is a Senior Engineering Fellow at Honeywell, where he designs ICs. He holds numerous patents and previously won the much-coveted and prestigious *EDN* magazine's annual Best Design Idea award.

This circuit is a high-speed, accurate, cascoded current source, and has a rather unique design, as you can see. Vaguely reminiscent of Widlar's design, this too provides an output that is a fraction of the input current. Unlike other current sources, this one has a flat regulated current output, regardless of the circuit's input voltage level. In this design, two peaking current sources are welded together—an NPN sink together with a PNP source—to form a cascode. It acheives a low temperature coefficient, by virtue of using aluminum resistors (3333 ppm/°C), which first-order compensate the circuit. Other types of resistors may be used, which include solid carbon, copper, gold, or silver, all of which have similar tempcos.

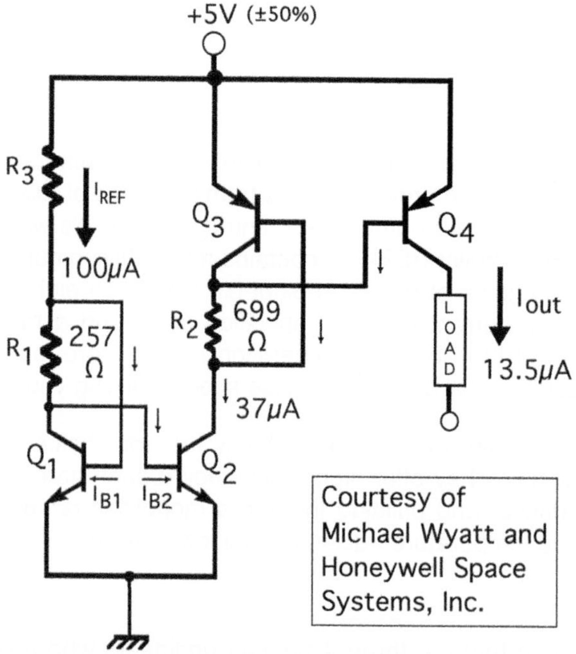

Figure 4.28. *The Wyatt Cascode Peaking Current Source. This design reduces the input current to a lower level, and combines high accuracy with a constant output, and temperature compensation.*

In this design, the following equations apply:

$$R_1 = \frac{V_T}{I_{IN}} \quad \text{(Eq.4.65)}$$

$$I_{IN} = \frac{V_T}{R_1} \quad \text{(Eq.4.66)}$$

where V_T is 25.7 mV, the transistor's theoretical thermal voltage. Resistor R_2 can be calculated by:

$$R_2 = e^1 \times R_1 \quad \text{(Eq.4.67)}$$

and the output current can be approximated within 1% or 2% by:

$$I_{OUT} = I_{IN} \times e^{-2} \qquad \text{(Eq.4.68)}$$

As an example, we will create a 13.5-μA current source using a supply voltage of 5 volts DC and an input current of 100 μA. The NPN pair, which forms the first stage, reduces the 100 μA down to 37 μA at the collector of Q_2. This current is the input for Q_3 of the PNP pair, which again reduces the current down, so that Q_4's collector outputs a constant current of 13.5 μA. We can show this by the following:

$$R_1 = \frac{25.7mV}{100\mu A} = 257\Omega \qquad \text{(Eq.4.69)}$$

$$I_{IN} = \frac{25.7mV}{257\Omega} = 100\mu A \qquad \text{(Eq.4.70)}$$

$$R_2 = e^1 \text{ x } 257\Omega = 698.6\Omega \qquad \text{(Eq.4.71)}$$

and the output will be:

$$I_{OUT} = 100\mu\text{A} \times e^{-2} = 0.0001 \text{ x } 0.13533 = 13.53\mu A \qquad \text{(Eq.4.72)}$$

The output is perfectly flat with an input voltage range of between 2 and 10 volts, and with an input current range of 50 μA and 200 μA, over a temperature range of 0°C to 100°C—certainly an impressive design. The Wyatt current source can be optimized by building it using a monolithic complimentary quad device, along with those aluminum resistors. If an uncompensated (tempco) version is acceptable, then regular low-tempco, low-tolerance, metal-film resistors can be used instead. The Wyatt current source is the least known of the Widlar-Wilson-Wyatt trio, but it provides even sharper current regulation than other (non-1:1) types, not unlike the performance of a JFET cascode (see Chapter 6 for more on that).

4.6 Multiple current mirrors

There are many occasions when one needs to build a current mirror with more than one output, as we have looked at up until now. Applications commonly include multi-stage amplifier biasing and high-impedance active loads. For simplicity's sake we will refer to these as multiple current mirrors. In the basic NPN current mirror shown in Figure 4.29, Q_1 is connected as a diode to ground and supplied with a fixed collector voltage and current via resistor R. Because Q_1's collector and base are shorted, and the emitter is grounded, the base is clamped via the V_{BE} at approximately 0.65 volt above ground. If Q_1's V_{BE} is now impressed across the parallel transistors Q_2, Q_3, and Q_4 (assumed to have identical V_{BE}s), then in theory the initial current I_{REF} is mirrored by the others, thus:

$$I_{REF} = I_2 = I_3 = I_4 \qquad \text{(Eq.4.73)}$$

That's the theory, but in practice as these additional transistors are added, an error term is introduced, which is caused by gain-dependent base-currents in Q_2, Q_3, and Q_4, which reduces I_{REF} by as much as 5%. This current error now adds to any errors caused by V_{BE} mismatches. Unfortunately, Widlar's technique of adding a small resistor in the base of the reference transistor for beta-compensation does not work well with these multiple mirrors. The relevant equations are included in the diagram. (The term *n* refers to the number of mirror channels. The diagram shows three, so here *n* = 3.)

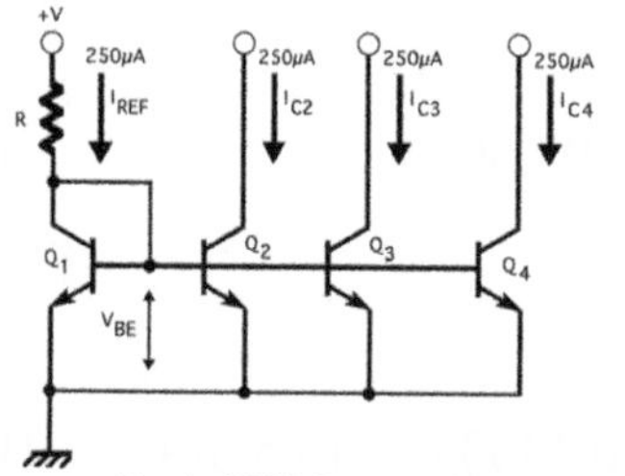

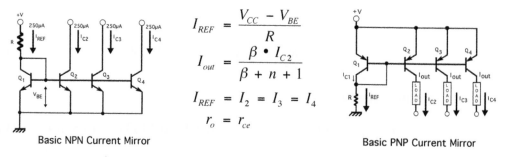

Basic NPN Current Mirror

$$I_{REF} = \frac{V_{CC} - V_{BE}}{R}$$

$$I_{out} = \frac{\beta \bullet I_{C2}}{\beta + n + 1}$$

$$I_{REF} = I_2 = I_3 = I_4$$

$$r_o = r_{ce}$$

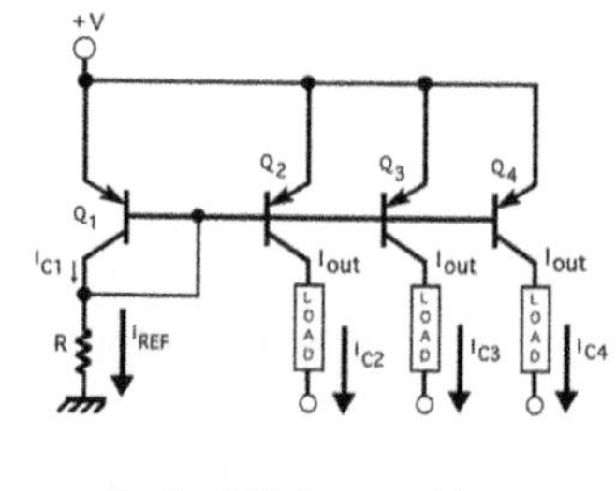

Basic PNP Current Mirror

Figure 4.29. Examples of multiple current mirrors.

Linear IC manufacturers frequently refine their designs in order to improve specs, boost yields, and match new processes. Despite this long experience with bipolar designs over the course of more than four decades, manufacturing BJTs of any type or in any process with exactly matched h_{FE}s and V_{BE}s is still all but impossible. Only *precision matched-pairs* come closest to this ideal. (While various companies make duals/quad devices, they are not true matched-pairs. In the United States, only National Semiconductor and Analog Devices make such products.) Long ago, however, Bob Widlar found that by incorporating small-value resistors (typically in the range of 100 Ω to 1.5 KΩ), in the emitters of the transistors reduces the sensitivity of the output currents to these small V_{BE} mismatches—a technique referred to as *emitter degeneration*. So, to reduce the V_{BE} errors in the multiple current mirror, adding small-value emitter resistors is recommended, as shown in Figure 4.30. It will definitely help.

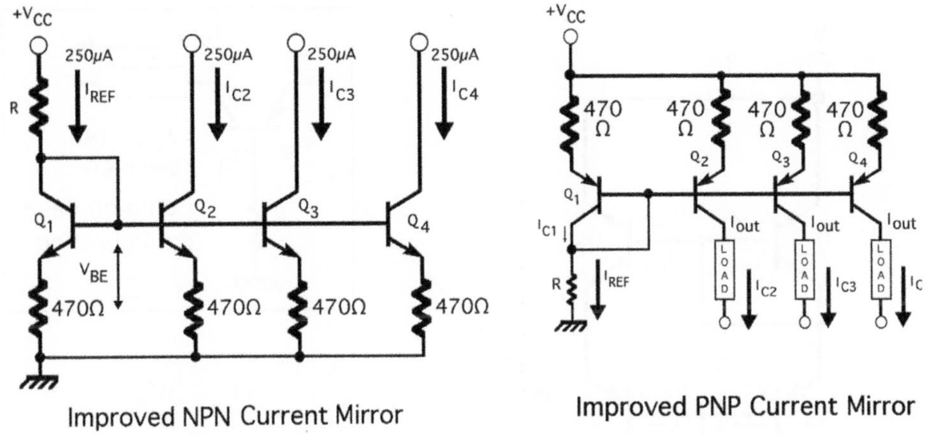

Figure 4.30. *Emitter degeneration helps compensate for small V_{BE} mismatches.*

For more than two decades, manufacturers have used laser-trimming techniques (most notably for adjusting op amp offset voltages) to precisely trim the values of the bias resistors. Precision Monolithics Inc. (now a part of Analog Devices) was a pioneer in this technique known as *zener zapping*. It automatically trims V_{OS} at the wafer level, by using a computer-controlled laser to effectively short one or more of a string of zener diodes, and thereby reduces the offset voltage from the millivolt level to less than 100 μV. By using computer control, this process takes less than 1 second per device on the production line. Applying similar techniques to current sources would result in much improved accuracy. Any of the current mirrors shown in this chapter, whether NPN or PNP, can include emitter degeneration and could have their emitter resistors precisely trimmed, if you were to use the circuit as part of a custom-made ASIC.

4.7 Cascode current mirrors

Although the current mirrors depicted in Figures 4.29 and 4.30 are easy to work with and give adequate performance, they lack much output impedance, or sharp regulation, and are not very good at high frequency either. However, better regulation can be achieved by using two similar devices in a cascode configuration. An example of a PNP cascode current mirror is shown in Figure 4.31. The figure shows a PNP cascode, but the circuit may be configured as an NPN cascode. A monolithic quad device will be perfect for this application. Cascoding generally improves high-frequency operation and provides even greater output impedance (Z_{OUT}). It also permits higher-voltage operation and compliance and appreciably reduces the small-signal output conductance (h_{oe}).

You have probably noticed that at first glance this looks exactly like the Full Wilson circuit (shown in Figure 4.27), but upon closer inspection you will see it is not. In the Full

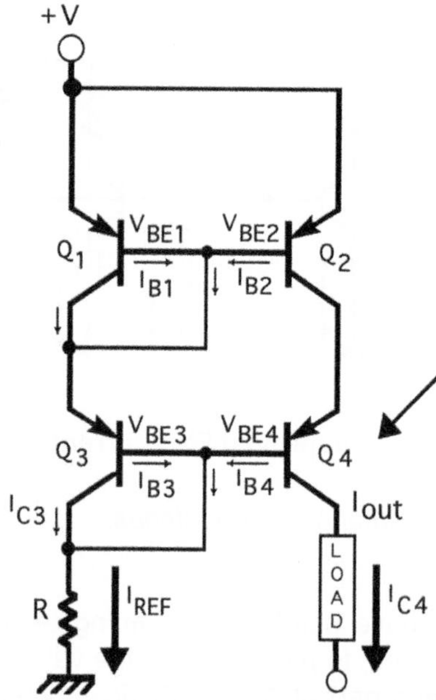

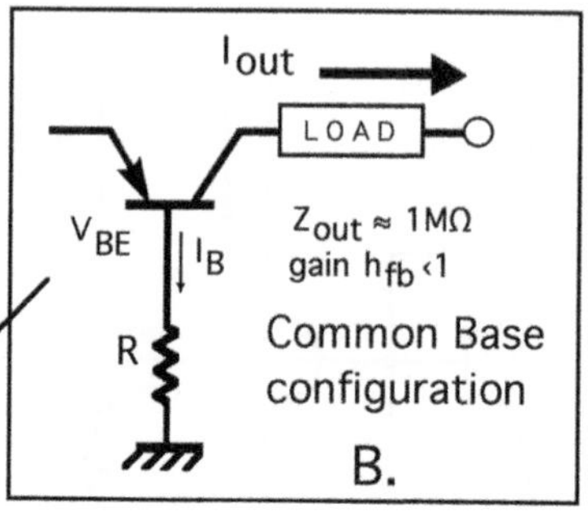

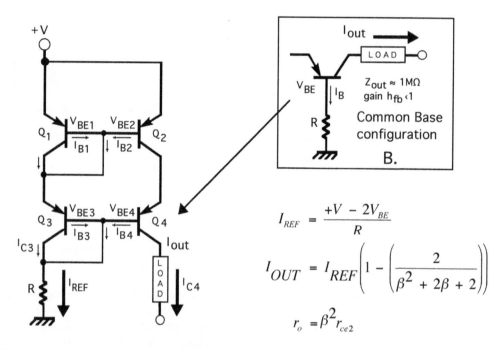

$$I_{REF} = \frac{+V - 2V_{BE}}{R}$$

$$I_{OUT} = I_{REF}\left(1 - \left(\frac{2}{\beta^2 + 2\beta + 2}\right)\right)$$

$$r_o = \beta^2 r_{ce2}$$

A. The PNP Cascode Current Source

Figure 4.31. *The Cascode configuration provides higher frequency response and maximum output impedance.*

Wilson, the pair of upper base terminals go to the right. With the cascode, the same two bases go to the left, and in this circuit cascoding buffers the current source from the load. As you will recognize, this is a basic current mirror, but with a common-base PNP transistor inserted into each collector leg (see Figure 4.31B). While the common-base configuration provides no current gain ($h_{fb} = < 1$), it does provide a very high output impedance (Z_{OUT}), typically of more than 10 MΩ. The common-base transistor makes V_{CE} relatively constant for the mirror transistors but greatly increases the output resistance. This means that while the emitter voltage of Q_4 remains constant, variations in the load voltage are accommodated by Q_2's V_{CE}. As a result, the voltage drop across the current source remains constant. Here the reference current (I_{C3}) is regulated by Q_3, and R sets the desired current, as previously discussed. In early development, the resistor R could be substituted by a multiturn potentiometer. This would allow easy adjustment of the desired current level, which could range from a few μA to many milliamps. (Because its tempco will not be as good as a 0.1% tolerance, 50-ppm precision resistor, it should be replaced by a quality fixed resistor.) The cascode circuit provides an internal circuit impedance of more than 10 MΩ at current levels less than 1 mA. It can be used equally well in circuits of more than 40 volts or down to about 3 volts, depending on one's exact choice of transistors. Because of its high impedance and reduced Miller capacitance, it also has very good high-frequency response. The basic equations for configuring the cascode circuit are shown here. It is assumed that the Q_1 and Q_2 pair, as well as the Q_3 and Q_4 pair, are all matched for V_{BE} and gain.

The following determines either the reference current, I_{REF}, or the value for resistor R:

$$I_{REF} = \frac{+V - 2V_{BE}}{R} \quad \text{(Eq.4.74)}$$

$$R = \frac{+V - 2V_{BE}}{I_{REF}} \quad \text{(Eq.4.75)}$$

The output current, I_{C2}, is determined by:

$$I_{OUT} = I_{REF}\left(1 - \left(\frac{2}{\beta^2 + 2\beta + 2}\right)\right) \quad \text{(Eq.4.76)}$$

The output resistance (r_o) for this four-transistor cascode current source is given by the internal collector-to-emitter resistance (r_{ce}), derived from the BJT's small-signal model, times beta squared. This accounts for the very-high-output resistance, which can range from about 1 MΩ to more than 100 MΩ:

$$r_o = \beta^2 r_{ce2} \quad \text{(Eq.4.77)}$$

where again r_{ce} is determined by:

$$r_{ce} = \frac{V_{CE}}{I_C} \quad \text{(Eq.4.78)}$$

This inexpensive, five-component circuit can be used as either a precision current source or sink, because the load can be connected to either terminal. Alternately, one can build an NPN cascode configured similarly. Even lower h_{oe} (output conductance) for the same current is achieved by the cascode, as a result of degenerative feedback. Here the circuit's output conductance (g_o, measured in either μS or μmhos) is much less than the h_{oe} of the single (noncascode) BJT circuit—at least 10 times lower. You can expect the cascode circuit's output conductance to be less than 20 μS. Regulation is about the same as the Wilson cells and will also have a really sharp knee, versus the softer knee of the other noncascode single BJT's circuits. The two big advantages with this circuit over most others is its much higher output resistance, and as a result its good high-frequency response.

Another example of a PNP cascode current mirror is shown in Figure 4.32. Here the bases of the cascode transistors are driven from a precision bandgap voltage reference, which provides a constant voltage and added temperature stability. It also provides multiple outputs, high output impedance, and good high-frequency performance. The circuit could be built using two PNP quad arrays. Transistors Q_1 to Q_4 could comprise one package, while the remaining Q_5 to Q_7 are the second package.

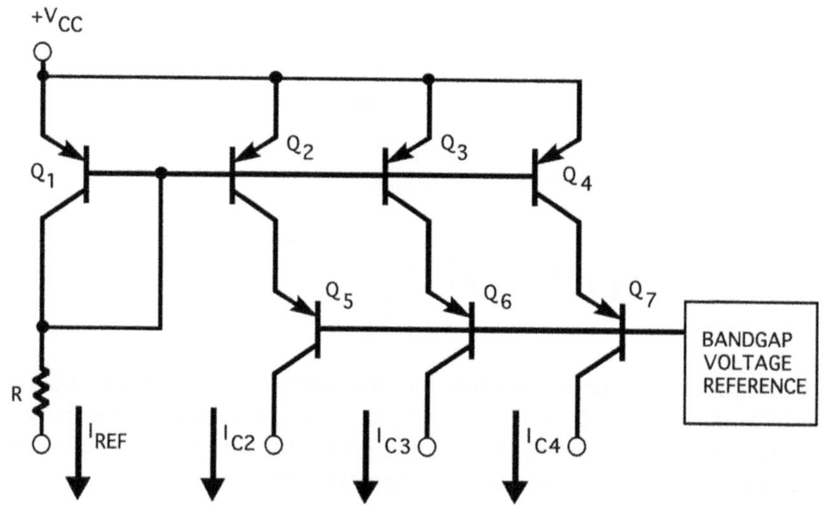

Figure 4.32. Multiple cascode current mirrors.

The following equation determines either the reference current, I_{REF}, or the value of the resistor R:

$$I_{REF} = \left(\frac{V_{CC1} - V_{BE1}}{R}\right) \qquad \text{(Eq.4.79)}$$

$$R = \left(\frac{V_{CC1} - V_{BE1}}{I_{REF}}\right) \quad \text{(Eq.4.80)}$$

In this circuit the output current, I_{C2}, is determined by:

$$I_{OUT} = \frac{\beta I_{REF}}{\beta + n + 1} \quad \text{(Eq.4.81)}$$

To further illustrate the importance of the PNP multiple cascode circuit, Figure 4.33 shows the biasing and input circuitry for a typical bipolar voltage-feedback op amp, using a PMOS transistor pair input stage. As you will recognize, the input stage relies on a PNP multiple cascode, as well as an NPN current mirror using emitter degeneration.

4.8 Current scaling

With the exception of the Widlar and the Wyatt current sources, up until now we have looked at current sources that generate input-to-output currents in a 1:1 ratio, but it is often necessary to generate integer multiples or fractions on an initial current. For the IC designer this is accomplished in the design by creating transistors with multiple emitters, or with bulk emitter area ratioed in either greater or lesser amounts. In this way a current can be made, say, 10 times or one-tenth of the initial input current. It is sometimes depicted in a manufacturer's circuit schematic as a number written next to the particular emitter (see Figure 4.22D). It is sometimes shown, for example, as 10E or 10A.

For the board-level designer, generating multiples or fractions of an initial current can be accomplished in different ways. One way to do this is by paralleling individual transistors as shown in the following circuits, in what is known as *current scaling.* The two circuits shown in Figure 4.34 are simple two times multipliers, which multiply the reference current (I_{REF}).

In either NPN or PNP multiplier circuits, the reference current, I_{REF}, is determined by:

$$I_{REF} = \left(\frac{V_{CC1} - V_{BE1}}{R}\right) \quad \text{(Eq.4.82)}$$

Also because the reference current is composed of both Q_1's collector current and the three base currents, then:

$$I_{REF} = I_{C1} + 3I_B \quad \text{(Eq.4.83)}$$

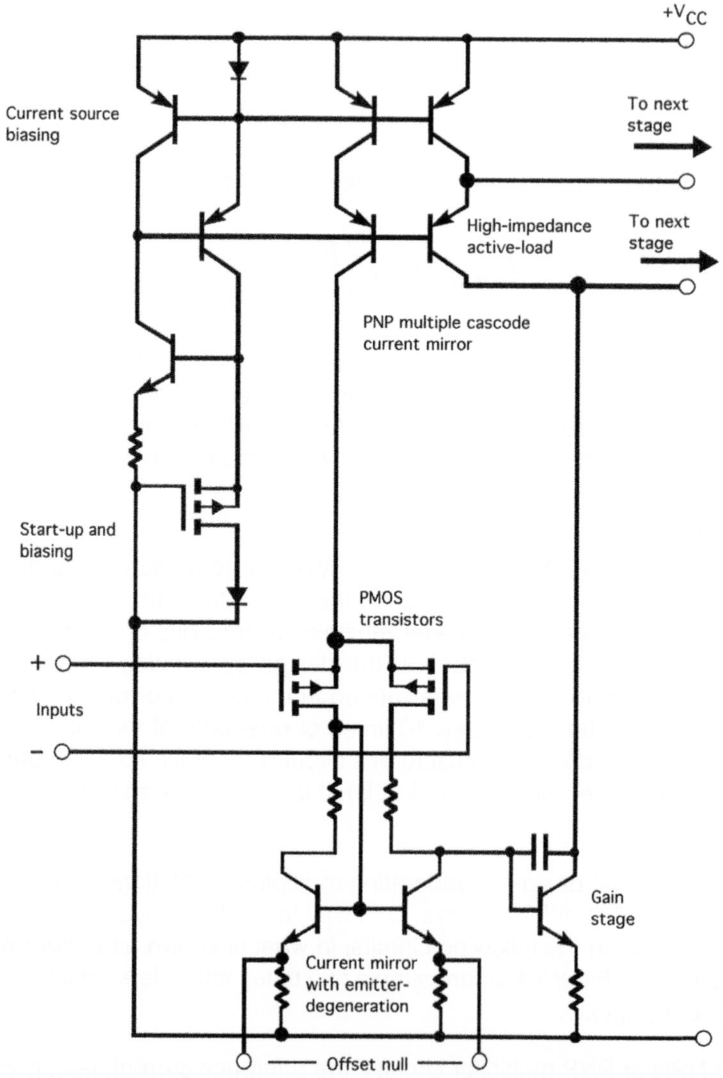

Figure 4.33. Example showing how cascoded current sources and a current mirror are used for active loads and biasing in a typical bipolar-FET voltage-feedback op amp.

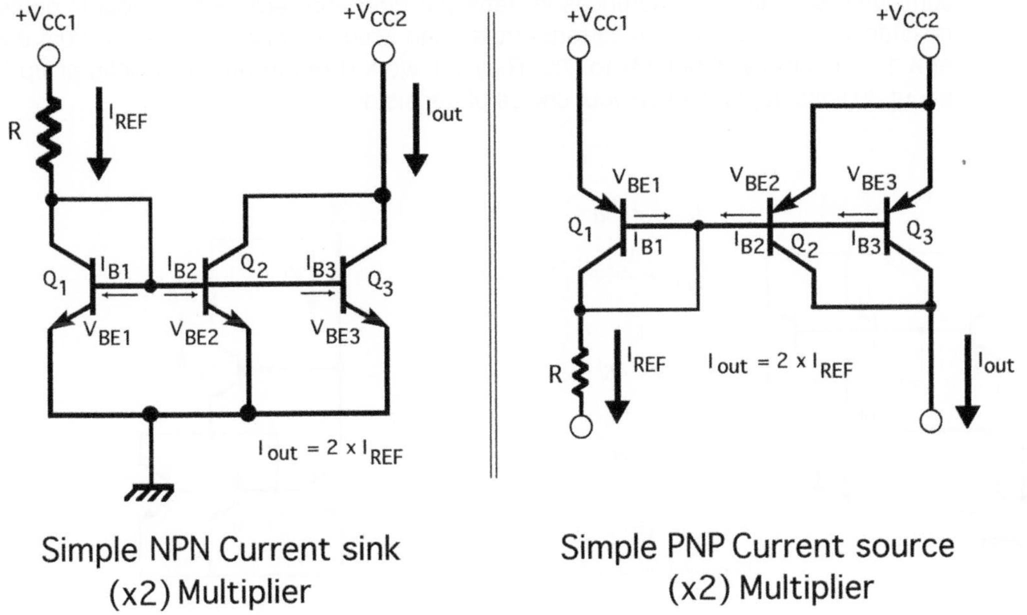

Figure 4.34. Examples of current multipliers.

Because the I_{REF} current is now mirrored by parallel-connected Q_2 and Q_3, the output current is therefore doubled. The output current, I_{C3}, is determined by:

$$I_{OUT} = 2\left(\frac{\beta}{\beta + 2}\right)\left(\frac{V_{CC1} - V_{BE1}}{R}\right) \quad \text{(Eq.4.84)}$$

$$\therefore I_{C3} \cong 2I_{C1} \quad \text{(Eq.4.85)}$$

The output resistance (r_o) for this configuration is given by the internal collector-to-emitter resistance (r_{ce}) of Q_3, derived from the BJT's small-signal model:

$$r_o = r_{ce3} \quad \text{(Eq.4.86)}$$

where again r_{ce} is determined by:

$$r_{ce} = \frac{V_{CE}}{I_C} \quad \text{(Eq.4.87)}$$

Overall, the circuits work well and are simple to use, but remember that they are somewhat sensitive to differences in transistor gain and have only a modest output resistance. Because of base current errors, their accuracy may be reduced (from that of a 1:1 circuit) by about 1% to 2%. They will work down to around 2 volts or up to about 30 volts depending on your choice of transistors.

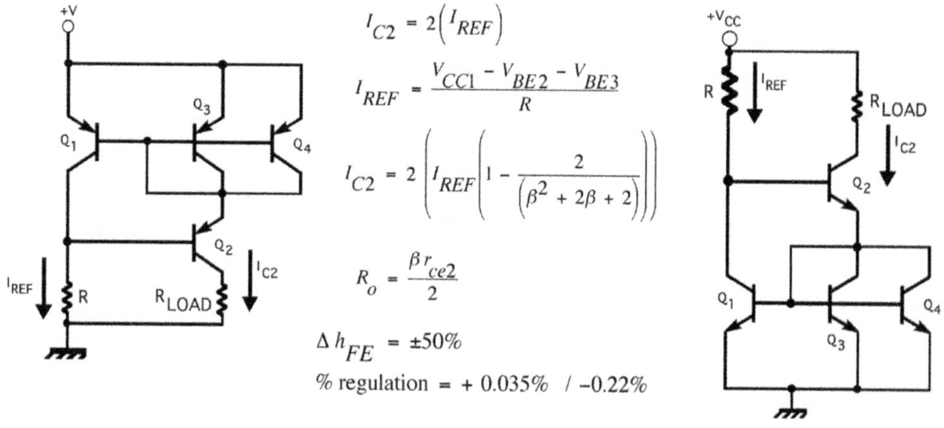

$$I_{C2} = 2\left(I_{REF}\right)$$

$$I_{REF} = \frac{V_{CC1} - V_{BE2} - V_{BE3}}{R}$$

$$I_{C2} = 2\left(I_{REF}\left(1 - \frac{2}{\left(\beta^2 + 2\beta + 2\right)}\right)\right)$$

$$R_o = \frac{\beta r_{ce2}}{2}$$

$$\Delta h_{FE} = \pm 50\%$$

$$\%\ \text{regulation} = +0.035\% \ / -0.22\%$$

A.
A modified basic Wilson PNP current source forms an effective x 2 multiplier.

B.
A modified basic Wilson NPN current sink forms an effective x 2 multiplier.

Figure 4.35. Examples of modified Wilson current multipliers.

For greater accuracy, the two circuits shown in Figure 4.35 (which are modified Basic Wilsons) would also provide two times multiplication, but with higher output resistance. The relevant equations are shown in the diagram. To illustrate how they work, let's design a 424-μA current source, with a 212-μA input. We will be using a quad device such as an MMPQ-3904 (NPN) or an MMPQ-3906 (PNP), and assume that all of the V_{BE}s and h_{FE}s are identical. The supply voltage will be 6 volts. Therefore:

$$V_{CC1} = 6V \quad \text{(Eq.4.88)}$$
$$V_{BE1} = V_{BE3} = V_{BE4} = 0.65V$$
$$\beta_{Q1} = \beta_{Q3} = \beta_{Q4} = 50$$
$$I_{B1} = I_{B3} = I_{B4} = 10\mu A$$
$$I_{REF} = 212\mu A$$
$$I_{OUT} = 424\mu A$$

We will begin by determining the resistor setting value (R), so that:

$$R = \frac{6V - 0.65V - 0.65V}{212\mu A} = 22.17K\Omega \quad \text{(Eq.4.89)}$$

Because the I_{REF} current is now mirrored by parallel-connected Q_3 and Q_4, the output current is therefore doubled. The output current ($I_{C2} = I_{OUT}$) is determined by:

$$I_{OUT} = 2\left(212\mu A\left(1 - \left(\frac{2}{50^2 + 2 \bullet 50 + 2}\right)\right)\right) = 423.7\mu A \quad \text{(Eq.4.90)}$$

The minimum output resistance (r_o) is approximately:

$$r_o = \frac{55000}{2} = 27.5K\Omega \quad \text{(Eq.4.91)}$$

Because these current sources are a modification of the Basic Wilson, they are much less sensitive to gain mismatches (beta-compensated) and provide excellent regulation and reasonable output resistance. Let's calculate the difference in output for a ±50% change in beta. For a 50% reduction in gain, this would mean that the gain is now 25, thus:

$$I_{OUT} = 2\left(212\mu A\left(1 - \left(\frac{2}{25^2 + 2 \bullet 25 + 2}\right)\right)\right) = 422.74\mu A \quad \text{(Eq.4.92)}$$

For a 50% increase in gain, this would mean that the gain is now 75, so that:

$$I_{OUT} = 2\left(212\mu A\left(1 - \left(\frac{2}{75^2 + 2 \bullet 75 + 2}\right)\right)\right) = 423.85\mu A \qquad \text{(Eq.4.93}$$

Considering that the nominal output current with a gain of 50 is 423.7 μA, this equates to a change in output of –0.22% for a 50% reduction in gain or a +0.035% for a 50% increase in gain. This is impressive regulation, all created by beta-compensation, but in practice because of base current errors, their accuracy may be reduced (from that of a 1:1 circuit) by about 1% to 2%. These circuits will probably work well down to about 2 volts or above 30 volts, if required.

Those were some examples of multiplying a current, but it's just as easy to *divide* the reference current. Consider this: if one paralleled the transistor normally handling the reference current, then the same reference current would flow, but now through two or more paralleled collectors (NPN) or emitters (PNP). Mirroring the current from either transistor will create an output that is just half of the reference current (i.e., divided by two). Figure 4.36 illustrates this point.

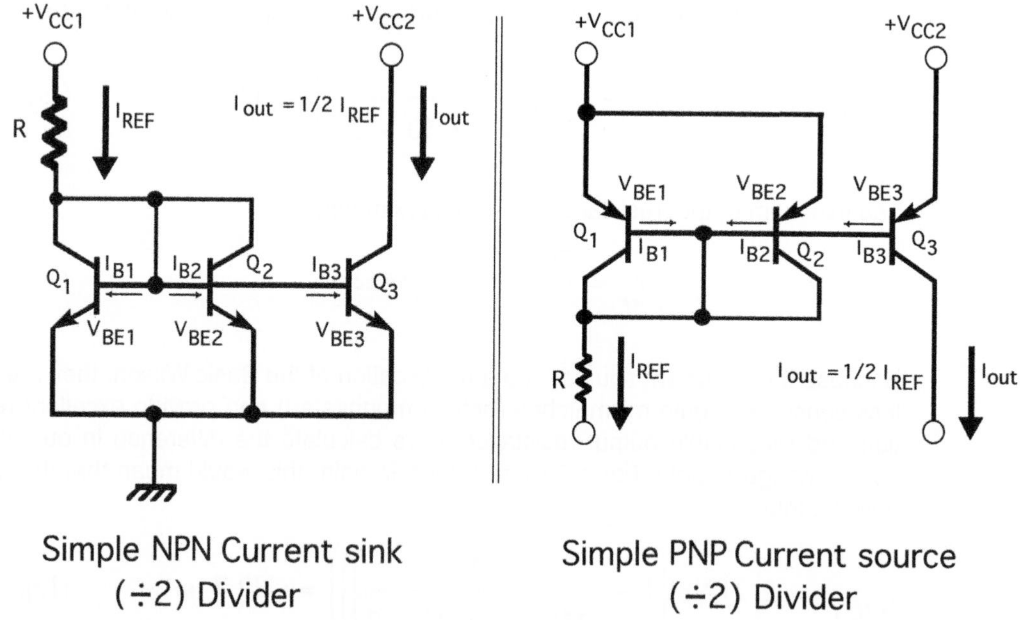

Figure 4.36. Examples of current dividers.

To determine the reference current, I_{REF}, in either the NPN or PNP divider circuit:

$$I_{REF} = \left(\frac{V_{CC1} - V_{BE1}}{R}\right) \quad \text{(Eq.4.94)}$$

Also, because the reference current is composed of both Q_1's collector current and the three base currents, then:

$$I_{REF} = I_{C1} + 3I_B \quad \text{(Eq.4.95)}$$

The output current, I_{C3}, is determined by:

$$I_{OUT} = \frac{\left(\frac{\beta}{\beta + 2}\right)\left(\frac{V_{CC1} - V_{BE1}}{R}\right)}{2} \quad \text{(Eq.4.96)}$$

$$\therefore I_{C3} \cong \frac{I_{C1}}{2} \quad \text{(Eq.4.97)}$$

The output resistance (r_o) for this configuration is given by the internal collector-to-emitter resistance (r_{ce}) of Q_3, derived from the BJT's small-signal model:

$$r_o = r_{ce3} \quad \text{(Eq.4.98)}$$

where again r_{ce} is determined by:

$$r_{ce} = \frac{V_{CE}}{I_C} \quad \text{(Eq.4.99)}$$

The two circuits work well and are simple to use, but they are sensitive to differences in transistor gain and have only a modest output resistance. In practice because of base current errors, their accuracy may be reduced (from that of a 1:1 mirror circuit) by about 1% to 2%. They will work down to around 2 volts or up to about 30 volts depending on the choice of transistors.

As with the previous multiplier circuits, we can use modified Basic Wilsons, which would also provide division-by-two, greater accuracy, and a higher output resistance. Two such circuits are shown in Figure 4.37. The relevant equations are shown in the diagram. As with the previous multiplication and division circuits, accuracy may be reduced (from that of a 1:1 mirror circuit), by about 1% to 2%. Either circuit can be easily built using a monolithic quad device, for optimum thermal considerations.

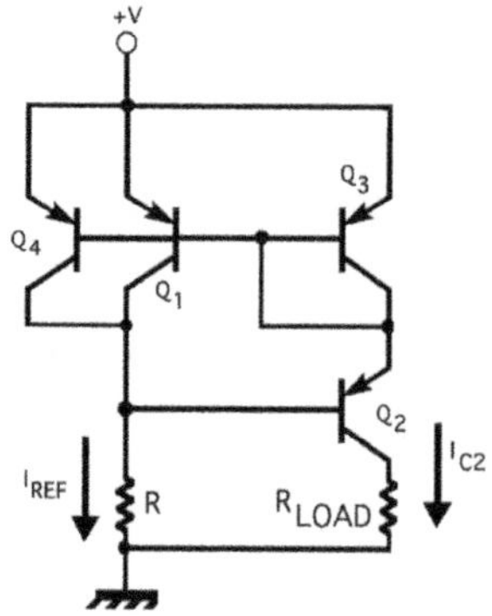

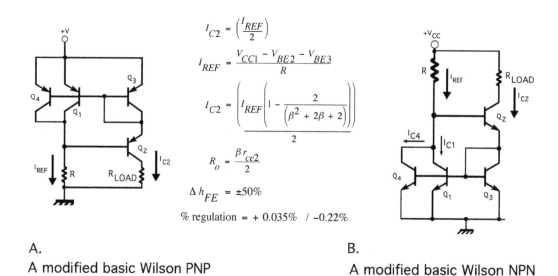

$$I_{C2} = \left(\frac{I_{REF}}{2}\right)$$

$$I_{REF} = \frac{V_{CC1} - V_{BE2} - V_{BE3}}{R}$$

$$I_{C2} = \frac{\left(I_{REF}\left(1 - \frac{2}{\left(\beta^2 + 2\beta + 2\right)}\right)\right)}{2}$$

$$R_o = \frac{\beta r_{ce2}}{2}$$

$$\Delta h_{FE} = \pm 50\%$$

$$\% \text{ regulation} = +0.035\% \; / \; -0.22\%$$

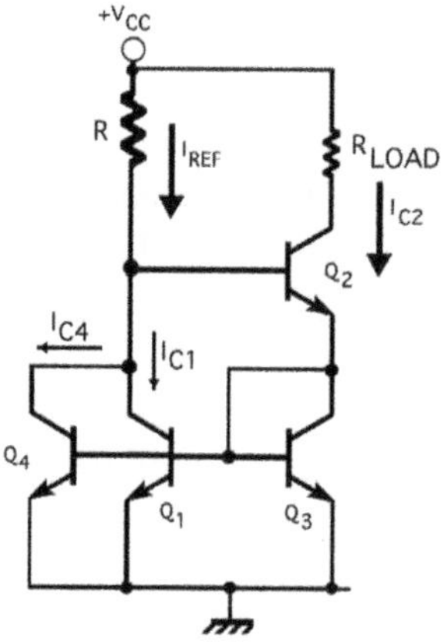

A.
A modified basic Wilson PNP current source forms an effective (÷2) divider.

B.
A modified basic Wilson NPN current sink forms an effective (÷2) divider.

Figure 4.37. Examples of modified Wilson current dividers.

Another way of performing multiplication or division of the reference current is by adding resistors. This was alluded to previously when looking at Widlar's circuits (see Figure 4.20). Now, if a resistor is inserted in series with the emitter of Q_2 in the divided circuit in Figure 4.38, the output current ratio between the two devices is no longer matched, but is divided and dependent on V_{BE} and the value of the resistor. This is also true for the multiplier circuit in which the reference current is multiplied, by putting R_2 in series with Q_2's emitter. These circuits are reasonably resistant to changes in the supply voltage and have a medium output impedance, although they do have a somewhat higher temperature coefficient than some of the other circuits we have looked at.

4.9 Modified current sources and example applications

For the rest of the chapter we will look at some applications, as well as some modified current sources, which may prove useful in one of your designs.

4.9.1 Running the current source from split power supplies

It's often necessary to run the current source in a split power supply situation. Figure 4.39 shows two examples: one a simple NPN 1.49-mA current sink and the other a simple PNP 3.47-mA current source. The relevant equations are shown, which apply to either configuration. Make sure that the combined +V and –V supply combination is a lot less than the breakdown voltage rating for the transistors you use. For higher out-

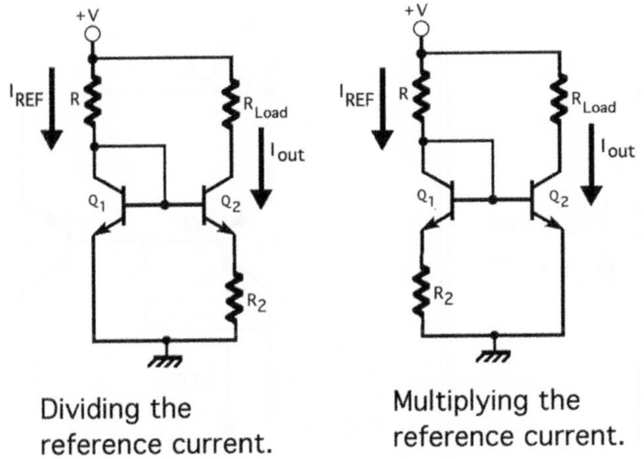

Figure 4.38. *Adding an emitter resistor provides an alternative method.*

put impedance, use either type of Wilson current mirror, or the cascode, and connected to the supply rails in a similar manner.

4.9.2 Improving power supply rejection

In a multiple-output current mirror, such as previously described in Figure 4.28, if greater power supply rejection is needed, then the circuit in Figure 4.40 can be used. For a 75% change in supply voltage, the output current will change only about 2%. Q_1 and Q_2 are both diode-connected and establish a two times V_{BE} potential at the bases of Q_3 and Q_4. This produces a potential of one times V_{BE} across the emitter resistors of Q_3 and Q_4, and causes current to flow in each collector. The current in R_2 and R_3 is I_{PTAT}, and the temperature coefficient (tempco) will be a little higher because of the 2 mV/°C change of V_{BE} with temperature. The output collector currents in Q_2 and Q_3 are determined by either transistor's V_{BE}, divided by its respective emitter resistor. Hence:

$$I_{OUT} = \frac{V_{BE}}{R_2} = \frac{0.7V}{467\Omega} = 1.5mA \qquad \text{(Eq.4.100)}$$

In some applications, using either a low-cost bandgap voltage reference IC (I_{OUT} = less than 10 mA) or a small linear regulator (I_{OUT} = less than 50 mA) nearby with bypass capacitors may be a much better solution, together with using a Wilson or cascode current source. This would significantly stabilize voltage levels, improve current regulation and accuracy, and boost the output impedance.

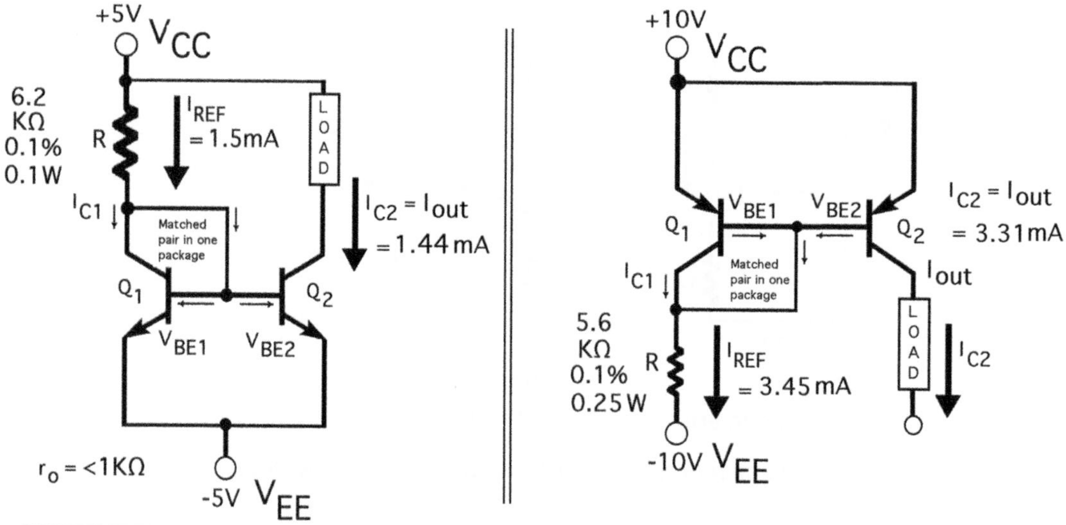

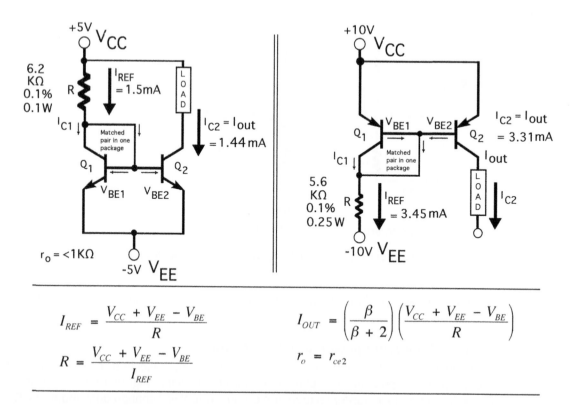

$$I_{REF} = \frac{V_{CC} + V_{EE} - V_{BE}}{R}$$

$$R = \frac{V_{CC} + V_{EE} - V_{BE}}{I_{REF}}$$

$$I_{OUT} = \left(\frac{\beta}{\beta + 2}\right)\left(\frac{V_{CC} + V_{EE} - V_{BE}}{R}\right)$$

$$r_o = r_{ce2}$$

Figure 4.39. Simple current sources running on split (±) supplies.

4.9.3 Alternative current divider

In this circuit, shown as Figure 4.41, the input reference current I_{REF} is mirrored by I_{C1}. This current is then divided by two. The upper three transistors (Q_1, Q_2, and Q_3) comprise a Basic Wilson current source, which is a cascode for the constant current divider. This is made up by the five lower transistors (Q_4, Q_5, Q_6, Q_7, and Q_8). For optimum performance, Q_1 and Q_2 should be matched, as well as Q_4 with Q_5, and then Q_6 with Q_7 and Q_8. This would require using a quad transistor array for Q_1, Q_2, and Q_3; a dual transistor array for Q_4 and Q_5; and another quad package for Q_6, Q_7, and Q_8. The circuit would have precision, good regulation, a very high output impedance, and create two equal output currents. The circuit operates from around 3.6 volts up to the limiting voltage of the transistors.

4.9.4 Modified three-transistor mirror-source

The circuit in Figure 4.42 is a modified PNP version of the three-transistor current sink we looked at earlier in this chapter, plus an extra mirror transistor. The equations for this are shown in the diagram. The output current (I_{OUT}) is calculated differently

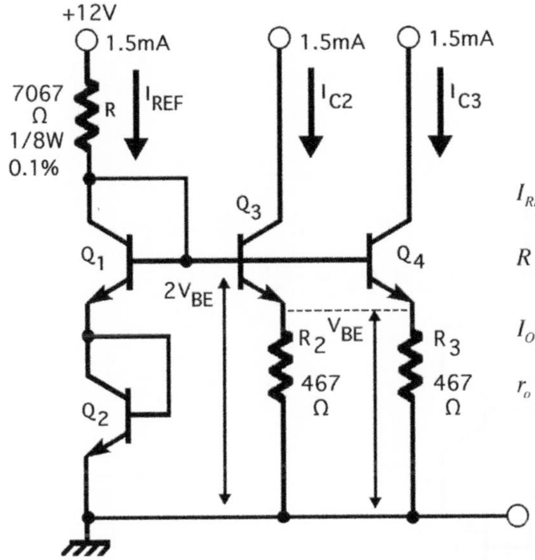

$$I_{REF} = \frac{+V_{CC} - V_{BE1} - V_{BE2}}{R}$$

$$R = \frac{+V_{CC} - V_{BE1} - V_{BE2}}{I_{REF}}$$

$$I_{OUT2} = \frac{V_{BE}}{R_2} \text{ or } I_{OUT3} = \frac{V_{BE}}{R_3}$$

$$r_o = r_{ce} + R_2$$

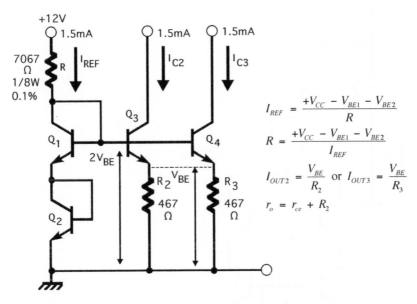

Figure 4.40. *Improved NPN current mirror with enhanced power-supply rejection.*

because it uses two output transitors (Q_3 and Q_4; thus n = 2). Normally one would likely incorporate small-value resistors in the emitters of the three upper transistors (Q_1, Q_3, and Q_4) for compensating any V_{BE} mismatches. If the application used a prime-grade, precision matched-pair instead, with a guaranteed maximum ΔV_{BE} of less than 1 mV, those emitter resistors would not be needed, in which case the error reduction would be proportional to beta^2. The circuit may be slow to turn off, because it has no base turn-off current. One way to remedy this is by adding either a reverse-biased diode or a resistor, as shown in the dotted line. In the case of the resistor, it may degrade accuracy somewhat.

4.9.5 Current source linearly charges capacitor in VCO

A current source is frequently used to charge a capacitor, because the constant current action generates a linear ramp across the capacitor. Figure 4.43 shows such a circuit. When the charge and discharge currents are identical, then the output waveform will also be symmetrical. When integrated into part of a timer or oscillator IC, this linear charging and discharging helps provide accuracy and repeatability. Depending on what kind of oscillator circuit the capacitor is part of, the output waveform may be a ramp, triangle, sine, square wave, and so on. The voltage-sensing circuit normally has an upper and a lower threshold. Assuming a constant charging current, then once the voltage level on the charging capacitor reaches the upper threshold, this triggers the circuit's output to change state. This turns the capacitor charging circuit off, while turning the discharge circuit on. The capacitor now discharges at a constant current. Once

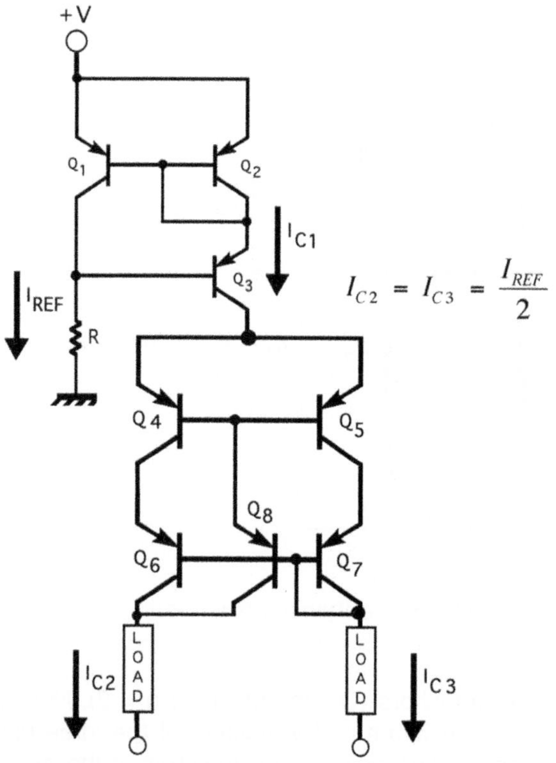

Figure 4.41. *A novel current divider.*

the discharge voltage level reaches the lower threshold, this triggers the output to change state again, and the process repeats.

In this circuit, transistors Q_1 and Q_2 form a composite NPN/PNP current source. Although it is simple, it is effective; it has low output capacitance, and thus offers a high bandwidth. The control voltage at the base of Q_1 establishes the current I_{REF} into the current mirror comprised of Q_3 and Q_4. Transistor Q_5 is simply used as a switch by the flip-flop's output, which controls the charge/discharge cycle. When Q_5 is off, the capacitor is linearly charged by the constant current through Q_2 and steering diode D_2. It may be expressed by the following equations:

$$Q = It \text{ ; in coulombs} \qquad \text{(Eq.4.101)}$$

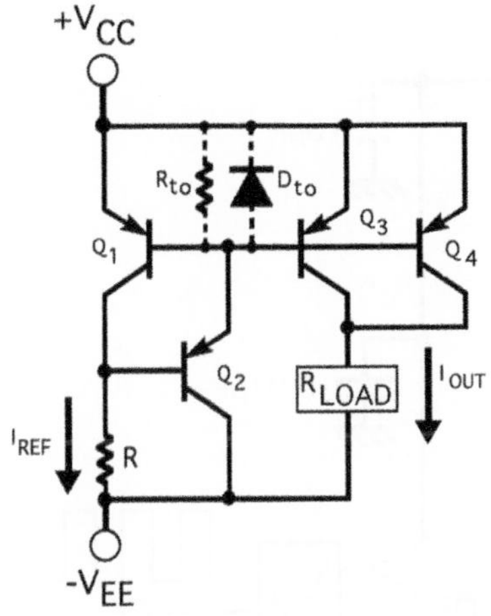

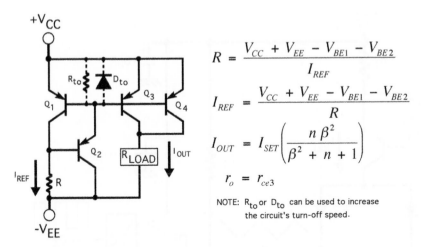

$$R = \frac{V_{CC} + V_{EE} - V_{BE1} - V_{BE2}}{I_{REF}}$$

$$I_{REF} = \frac{V_{CC} + V_{EE} - V_{BE1} - V_{BE2}}{R}$$

$$I_{OUT} = I_{SET}\left(\frac{n\,\beta^2}{\beta^2 + n + 1}\right)$$

$$r_o = r_{ce3}$$

NOTE: R_{to} or D_{to} can be used to increase the circuit's turn-off speed.

Figure 4.42. Modified 3-transistor mirror-source.

$$I = \frac{Q}{t} \text{ ; in Amps} \qquad \text{(Eq.4.102)}$$

$$I = C\frac{\Delta V}{\Delta T} \text{ ; in Amps} \qquad \text{(Eq.4.103)}$$

Once the voltage across the capacitor has reached the comparator's upper trip-point (established by the voltage between R_2 and R_3), the flip-flop changes state and switches on Q_5. The I_{REF} current via Q_1 and Q_2 establishes the current into the current mirror Q_3 and Q_4. With D_2 now effectively reverse-biased, the capacitor discharges at a constant rate via Q_4 and Q_5. Once the lower threshold is reached (established by the voltage between R_3 and R_4), the flip-flop changes state again, and switches Q_5 off. The oscillator's duty cycle is fixed because the charge and discharge currents are identical. The circuit offers good stability, on the order of 100 ppm/°C, and needs at least 3.5 volts to function.

4.9.6 Current source in a high-frequency laser transmitter

In the example shown in Figure 4.44, a diode laser is controlled by an incoming AC waveform, and the resulting modulated beam is then transmitted down a fiber-optic cable. In order to set the correct biasing for the laser, a simple PNP current mirror is used to set a 30-mA quiescent current. The combination of R_2 and P_1 set the refer-

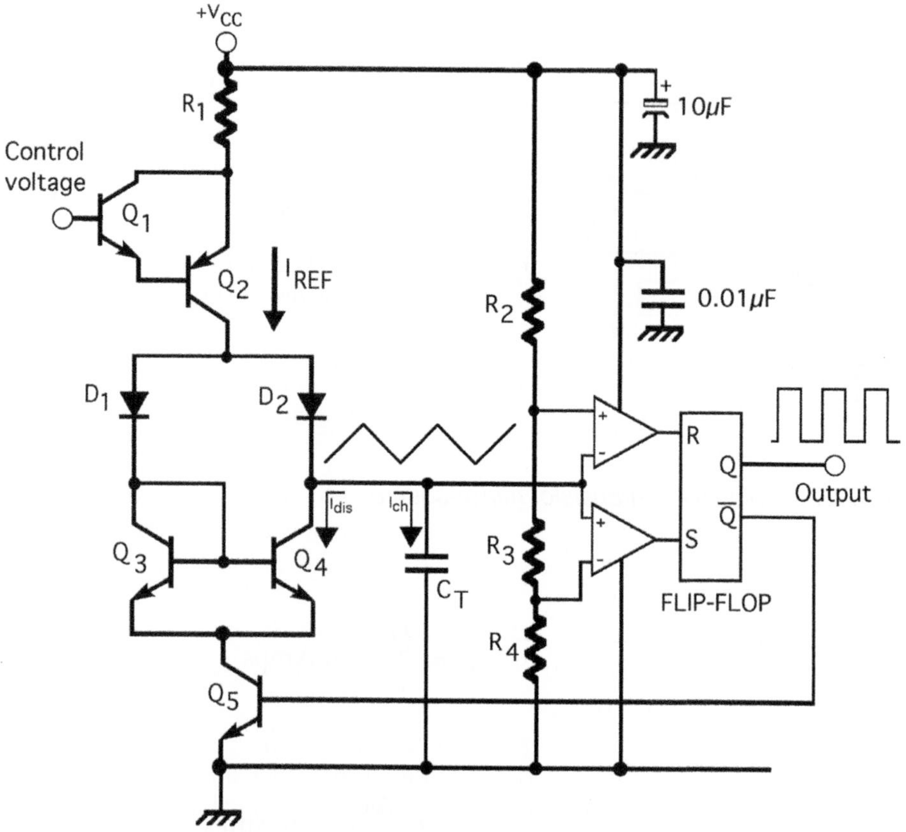

Figure 4.43. *This current mirror precisely controls the charge and discharge time of a voltage-controlled oscillator.*

ence current, I_{REF}. This current is mirrored by Q_2 as I_{C2}, and so powers the diode laser. Potentiometer P_1 allows adjustment for the optical power required, because this varies slightly from one diode laser to another. Diode D_1 guards the expensive laser diode against any reverse voltages.

4.9.7 Temperature-compensated current sink

Figure 4.45 shows one way of providing temperature compensation for a simple current sink or source. In this circuit, a quad matched array is used, so that differences in h_{FE}, V_{BE}, drift, and temperature effects are all minimized. It also uses a split power supply, although this could be a single positive and ground, or ground and negative supply. Here the collector currents for all four transistors are made identical, so that the output current I_{C4} mirrors the input current I_{C1}. The currents through R_1 and R_2 are equal, as are the currents through R_2 and R_5. The currents through R_3 and R_4 are equal, as are

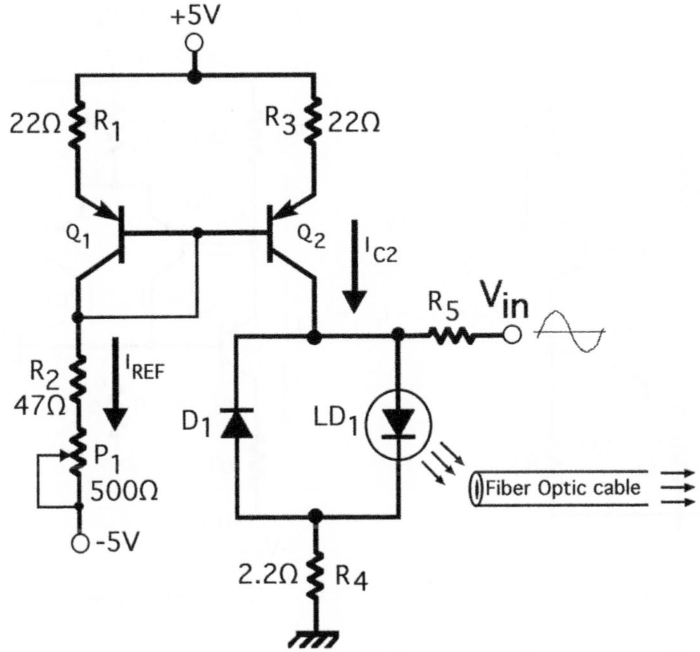

Figure 4.44. *This simple current source sets the quiescent current in this modulated laser transmitter.*

the collector currents I_{C2} and I_{C3} (which are also equal to the collector currents I_{C1} and I_{C4}). Transistors Q_2 and Q_3 both monitor the voltage on R_2 and R_5, repectively, and both compensate Q_1 and Q_4's V_{BE} temperature drift by creating a compensating temperature-dependent current in R_3. This action changes the base voltage of Q_1 and Q_4 so as to oppose their change in V_{BE}. As a result, a state of equilibrium is reached, where the current sink/source provides a constant current, despite changes in temperature. The design equations are shown in the diagram. Similar techniques are also used for current-limiting in some audio output stages and power supplies.

4.9.8 Compound current mirrors

It is common for designers, particularly IC designers, to create compound current mirrors, such as that shown in Figure 4.46. This circuit serves as an example and consists of a PNP quad array and an NPN quad array. Transistors Q_1 and Q_2 form a Basic Wilson current mirror, with I_{C1} mirroring I_{REF} The current I_{C1} is now mirrored by the Modified Wilson current source Q_5, Q_6, and Q_7, and produces mirrored current I_{C2}. Transistor Q_8 separately mirrors I_{C1}, producing I_{C3}, which feeds into a separate load. The two resulting currents I_{C2} and I_{C3} now flow into the mirrored-sink comprising Q_3 and Q_4. Because this combination multiplies I_{REF} by two, it serves as a mirror-sink for I_{C2} and I_{C3}. Diodes D_1 and D_2 steer the currents and guard against reverse-bias con-

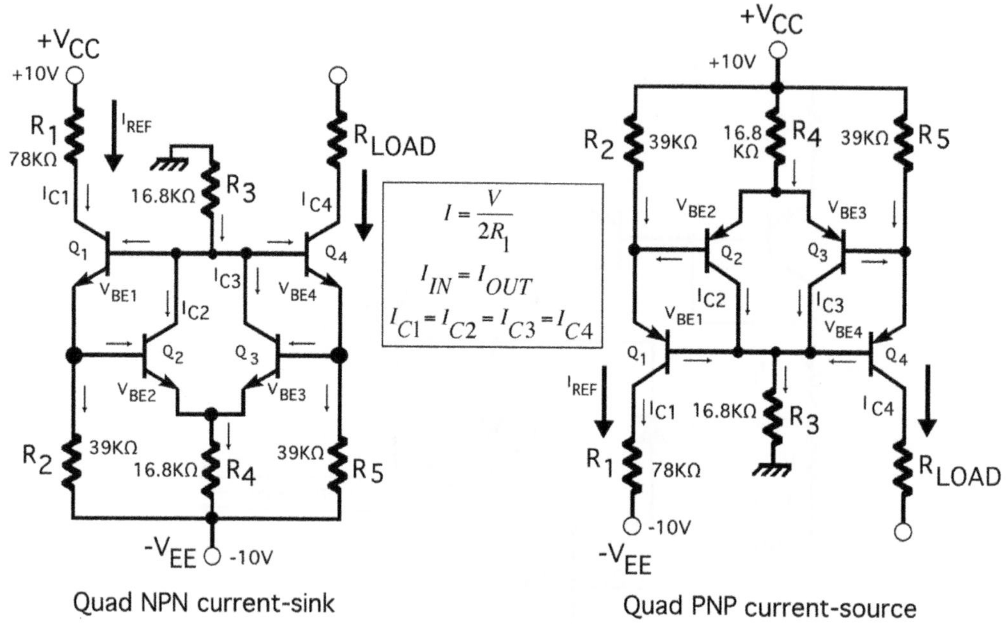

Figure 4.45. Temperature-compensated current sources.

ditions from the two loads. Higher accuracy will be attained if one uses a Wilson current source in place of Q_1 and Q_2. However, this will require more transistors and take up extra circuit board area. It comes down then to accuracy and cost versus size.

4.9.9 Current mirrors help DACs control oscillator frequency and duty cycle

In a machine-tool application, it is necessary to provide a variable drive frequency, with independent duty-cycle control. This circuit (shown as Figure 4.47) would be one of several motion-control circuits and part of a processor-based system. The frequency generation and duty-cycle control would use two 555 timer ICs. Frequency would range from 5 Hz to 250 Hz, and the duty cycle would be variable from 1% to 99%. One timer functions as an astable multivibrator, while the other is used as a one-shot. Rather than using resistors, two Full Wilson current mirrors are used, which improve accuracy, current regulation, impedance, and repeatability. Both mirrors provide constant currents for linearly charging the timing capacitors C_{T1} and C_{T2}. While both currents are regulated, they are also adjustable. This allows adjustment of both the frequency and the duty cycle. The use of Full Wilson current mirrors provides a very high impedance ($r_o = h_{FE} \times r_{ce}$) to either 555 timer circuit, as well as the DAC (both circuits are controlled by 8-bit DACs, which interface with the processor). The DACs set the reference currents for the current mirrors, which in turn duplicate the

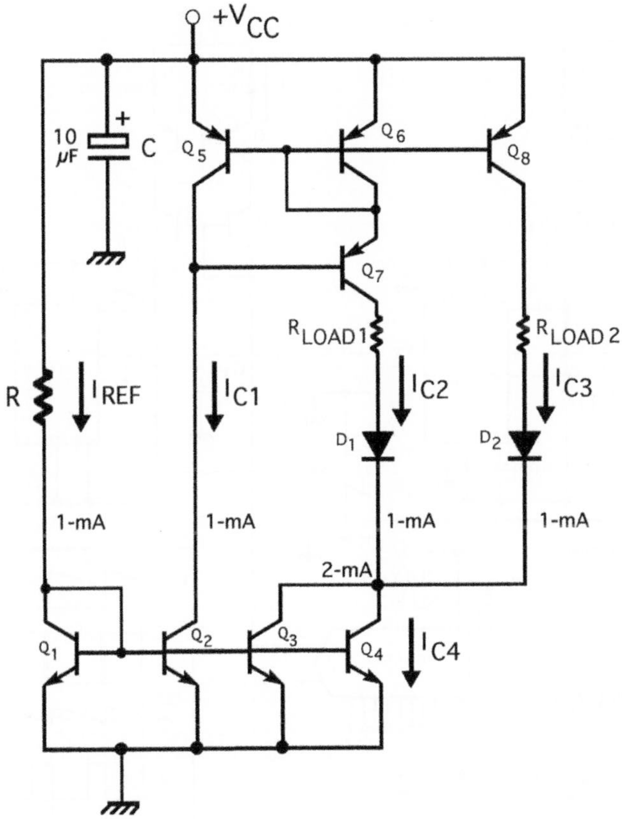

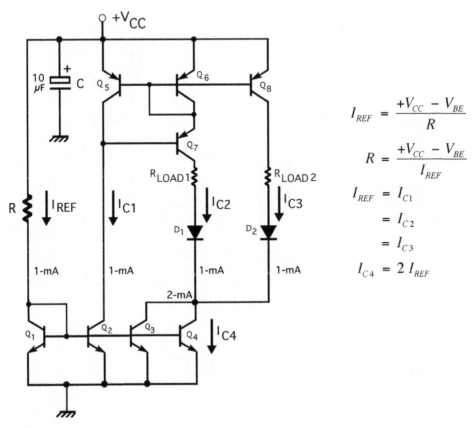

$$I_{REF} = \frac{+V_{CC} - V_{BE}}{R}$$

$$R = \frac{+V_{CC} - V_{BE}}{I_{REF}}$$

$$I_{REF} = I_{C1}$$
$$= I_{C2}$$
$$= I_{C3}$$
$$I_{C4} = 2\, I_{REF}$$

Figure 4.46. *Compound current mirrors.*

respective currents. One DAC circuit acts as the Frequency Select, while the other is the Duty-Cycle Select. The op amp acts as a buffer between I_{C1}'s RC timing components and I_{C2}'s high-impedance threshold input. The waveforms and equations are shown in the diagram. The design could use all bipolar devices or a mixture of bipolar and CMOS.

4.9.10 Using current sources as active loads

The circuit in Figure 4.48 shows a typical low-noise amplifier using active loads. There are many advantages in replacing the normal passive-resistive loads in front-end amplifiers with active loads in the form of current sources. Benefits include higher impedance, higher bandwidth, and increased precision. This is especially beneficial in low-noise front-end amplifiers, where the output impedance of the current source provides a high collector resistance, along with a small voltage drop. Noise levels in some precision amplifiers can reach below 1 nV/√Hz, if the circuit is designed well and uses matched transistors.

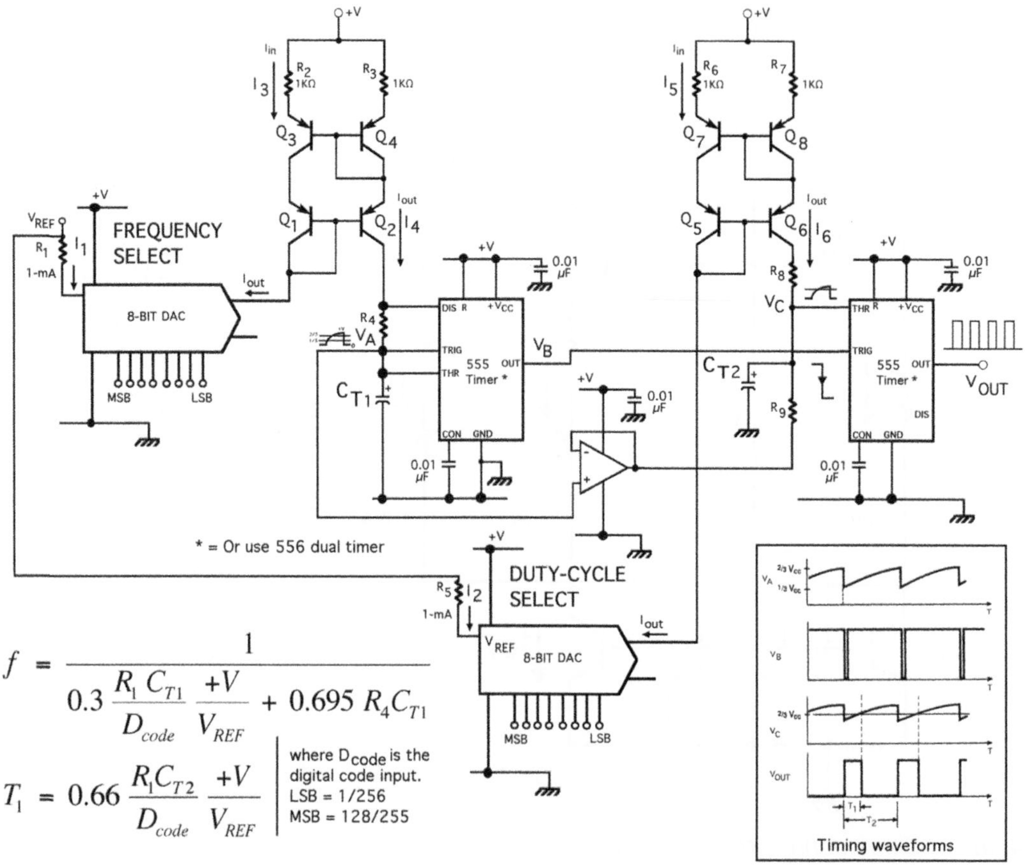

$$f = \frac{1}{0.3\,\dfrac{R_1 C_{T1}}{D_{code}}\,\dfrac{+V}{V_{REF}} + 0.695\,R_4 C_{T1}}$$

$$T_1 = 0.66\,\frac{R_1 C_{T2}}{D_{code}}\,\frac{+V}{V_{REF}}$$

Figure 4.47. Current sources help create a digitally controlled frequency generator, with separate control of frequency and duty cycle.

In this circuit, transistors Q_1, Q_2, and Q_3 form a simple current mirror, with Q_2 and Q_3 mirroring the reference current. The reference current for the multiple PNP current sources is set up by R_2. Transistors Q_4 and Q_5 are the low-noise amplifier pair. The collector current for each transistor is provided by each current mirror. These currents flow through both emitters and feed into the Basic Wilson NPN current sink, which provides excellent regulation. The reference current for the Wilson current sink is set up by R_1. In this example, the I_{REF2} reference current is set for 1 mA, while the I_{REF1} reference current is set for 2 mA. All of the mirror-sources and sinks use emitter-degeneration resistors (R_E = 470 Ω), to reduce differences in the transistors' V_{BE}s and to increase the output resistance. The effective load resistance of the differential pair is given as follows:

$$R_{L1} = R_{L2} = r_{ce4} \| \left(r_{ce2}\left(1 + g_m R_{E2}\right)\right) \qquad \text{(Eq.4.104)}$$

4.9.11 Modified current source squares the reference current

Previously in this chapter we looked at ways of multiplying and dividing the input reference current. The example depicted here in Figure 4.49 shows a means of providing a square-law relationship with a relatively low input current (I_{REF}). In this circuit, a simple NPN mirror (composed of transistors Q_1, Q_2, and Q_5) is used to provide a multiple mirror reference current, which is fed to the simple PNP mirror composed of transistors Q_3 and Q_4. As you probably noticed, the circuit is *not* connected quite as normal. A potentiometer is in series with R_1, which allows the input current to be changed, but then an added low-value resistor R_2 is connected in series with both NPN emitters. Additionally, the output is made at the junction of Q_4 and Q_5, via their respective collectors.

Normally with a 1:1 ratio current mirror, an exact duplicate of the input current will be produced at the output. If this were plotted on a graph, it would result in a straight diagonal line from zero (for I_{IN} vs. I_{OUT}). With this square-law circuit, because of the way it is configured, it produces an exponential curve at low currents (less than 400 μA). The transistors in this circuit are considered to have equal h_{FE}s, I_Bs, I_Cs, and V_{BE}s. As a result, the reference current is mirrored by several other currents, as follows:

$$I_{C1} = I_{C2} = I_{C3} = I_{C4} = I_{REF} \qquad \text{(Eq.4.105)}$$

The reference current is determined by:

$$I_{REF} = \left(\frac{V_{CC} - V_{BE1}}{R_{POT} + R_1 + R_2} \right) \qquad \text{(Eq.4.106)}$$

Because the base-emitter voltages are the same for this example, then:

$$V_{BE5} = V_{BE1} + R_2\left(I_1 + I_2\right) \qquad \text{(Eq.4.107)}$$

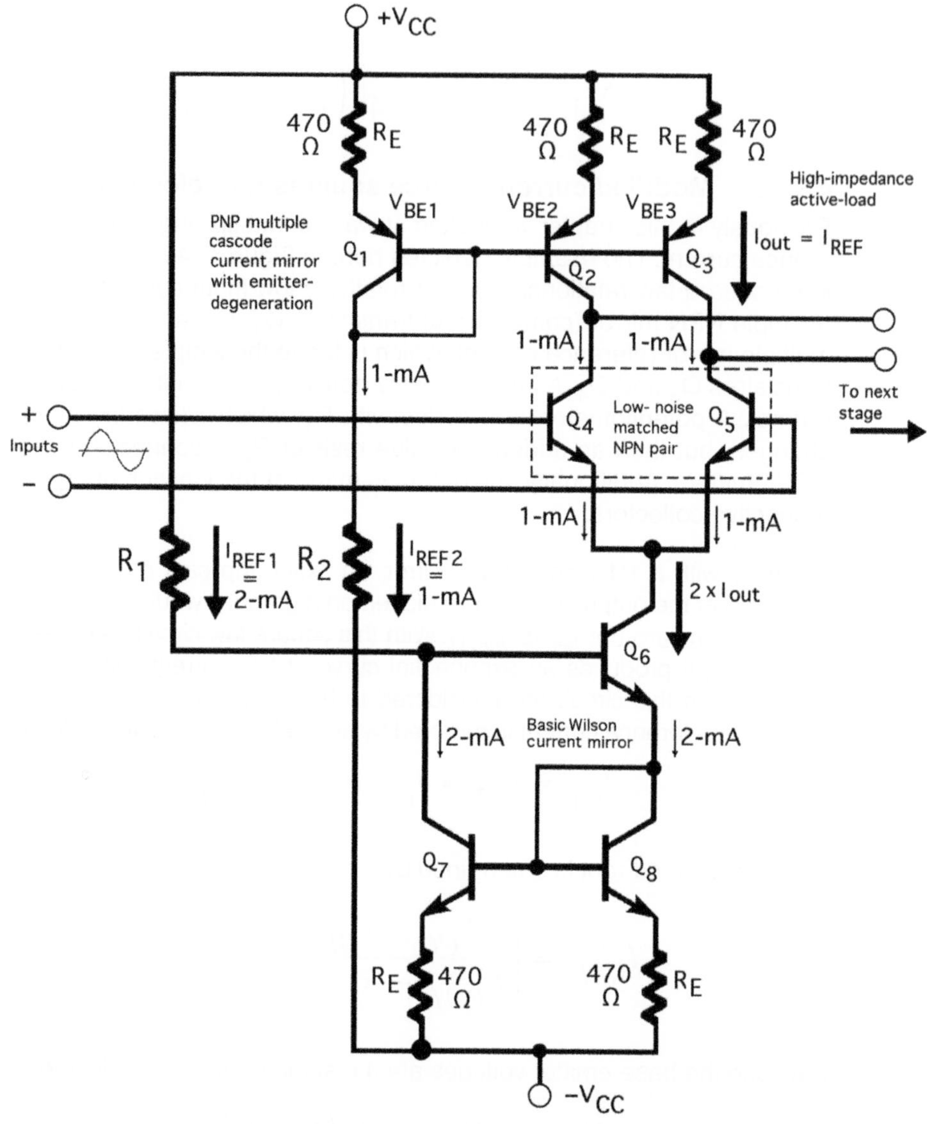

Figure 4.48. *Current sources are used as active loads in this front-end differential amplifier circuit.*

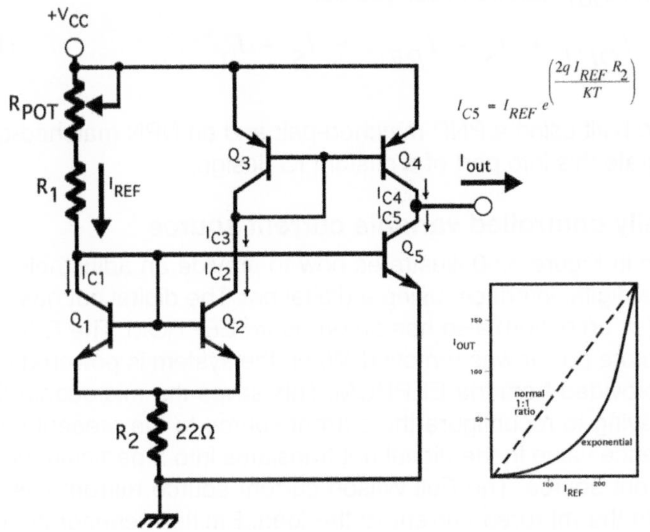

Figure 4.49. *Composite current mirrors provide an exponential output current.*

The ratio of collector currents between I_{REF} and I_5 can be found by:

$$\frac{I_{C5}}{I_{C1}} = e^{\frac{q\left(V_{BE5} - V_{BE1}\right)}{KT}} \qquad \text{(Eq.4.108)}$$

where: q is the charge on an electron (1.6×10^{-19} J);

K is Boltzmann's constant (1.38×10^{-23} J/°K);

T is the temperature (°Kelvin; 298°K = 25°C)

Transistor Q_5's collector current can be determined by:

$$I_{C5} = I_{REF}\, e^{\left(\frac{2q\, I_{REF}\, R_2}{KT}\right)} \qquad \text{(Eq.4.109)}$$

The output resistance for this configuration is fairly low and can be found by:

$$r_o = r_{ce} \qquad \text{(Eq.4.110)}$$

The output current, I_{OUT}, can be described as:

$$I_{OUT} = I_5 - I_{REF} = I_5 - I_4 \qquad \text{(Eq.4.111)}$$

The circuit can be built using a PNP matched-pair and an NPN matched-quad, unless you want to integrate this into part of a custom IC design.

4.9.12 Digitally controlled variable current source

The circuit shown in Figure 4.50 illustrates how to provide an adjustable/variable current source with a digital interface, using a digital pot. The digital pot has 256 settings, and in this case (as an option) also has an on-board EEPROM. The EEPROM retains the last setting before power was removed. When the system is powered up again, the same setting is provided from the EEPROM. This saves the microcontroller time and program steps having to reconfigure the current source to the present setting. In this circuit, the resistance value in the digital pot translates into a particular current setting (I_{SET}) to the current source. The Full Wilson current source mirrors this current precisely and outputs the mirrored current to the load. Emitter degeneration is provided by resistors R_1 and R_2 to compensate for V_{BE} mismatches, and the current mirror presents a high-output impedance to the load. Manufacturers of digital pots that have an on-board EEPROM include Analog Devices, Maxim/Dallas, and Intersil/Xicor.

4.9.13 High-pass filter's response is set by compound current mirrors

The circuit shown in Figure 4.51 functions as a high-pass filter for signals coming from a high-impedance source. It allows one to filter signals based on their slew rate, rather than their frequency, and is best suited as an input stage to a more complex filter. A compound set of current mirrors is used in this design and allows one to adjust the filter's slew-rate response, as well as to alter the signal's baseline voltage level. Being able to zero out any DC errors and offsets, by adjusting or restoring the baseline level, is really helpful. Additionally, being able to filter out the noise or unwanted parts from a signal helps improve the signal-to-noise (S/N) ratio. It allows one to measure and work with those signals with greater precision and accuracy. The voltage at amplifier A_2's V_{REF} input is responsible for this function.

The signal input appears at amplifier A_1's inverting input, is processed, and then is available as a buffered output at amplifier A_3's V_{OUT} pin. The slew-rate cutoff is directly adjustable by changing I_{REF} with the potentiometer R_{POT}, located in transistor Q_2's collector. This current, shown in the diagram as 1 mA, is then mirrored as I_{M1} (mirrored current) and I_{M2}. Current I_{M2} is responsible for charging the capacitor C (47 μF), whose purpose is to retain the baseline voltage setting. This voltage is regulated using feedback from amplifiers A_2 and A_4 and diode D_2, by using currents I_{M2}, I_{M3}, and I_{M4}. If the output voltage at V_{OUT} attempts to rise, the current through diode D_2 is reduced, which in turn reduces the voltage on capacitor C. Alternately, if the output voltage at V_{OUT} attempts to fall, then the current through diode D_1 is increased, which in turn

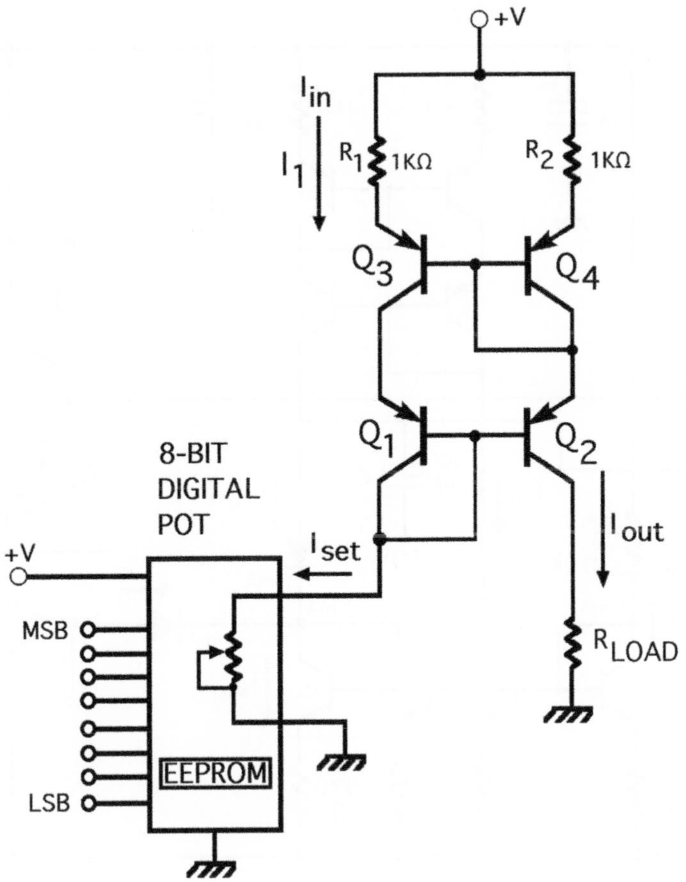

Figure 4.50. *Full Wilson current source controlled by a digital pot. The EEPROM retains setting.*

increases the capacitor's voltage to the level set by V_{REF}. Thus, it works like a voltage regulator or an AGC circuit.

Transistors Q_3 and Q_4 are part of the Q_1 and Q_2 mirror, but are configured as a current doubler (times two), because the two currents I_{M2} and I_{M3} are combined into I_{M4}, which flows through transistors Q_3 and Q_4. The two currents are required for symmetrical operation on AC signals. The circuit is probably best built using two quad arrays for the current mirrors and two dual op amps. Amplifiers A_1 and A_3 should both be quality precision types, while A_2 and A_4 need to be high-speed types, with a high slew rate (50 V/μS)—in fact, significantly higher than the highest signal frequency being filtered.

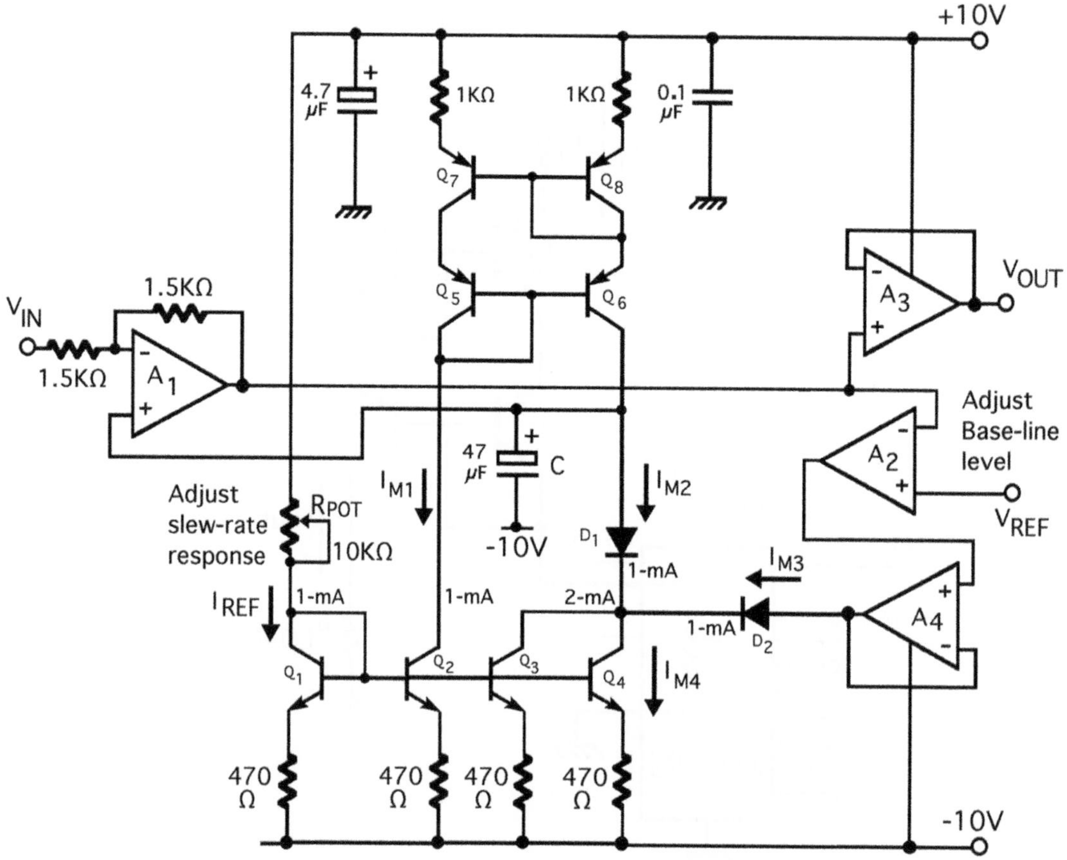

Figure 4.51. *Compound current mirrors help tailor this high-pass filter's response.*

4.9.14 Simple LED current sources

In Figure 4.52, some simple current sources can be created from the combination of a BJT and a visible red LED. The LED's forward voltage only serves as a voltage reference (not as a normal indicator) and approximately compensates the BJT over the temperature range. This is because the forward voltage of a red LED (V_{FWD} = 1.65 volt) and the forward base-emitter voltage (V_{BE} = 0.65 volt) of a transistor are well known and predictable. This is because of the relationship:

$$\frac{\Delta V_{BE}}{\Delta T} \cong \frac{V_{FWD}}{\Delta T} \cong -2.2mV/°C \qquad \text{(Eq.4.112)}$$

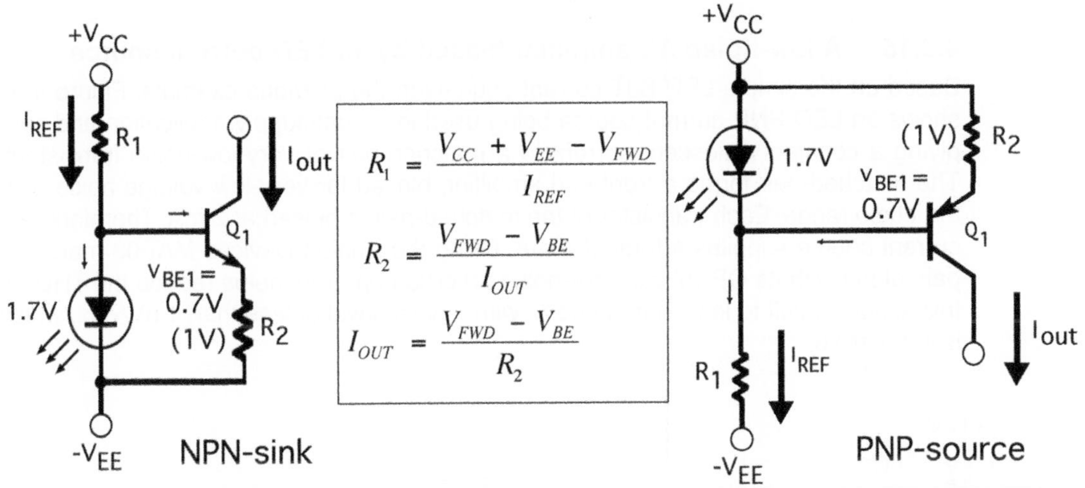

Figure 4.52. *Using the forward voltage (V_{FWD}) of a red LED to temperature-compensate the transistor's V_{BE}.*

The voltage difference (≈ 1 volt) is dropped across resistor R_2, which in turn provides a temperature-stabilized current at Q_1's emitter. Other color LEDs may be used for the same purposes, but they mostly all have higher bandgap voltages and thus forward voltages as shown in Table 4.4. Of all these the most practical and the cheapest is the standard red type.

LED COLOR	FORWARD VOLTAGE $V_{(fwd)}$
Infra-Red (IR)	1.40V
Red	1.65V
High-efficiency red	2.20V
Pink	2.00
Orange	2.10
Yellow	2.20
Green	2.20
High-efficiency green	2.30
Blue	3.10
The forward voltage is measured @ 10-mA.	

Table 4.4 LED color vs. forward voltage.

4.9.15 A low-noise AC amplifier biased by an LED current source

Based on the simple LED-BJT current sources in the previous example, Figure 4.53 shows an LED-PNP current source being used in a dual-supply application and supplying a constant quiescent current to a matched-pair of very-low-noise transistors. The matched-pair forms a front-end amplifier, biased for very-low-voltage noise over the audio range. Each transistor of the matched-pair is biased at 2 mA. Therefore, the current source supplies a total of 4 mA. Using the Analog Devices MAT-03 matched-pair, along with its OP-27 very-low-noise precision op amp, helps reduce the THD for this simple circuit to less than 0.005%, with a noise level of less than 1 nV/√Hz, which is impressive.

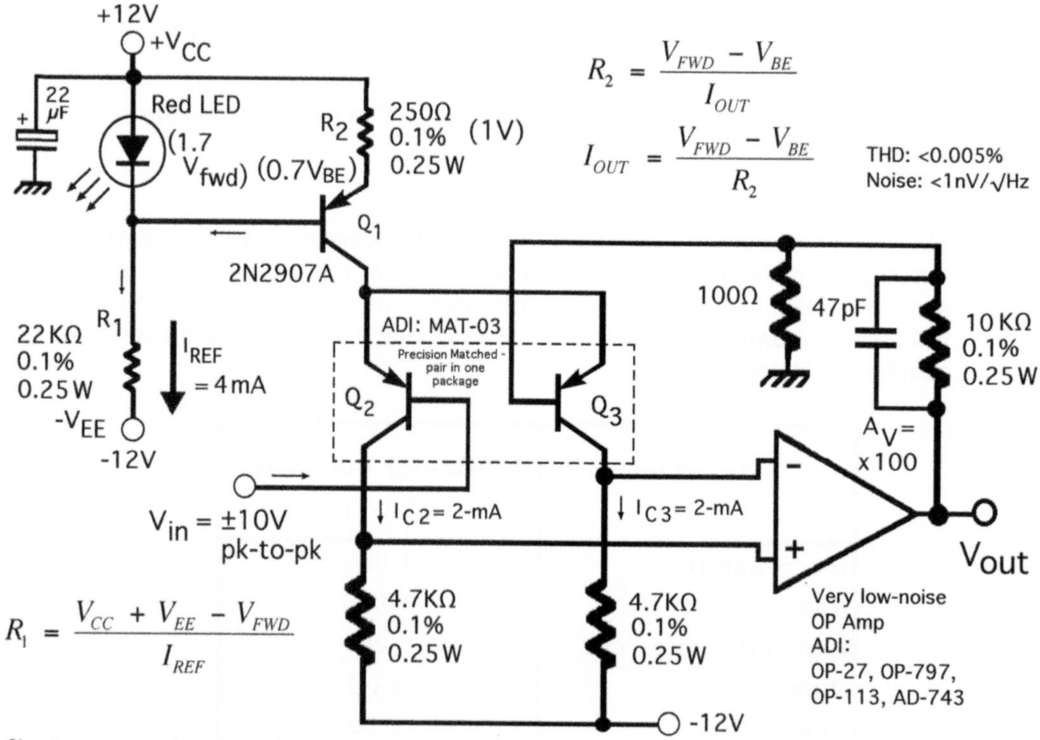

$$R_2 = \frac{V_{FWD} - V_{BE}}{I_{OUT}}$$

$$I_{OUT} = \frac{V_{FWD} - V_{BE}}{R_2}$$

$$R_1 = \frac{V_{CC} + V_{EE} - V_{FWD}}{I_{REF}}$$

Circuit courtesy of Analog Devices, Inc.

Figure 4.53. This simple LED-referenced current source is used at the input stage of this low-noise AC amplifier, where each of the matched transistor pair is fed a constant 2 mA to reduce voltage noise.

4.9.16 A composite BJT-JFET current source with very high output impedance

If you wanted an extremely high output impedance, together with really sharp current regulation, then the circuit in Figure 4.54 would be a good possibility. In this 28-volt avionics circuit, the output current is set at 125 μA, provided by the Full Wilson PNP current mirror at the top, together with a pair of JFETs in a cascoded configuration. The Wilson current mirror provides accuracy and beta-compensation, while the cascoded complimentary JFETs sharpen regulation and boost the output impedance. The circuit's output impedance (g_o) is greater than 5 GΩ, because of the JFET cascode's very low output conductance. In fact, the output conductance for the circuit is about 0.001 μS (1 nSec). A BJT cascode mirror-sink is used, rather than a Full Wilson, because it adds even higher output resistance. The two capacitors are simply for stability, and the output current is determined by R_{SET}, R_3, and R_4. Emitter degeneration is provided by resistors R_1, R_2, R_5, and R_6, which compensate for differences in the NPN transistors' V_{BE}s. This circuit is probably best built using two quad arrays, together with two discrete JFETs, which could be either through-hole or surface-mount components. All resistors should be quality metal-film, one-eighth watt, with a 0.1% tolerance or better.

4.9.17 A composite BJT-MOSFET high-power current source

Although most current sources we have looked at are in the low microamp to low milliamp range, there may be occasions when one needs a power circuit, delivering amps. The circuit shown in Figure 4.55 uses a combination of a JFET current regulator diode (CRD1), NPN transistors, and resistors in order to set the power MOSFET's drain current. CRD1 provides the operating bias for transistor Q_1, irrespective of power supply fluctuations. The transistor is part of a Widlar bandgap voltage reference, whose temperature-stable 1.22-volt output is used as a reference voltage across the power MOSFET's source resistor, R_S. This forces the MOSFET to pass constant drain current irrespective of its drain-to-source voltage. The source resistor should be a precision, low-tempco, wirewound type, with its wattage dependent on the maximum current through it (+30% margin). The three NPN transistors that make up the bandgap voltage reference must be part of a monolithic array (i.e., CA3045, CA3046, MMPQ3904, MMPQ2222A), so that they share the same temperature characteristics. CRD1 was chosen because at 430 μA it has a flat temperature characteristic yielding 0 ppm/°C, and provides a stable operating bias for the NPN transistor, Q_1. As a result, this design provides a stable medium-power (1.22-Amp) current sink. In this design, the power MOSFET uses a TO-220 package, which should be mounted on a good-quality heatsink. In some designs, you may want to consider heatsinking the power resistor too.

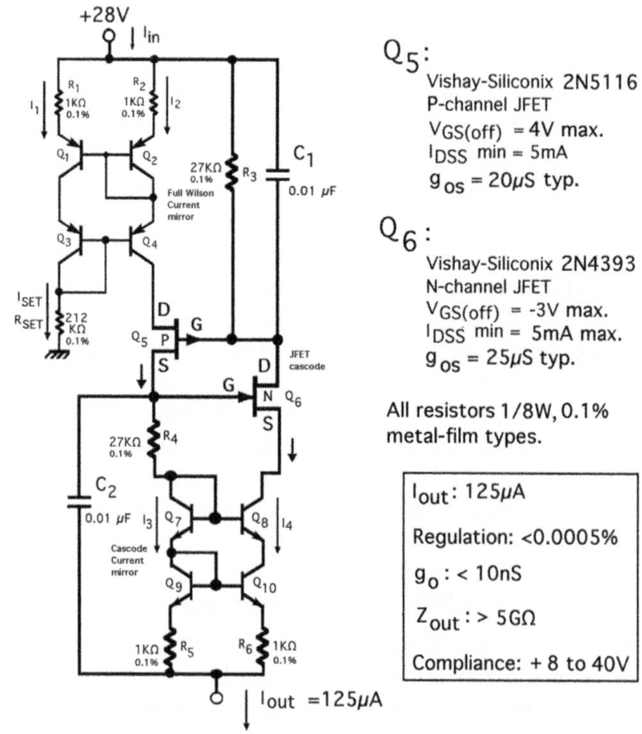

Figure 4.54. *Composite cascoded BJT-JFET current source provides excellent regulation and ultra high output impedance.*

4.9.18 A DAC-controlled current pump uses a Wilson current source

The circuit shown in Figure 4.56 functions as a bipolar current pump, which both sources and sinks current to the load. This design combines a Basic Wilson current mirror together with an 8-bit multiplying DAC. The venerable industry-standard Analog Devices DAC-08 (8-bit) has complementary current outputs, with a range of ±2 mA full-scale, and having a typical drift of about ±10 ppm/°C. The DAC's digital code ranges from 0000 0000 to 1111 1111, which corresponds to an output current range of –1.992 mA to +1.992 mA. The current source maintains linearity within one LSB (which is 15.6 μA). In this design, the circuit uses split power supplies of ±15 volts and a +10-volt reference voltage. The external voltage reference shown in the diagram as V_{REF} is 10 volts. This sets the I_1 reference current through resistor R_1 at 2 mA. Hence the DAC's outputs range from 0 to +1.992 mA or 0 to –1.992 mA. The processor sends the appropriate digital code to the DAC, whose output current changes to that value. This sets the current mirror's reference current through Q_1. Transistor Q_3 mirrors this current, which is regulated and buffered by Q_2. Diode D_1 matches the V_{CE} across Q_1 and Q_3, keeping the voltage constant across Q_3. Diodes D_2 and D_3 form a Baker

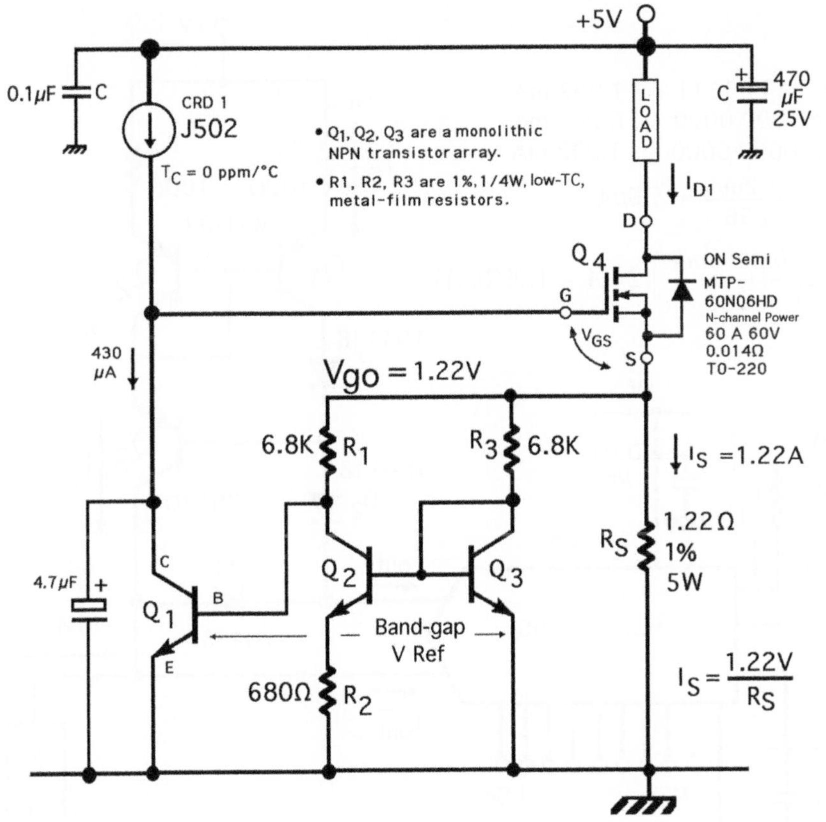

Figure 4.55. *A high-power BJT/Power MOSFET current sink using a Widlar band gap voltage reference for improved temperature stability and current regulation.*

clamp, preventing Q_2 from turning off, and so improving the overall switching speed of the circuit. Because D_3 also serves as a blocking diode between the DAC's output and the load, it is subject to very fast reverse voltages. Although the other diodes shown in the circuit can be regular 1N4148 switching diodes, D_3 should be a fast-recovery diode, such as a 1N4937 or similar (1 Amp; 600 volt; trr = 200 nS).

As you have seen in this chapter, some of the most precise current sources can be built with BJTs. Currents can be easily matched, multiplied, divided, scaled, or inverted in various ways. The best current sources are the Wilson, Widlar, Wyatt, and cascodes. When you consider how far the junction transistor has come since Dr. Shockley invented it in the late 1940s and Bob Widlar showed us how to use it in the '60s and '70s, how it has evolved and changed society, it has to be one of the most useful devices ever invented in the history of humankind. Invented and developed in America, transistors are now made in dozens of countries around the world. The BJT

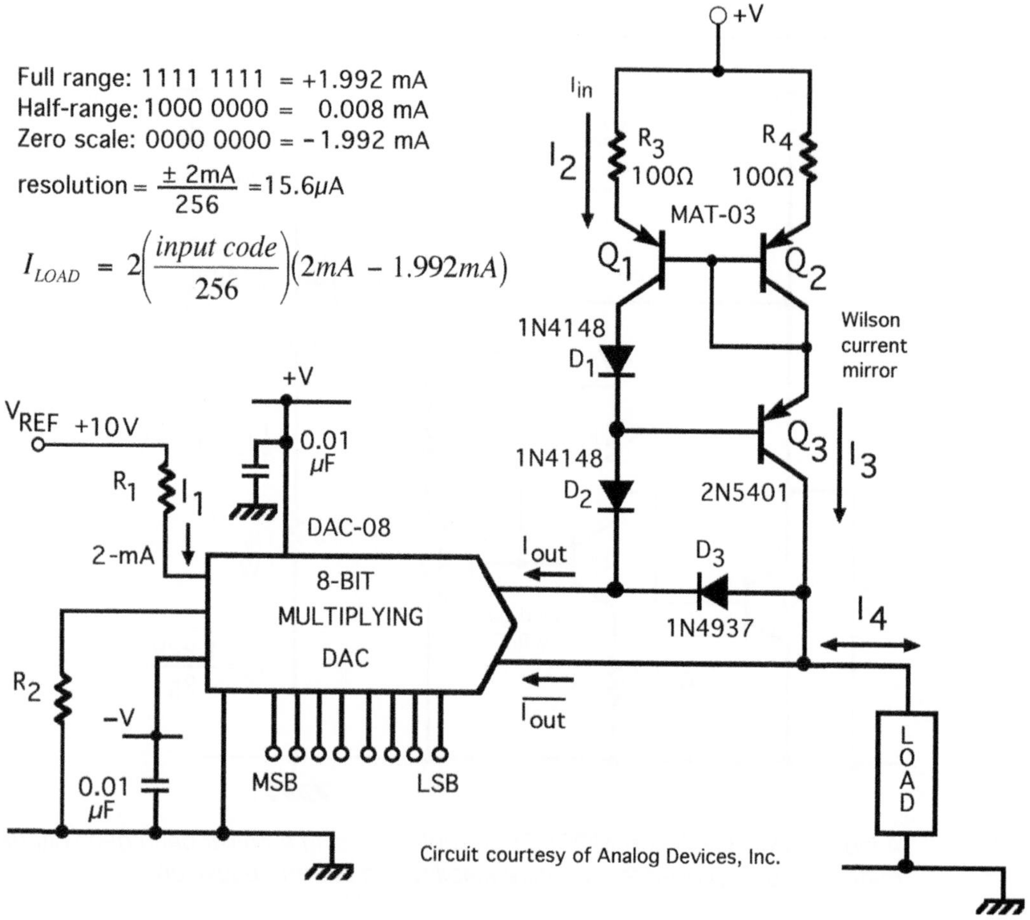

Figure 4.56. *A digitally controlled bipolar current pump.*

will surely be a major part of our electronics design kit for a few more decades to come, and with it the current source.

Chapter 5

Using Precision Matched-Pairs, Duals, and Quads

5.1 Precision BJT matched-pairs

As we learned from the previous chapter, some of the most accurate BJT-based current sources that the designer can create at the circuit board level use precision matched-pairs and quads. These are available from Analog Devices, National Semiconductor, and second-source specialists Linear Integrated Systems. Precision matched monolithic BJT pairs and quads represent the highest quality and precision available in a silicon transistor. All three manufacturers make NPN matched-pairs. Analog Devices and Linear Integrated Systems also offer PNP matched-pairs. Additionally, Analog Devices makes a quad NPN device. Overall, Linear Integrated Systems provides a wider range of different package options than any other manufacturer. Some of the company's matched-pairs are super-gain transistors (not Darlingtons), with h_{FE}s upward of 1000. Many of the BJT circuits shown in Chapter 4 can be optimized by using any of these high-performance matched-pairs and quads.

What makes precision matched-pairs so special? Unlike dual transistors or multiple transistors that are packaged as an array, matched-pairs have specifications far exceeding those of high-performance discretes or a typical multitransistor array. Precision matched-pairs rely on a specially designed paralleled transistor structure (not unlike today's power MOSFET structure, which also uses many paralleled cells in order to create one single FET device with very low on-resistance). With the precision matched-pair/quad BJT, a similar technique dramatically reduces the base-emitter bulk resistance to levels typically less than an ohm (this resistance is often more than 50 Ω in a regular discrete transistor). They also receive special processing, in order to achieve these exotic characteristics. This includes a multiple passivation process, which stabilizes the important parameters over a wide temperature range, reduces leakage currents, and enhances the isolation between individual transistors. These parameters are also stabilized over a range of collector currents, which typically span from one microamp to several milliamps. Additionally, each transistor is individually tested against its published data sheet specifications over the full operating temperature range, and those characteristics, such as noise levels, gain matching, and offset voltage drift, all come close to matching the specs for the ideal transistor.

To illustrate this concept further, the best NPN matched-pairs have V_{BE}s that are matched to better than 50 μV (referred to as ΔV_{BE} or V_{OS}) and high h_{FE}s that are

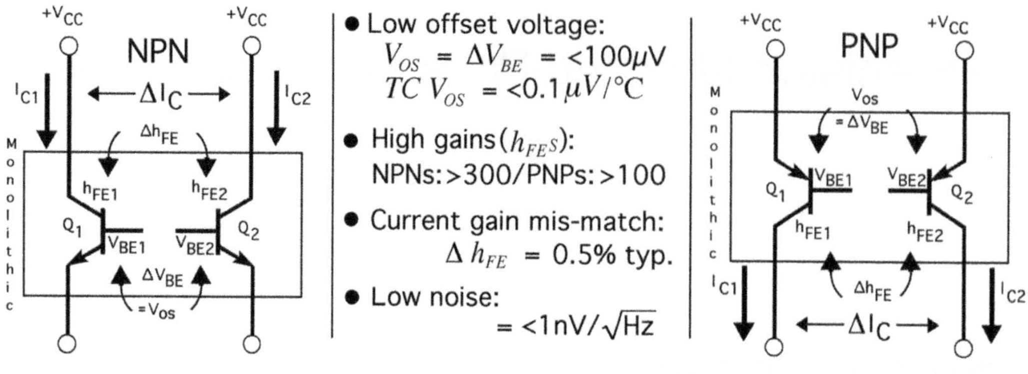

Figure 5.1. *An example of excellent transistor matching using precision matched pairs.*

matched to better than 0.5%. They have offset voltage drifts of less than 0.1 μV/°C, offset voltage CMRRs ($\Delta V_{OS}/\Delta V_{CB}$) that are better than 120 dB, and ultra-low base-emitter bulk resistance (r_{be}) of less than 0.5 Ω.They also feature very low noise levels, typically less than 1 nV/√Hz (see Figure 5.1). National Semiconductor's venerable LM194/394, for example, has specs that include a typical offset voltage of 25 μV, h_{FE} matching typically 0.5%, offset voltage drift less than 0.1 μV/°C, and a wide operating range of between 1 and 40 volts. Operation is specified at room temperature and over a range of either –55°C to +125°C (LM194) or –25°C to +85°C (LM394).

Analog Devices' MAT02 has specs that even improve on this, over the full military temperature range. In addition, it provides a quad NPN product, the MAT04, as well as lower-cost devices, the SSM2210 (NPN) and SSM2220 (PNP). See Table 5.1 for more specifics. When it comes to high-gain matched PNP pairs, Analog Devices' MAT03 has offsets of 100 μV max, while their lower-cost SSM2220 has offsets of 200 μV max. Unlike discrete PNPs, these devices have good h_{FE}s, as well as very low r_{bes}, and ultra-low noise. Linear Integrated Systems offers higher gain devices, but with higher V_{OS} (i.e., their LS352 has an h_{FE} of greater than 200 and a V_{OS} max of 500 μV). Linear Integrated Systems also offers some useful log-conformance, matched NPN and PNP duals in a choice of several different packages.

Unlike normal discrete or multiple transistors, which often require beta-compensation for h_{FE} mismatches, or emitter-degeneration resistors to compensate for differences in V_{BE}s, precision matched-pairs usually do not. In fact, these types of compensation may be completely eliminated in many applications. Because of their far superior specs, precision matched-pairs are ideal for use as differential input stages for high-gain instrumentation amplifiers, squaring and square root circuits, log/antilog amplifiers, converters, multiplier and divisor circuits, professional audio equipment, and precision current sources. With the exception of Analog Devices' MAT01 (an NPN pair), the other products (i.e., MAT02, 03, 04, and LM194/394) all include internal clamp

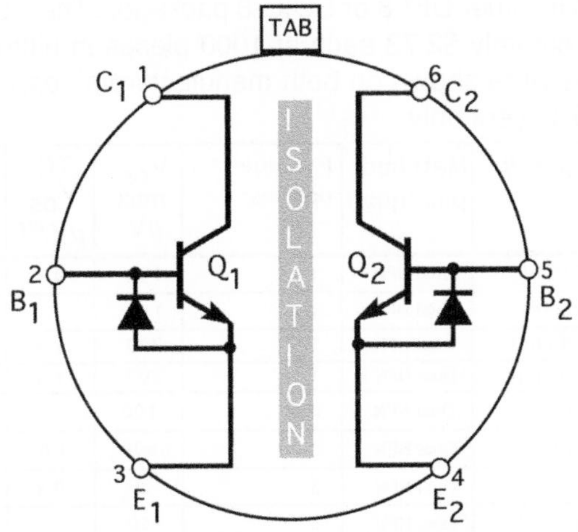

Figure 5.2. *Pin-out for National Semiconductor's LM194/LM394 and Analog Devices MAT02 dual monolithic transistors in a 6-lead, insulated header, metal-can package. Note that the substrate is connected to the case.*

diodes between the emitter and base junction of each transistor (see Figure 5.2). This feature serves a dual purpose: (1) it clamps the substrate to the most negative emitter and ensures complete isolation between the devices, and (2) the diodes help prevent degradation caused by reverse-biased emitter current, which could otherwise cause damage, instability, and drift of some of the key matching parameters.

Being precision devices, these matched-pairs and quads are intended for high-performance instrumentation. One could expect to pay a very high price for this level of quality, but many devices are inexpensive and affordable. Others are aimed at MIL/aerospace-type applications and are priced accordingly. National Semiconductor's LM194 and LM394B/C are available in an isolated header six-lead TO-5 metal can. The LM194 is identical to the LM394 except for tighter electrical specifications and a wider temperature range. Because the LM194 is a military/aerospace device, it is also the most costly of all the precision matched-pairs ($17.50 each), whereas the LM394 devices are considerably less expensive (ranging from approximately $3.30 to $5.20 each).

Analog Devices' MAT01/02/03 products are also available in an isolated header six-lead metal can. The MAT01 costs between approximately $5.60 (GH) to $9.30 (AH) each for 100 pieces. The MAT02, with its improved specs, costs between approximately $6.00 (FH) to $11.65 (EH) each at 100 pieces. The MAT03, the PNP matched-pair, costs between approximately $7.33 (FH) to $13.98 (EH) each at 100 pieces. The MAT04, the NPN matched-quad device, is considerably less expensive and costs approximately $3.95 each at 100 pieces for either the 14-pin DIP (FP) or the SOIC14

(FS) device. The SSM2210 (a matched NPN pair) costs approximately $2.24 each at 1000 pieces in either DIP-8 or SOIC-8 packages. The SSM2220 (a matched PNP pair) costs approximately $2.73 each at 1000 pieces in either DIP-8 or SOIC-8 packages. These prices were shown on both manufacturers' respective Web sites but are applicable to U.S. buyers only.

Mfr.	Part #:	Matched pair/quad	Package options:	V_{OS} max µV	TC V_{OS} µV/°C	h_{FE} min	Δh_{FE} Max. %	e_n typ. nV/$\sqrt{Hz}$	Op. temp. range
NS	LM 194H	Dual NPN	1	100	0.3	300	2	1.8	MIL
NS	LM 394H	Dual NPN	1	150	1	200	4	1.8	Ind
NS	LM 394BH	Dual NPN	1	200	1.5	150	5	1.8	Ind
NS	LM 394CH	Dual NPN	1	200	1.5	150	5	1.8	Ind
AD	MAT 01AH	Dual NPN	3	100	0.5	500	3	6.1	MIL
AD	MAT 01GH	Dual NPN	3	500	1.8	250	3	6.1	MIL
AD	MAT 02EH	Dual NPN	3	50	0.3	400	2	0.9	Ind
AD	MAT 02FH	Dual NPN	3	150	1	300	4	0.9	Ind
AD	MAT 03EH	Dual PNP	3	100	0.5	80	3	0.7	Ind
AD	MAT 03FH	Dual PNP	3	200	1	60	6	0.7	Ind
AD	MAT 04FP	Quad NPN	5	400	2	300	4	1.8	Ind
AD	MAT 04FS	Quad NPN	6	400	2	300	4	1.8	Ind
AD	SSM 2210P	Dual NPN	2, 8	200	1	300	5	1.0	Ind
AD	SSM 2210S	Dual NPN	2, 8	200	1	300	5	1.0	Ind
AD	SSM 2210SZ	Dual NPN	2, 8	200	1	300	5	1.0	Ind
AD	SSM 2220P	Dual PNP	2, 8	200	1	80	2	1.0	Ind
AD	SSM 2220S	Dual PNP	2, 8	200	1	80	2	1.0	Ind
LIS	LSIT 120A	Dual NPN	2, 3, 4, 7, 8	400	3	200	2	I	Ind/MIL
LIS	LSIT 130A	Dual PNP	2, 3, 4, 7, 8	400	3	200	2	I	Ind/MIL
LIS	LS 301	Dual NPN	2, 3, 4, 7, 8	1000	5	2000	5	I	Ind/MIL
LIS	LS 302	Dual NPN	2, 3, 4, 7, 8	1000	5	1000	5	I	Ind/MIL
LIS	LS 303	Dual NPN	2, 3, 4, 7, 8	1000	5	2000	5	I	Ind/MIL
LIS	LS 310	Dual NPN	2, 3, 4, 7, 8	1000	2	150	10	I	Ind/MIL
LIS	LS 311	Dual NPN	2, 3, 4, 7, 8	1000	5	150	5	I	Ind/MIL
LIS	LS 312	Dual NPN	2, 3, 4, 7, 8, 9	500	2	200	5	I	Ind/MIL
LIS	LS 313	Dual NPN	2, 3, 4, 7, 8	1000	5	400	5	I	Ind/MIL
LIS	LS 318	Dual NPN #	2, 3, 4, 7, 8	400	5	150	5	I	Ind/MIL
LIS	LS 351	Dual PNP	2, 3, 4, 7, 8	1000	10	150	5	1.5	Ind/MIL
LIS	LS 352	Dual PNP	2, 3, 4, 7, 8, 9	500	2	200	5	1.5	Ind/MIL
LIS	LS 358	Dual PNP #	2, 3, 4, 7, 8	400	5	100	5	I	Ind/MIL

NOTES:
NS = National Semiconductor AD = Analog Devices LIS = Linear Integrated Systems
Package options:
1 = 6-lead, isolated header, TO-5 metal can; 2 = 8-pin plastic DIP; 3 = 6-lead, isolated header, TO-78 metal can; 4 = 6-lead, isolated header, TO-71; 5 = 14-pin plastic DIP; 6 = 14-pin SO; 7 = SOIC-8; 8 = SOT-23/8pin; 9 = SOT-23/6pin; # = Log conformance

Table 5.1 BJT precision matched-pairs and quads.

5.2 Quality dual transistors

There was a time at the beginning of the electronics era (early 1960s) that dual transistors were readily available and were even used to build early modular op amps. Major vendors of the day included Fairchild Semiconductor, Union Carbide, Transitron, and Motorola Semiconductors. Quality dual transistors, unlike most of the general-purpose duals and quads that we will look at next, are matched, but to a lower degree than the ultra-precision devices we looked at previously. They are useful in development work and in lower accuracy applications. These monolithic devices share virtually identical characteristics and are also thermally matched, which is an important factor. As stated previously in Chapter 4, accuracy can always be significantly improved by adding relatively low-value resistors (100 Ω to 1.5 KΩ), inserted in either the uppermost emitters of PNP devices (see Figure 5.3A) or the lowest emitters of NPN devices. They can also be beta-compensated by incorporating a low-value resistor (typically between 22 Ω and 100 Ω) in the base of the reference transistor, Q_1, as shown in Figure 5.3B. In either case, the resistors should be quality metal-film types, ±1% tolerance or better, with a low tempco (less than 100 ppm). Some excellent current sources can be created using quality matched dual transistors, *if* either or both of these two techniques are employed, as shown in Figure 5.3.

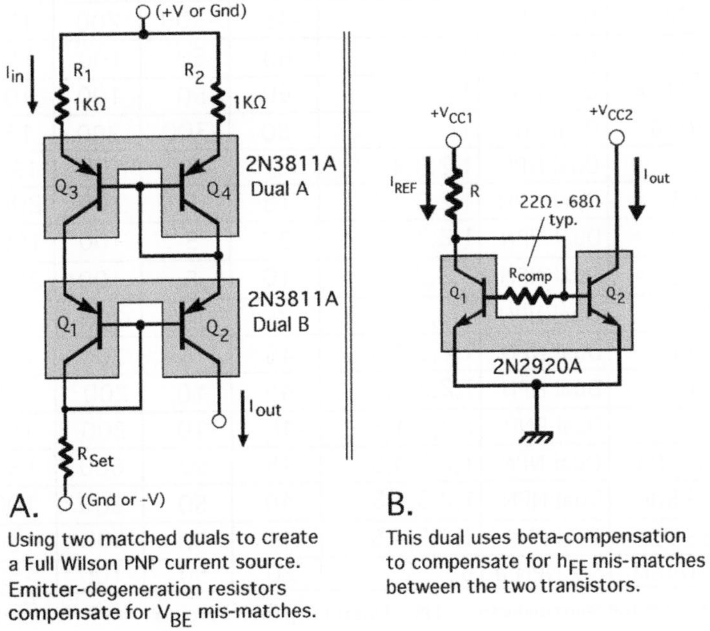

Figure 5.3. Examples showing how to improve matched duals for use as a current source (A) or a current sink (B).

Quality dual transistors are fairly uncommon, however. Most manufacturers today prefer to offer general-purpose duals and quads, and so do not publish specifications that includes the key matching data between the transistors (such as ΔV_{BE} or V_{OS} max, TC V_{OS}, and Δh_{FE}). Those particular specifications are shown with the precision matched-pairs and quads, and require that each transistor in a pair or quad be individually tested against its published data sheet specs. This adds cost and results in a more expensive device. A quality dual transistor does include most of those specs, although not as high performance (where a ΔV_{BE} might be 500 μV vs. 50 μV for a precision matched-pair). A general-purpose dual/quad does not include those specs at all. Two manufacturers that do make quality matched dual transistors are Central Semiconductor and Linear Integrated Systems. Notice the key matching specifications, ΔV_{BE} and Δh_{FE}. Table 5.2 shows some of these products.

Mfr.:	Part #:	Dual/quad NPN/PNP	Package options:	V_{CBO} (V)	I_C (mA)	f_t min (MHz)	h_{FE} min	Matching Δh_{FE} max. %	Matching ΔV_{BE} max. (mV)
CS	2N2915A	Dual NPN	1	45	30	60	60	10	1.5
CS	2N2916A	Dual NPN	1	45	30	150	60	10	1.5
CS	2N2919A	Dual NPN	1	60	30	60	60	10	1.5
CS	2N2920A	Dual NPN	1	60	30	60	150	10	1.5
CS	2N3727	Dual PNP	1	-45	-300	200	135	10	2.5
CS	2N3810A	Dual PNP	1	-60	-50	100	150	5	1.5
CS	2N3811A	Dual PNP	1	-60	-50	100	100	5	1.5
CS	2N4016	Dual PNP	1	-60	-300	200	135	10	2.5
LIS	IT124 *	Dual NPN	1,2,3,4	2	10	100	1500	10	5
LIS	LS301 *	Dual NPN	1,2,3,4,5	18	5	100	2000	5	1
LIS	LS302 *	Dual NPN	1,2,3,4,5	35	5	100	1000	5	1
LIS	LS303 *	Dual NPN	1,2,3,4,5	10	5	100	2000	5	1
LIS	LS310	Dual NPN	1,2,3,4,5	25	10	200	150	10	3
LIS	LS311	Dual NPN	1,2,3,4,5	45	10	200	150	5	1
LIS	LS313	Dual NPN	1,2,3,4,5	45	10	200	400	5	1
LIS	LS351	Dual PNP	1,2,3,4,5	-45	-10	200	150	5	1
LIS	LS3250A	Dual NPN	1,2,3,4,5	45	50	600	150	10	2
LIS	LS3250B	Dual NPN	1,2,3,4,5	40	50	600	100	10	5
LIS	LS3550A	Dual PNP	1,2,3,4,5	-45	-50	600	150	10	2
NS	LM3046M	5 x NPN #	6	20	50	>100	>50	10	5

NOTES: CS = Central Semiconductor LIS = Linear Integrated Systems NS = National Semiconductor

Package options:

1 = 6-lead, TO-78 metal can; 2 = 6-lead, TO-71 metal can; 3 = 8-pin plastic DIP; 4 = SOIC-8;

5 = SOT-23/6-pin ; 6 = SOIC-14 * = super-beta, v.low-voltage type # = 3 x NPN + a differential pair

Table 5.2 Closely matched dual NPN and PNP transistors.

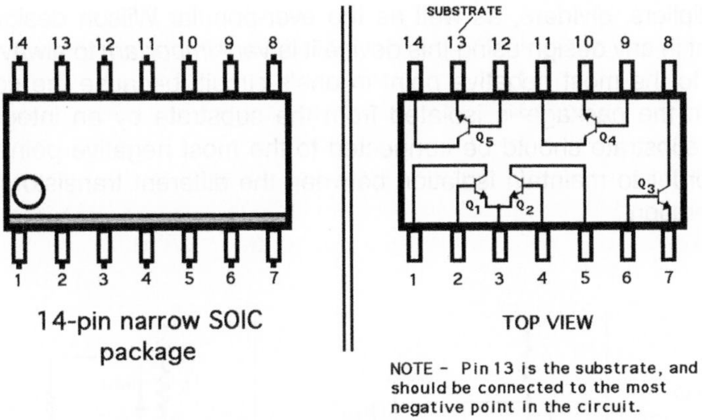

Reprinted with permission from National Semiconductor Corp.

Figure 5.4. *National Semiconductor LM3046M, a 5 × NPN transistor monolithic array, with matched V_{BE}s.*

Back in the 1970s and early '80s, RCA Solid State developed and introduced a product line in dual-in-line packages, known as the CA-3000 series. These became the forerunners of most other BJT arrays/hybrids from various companies and were used mostly in front-end amplifier applications. (RCA was after all primarily a TV/radio manufacturer, and so the line was probably developed for use in such equipment.) Some devices were second-sourced by Motorola's Semiconductor Group and National Semiconductor, among others. When RCA's discrete and analog portfolio was sold off to Harris Semiconductor, it too made them. Today, however, they are no longer available, which is a pity because some of the products were closely matched and could easily be used for current source applications (e.g., CA3018A and CA3146 both had a maximum ΔV_{BE} of only 2 mV). Only National Semiconductor continues to produce one of those products, the LM3046 (a 14-pin SOIC shown in Figure 5.4), which contains an NPN differential pair as well as three general-purpose NPNs.

Although at first glance only the differential pair appears to have matched V_{BE}s, actually when one examines the data sheet carefully, all five transistors have their V_{BE}s matched to within ±5 mV of one another (typically 0.45 mV), and can therefore be used to create some inexpensive current sources. Because the device has a maximum collector current rating of 50 mA for each transistor, typical current (sink) levels can range from about 2 μA to about 10 mA, without heat dissipation becoming a major factor.

In some volume applications that may be noncritical or require only medium accuracy, the LM3046M can be useful, particularly because it is a surface-mount package. The best part is that the U.S. price at 1000 pieces is $0.45 each, which is extremely economical. Figure 5.5 shows some common current-sink designs that can be built using the device. Because of its NPN format, these all take the form of common mirror-sinks, multipliers, dividers, as well as the ever-popular Wilson designs. Remember, though, that in any design using this device it is *very important* to always tie pin 13 (the substrate) to the most negative point in one's circuit, because the collector of each transistor in the package is isolated from the substrate by an integral diode. As a result, the substrate should be connected to the most negative point in the external circuit, in order to maintain isolation between the different transistors and to ensure normal operation.

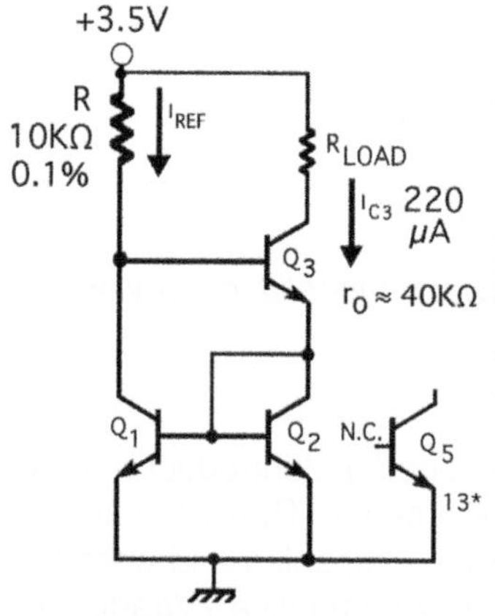

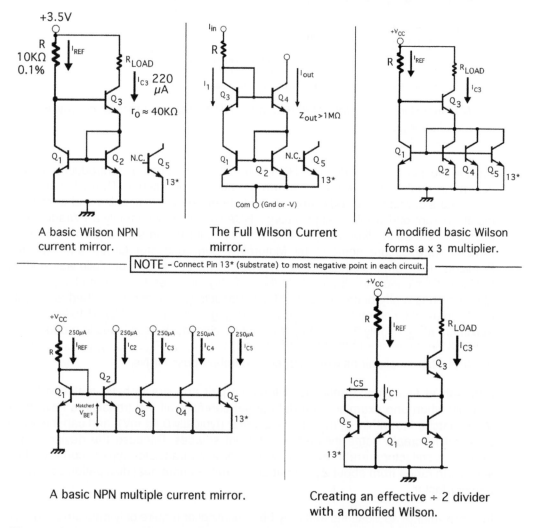

A basic Wilson NPN current mirror.

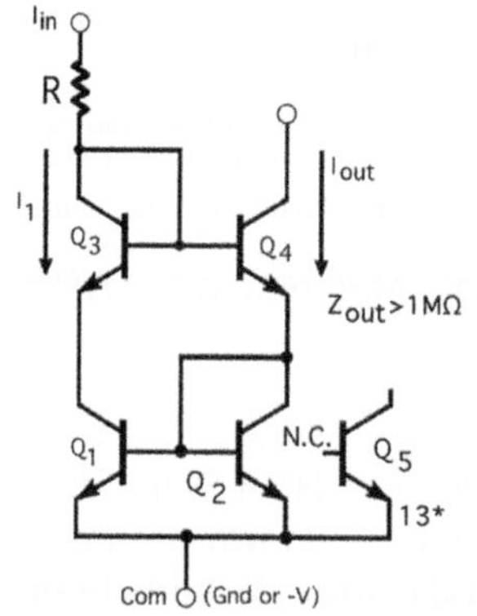

The Full Wilson Current mirror.

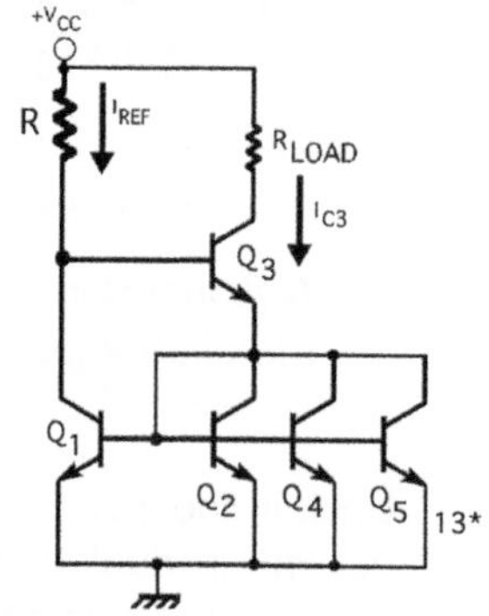

A modified basic Wilson forms a x 3 multiplier.

NOTE - Connect Pin 13* (substrate) to most negative point in each circuit.

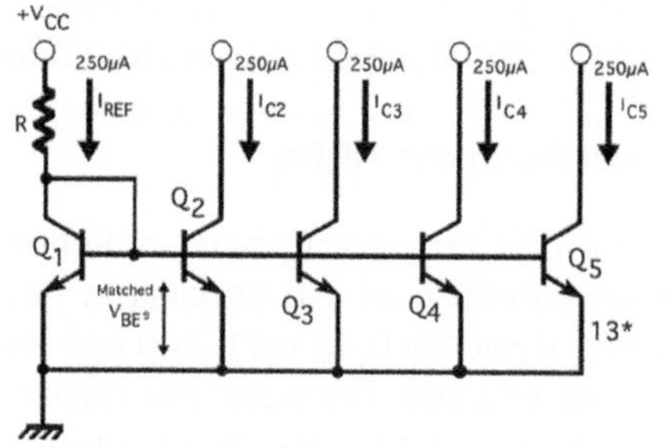

A basic NPN multiple current mirror.

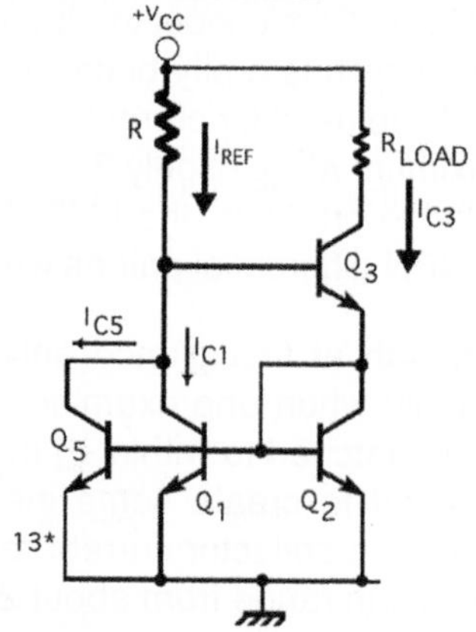

Creating an effective ÷ 2 divider with a modified Wilson.

Figure 5.5. LM3046M current source applications.

5.2.1 General-purpose BJT duals and quads

Today various general-purpose BJT duals and quads are readily available in the marketplace. Although most are not critically matched to the same degree as we have seen up until now, they are (usually) built with adjacent die from the same wafer and so are at least thermally matched. (Devices of the same type, but from different wafers, may be regarded as hybrids and are not a good choice for using in current source designs. They would be only slightly better than using random discrete transistors, which we have learned previously are not a good choice.) Therefore, general-purpose BJT duals and quads are useful in development work or in noncritical or low-accuracy applications, but remember that because these products are not matched, they will require V_{BE} and/or h_{FE} compensation for any semiprecision current source designs.

Dual transistors are available from Central Semiconductor, Crystalonics, Diodes Inc., Fairchild Semiconductor, Linear Integrated Systems, Microsemi, ON Semiconductors, Philips Semiconductors, and Vishay Semiconductors. They are available in different packages, including surface-mount, and in different temperature ranges. The largest makers of duals include Central Semiconductor, Fairchild Semiconductor, and Linear Integrated Systems. Crystalonics makes various former and discontinued Motorola Semiconductor parts, including some that are MIL/JAN-qualified devices. Microsemi also offers an impressive range of MIL/JAN-qualified devices in various packages. Common dual transistors offered by several manufacturers include 2N3811, IT130, and 2N4937 (all PNP pairs), as well as 2N2920, IT120, and 2N5794 (all NPN pairs). Table 5.3 gives an overview of the manufacturers and the types of products they offer.

Figure 5.6. *Just a few of the packages that have been used for the popular 2N3906, which include metal, plastic, and surface-mount types.*

Manufacturer:	Ultra-precision, matched duals and quads			Matched duals and quads			General purpose duals and quads		
	Metal-can /leaded	DIL/ leaded	Surface-mount	Metal-can /leaded	DIL/ leaded	Surface-mount	Metal-can /leaded	DIL/ leaded	Surface-mount
Analog Devices	√	√	√						
Central Semiconductor				√		√	√		√
Crystalonics							√	√	√
Diodes Inc.									√
Fairchild Semiconductor									√
Linear Int. Systems	√	√	√	√	√	√	√	√	√
Microsemi							√	√	√
National Semiconductor	√					√			
ON Semiconductor								√	√
Philips Semiconductors								√	√
Vishay Semiconductors								√	√

NOTES:

1. Ultra-precision matched devices are specified on their data sheet with ΔV_{BE}, Δh_{FE}, and TC_{VOS}.
2. Matched devices are specified on their data sheet with ΔV_{BE}, and /or Δh_{FE}.
3. General purpose devices do not have ΔV_{BE}, Δh_{FE}, or TC_{VOS} specified on their data sheet.
4. Some devices in all three categories are qualified to MIL/JAN standards.
5. Some general purpose quads are complementary, containing both a pair of NPN and PNP transistors.

Table 5.3 Manufacturers of dual and quad BJT transistors.

Quad transistors are available from Fairchild Semiconductor, Central Semiconductor, Crystalonics, ON Semiconductor, Diodes Inc., Microsemi, and others. Most products contain either all-NPN or all-PNP devices. Some examples include the PQ-3904 (four NPNs with similar specs to a 2N-3904) and the PQ-3906 (four PNPs with similar specs to a 2N-3906). Some are dual-complementary (two NPN and two PNP in the same package), such as the PQ-6002 and the PQ-6502 (two NPNs and two PNPs with specs based on 2N3904/3906), which can provide a simple current source and current sink in one package. A sample of some of these various devices are listed in Table 5.4.

In summary, some of the best and most accurate current sources that can ever be created at the circuit board level use well-proven BJT technology. Monolithic matched-pairs and quads represent the highest quality and precision available, while lower-priced matched-duals can also produce some excellent current sources. Even general-purpose duals and quads can produce good, stable current sources, so long as one remembers to include the necessary compensation. For many engineers designing with BJTs, with which they are more familiar, is easier than with other technologies. This will probably remain so for a long time to come.

Part No.:	NPN/ PNP/ Comp.	Dual	Quad	V_{CEO} min (V)	V_{CBO} min (V)	I_C max (mA)	I_{CBO} max (nA)	h_{FE} min @ 1mA	F_T min (MHz)	
2222/A	NPN	√	√	40	75	1000	10	50	300	
2369	NPN		√	15	40	200	400	40		
2484	NPN		√	40	60	50	20	300	50	
2907/A	NPN	√	√	60	60	800	20	100	200	
3799	PNP		√	-60	-60	-50	20	200	60	
3904	NPN	√	√	40	60	200	50	70	300	
3906	PNP	√	√	-40	-40	-200	50	80	250	
5401	PNP	√		-150	-160	-200	50	50	100	
5551	NPN	√		160	180	200	50	80	100	
6002	Compl		√	30	60	500	30	50	200	
6502	Compl		√	30	60	500	30	50	200	
6600	Compl		√	45	60	50	10	150	50	
6700	Compl		√	40	40	200	50	50	200	
7043	NPN		√	250	250	500	100	25	50	
7051	Compl		√	150	150	500	100	25	50	
7093	PNP		√	-250	-250	-500	250	25	50	
BC846	NPN	√		65	80	100	15	150	100	
BC847	NPN	√		45	50	100	15	150	100	
BC848	NPN	√		30	30	100	15	150	100	
BC856	PNP	√		-65	-80	-100	15	150	100	
BC857	PNP	√		-45	-50	-100	15	150	100	
BC858	PNP	√		-30	-30	-100	15	150	100	

NOTES:

1. The numeric part numbers shown above are "generic". For example the "3906" may be prefixed with alpha characters such as "2N/PN-3906" (single),"FMB-3906" (a dual), or "MMPQ-3906" (quad). In addition to identifying the manufacturer, and type (dual or quad), the prefix often denotes the package type. For example, "MMPQ-2907A" is a quad PNP in a SOIC-16 package, while "FMB-2907A" is a dual PNP in a SOT-6 package, and "FFB-2907A" is a dual PNP in a SC-70 package.

2. The devices listed are available from multiple sources, and in various different packages.

3. "Comp" shown in column 2, denotes a complementary pair. The quad devices shown contain a pair of NPN and a pair of PNP devices. One such quad may be used to build a simple current source and sink.

Table 5.4 Common transistors available as dual and quads.

Chapter 6

Using JFETs and CRDs to Create Current Sources

6.1 The JFET paved the way

Just as BJTs can be used to create current sources, so too can field-effect transistors. The designer can readily utilize *junction* FETs (JFETs) and complementary metal oxide semiconductor (CMOS) FETs in low current-source applications. Alternately, one can use double-diffused MOS (DMOS) FETs for higher current requirements. However, these types of devices are often overlooked by the OEM board-level designer, because of their somewhat unfamiliar terminology, as opposed to the wide-scale popularity of the bipolar junction transistor (BJT). Actually, use of the FET can often render a simpler and less costly circuit, if one understands some of the critical parameters involved, rather than using more expensive solutions. These might include using op amps with voltage references or matched bipolar transistors, as discussed previously. Although it is a little more complex than is depicted here, because of process differences (e.g., silicon gate, metal gate, vertical, double-diffused), the basic FET family tree is shown in Figure 6.1. Not all FET versions are available or easy to manufacture. For example, you can see from the diagram that JFETs only function in the depletion mode, and P-channel depletion-mode MOSFETs are unavailable.

The basic theoretical principles of the FET were outlined in various papers published in the late 1920s by a German-born physicist, Dr. Julius E. Lilienfeld. He was granted a patent by the U.S. Patent Office in 1930. Some 17 years later, Dr. Shockley and his colleagues at AT&T's Bell Labs further theorized on this device (the unipolar transistor, as it was originally called), and in the course of their research with the FET, won the Nobel Prize for Physics in 1956, for inventing the bipolar transistor. After patenting the BJT, Dr. Shockley went on to publish various theoretical papers on other semiconductor devices, and in 1952 he invented and patented the JFET, although he did not actually make one. It had been some 22 years since Dr. Lilienfeld first theorized on an FET-like device.

For several years after, the FET languished only as a lab curiosity, during which time the epitaxial process was developed by Dr. Ross's group at Bell Labs. Then in 1959, Dr. Jean Hoerni, a co-founder of Fairchild Semiconductor, developed and patented the semiconductor industry's planar process and designed the first modern-day planar epitaxial JFETs. (Dr. Hoerni subsequently went on to co-found Teledyne Amelco, which also became a major source of JFETs, then later he co-founded Intersil, a major

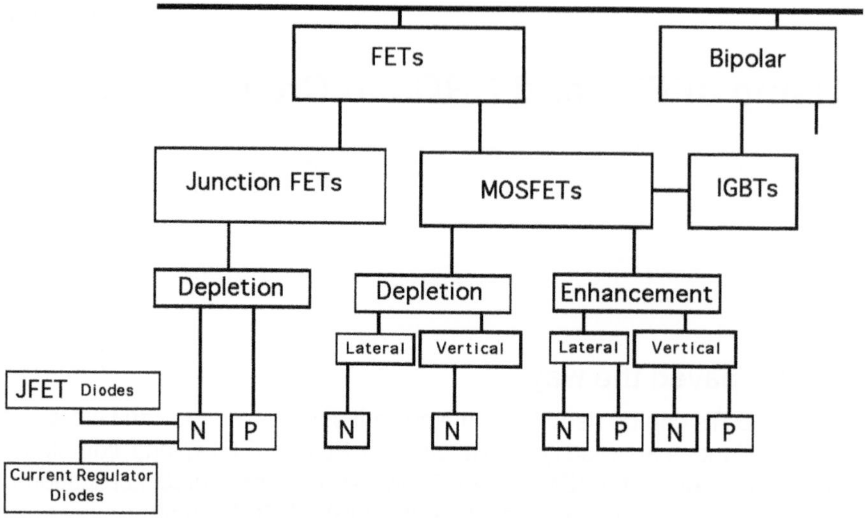

Figure 6.1. The FET family tree.

analog CMOS pioneer.) In the early 1960s, Siliconix, Amelco, and Crystalonics were some of the first companies in the world to specialize and manufacture JFETs in volume. One of Siliconix's co-founders, Bernie Murphy, was a former Bell Labs researcher who had invented the important "buried collector" structure, which has been indispensable to manufacturing analog ICs.

Besides Fairchild Semiconductor, Siliconix, Crystalonics, and Teledyne Amelco, other notable JFET pioneers included National Semiconductor, Solitron Devices, Motorola Semiconductor, Texas Instruments, Philips Semiconductor, and Union Carbide. Some of these companies no longer make discretes or JFETs; only a few of these original pioneers are still even selling FETs. They should all be congratulated, though, because without their bold efforts we would not have CMOS, DMOS, gallium-arsenide FETs, silicon power MOSFETs, or IGBTs, all of which now dominate in various market segments of present-day electronics. By today's standards these early devices were fairly raw, but the limitations on these pioneers was only in the available technology of their time—namely in processing, precision lithography, and diffusion techniques.

Although both BJTs and FETs are used as switches, amplifiers, and current sources, the JFET can also be used in its simplest form as a voltage-controlled resistor. The JFET has many diverse applications (see Figure 6.2), which exploit its various characteristics, when compared with the BJT. Besides voltage-controlled resistors, they include low-noise, differential, and high input-impedance (Z_{in}) amplifiers, audio and AGC amplifiers, video and RF amplifiers, analog switches, RF mixers, H-F oscillators, and current sources. In the design of many of today's analog ICs, JFETs are commonly used for circuit start-up, biasing, and front-ends in bipolar-FET op amps.

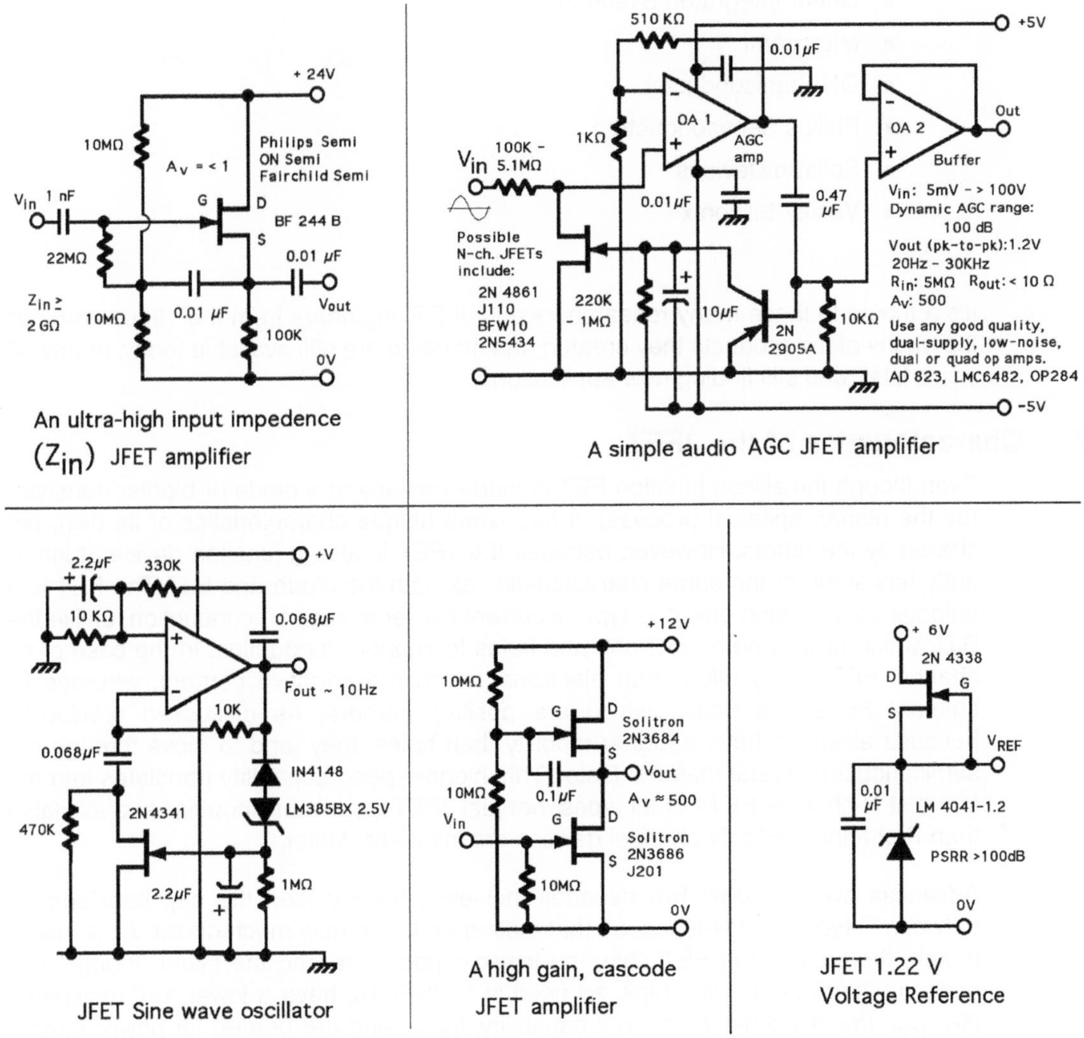

Figure 6.2. Common JFET applications.

In the United States, the following semiconductor manufacturers make silicon JFETs:

- Calogic
- Central Semiconductor
- Crystalonics
- Fairchild Semiconductor
- InterFET

- Linear Integrated Systems
- Microsemi
- ON Semiconductor
- Philips Semiconductors
- Solitron Devices
- Vishay-Siliconix

It's a tribute to those many researchers and JFET engineers from the 1960s and '70s that many of the products they created and modeled are still available today, nearly 40 years later, and still finding new applications.

6.2 Characteristics of the JFET

Even though the silicon junction FET is made similarly to a diode or bipolar transistor (by the planar, epitaxial process), it has some unique characteristics of its own, not shared by the others. However, because the JFET is also a junction device, it inherently has some of the same characteristics as both the diode and BJT. The FET is a unipolar device using only one type of current carrier to support conduction, unlike the BJT, which utilizes both electrons and holes to support conduction. In the case of N-channel FETs, they utilize free electrons, which are negative carriers, whereas P-channel FETs use holes, which are positive carriers. As discussed previously, because electrons have a higher mobility than holes, they tend to move through the semiconductor crystal material faster. This higher-speed capability translates into the fact that N-channel FETs of all types, not just JFETs, are more popular and available than P-channel FETs; N-channel devices are invariably faster.

Advances over the past two decades, however, have closed the gap considerably between P-types and N-types, so that their speeds are now much closer. As a rule of thumb, the smaller the FET chip, the lower capacitance and the faster, though less powerful, it will be. Large chips, as used in MOSFETs, have a lower "on" resistance $R_{DS\ (on)}$, thereby a higher current capability, $I_{D(on)}$, and are geared for power switching. They tend to have higher input and output capacitance and are therefore slower. A JFET's input capacitance (C_{iss}) is typically less than 10 pF and remains constant up to about 1 GHz. For a small MOSFET, this may be more than 30 pF, but with a power MOSFET, it can be more than 1000 pF. The output capacitance (C_{oss}) for the JFET and small MOSFET are typically less than 30 pF, while that of the power device may be 10 or 20 times higher. As a result, JFETs and small DMOS FETs are usually much faster than power MOSFETs. Varactor FETs exploit the device's unique capacitance characteristics, have two separate gates, and are used for RF applications.

Individual JFETs are typically three-terminal devices like BJTs, with the *gate*, *drain*, and *source* corresponding to the BJT's base, collector, and emitter (see Figure 6.3).

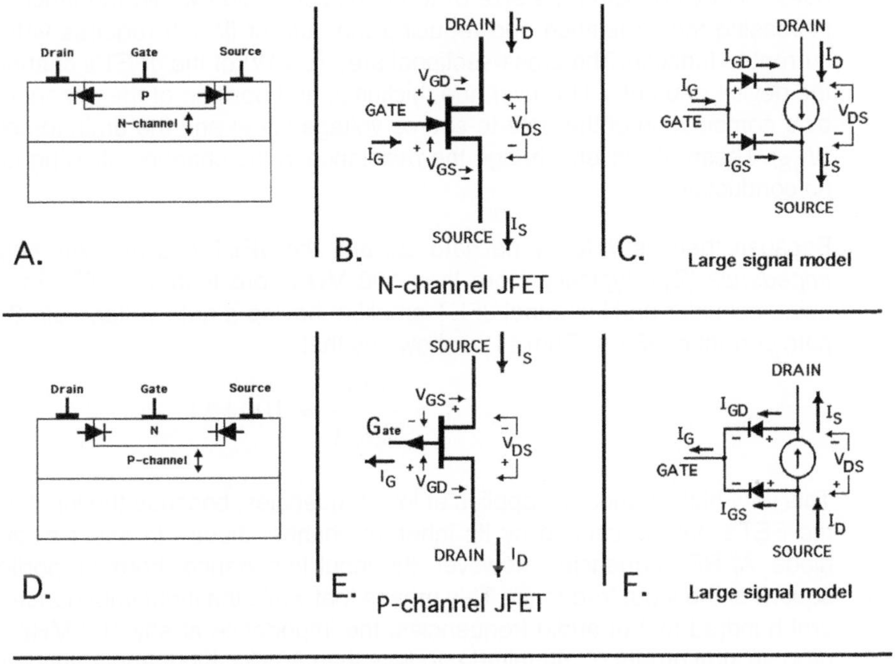

Figure 6.3. *Showing the structure of the JFET, as well as its various voltages, currents, and polarities. Notice the two-channel diodes that provide forward and reverse bias.*

Besides individual discretes, some FETs are also available as matched-pairs, duals, and arrays, in standard metal, plastic, and surface-mount packages.

Another unique characteristic of the FET is that it is voltage-controlled, whereas the bipolar transistor is current-controlled. To control current passing through any type of FET, one uses a voltage at its gate, whereas with the bipolar transistor a combination of base voltage and current is used. The N-channel JFET is a *normally-on* device, until the gate-to-source voltage (V_{GS}) becomes negative enough to reverse-bias the gate channel diode, and so turn it off. The point where this occurs is known as $V_{GS\ (off)}$, which we will look at in more detail shortly. The P-channel JFET is also a *normally-on* device, until the gate voltage becomes positive enough to reverse-bias the gate channel diode, and thereby turn it off. In contrast, the bipolar transistor (both NPN and PNP) is a normally-off device until sufficient voltage (V_{BE}) and base current (I_B) are provided to turn it on.

N- and P-channel JFETs work only in the *depletion mode* (as shown in Figure 6.3A and D). Other types of FETs can operate in either the depletion mode or the enhancement mode, which we will cover in a later chapter. The depletion mode operates by a gate voltage depleting or shutting off the majority current carriers in the channel. It does this by changing the size of the depletion region within the junction area, thus increasing the resistance and reducing the current flow. It requires virtually no gate current to function. The cross-sectional area (L × W) of the JFET's channel is fixed by the device geometry. However, the thickness and position of the channel is controlled by a combination of the gate-to-source voltage (V_{GS}) and the drain-to-source voltage (V_{DS}). These effectively change the resistance in the channel, allowing full, partial, or no conduction.

Because there is virtually no gate current, the JFET has an extremely high *input impedance* (Z_{in}), typically more than 100 MΩ (more than 1×10^8). To illustrate this point, consider an N-channel JFET amplifier having a gate voltage of –2.5 volt and a gate current of 25 nA. Ohm's law shows us that:

$$R = \frac{2.5V}{25nA} = 100M\Omega \qquad \text{(Eq.6.1)}$$

This very high impedance applies at low frequencies, because the input impedance of the FET's gate is caused by its inherent channel diode—in effect a reverse-biased diode. At RF frequencies, however, the input impedance drops proportionally to the square of the input frequency. This means that while the input impedance may be several hundred MΩ at audio frequencies, the impedance at say 100 MHz will probably drop several orders of magnitude, to less than 50 KΩ. Input impedance also depends on the value of any external gate-source resistance used in the gate circuit (which is usual when using the JFET as an amplifier). The lower the value of this external resistor, the higher the JFET's frequency response will be, and vice versa. This resistor (often labeled RG) is usually found between the gate and the source in JFET amplifier circuits. Its function is to set a predetermined input impedance to match the signal source and to ensure that the JFET's gate is at 0 volts with no input signal present. The output impedance of the JFET depends on its gain, frequency, internal source resistance, and the value of the external source load resistor. It can range from a few ohms to more than 1 MΩ. Higher current JFETs, such as many analog switches, have a much lower drain-to-source resistance, $R_{DS(on)}$, a lower internal source resistance, and so a very low output impedance.

Because JFETs are voltage-controlled, they have much lower equivalent *noise characteristics* (nV√Hz) than do most bipolars, which explains their popularity as low-level preamplifiers, as well as being used for the same reasons as op amp input stages. JFETs are also well suited to *high-frequency* operation, including RF/VHF/UHF, and have much better linearity than BJTs. Although JFETs have a comparatively lower voltage gain than bipolars, they have a much higher dynamic range than bipolars (typically better than 100 dB), so they can amplify smaller signals with lower noise or amplify large signals without introducing distortion.

Discrete JFET amplifiers fall into one of several different types: *high-frequency, low-leakage, low-noise,* and *general-purpose.* High-frequency types typically have very low capacitance and a high transconductance (g_{fs}), and often sacrifice other characteristics like gate leakage current and breakdown voltage for high frequency. Low-leakage types offer low capacitance, low transconductance, and leakage currents of less than 1 pA. Low-noise types also offer a low conductance (g_{os}), as well as reasonable breakdown voltages, leakage currents, and transconductance. General-purpose types offer many of the good characteristics of the other three types and usually have higher values of transconductance and noise than, say, low-leakage types. All four types have devices that are suitable as current sources, but the most suitable JFETs come from the low-noise amplifier variety.

A typical set of curves for an N-channel JFET showing drain current, drain-source voltage, various gate-source voltages, and other characteristics is shown in Figure 6.4. Notice first that three regions govern the JFET's operation: (1) the *ohmic* or *linear* region, (2) the *saturation* region, and (3) the *breakdown* region. The ohmic or linear region occurs at low drain currents and low drain-source voltages, below the point shown as "V_p" in Figure 6.4.

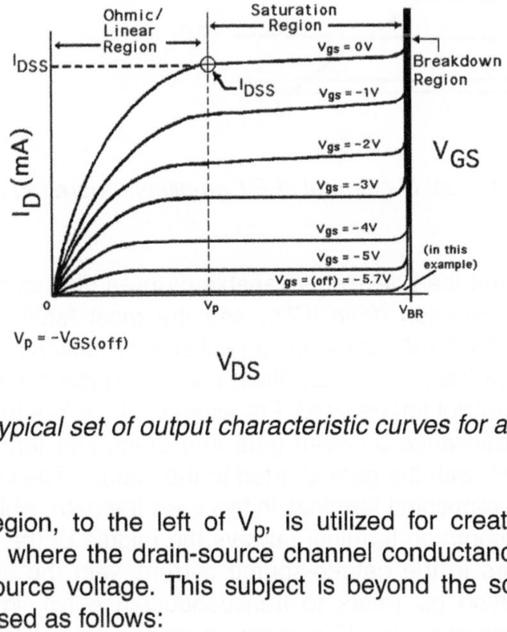

Figure 6.4. *A typical set of output characteristic curves for an N-channel JFET.*

This region, to the left of V_p, is utilized for creating voltage-controlled resistance (VCR), where the drain-source channel conductance is a near-linear function of the gate-source voltage. This subject is beyond the scope of this book, but it may be expressed as follows:

$$I_D = \frac{2I_{DSS}}{V_{GS(off)}^2} V_{DS}\left(V_{GS} - V_{GS(off)} - \frac{V_{DS}}{2}\right) \quad \text{(Eq.6.2)}$$

With a JFET amplifier, its DC performance depends on its optimum bias point (Q-point) within the particular circuit, and it must be correctly biased with respect to the supply voltages (see Figure 6.5). As mentioned, the JFET is composed of two junctions, as shown previously in Figure 6.3A and D. In the N-channel JFET, both are forward-biased junctions, which are similar in many respects to the diode we looked at in a previous chapter. In actual use, though, the polarity of the voltage at the gate reverse-biases these two junctions, in that the V_{GS} ranges from full conduction (V_{GS} = 0 volt) to some negative amount of several volts to turn it off— $V_{GS\ (off)}$. With the P-channel JFET, the gate voltage must become positive enough to reverse-bias the gate channel diode, and thereby turn it off. Again, turning the P-channel JFET full on and conducting occurs at V_{GS} = 0 volt. Thus the voltage at the gate of the P-channel device typically ranges from 0 volt to a more positive level of several volts.

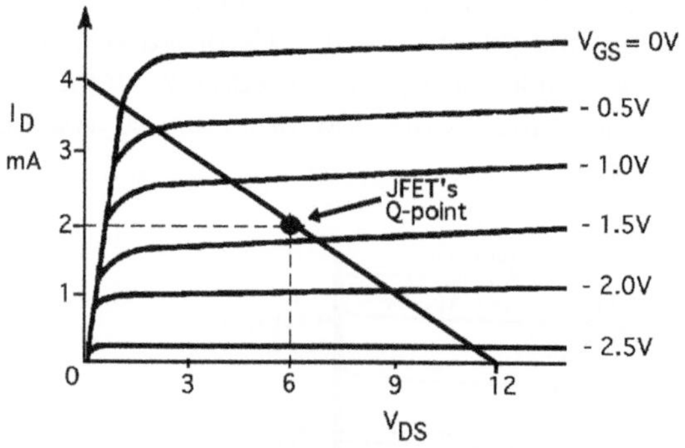

Figure 6.5. *The Q-point for a typical N-channel JFET amplifier, operating at 12 volts at 4 mA.*

The JFET, being a three-lead device, is capable of three distinct modes of operation: *common gate* (CG), *common drain* (CD), and the most familiar—*common source* (CS). In addition, the JFET has some unique subscript notations that reference either the connection, current flow, or test condition used. In some situations there may be one, two, or three subscript letters used. For example, I_G refers to gate current, while C_{GD} refers to the capacitance between gate and drain. The term I_{DSS} refers to the drain-to-source current, with the gate shorted to the source. The last letter ("S") refers to the condition of the remaining terminal, in this case the gate, which is designated as being shorted to the common terminal (always the middle of the three subscript letters). Thus, I_{GSS} refers to the gate-to-source current, with the drain shorted to the source. The abbreviation g_{fs} refers to transconductance (g), in this case forward transconductance, designated by "f," in common source mode, designated by the "s." Although not a complete study, some relevant JFET equations and models are given in Appendix A. Here we look at some of the JFET's more important and relevant characteristics as regards potential current sources.

One important JFET characteristic is the gate-to-source breakdown voltage, $V_{(BR)GSS}$. This is measured with the JFET's drain shorted to the source, at a specified gate current. This parameter is invariably specified at the beginning of a JFET's data sheet. In normal operation, though, the V_{GD} is much higher than the V_{GS}. However, JFETs are manufactured according to a particular process, much of which dictates the breakdown voltage of the device. As a result, JFETs normally use the *same value for both* $V_{(BR)GSS}$ and $V_{(BR)GDS}$, although the latter is not usually specified other than in the absolute maximum ratings for V_{GD}. Actually, this gate-to-drain voltage is responsible for most of a JFET's tiny leakage current (I_{GD}), which like any P-N junction increases with either increased voltage or temperature.

Although the JFET is controlled by a voltage at its gate, there is actually a tiny current involved called the *gate operating current* (I_G) typically in the range of 1–100 pA, at room temperature. This current applies when the drain current is flowing and rises rapidly with an increase in either temperature or V_{DG}. As the V_{DG} is increased, typically upwards of about 20 volts, the gate current rises exponentially, caused by a type of avalanche breakdown at the junction. This is often referred to as the gate current's *breakpoint.* Another current referred to in many course books and texts is the *gate-to-source reverse leakage current* (I_{GSS}). It is also specified on many JFET data sheets. The last letter indicates that the drain is shorted to the source. It is a very small leakage current usually measured in pico-amps, which applies at a specified gate-to-source voltage. It is composed of two currents: the gate-to-source (I_{GS}) and the gate-to-drain (I_{GD}) currents. This I_{GSS} leakage current approximately doubles for every 10°C rise in temperature, as shown in Figure 6.6.

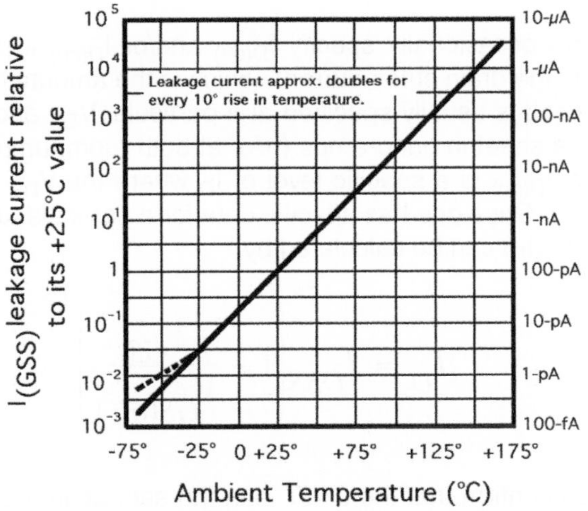

Figure 6.6. Example of a JFET's junction leakage current, as a function of temperature.

With a JFET, the maximum limiting current that can flow between the drain and source is known as the *drain saturation current* (I_{DSS}). This occurs at a particular V_{DS}, when the V_{GS} is at zero. It is clearly seen in Figure 6.4, where the drain current increases linearly, then begins to be pinched off or limited at the knee of the curve. This particular set of curves, which displays the drain current (I_D) versus the V_{DS}, with applied V_{GS}, is normally referred to as a JFET's *output characteristic.* It is depicted in most JFET data sheets and can be easily measured, using the circuit shown in Figure 6.7. With a JFET, this current is typically less than 50 mA, although some devices are capable of higher current.

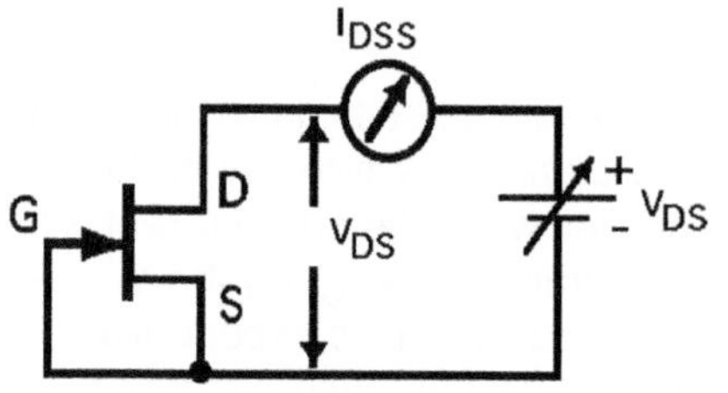

Figure 6.7. *I_{DSS} measurement for an N-channel JFET.*

At the low end of the I_{DSS} scale, some JFETs can only work below 1 milliamp. In any event, I_{DSS} may typically range over 3:1 for similar devices, so it is important that one considers both the minimum and maximum ratings/curves in the range. I_{DSS} is also temperature sensitive and has a small negative temperature coefficient of approximately –0.5%/°C.

Some manufacturers occasionally specify $I_{D(off)}$ and/or $I_{D(on)}$ in their JFET product data sheets. $I_{D(off)}$ is the drain cutoff current, which is the amount of leakage between the drain and source. It is usually specified at a particular V_{DS} and V_{GS}. $I_{D(off)}$ is usually shown on a data sheet in nano-amps (nA), at both room temperature and at elevated temperatures. $I_{D(on)}$ is a specific level of I_D where the V_{DS} is greater than the pinch-off voltage (V_p). Once one has specific data for a particular JFET's $V_{GS\ (off)}$ and I_{DSS}, its drain current (I_D) can be calculated by:

$$I_D = I_{DSS}\left(1 - \frac{V_{GS}}{V_{GS(off)}}\right)^2 \qquad \text{(Eq.6.3)}$$

As an example, let's confirm that the JFET is in the saturation region and at full conduction, where I_{DSS} is at its maximum. Let's assume for this example that the JFET's I_{DSS} equals 3.6 mA, its $V_{GS\ (off)}$ equals –2.65 volts, its V_{GS} equals 0 volt, the test circuit is as shown in Figure 6.7, and this test is conducted at room temperature.

$$I_D = 0.0036\left(1 - \frac{0V}{2.65V}\right)^2 = 3.6\text{mA, ie. } I_{DSS} \text{ value} \qquad \text{(Eq.6.4)}$$

Seeing as the key parameter V_{GS} equals 0 volt, this confirms that the JFET is at full conduction.

Another important characteristic is $V_{GS\ (off)}$, the *gate-to-source cutoff voltage* necessary to turn the FET off. The V_{GS} for an N-channel JFET ranges from 0 volt for full conduction to some negative amount of typically several volts to turn the device off. This is shown in Figure 6.4, and we just saw it used in equation 6.4. Some JFETs have a remarkably low $V_{GS\ (off)}$ value. For example, a 2N4338 has a maximum –1 $V_{GS\ (off)}$, while a J202 has a maximum $V_{GS\ (off)}$ of –1.5 volt. At the other end of the scale, the popular Vishay-Siliconix general-purpose 2N3819 JFET has a maximum –8 $V_{GS\ (off)}$, while a Crystalonics high-voltage (300 V_{DGO}) 2N5543 has a –15 $V_{GS\ (off)}$. For some applications, where high supply voltages are involved (more than 20 to 90 volts), higher $V_{GS\ (off)}$ devices like these can be indispensable.

For a P-channel JFET, $V_{GS\ (off)}$ ranges from 0 volt for full conduction to some positive amount of several volts. For example, a J176 has a $V_{GS\ (off)}$ of +4 volt (remember it's a P-channel device, so it has a positive gate-to-source bias). With either type, this occurs when the V_{DS} is equal to or greater than the the pinch-off voltage (V_p), so that only V_{GS} controls conduction through the channel. So with an N-channel JFET, while the value of V_p is some relatively low positive value, the value of $V_{GS\ (off)}$ will be the same value, but of opposite polarity. Today most manufacturers and circuit designers prefer using the term $V_{GS\ (off)}$ rather than V_p, and this is reflected in most JFET product data sheets. It is important for the circuit designer to know the knee of the curve and the minimum V_{DS} necessary to reach the pinch-off point. The graph shown in Figure 6.4 clearly shows the complete I_D versus V_{GS} curves (transfer curves), from V_{GS} equals 0 through $V_{GS\ (off)}$, for a typical N-channel JFET. However, the data sheet for your chosen device may not necessarily portray the data in exactly the same manner, and you might have to resort to measuring devices and plotting your own curves. These transfer curves are depicted in most JFET data sheets, and the circuit shown in Figure 6.8 shows how this is measured.

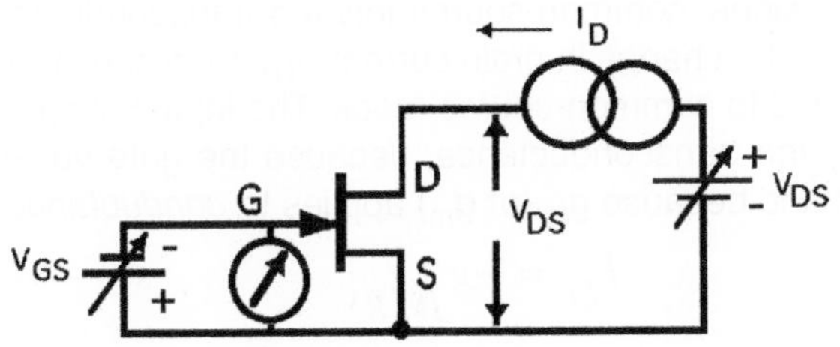

Figure 6.8. $V_{GS(off)}$ *measurement for an N-channel JFET.*

As you can see in Figure 6.9, I_{DSS} and $V_{GS\ (off)}$ both shift with temperature. Actually, $V_{GS\ (off)}$ has a negative temperature coefficient of approximately –2 mV/°C.

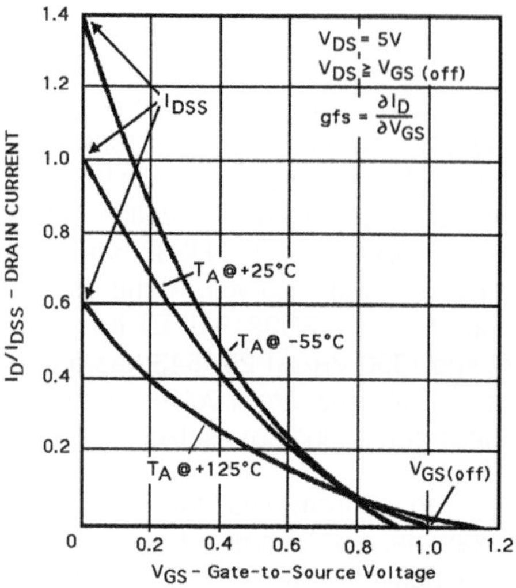

Figure 6.9. *Typical JFET transfer versus temperature curves.*

Most JFET data sheets specify $V_{GS\ (off)}$ with typical and maximum values, at a particular V_{DS} and drain current. Some devices also have a minimum value of $V_{GS\ (off)}$ specified. For example, the popular 2N4416A is specified with a V_{DS} equals 15 volts and I_D equals 1 nA, and has a typical $V_{GS\ (off)}$ of –3 volts, a minimum of –2.5 volts, and a maximum of –6 volts. In some JFET circuits, particularly when two devices are used together, such as in a cascode circuit, one FET may need to have a higher $V_{GS\ (off)}$ than the other. Because $V_{GS\ (off)}$ can typically range over 3:1 or 4:1 for similar devices, it is important that one considers both its minimum and maximum ratings in the range for your application.

Another important characteristic of the JFET is its *transconductance/forward conductance* (g_{fs}) or small-signal common-source forward transconductance (g_m or Y_{fs}). G_{fs} measures the effect of a change in drain current (I_D), for a specific change in gate voltage (V_{GS}), referenced to common-source mode. The input voltage and the output current are related by the transconductance, because the gate voltage is *transferred* to the source current and because g_{fs} (or g_m) applies to *conductance.* Thus,

$$I_D = -g_{fs} V_{gs} \qquad \text{(Eq.6.5)}$$

You probably recognize this from the JFET's small-signal model, part of which is shown here in Figure 6.10, and more fully in Appendix A7.

The transconductance value can be found by either of the following equations:

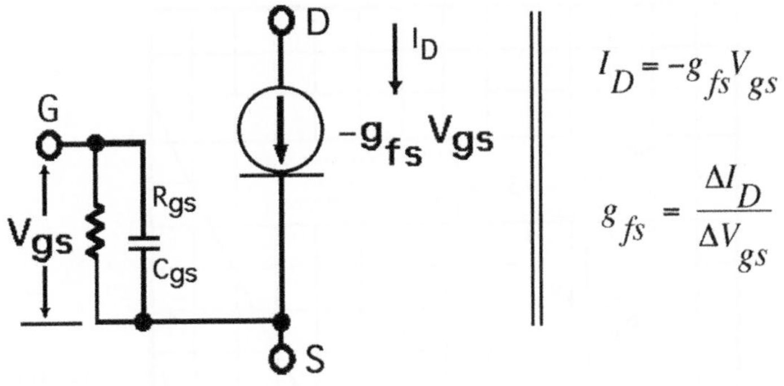

Figure 6.10. *The JFET's small-signal transconductance model.*

$$g_{fs} = \frac{\Delta I_D}{\Delta V_{gs}} \mid V_{DS} = \text{constant} \qquad \text{(Eq.6.6)}$$

$$g_{fs} = \frac{2I_{DSS}}{V_{GS(off)}}\left(1 - \frac{V_{GS}}{V_{GS(off)}}\right) \qquad \text{(Eq.6.7)}$$

In practical terms, transconductance is a measure of the JFET's gain, or ability to amplify, and is a figure of merit. On a graph it appears as a steep or shallow slope, as shown in Figure 6.11, and is usually referred to in milli-Siemens (mS). The old value was measured in μmhos (ohms reversed), but most JFET manufacturers now specify g_{fs} in mS or μS as standard. G_{fs} is usually measured as milliamps per volt (mA/V). In the example shown in Figure 6.11, the JFET's V_{DS} is held constant at 5 volts, while the I_D current's slope is measured between 5 and 8 mA, and 1 to 2 V_{GS}. This results in a g_{fs} of 3 mA/V or 3 mS. While transconductance depends on both the V_{DS} and the V_{GS}, it mostly depends on the latter. It is highest when V_{GS} equals 0, where I_{DSS} will be at its maximum, and lowest when V_{GS} equals $V_{GS\ (off)}$ and I_{DSS} equals 0. It also depends on chip area and gate oxide thickness. G_{fs} is specified on some data sheets with a minimum or maximum value; others specify only a typical value. Be aware how it is measured though. The g_{fs} value given in most manufacturers' data sheets is measured at full conduction (I_{DSS}), where g_{fs} is at maximum. Others also specify it with a particular V_{DS} and at a 1-KHz frequency.

Yet another important characteristic of the JFET is its *output impedance* (Z_{OUT}), which can range from a few ohms to more than a Meg ohm. This depends on the JFET's

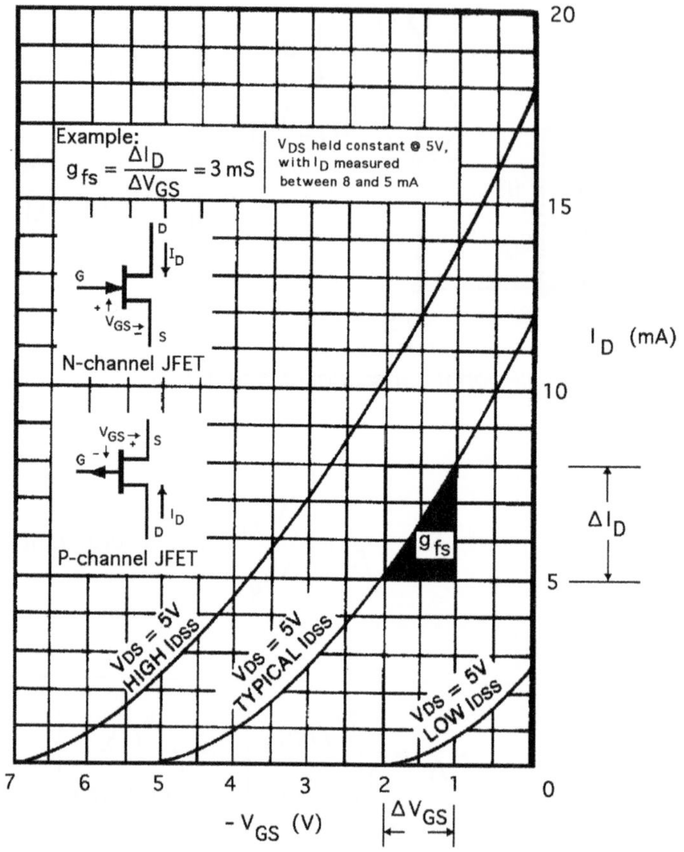

Figure 6.11. A JFET's transconductance (g_{fs}) characteristics.

gain (g_{fs}), input frequency, internal source resistance, and the value of the external source load resistor. This often just approximates to the value of its internal source resistance (r_S), compared with the value of the external source resistor (R_S), which is usually much larger. This internal source resistance forms a voltage divider with any external source load, such that:

$$r_S = \frac{1}{g_{fs}} \qquad \text{(Eq.6.8)}$$

and the output impedance is given by:

$$Z_{out} = \frac{r_S R_S}{r_S + R_S} \qquad \text{(Eq.6.9)}$$

where r_s is the JFET's internal source resistance; g_{fs} is the transconductance or gain; Z_{OUT} is the output impedance in ohms; and R_s is the value of the external source resistor. As an example, let's obtain an approximate value of output impedance for several popular N-channel JFETs, all with the same source resistance of 2.2 KΩ (Figure 6.12). Their typical g_{fs} specs are as follows:

2N 5434 = 17 mS

2N 4867A = 1 mS

2N 3819 = 5 mS

2N 4416A = 6.5 mS

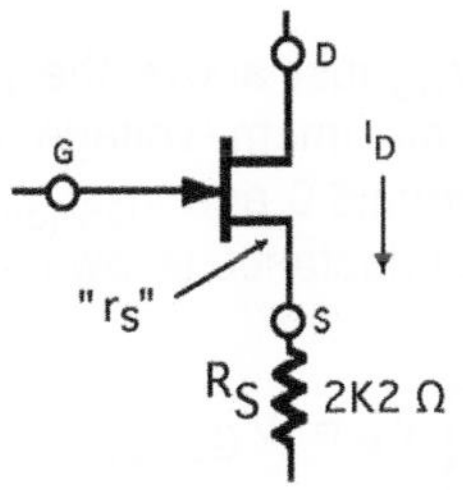

Figure 6.12. Example N-channel JFET used to calculate $Z_{(out)}$, the output impedance.

Using equations 6.8 and 6.9, the results are:

2N 5434	r_S = 58.82 Ω;	Z_{OUT} = 57.5 Ω
2N 4867A	r_S = 1 KΩ;	Z_{OUT} = 688 Ω
2N 3819	r_S = 200 Ω;	Z_{OUT} = 183 Ω
2N 4416A	r_S = 154 Ω;	Z_{OUT} = 143 Ω

As you can see, the output impedance is fairly close to the value of the internal source resistance, which in turn depends on the transconductance. We are going to cover two more important characteristics of the JFET shortly. These are the *output conductance* and the zero temperature coefficient. For now, though, we have probably covered the most important JFET basics.

6.3 Using the JFET as a current source

Another unique characteristic of JFETs, particularly in relation to current sources, is the extremely low output conductance (g_{os}, y_{os}, and $Re_{(yfs)}$) of some low-noise JFET amplifiers, when operating in their saturation regions. Besides low-noise JFETs, other potential devices exist from the different types of JFET amplifiers and switches. This is shown in a table later in this chapter. Although most JFETs do not make good current sources, those with low conductance usually do. This applies equally to both N- or P-channel devices, although N-channel devices predominate. Conductance is a DC characteristic and the reciprocal of resistance where:

$$R = \frac{1}{G} \quad \text{and} \quad G = \frac{1}{R} \qquad \text{(Eq.6.10)}$$

$$\text{(Ohms)} \qquad \text{(Siemens / mhos)}$$

It follows then that a 100-Ω resistor has a conductance of 1/100th or 10 mS, while a 100-KΩ resistor has a conductance of 1/100th K or 10 μS.

In summary, the saturation region is the V_{DS} just above the pinch-off voltage (V_p), extending horizontally to just below the maximum voltage rating of the device, $V_{(BR)DSS}$, and vertically between the V_{GS} equals 0 and $V_{GS\ (off)}$ limits. Whenever the JFET is operated in this region, its output conductance is low (see Figure 6.13).

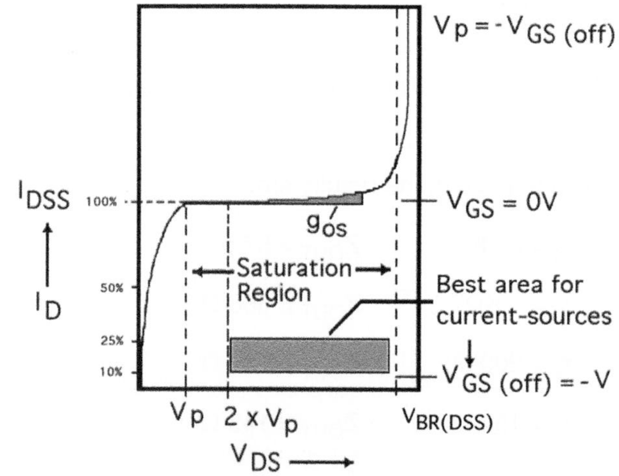

Figure 6.13. *N-channel JFET output conductance (g_{os}) characteristics.*

The quality of the current regulation is strongly dependent on output conductance, though, which is in turn closely related to the drain current (I_D). The lower the drain current, the lower the conductance, and the better the regulation will be. Some JFETs have extremely low values of output conductance, often less than 10 μS (micro-Siemens), which make them excellent candidates as current sources (i.e., 2N 4338 = N-channel: g_{os} = 5 μS max, $V_{GS\ (off)}$ = –1 V, I_{DSS} = 0.6 mA; 2N 4867 = N-channel: g_{os} = 1.5 μS max, $V_{GS\ (off)}$ = –2V, I_{DSS} = 1.2 mA). Others have output conductance values in the hundreds of micro-Siemens and higher, making them mostly unsuitable for use as current sources.

As previously stated, the JFET's maximum limiting current (I_{DSS}) occurs when the V_{GS} is at 0 volts, at a particular V_{DS}. Under these conditions, the output conductance will be at its maximum ($g_o = g_{os}$ at I_{DSS}). The simplest current source therefore can be made with the JFET's gate tied directly to its source, as shown in Figure 6.14. The circuit's output conductance (g_o) will be approximately equal to the g_{os} specs for that particular JFET. However, the overall performance using this configuration is poor, because the drain current will be at full conduction and will be whatever the I_{DSS} value of this particular JFET happens to be. This is not a very good solution, but analog IC makers often use this same configuration as part of their start-up bias circuitry, where only a current limit (rather than a current source or sink) is needed. So really this I_{DSS} configuration need only be considered here as a starting point.

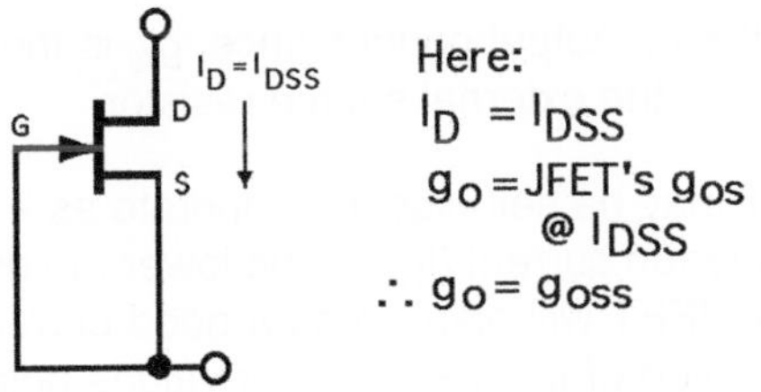

Figure 6.14. The simplest JFET current source uses the I_{DSS} value.

Actually the best regulation performance is obtained when using a JFET that is biased well below its I_{DSS}, at between say 5% to 33% of I_{DSS}. To accomplish that, a resistor is inserted into the source connection, to provide the necessary V_{GS} biasing (see Figure 6.15). At first glance this looks similar to a JFET source-follower circuit, but the gate is tied to the bottom end of the resistor (R_S), so that a feedback voltage controls the gate and establishes a state of equilibrium, keeping the current constant despite changes in drain voltage. The output conductance (g_{os}), which is a function of the operating point of the JFET, decreases linearly with I_D, because the JFET is biased closer to $V_{GS\,(off)}$. As a result, the circuit's output conductance (g_o) can be found by either of the following equations:

$$g_o = \frac{g_{os}^{\;2}}{1 + R_S\left(g_{os} + g_{fs}\right)} \qquad \text{(Eq.6.11)}$$

$$g_o \approx \frac{g_{os}}{1 + R_S g_{fs}} \qquad \text{(Eq.6.12)}$$

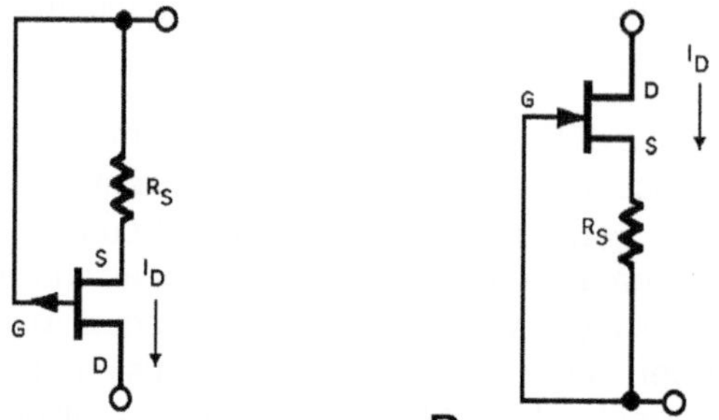

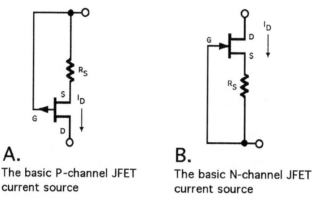

A. The basic P-channel JFET current source

B. The basic N-channel JFET current source

Figure 6.15. *Basic Wilson current mirrors.*

where g_{os} is the JFET's output conductance; g_{fs} is the value of its transconductance, and R_S is the value of the external source resistor.

Although the JFET may be self-biased to operate as a constant current source at any level below its saturation current (I_{DSS}), the lower the drain current, the tighter the regulation will be. The JFET will operate as a good current source whenever the V_{DS} is twice or more the value of the $V_{GS\ (off)}$. For values below two times $V_{GS\ (off)}$, conductance will be significantly higher and regulation will be degraded. Where $V_{GS\ (off)}$ and I_{DSS} are both known, then the approximate V_{GS} bias voltage for a particular drain current (I_D) can be found by the following:

$$V_{GS} = V_{GS(off)}\left(1 - \sqrt{\frac{I_D}{I_{DSS}}}\right) \qquad \text{(Eq.6.13)}$$

and also :

$$I_D = I_{DSS}\left[1 - \left(\frac{V_{GS}}{V_{GS(off)}}\right)\right]^2 \qquad \text{(Eq.6.14)}$$

The bias resistor (R_S) required between the gate and the source can be found by applying Ohm's law, so that:

$$R_S = \frac{V_{GS}}{I_D} \qquad \text{(Eq.6.15)}$$

As an example, if we assume these values:

$I_D = 100\ \mu A$

$I_{DSS} = 2.5\ mA$

$V_{GS} = -4$ volt

Then the value of R_S would be 40 KΩ.

A change in either the supply voltage or the load impedance will cause only a small change in current as a result of a low output conductance (g_{os}), and is shown by:

$$\Delta I_D = \Delta V_{DS}\, g_{os} \qquad \text{(Eq.6.16)}$$

In our example, if the particular JFET device had a g_{os} value of 50 μS, and a change in V_{DS} of 25 mV occurred, then the output current would change by 1.25 μA (1.25%). If we chose another device, having a significantly lower value of g_{os} (i.e., 5 μS), the same 25-mV change in V_{DS} would yield a change of only 0.125 μA (0.12%). In practical terms, because g_{os} can range from less than 1 μS to more than 500 μS, according to the JFET type, the dynamic impedance can be more than 1 MΩ to less than 2 KΩ, which corresponds to a current stability range of less than 1 μA/V to more than 1 mA/V. In this example, the dynamic impedance is more than 200 KΩ.

In many current source applications, it may be necessary to bias the JFET at or near its *zero temperature coefficient point* (zero TC, TC0, or zero tempco). This is a unique point at which the drain current varies least (I_{DZ}), with changes in temperature (either large or small ΔT), and which provides added thermal and circuit stability. You can see an example of the zero TC point in Figure 6.16. Actually, I_{DZ} will change slightly from one device to another, because each device will have a slightly different value of saturation current (I_{DSS}), and I_{DZ} will change proportionally to that. The theoretical value of I_{DZ} can be found by:

$$I_{DZ} \cong I_{DSS}\left(\frac{0.63}{V_{GS(off)}}\right)^2 \qquad \text{(Eq.6.17)}$$

The necessary gate-to-source bias voltage for zero TC would then be:

$$V_{GS(0TC)} \cong V_{GS(off)} - 0.63V \qquad \text{(Eq.6.18)}$$

One will also need to determine the value of the JFET's source resistance (R_S), and one can do that by drawing a load-line from a convenient fixed reference point (i.e., 1 V and 1 mA = 1 KΩ), so that the load-line intersects as closely as possible with or near

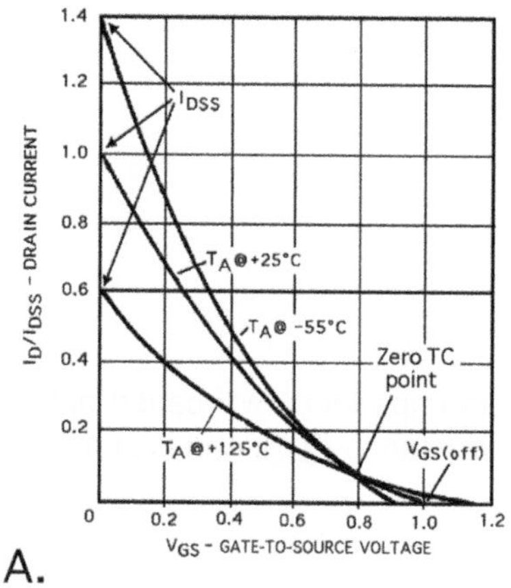

A.

Transfer characteristic, showing the zero TC point at the intersection of the three temperature curves.

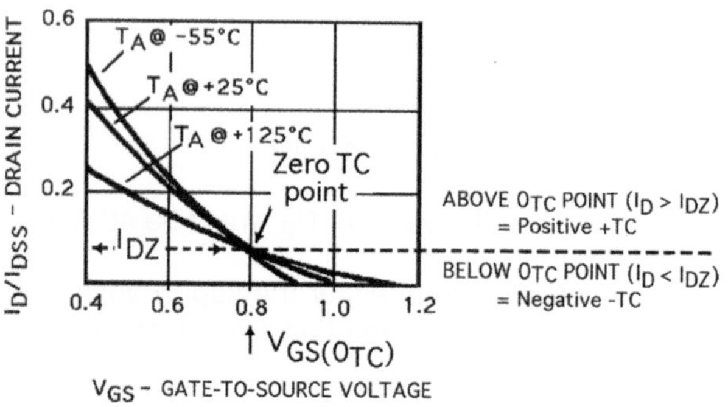

B.

Close-up view of the zero TC point, showing the I_{DZ} current, and $V_{GS(0TC)}$. Notice how the -TC changes to a +TC at the zero TC point.

Figure 6.16. Examples of a JFET's Z_{TC} point.

the zero TC point, then back to the origin at 0. Where it intersects with the curves will determine the V_{GS} and I_D.

$$R_{S\,(0TC)} = \frac{V_{GS(0TC)}}{I_{DZ}} \qquad \text{(Eq.6.19)}$$

Knowing where the zero TC point is allows one to tailor the circuit to have a positive tempco, by operating the drain current *above* I_{DZ}, or to have a negative tempco, by operating the drain current *below* I_{DZ} (see Figure 6.16B). Normally one will have to build a circuit, then test it under extended temperature conditions (in an accurate oven), logging the results on a graph. Once you have some data recorded and graphed, you can then compare it with your initial design calculations. This is not the kind of project one can do using SPICE models, which for JFETs are usually approximations at best.

As a design example, we will first determine the zero TC point for a 2N 4340 (a low-noise N-channel JFET), and then establish what the appropriate load resistor would be to keep the FET current source at that zero TC point. In this example, the JFET has a $V_{GS\,(off)}$ equals –1.5 volt, an I_{DSS} minimum equals 1.2 mA, and it is operated from a V_{DS} equals 10 volts. Looking at its transfer characteristic graph (Figure 6.17), we can visually confirm that the intersection of the three temperature curves (the zero TC point) occurs at approximately 0.84 volt and 0.2 mA. Using equations 6.17, 6.18, and

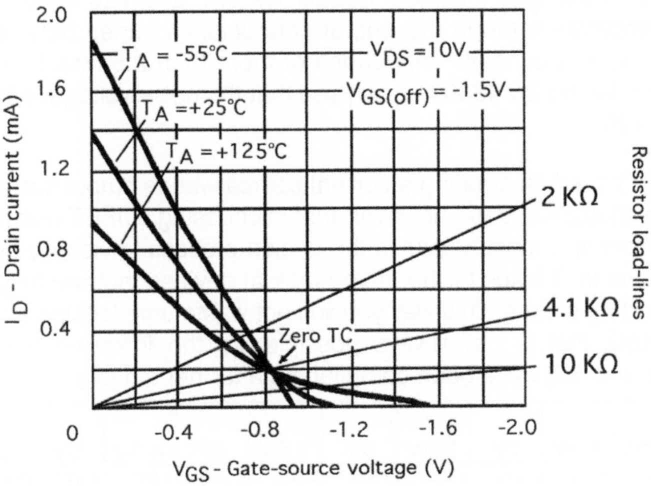

Figure 6.17. *Example showing 2N4340 N-channel JFET's zero TC point and three load-lines.*

6.19, we can determine from the $V_{GS\ (off)}$, I_{DSS}, and V_{DS} values given what the zero TC values for V_{GS}, I_{DZ}, and R_S are:

$$V_{GS(0TC)} \cong 1.5 - 0.63 = 0.87\ V \qquad \text{(Eq.6.20)}$$

$$I_{DZ} \cong 0.0012 \left(\frac{0.63V}{1.5V}\right)^2 = 212\mu\text{A} \qquad \text{(Eq.6.21)}$$

$$R_{S\ (0TC)} = \frac{0.87V}{212\mu A} = 4.1\ \text{K}\Omega \qquad \text{(Eq.6.22)}$$

Comparing the graph with the calculations, you will see that this is fairly close, although one should remember that $V_{GS\ (0TC)}$ and I_{DZ} are both *theoretical* (approximate) values, much as the reverse saturation current (I_S) is with the P-N junction. Looking at the three load-lines shown in Figure 6.17, the 2-KΩ load-line is too high and is above the zero TC point, thereby having a negative TC. It has an approximately 45-degree slope as it intersects with the three temperature curves and would drift about 80 μA over the temperature range, which is unacceptable. The 10-KΩ load-line is on the low side and is below the zero TC point, thereby having a positive TC. Although it has a more shallow slope, it would still drift about 25 μA. The 4.1-KΩ load-line intersects exactly at the zero TC point (4.1 KΩ) and would drift less than 5 μA, over the full military temperature range. We can also determine from the data sheet the approximate output conductance from the device's "Output Conductance vs. Drain Current" graph. This shows that for a drain current of approximately 0.2 mA (with V_{GS}

(off) equals −1.5V, I_{DSS} minimum equals 1.2 mA, and V_{DS} equals 10 volts), over the military temperature range the output conductance varies between 0.3 and 0.4 μS. This would give a dynamic impedance of more than 2.5 MΩ. From an adjacent data sheet graph for the 2N 4340, we can see that the noise voltage is in fact greater than 8 nV√Hz at 1 KHz.

All of this is based on creating a current source whose output current is at the zero TC point. This should illustrate, however, that in choosing a JFET and knowing the desired output current, it is most helpful in the design process to visually determine where the zero TC point is. It helps further if you look at devices that are fully characterized and with supporting graphs. This way you will not waste time looking at unsuitable devices. Unfortunately, that is not always possible, so the following reference charts (see Tables 6.1 and 6.2) have been compiled as a further aid to you.

VGS(off): (V)	1V	1.25V	1.5V	2V	2.5V	3V	3.5V	4V	4.5V	5V	6V	8V	10V
IDSS: (mA) 0.1	0.37V 39 μA	0.62V 25 μA	0.87V 17.6 μA	1.37V 10 μA	1.87V 6.35 μA	2.37V 4.4 μA	2.87V 3.2 μA	3.37V 2.5 μA	3.87V 1.96 μA	4.37V 1.58 μA	5.37V 1.1 μA	7.37V 0.62 μA	9.37V 0.39 μA
0.5	0.37V 200 μA	0.62V 127 μA	0.87V 88 μA	1.37V 50 μA	1.87V 32 μA	2.37V 22 μA	2.87V 16.2 μA	3.37V 12.4 μA	3.87V 9.8 μA	4.37V 7.9 μA	5.37V 5.5 μA	5.37V 3.1 μA	9.37V 1.98 μA
1	0.37V 400 μA	0.62V 254 μA	0.87V 176 μA	1.37V 99 μA	1.87V 63 μA	2.37V 44 μA	2.87V 32.4 μA	3.37V 25 μA	3.87V 19.6 μA	4.37V 15.8 μA	5.37V 11 μA	5.37V 6.2 μA	9.37V 3.96 μA
2	0.37V 793 μA	0.62V 508 μA	0.87V 352 μA	1.37V 198 μA	1.87V 127 μA	2.37V 88 μA	2.87V 64 μA	3.37V 50 μA	3.87V 39 μA	4.37V 31.7 μA	5.37V 22 μA	5.37V 12.4 μA	9.37V 7.93 μA
2.5	0.37V 992 μA	0.62V 635 μA	0.87V 441 μA	1.37V 249 μA	1.87V 158 μA	2.37V 110 μA	2.87V 81 μA	3.37V 62 μA	3.87V 49 μA	4.37V 39.6 μA	5.37V 27.5 μA	5.37V 15.5 μA	9.37V 9.92 μA
3	0.37V 1.2 mA	0.62V 762 μA	0.87V 529 μA	1.37V 298 μA	1.87V 191 μA	2.37V 132 μA	2.87V 97 μA	3.37V 74.4 μA	3.87V 58.8 μA	4.37V 47.6 μA	5.37V 33 μA	5.37V 18.6 μA	9.37V 11.9 μA
4	0.37V 1.58 mA	0.62V 1.01 mA	0.87V 705 μA	1.37V 397 μA	1.87V 254 μA	2.37V 176 μA	2.87V 129 μA	3.37V 99 μA	3.87V 78 μA	4.37V 63.5 μA	5.37V 44 μA	5.37V 24.8 μA	9.37V 15.8 μA
5	0.37V 2 mA	0.62V 1.27 mA	0.87V 882 μA	1.37V 496 μA	1.87V 317 μA	2.37V 220 μA	2.87V 162 μA	3.37V 124 μA	3.87V 98 μA	4.37V 79 μA	5.37V 55 μA	5.37V 31 μA	9.37V 19.8 μA
10	0.37V 4 mA	0.62V 2.54 mA	0.87V 1.76 mA	1.37V 1 mA	1.87V 635 μA	2.37V 441 μA	2.87V 324 μA	3.37V 248 μA	3.87V 196 μA	4.37V 159 μA	5.37V 110 μA	5.37V 62 μA	9.37V 39.6 μA
25	0.37V 10 mA	0.62V 6.35 mA	0.87V 4.41 mA	1.37V 2.48 mA	1.87V 1.58 mA	2.37V 1.1 mA	2.87V 810 μA	3.37V 620 μA	3.87V 490 μA	4.37V 397 μA	5.37V 275 μA	5.37V 155 μA	9.37V 99 μA
50	0.37V 20 mA	0.62V 12.7 mA	0.87V 8.82 mA	1.37V 4.96 mA	1.87V 3.17 mA	2.37V 2.2 mA	2.87V 1.62 mA	3.37V 1.24 mA	3.87V 980 μA	4.37V 794 μA	5.37V 551 μA	5.37V 310 μA	9.37V 198 μA
75	0.37V 29.7 mA	0.62V 19 mA	0.87V 13.2 mA	1.37V 7.44 mA	1.87V 4.76 mA	2.37V 3.3 mA	2.87V 2.43 mA	3.37V 1.86 mA	3.87V 1.47 mA	4.37V 1.19 mA	5.37V 826 μA	5.37V 465 μA	9.37V 297 μA
100	0.37V 39.7 mA	0.62V 25.4 mA	0.87V 17.6 mA	1.37V 9.92 mA	1.87V 6.35 mA	2.37V 4.4 mA	2.87V 3.24 mA	3.37V 2.48 mA	3.87V 1.96 mA	4.37V 1.58 mA	5.37V 1.1 mA	5.37V 620 μA	9.37V 397 μA

NOTES:

1. Select the nearest Vgs(off) for your JFET along the top horizontal axis, then select the nearest IDSS, from the left vertical axis. Where that column and row intersect, read off the resulting Vgs(OTC) and IDZ values.
2. Alternatively, select the nearest Vgs(OTC) and IDZ combination from within the cells, then read the corresponding vertical Vgs(off) and horizontal IDSS values.
3. These values are approximate only.

Table 6.1 Vgs(OTC) and IDZ calculator

Table 6.1 allows you to quickly determine the corresponding $V_{GS\ (0TC)}$ and I_{DZ} values for many combinations of $V_{GS\ (off)}$ and I_{DSS}. You can also use this in reverse, so that you can find the corresponding $V_{GS\ (off)}$ and I_{DSS} values for various combinations of $V_{GS\ (0TC)}$ and I_{DZ}. Remember these are approximate values only, but they should be close enough.

Table 6.2 allows you to quickly determine a suitable device and includes the most important characteristics as far as potential JFET current sources are concerned. This table lists various N-channel devices with a $V_{GS\ (off)}$ value of less than –3.6 volts and P-channel devices with a $V_{GS\ (off)}$ value of less than +4.1 volts. These are mostly low-noise devices from 10 different manufacturers, and all with low values of output conductance (g_{os} less than 80 μS). Both tables are easy to use and should save you some time, by revealing either the approximate values or a commonly available device you may need in your design.

Manufacturers and circuit designers commonly portray a product's characteristics by using the temperature coefficient as either a percentage or in parts per million. In relation to a current source, the temperature coefficient of the output current ($\alpha\ I_{O(TC)}$) is the ratio of change in output current to the change in temperature (e.g., 20°C to 85°C). This is usually expressed as a *percentage* (%) of the output current (i.e., 0.0015%), and is the average value for the total temperature change. You can determine the tempco from your data by the formula in equation 6.23:

$$\alpha I_{O\,(TC)} = \pm \left(\frac{I_O @ T_2 - I_O @ T_1}{I_O @ 25°C} \right) \left(\frac{100\%}{T_2 - T_1} \right) \qquad \text{Eq.6.23}$$

For example, some current measurements gave us the following results: I_o at 25°C = 502 μA; I_o at 20°C = 500 μA; I_o at 85°C = 512 μA. The temperature range of interest is +20°C to +85°C. Using equation 6.23 would yield:

$$\left(\frac{512\mu A - 500\mu A}{502\mu A} \right) \left(\frac{100}{65} \right) = 0.03677\% \qquad \text{Eq.6.24}$$

You can also express the tempco as a value in *parts per million per degree Celsius* (ppm/°C), by using the formula in equation 6.25:

$$\alpha I_{O\,(TC)} = \left(\frac{I_O @ T_2 - I_O @ T_1}{I_O @ 25°C} \right) \left(\frac{1 x 10^6}{T_2 - T_1} \right) \qquad \text{Eq.6.25}$$

In the previous example, the output current's tempco could also be expressed as 368 ppm/°C, suggesting an accurate current source.

Part #:	N / P type:	Vgs(off) max: V	VBR(gss) min: V	Idss mA min:	Gos typ µS:	Gfs typ mS:	Typ. equiv. noise: nV√Hz	Pkg:	Manufacturers:
2N 3574	P	2	25	-0.15	10	0.2		TO-72	Sol
2N 3575	P	4	25	< -1	20	0.3		TO-72	Sol
2N 3578	P	4	20	-1	15	1.2		TO-18	Sol
2N 3685/A	N	- 3.5	-50	1	25	2		TO-72	Sol
2N 3686/A	N	- 2	-50	0.4	10	1.5		TO-72	Sol
2N 3687/A	N	- 1.2	-50	1.2	5	1		TO-72	Sol
2N/MPF 3821	N	- 2	- 50	0.5	10	3	100	TO-72	IF, ON
2N/PN/FN/SST/ LS/MMBF 4117/A	N	- 1.8	- 40	0.09 mA	3 µS	0.08 mS	1.5 nV	TO-18/TO-72/ SOT-23	C, F, IF, LIS, Sol,VS
2N/PN/FN/SST/ LS/MMBF 4118/A	N	- 3	- 40	0.24 mA	3 µS	0.15 mS	1.5 nV	TO-18/TO-72/ SOT-23	C, F, IF, LIS, Sol,VS
2N 4338	N	- 1	- 30	0.2	5	0.5	6	TO-18/TO-72	C, CS, IF, Sol, VS
2N 4339	N	- 1.8	- 50	0.5	15	2	6	TO-18	C, CS, IF, LIS, Sol, VS
2N 4340	N	- 3	- 50	1.2	30	2.5	6	TO-18	C, CS, IF, Sol, VS
2N/PN/SST 4393	N	- 3	- 40	5 mA	25 µS	6.0 mS	3.0 nV	TO-18/92/SOT-23	Ph, VS
2N 4867/A	N	- 2	- 40	0.4	1.5	1	3	TO-18/TO-72	IF, Sol, VS
2N 4868/A	N	- 2	- 40	1	1.5	2	6	TO-72	IF, LIS, Sol
2N 5020	P	1.5	25	- 0.3	20	3		TO-18	CS, IF, Sol
2N 5021	P	2.5	25	- 1	20	4		TO-18	IF, Sol
2N 5116	P	4	30	- 5	20	4.5	20	TO-18	CS, Ph, VS
2N 5265	P	3	60	-0.5	75	1		TO-72	Sol
2N 5266	P	3	60	-0.8	75	1.5		TO-72	Sol
2N 5358	N	- 3	- 40	0.5	10	2		TO-72	Sol
2N/SST/MMBF 5484	N	- 3	- 25	1 mA	50 µS	4.0 mS	10 nV	TO-92/SOT23	C, CS, F, Ph, VS
2N 5716	N	- 3	- 40	0.05	25	1		TO-92	Sol
2N 6451/2	N	- 3.5	- 20	5	50	20	5	TO-72	IF
2N 6550/CM 860	N	- 3	- 20	10	50	40	10	TO-46/TO-72	Cr
CM 697	N	- 3	- 25	30				TO-46	Cr
J 113A	N	- 3	- 35	8	25	6	4	TO-92	F, IF, LIS, ON, Ph,VS
J/MMBFJ/SST 177	P	4	30	-1.5	20	4.5	20	TO-92/SOT-23	C, F, IF, LIS, ON, Ph, Sol, VS
J/SST/MMBF J 201	N	- 1.5	- 40	0.2	1	0.5	6	TO-92/SOT-23	C, F, IF, LIS, ON, Sol, VS
J 230	N	- 3	- 40	0.7	1.5	2	8	TO-92	IF
J305	N	- 3	- 30	1	30	3	10	TO-92	VS
KK 3685	N	- 3.5	-50	1	25	2		TO-92	Sol
KK 3686	N	- 2	-50	0.4	10	1.5		TO-92	Sol
KK 3687	N	- 1.2	-50	0.1	5	1		TO-92	Sol
KK 4304	N	- 1	- 30	0.5	50	1		TO-92	Sol
KK 5033	P	2.5	20	-0.3	20	5		TO-92	Sol

MANUFACTURERS' KEY:
C = Calogic; CS = Central Semi; Cr = Crystalonics; F = Fairchild Semi; IF = InterFet; LIS = Linear Int. Systems; ON = ON Semi; Ph = Philips Semi; Sol = Solitron Devices; VS = Vishay-Siliconix.

* **NOTE:** Table shows both P-channel, and N-channel JFET devices with low $V_{gs(off)}$. N-ch. are < 3.6V, and P-ch. are < 4.1V. All are low g_{os} < 80 µS devices.

Table 6.2 Low G_{os} and low Vgs(off) single JFETs

In summary, the lower the drain current and the lower the output conductance, then the tighter the regulation will be. The closer the source resistor load-line intersects the zero TC point, the less the current source will drift over the intended temperature range, and the more stable and reliable your circuit will be. You can determine just how accurate your current source is by applying equations 6.23 and 6.25.

As another design example, let's create a simple JFET current source, in some easy steps. We can use the following checklist (see Table 6.3) to aid in our search for the best device(s) for the design. While the column headings are self-explanatory, the one titled "0_{TC} point" allows you to record the approximate $V_{GS\ (0TC)}$ and I_{DZ} values from the data sheet transfer characteristic graphs.

Manufacturer:	Part No.:	Pkg:	N or P type:	I_{DSS} min/ max: (mA)	$V_{GS(off)}$ min/ max: (V)	$V_{BR(dss)}$ min/ max: (V)	G_{os} typ: (μS)	0_{TC} point @ V_{GS} and I_{DZ}: (V/mA)

Table 6.3 JFET Current Source Design Check List

1. Determine (1) the desired constant current level, (2) the supply voltage, (3) the load requirements, and (4) current regulation necessary for your circuit, together with the applicable ±% tolerances. (Current regulation is usually expressed as a percentage, as in this example: I_D = 200 μA, tolerance = 1% = 2 μA).

$$\left(\frac{2\mu A}{200\mu A}\right) \text{x } 100\% = 1\% \text{ regulation} \qquad \text{(Eq.6.26)}$$

2. Does the design involve limited space or other mechanical constraints? This may apply to what kind of JFET package would best suit your design: TO-18, TO-92, SOT-23 (SMD), etc.
3. What polarity of JFET, N-channel, or P-channel will be required?
4. Make the constant current/drain current (I_D) less than 33% of I_{DSS} max. The lower the I_D, the tighter the regulation will be.
5. Either pick a device from the listing in Table 6.2, low G_{os} and low $V_{GS\ (off)}$, or look through a JFET catalog. Pick a device with an appropriate I_{DSS} value. Record this data on the checklist.
6. Now choose a device with a $V_{GS\ (off)}$ that matches your design (negative value $-V_{GS\ (off)}$ for N-channel devices, or a positive value $+V_{GS\ (off)}$ for P-channel devices). Record this data also on the checklist.
7. In your circuit, ensure that the JFET's V_{DS} is two times or more the value of the $V_{GS\ (off)}$, or else if V_{DS} is less than $V_{GS\ (off)}$, the JFET will fall out of saturation. Determine both the *minimum* and *maximum* supply voltage (V_{DS}) across the JFET. Is the maximum supply voltage less than its breakdown voltage? Remember too that the voltage compliance across the JFET is from $V_{GS\ (off)}$ to its breakdown voltage $V_{(BR)GSS}$.
8. Next pick a suitable low g_{os} device (with a low-noise value, e.g., less than 25 nV√Hz) from the listing in Table 6.2 or from a JFET catalog. Record this data on the checklist.
9. You should now have recorded a short list of possible devices.
10. This step involves looking at the graph of the transfer curve (I_D vs. V_{GS}) for each JFET on your checklist. Look at either the printed data sheet graph or the associated PDF file, if downloaded from a Web site. If a chosen device does *not* show a transfer curve, then you should move down the list and find another one that does. As shown previously, the transfer curve is a *very important* element in the design sequence, because it allows one to physically see just how close one's intended design is going to be in relation to the zero TC point. A zero TC means that there will be virtually no change or drift in the current through the JFET, over the intended temperature range. This is an ideal situation and should be one's goal (see Figures 6.16 and 6.17). From the JFET's transfer curve, record the approximate $V_{GS\ (0TC)}$ and I_{DZ} values in the checklist's right-hand column labeled zero TC point.
11. Relative to the zero TC point, see where your desired current level (in μA or mA) occurs (above it yields a negative TC, while below it yields a positive TC).

12. Draw a resistor load-line on the graph that intersects at the V_{GS} value arrived at in steps 6 and 7, and at your desired current level. Ideally the load-line should be close to or intersect with the zero TC point. If not, try another device where this happens, or consult Table 6.1 for an approximate value of $V_{GS\ (0TC)}$ and I_{DZ}. If the intersect point is *not* close to the zero TC point, the current source will drift with temperature.
13. Determine the value of the JFET's source resistor (R_S). Make sure that you use a metal-film type, or a precision wirewound resistor, and that its tolerance (e.g., ±0.1%, ±0.01%), wattage rating ($P_{diss} = V_{drop} \times I_D$), and tempco are all within your design limits. The tempco should be as low as possible, so that it does not degrade your circuit (i.e., ±0.001% = 10 ppm/°C), and it should have a low thermal resistance (higher power rating than needed).
14. Build and double-check your circuit's connections, *before* applying power.
15. Now test the circuit, and after some initial evaluation, decide whether you need to adjust the current level, and thereby change the value of R_S. You should test the performance of the circuit by changing the value of V_{DS} and noting any change in drain current (ΔI_D). You will probably need to graph the results. You may decide that over the intended temperature range, the circuit requires a change in its tempco. You can determine what the circuit's tempco is in ppm/°C or as a percentage by using equations 6.23 and 6.25. By employing the techniques previously discussed, you can design the circuit to compensate for the load or following circuitry, by having either a positive, a negative, or even a zero tempco.
16. Once you have implemented your modifications, you should retest the circuit over the intended temperature range. Recalculate the new tempco, and see how that has changed.
17. Build a duplicate circuit, but now using the same modifications you made previously. How close is its performance to the prototype built in steps 1 to 17? You may need to average the value of R_S for a mass-production situation.

6.4 The JFET cascode current source

Although performance is good using just one FET, even better regulation can be achieved by using two N-channel JFETs in a *cascode* configuration, as shown in Figure 6.18B. Cascoding improves high-frequency operation, provides even greater output impedance (Z_{out}), increases high-voltage operation and compliance, and reduces output conductance. In this circuit, cascoding buffers the current source Q_1 from the load, by using Q_2. This means that while the source voltage of Q_2 remains constant, variations in the load voltage are accommodated by Q_2's drain. As a result, the voltage drop across the current source remains constant. Here the drain current (I_D) is regulated by Q_1 and R_S, so that effectively V_{DS1} equals $-V_{GS2}$. The current source is set up as previously discussed. Both JFETs must be operated with adequate V_{DG}, so that V_{DG} is ideally more than twice $V_{GS\ (off)}$, or else the circuit's output conductance will increase significantly. JFET Q_2 needs to have a higher $V_{GS\ (off)}$ than Q_1. R_S sets the desired current, as previously discussed, but in early development it could be substituted by a multiturn potentiometer. This would allow easy adjustment of the desired current level and could range from a few μA to several milliamps.

In the cascode circuit shown in Figure 6.18B, it provides an internal circuit impedance of more than 10 MΩ at current levels less than 1 mA. It can be used equally well in circuits of more than 40 volts or down to about 6 volts, depending on one's exact choice of $V_{GS\ (off)}$. Capacitor C_1 is just included for circuit stability. This inexpensive four-component circuit can be used as either a precision current source or sink, because the load can be connected to either terminal. Even lower g_{os} for the same current is achieved by the cascode, because of degenerative feedback, and the circuit's lower output conductance (which in turn is caused by the combined forward transconductance). Here the circuit's output conductance (g_o) is much less than the g_{os} of the single JFET circuit (Figure 6.18A)—about a *hundred* times lower. (This is proven by the two equations included in the figure.) You can also expect the cascode circuit's output impedance to be more than 10 MΩ. Regulation is about 100 times better, as shown in Figure 6.18C; notice the really sharp knee versus the softer knee of the single JFET's curve.

Using JFETs such as 2N4340/41 in the cascode circuit guarantees extremely low noise, typically less than 10 nV√Hz. For low-voltage and low-current applications, consider using these N-channel devices: 2N4118 ($V_{GS\ (off)max}$ equals –1.8 volt and I_{DSSmax} equals 0.09 mA); 2N4119 (–3 volt/0.24 mA); 2N4338 (–1 volt/1.5 mA); 2N4339 (–1.8 volt/0.6 mA); 2N4867 (–2 volt/3 mA); J202 (–1.5 volt/1 mA). For low voltages around, say, 3.6 volts, one could probably use a combination of 2N4338 and 4339, and below that, around, say, 2.25 volts, a single current source using a 2N4338 can provide a good solution. Selecting devices might be necessary in certain situations, where a lower actual $V_{GS\ (off)}$ might be required, thus leading to lower voltage operation down to about 1.5 volts. A few low-voltage P-type devices also exist that include J201 (2.25 volt/20 mA), 2N3574 (2 volt/0.38 mA), 2N5265 (3 volt/1 mA), 2N5266 (3 volt/1.6 mA), and KK5053 (2.5 volt/3.5 mA), although none provide as low a voltage as the N-channel JFETs previously noted.

In some instances, you may be tempted to use a JFET *matched-pair* instead. However, monolithic matched-pairs are only available as N-channel devices and are usually more expensive. Although thermal tracking is excellent (because both devices share the same substrate), in most cases overall performance is *not* as good as when using two discrete N-channel devices. One should definitely *avoid* monolithic pairs, in which two intrinsic diodes are formed in manufacture (such as U401, U421, 2N3598, 2N5196, SST440/1), because some biasing schemes may inadvertently forward-bias

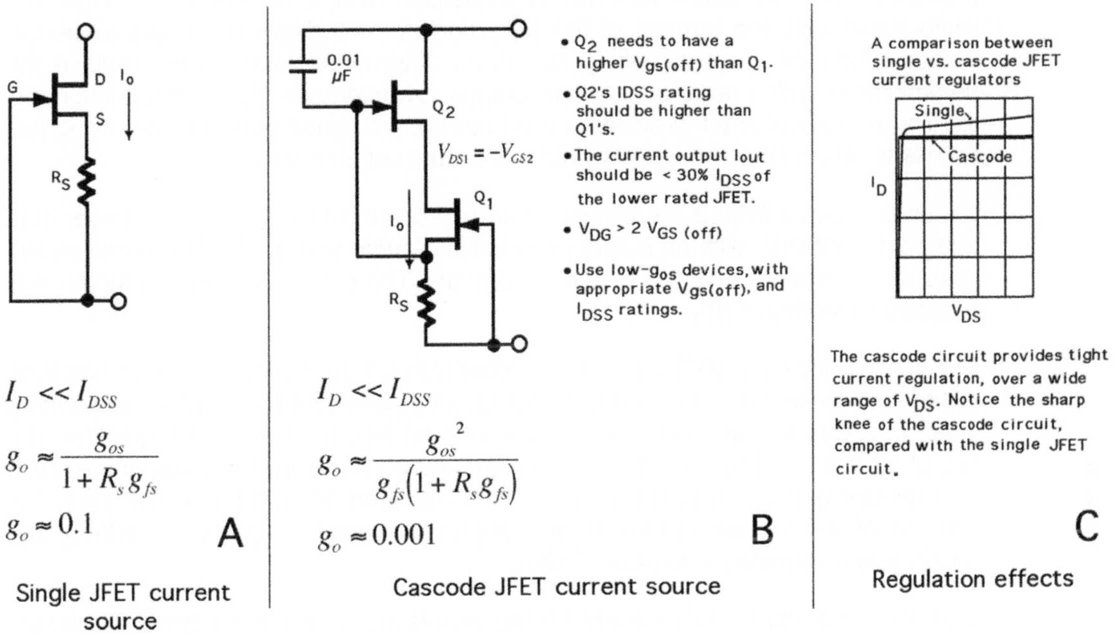
G
D
S
Io
RS
ID << IDSS
go ≈ gos / (1 + Rs gfs)
go ≈ 0.1
A
Single JFET current source
0.01 µF
Q2
VDS1 = −VGS2
Q1
Io
RS
ID << IDSS
go ≈ gos^2 / (gfs(1 + Rs gfs))
go ≈ 0.001
B
Cascode JFET current source
• Q2 needs to have a higher Vgs(off) than Q1.
• Q2's IDSS rating should be higher than Q1's.
• The current output Iout should be < 30% IDSS of the lower rated JFET.
• VDG > 2 VGS (off)
• Use low-gos devices, with appropriate Vgs(off), and IDSS ratings.
A comparison between single vs. cascode JFET current regulators
Single
Cascode
ID
VDS
The cascode circuit provides tight current regulation, over a wide range of VDS. Notice the sharp knee of the cascode circuit, compared with the single JFET circuit.
C
Regulation effects

Figure 6.18. *Comparing the single and cascode JFET current source.*

one of these diodes. You may even be tempted to use a two-chip *matched-pair*, such as U440/1 ($V_{GS\ (off)}$ equals 6 volts, I_{DSS} equals 30 mA); 2N5564/5/6 (–3 volts/30 mA); 2N5911/2 (–5 volts/40 mA). However, you should consider that both JFETs will have *very* close specs, because they were designed for differential-pair/front-end amplifier tasks (as shown later in Figure 6.29). This would not work well for a JFET cascode current source, because the upper device—the cascode—needs both a higher $V_{GS\ (off)}$ *and* a higher I_{DSS} than the lower one—the current regulator.

Alternately, they may have too close a $V_{GS\ (off)}$, which could also cause circuit problems. Even if you hand-selected devices for your lab prototypes or preproduction circuits, if the devices were unwittingly reversed during assembly at the circuit board level, it could prevent proper current regulation and your overall circuit from functioning properly. Using *two* separate devices will always yield the best performance, and these are easily implemented in either throughhole or surface-mount devices. Even a single JFET current source will be a better, safer, more reliable current source than a *mis-matched* matched-pair. In fact, the only time to consider a dual or matched-pair would be if you needed *two* separate, *single* JFET current sources in the same circuit. Each would work equally well as any of the single JFET current sources described.

Some circuit examples of discrete JFETs being used as current sources are shown in Figure 6.19. Circuit A shows a pair of JFET current sources: one N-channel, the other

P-channel. The N-channel example is adjustable over a 500-μA range. However, while convenient, the tempco of the potentiometer will degrade the overall performance. Once the desired setting has been determined, one should replace the potentiometer with a good-quality, low-tempco, low-tolerance (0.1%) metal-film resistor. The P-channel JFET provides a fixed current with good performance, using just two parts, which could both be tiny surface-mount components.

Circuit B shows a fixed 2-mA cascode current source, which can provide better than 300-ppm regulation, with an output impedance of more than 5 MΩ. The circuit shown uses two Crystalonics TO-46 metal-can devices. The circuit's output conductance is probably lower than 1 μS.

Circuit C combines a JFET cascode together with a high-quality monolithic precision voltage reference. The reference IC keeps Q_1's V_{GS} constant over a wide temperature span, to create a high-quality current source with less than 50-ppm regulation. The output current is set by the reference voltage output divided by the value of the precision resistor (a 0.1% or better tolerance and less than 50-ppm metal-film type). The addition of the voltage reference also improves overall temperature stability and boosts output impedance to about 1 GΩ.

One or two additional JFET current source circuits are shown in this book, which may be helpful in your designs. One example is the JFET-DMOS cascode current source circuit shown in Chapter 7's Figure 7.11, which allows a much higher voltage range (5–450 volts) than an all-JFET cascode. Another can be seen in Figure 15.35 of chapter 15, which shows some ultra high-precison current sources using Intersil's state-of-art FGA™ voltage references.

6.5 JFET current regulator diodes

The current regulator diode (CRD) is equivalent to the single JFET current source, as depicted in Figure 6.20, but it is fully monolithic, with its geometry specially designed and enhanced for precise current regulation. The electronic symbol for the JFET CRD is shown on the right.

JFET current regulators were first introduced in the late 1960s, and they have been popular and readily available ever since. Major vendors at the time included Siliconix, Motorola Semiconductors, Teledyne-Crystalonics, and Fairchild Semiconductor. Many of the devices were granted MIL-classification (JAN, JAN-TX, etc.) and were part of a vendor's MIL product portfolio. However, in the late '80s and early '90s, there were numerous cuts in U.S. military programs, and as a result some vendors either sold off parts of their MIL business or discontinued manufacture of many discrete MIL product lines. For that reason, neither Fairchild Semiconductor nor ON Semiconductor (formerly Motorola's discretes division) makes CRDs today, although they pioneered them long ago. CRDs are still popular and readily available in the United States from the following major sources:

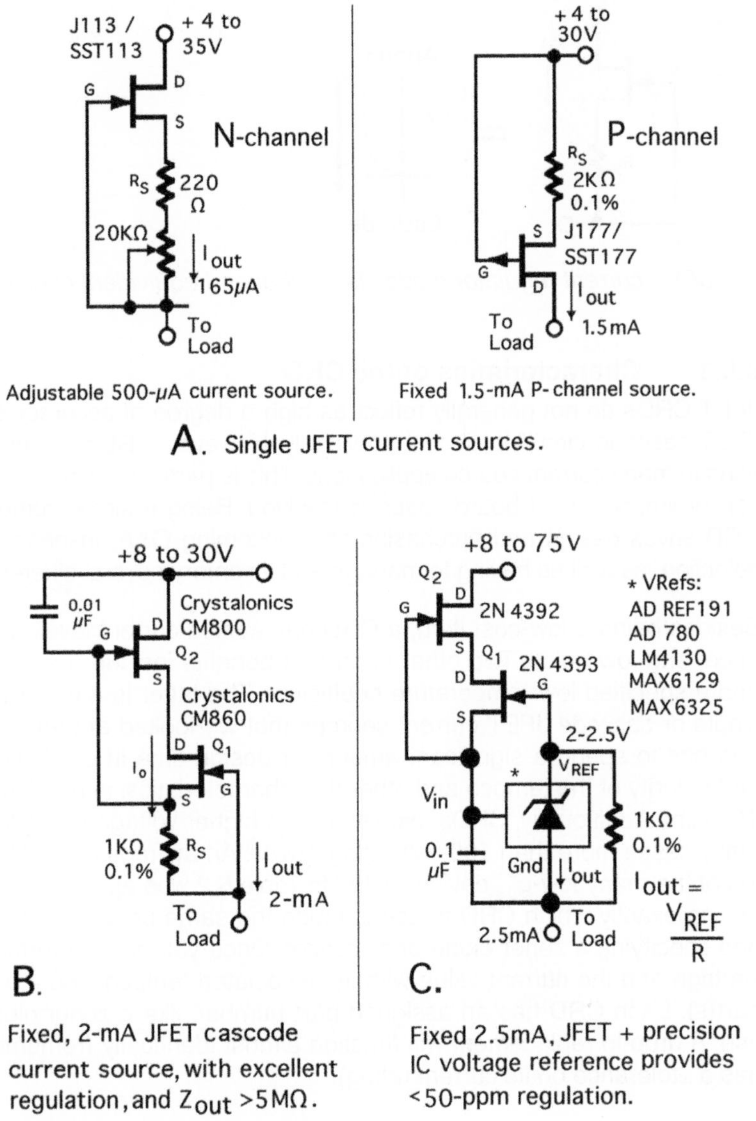

Figure 6.19. *Examples of JFET current sources.*

- Calogic
- Crystalonics
- Linear Integrated Systems
- Solitron Devices
- Central Semiconductor
- InterFET
- Microsemi
- Vishay-Siliconix

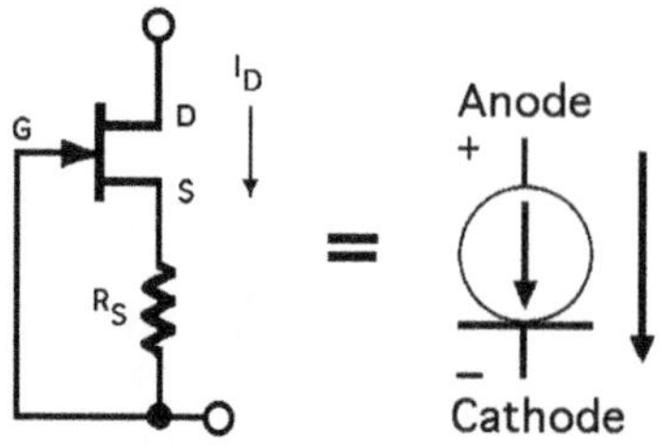

Figure 6.20. *JFET current regulator diode: its symbol and equivalent circuit.*

6.5.1 Characteristics of the CRD

JFET CRDs do not generally reflect as high a degree of accuracy or regulation as the JFET cascode circuit looked at previously. However, CRDs do remain attractive solutions to many current source applications. This is particularly true where volume production or limited circuit board space is involved. Being a single-component solution, the CRD saves design and purchasing time, incoming Q&A, inspection, and part-by-part selection, as well as having to manage and inventory several different components.

Besides being a low-cost item, a CRD has a fixed current level, a high output impedance, and low noise. Two other important benefits include a small two-lead package and a specified low temperature coefficient. The latter is a major advantage over the single or cascode JFET current sources that we looked at previously, because there one has to spend a significant amount of design time in determining the magnitude and polarity of the tempco and other drift characteristics, in relation to the JFET's zero TC point. Additionally, CRDs usually have a higher voltage rating than most JFETs, in many cases more than 100-volt rating. (*Note*: An average of the $V_{(BR)GSS}$ voltage rating of the many low-g_{os} devices listed in Table 6.2 was approximately 38 volts.) Determining exactly which CRD to use is much the same procedure as when designing in and specifying a zener diode or a resistor. Once you have determined the necessary wattage and the current value with its associated tempco, you are ready to order the part(s). Each CRD has an assigned part number, like a zener diode, so it is easy to use. A drop-in replacement will function almost identically (remember that each CRD has a ±tolerance on its current rating).

CRDs are monolithic two-terminal devices, specially designed for precise current regulation. This is achieved by modifying their internal structure, thus creating the equivalent of an internal resistor at the JFET's source (see Figure 6.21A). This internal bulk-source resistance sets the output current. The gate and source are connected together by metallization. The structure's equivalent can be described by either one of two models. The more popular model (Figure 6.21B) is a current generator in series with the parallel combination of an impedance and a capacitance. The second model (Figure 6.21C) is an older model and is of a current generator in parallel with a capacitance and a conductance. Some devices are further enhanced for higher voltage or

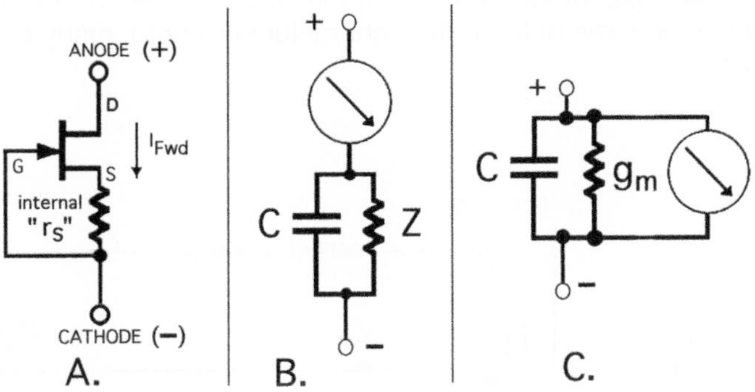

Figure 6.21. *CRD equivalent circuits.*

higher current operation (like some Crystalonics products, which can provide up to 50-mA constant current). Others are able to provide precise current regulation at very low voltages.

In fact, a somewhat overlooked feature of some of these devices is their ability to work down to approximately 1 volt. In Figure 6.22, you will notice that there are several devices in this example, which operate around the 1 volt at 1 mA level or less (there are others too not shown). In today's world, where an increasingly mobile society requires cell phones, digital cameras, MP-3 music players, pods, handheld PCs, and laptops, this unique low-voltage feature is attractive to manufacturers of battery-powered equipment, consumer items, and portable instrumentation.

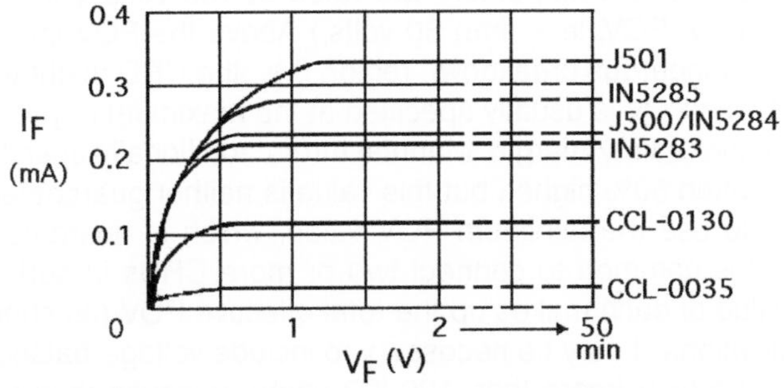

Figure 6.22. *Ultra-low-voltage operation with current regulator diodes.*

Being a two-electrode device, the CRD meets the industry standard for being a type of diode; in fact, the JEDEC prefix of "1N" confirms that (e.g., 1N5290). As a result, JFET terminology no longer applies, and diode terminology is used instead. Examples include *anode* instead of drain, *cathode* instead of source, *peak operating voltage* instead of $V_{(BR)GSS}$, *forward current* (I_F) instead of drain current (I_D), and so on. We will briefly review some of the CRD's characteristics here, many of which are shown in Figure 6.23.

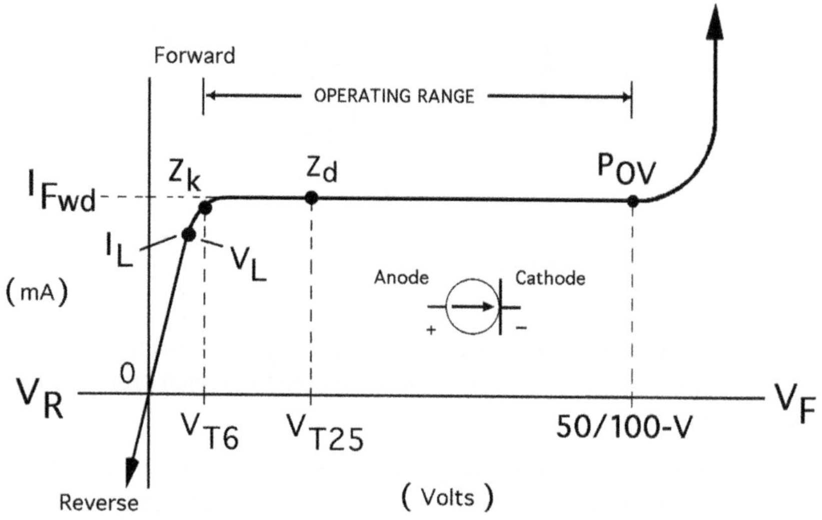

Figure 6.23. The current regulator diode's V-I characteristic curve.

The *peak operating voltage* (POV) simply refers to the maximum voltage that can be applied to the CRD, at 25°C room temperature. The POV is shown in Figure 6.23, and for many devices is 50 volts; others are rated at 100 volts. (Some very-high-current devices have a low POV, less than 30 volts.) Above the POV limit lies the uncharted and potentially dangerous breakdown region (i.e., the CRD might work fine, or it might self-destruct). The POV is usually specified at the maximum regulated current, with an added 10% higher safety margin. Manufacturers additionally specify a *typical* value of POV, which is often 50% higher, but this value is neither guaranteed nor tested. To be safe you should use the minimum POV value, which is guaranteed. In high-voltage applications, it is common to connect two or more CRDs in series, where the combined POV value of each makes up the total effective POV (as shown in Figure 6.24). In some applications, it may be necessary to include voltage-balancing resistors. Their values should be high (more than 100 KΩ each), because they effectively shunt the output resistance. Each CRD should be paralleled by a resistor, as shown in Figure 6.24. A similar technique, known as *current balancing*, is sometimes used when paralleling zener diodes.

Manufacturers refer to the minimum operating voltage as the *limiting voltage* (V_L); this is the *minimum* forward voltage (V_{Fwd}), required to ensure just 80% of the *minimum*

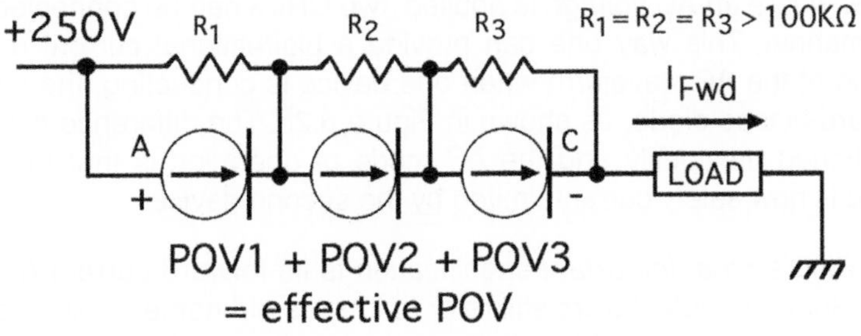

Figure 6.24. *Combining CRDs in series for high-voltage applications.*

forward current (I_{Fwd}). Some low-current CRDs have a very low limiting voltage of less than 1 volt, such as those devices shown in Figure 6.22.

Some data sheets refer to a reverse voltage (V_R), which is simply the value of a *reverse-bias* voltage that the device can withstand (with the cathode positive and the anode negative). It is usually measured at a low reverse current (I_R) of 1 mA and is approximately a diode drop of 0.7 to 0.8 volt. This is shown in Figure 6.25 and can be considered equivalent to the JFET's gate-drain channel diode. Exceeding the specified absolute maximum rating on the data sheet for reverse current may destroy the CRD. For example, J-500, SST-500, CR, and CRR series all have a 50-mA limit (assume that all other types are lower unless you confirm differently from the manufacturer).

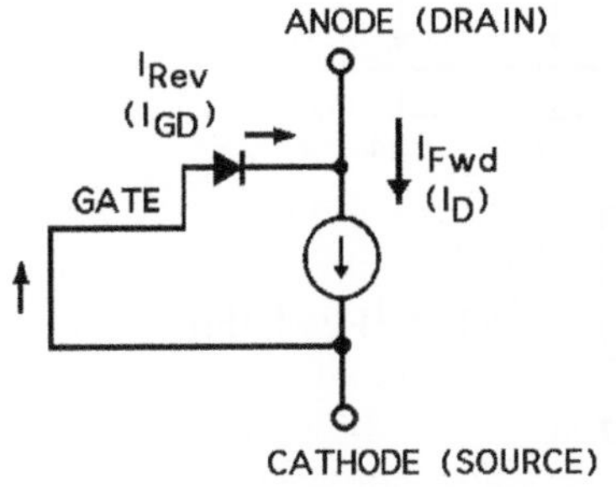

Figure 6.25. *Reverse-bias equivalent of the current regulator diode.*

In some designs, you may need to insert a protective diode in series with the anode or cathode, so that it is forward-biased in the same direction as the CRD's I_{Fwd}, if there is a possibility that the CRD may be inadvertently reversed. Doing so, however, could alter your circuit's temperature coefficient, because the diode's forward voltage (V_{fwd}) has a negative temperature coefficient of 2.2 mV/°C. You may need to choose another CRD with a different current value and associated tempco to compensate. In some cases where an AC voltage is applied, two CRDs can be connected in a series-opposing manner. This way one can provide a bidirectional current regulator. During the period of the AC waveform when one device is conducting, the other one is simply a forward-biased diode, as shown in Figure 6.25. The difference between the DC mode mentioned previously and the AC mode of operation is that the AC forward-biased diode is now safely current-limited by the second device.

The CRD's *most important* specification is its forward current (I_F) or regulating current. Some manufacturers still refer to it by its old name of pinch-off current (I_p). This is the current you are essentially designing around. In the data sheet it is specified at a particular test voltage (usually 25 volts), at room temperature. It is shown in Figure 6.23 as "VT_{25}," meaning at a test voltage of 25 volts. Manufacturers usually measure the forward current using a pulsed test, lasting a few milliseconds. As with most other semiconductors (e.g., MOSFETs, SCRs, diodes, BJTs), using a pulsed test lends itself well to automated test systems. The alternative is a slow (more than 60 seconds per device) steady-state method, which must use a large heatsink. The manufacturer usually specifies a typical value for I_F, as well as a minimum and a maximum. For premium-grade devices, the forward current is usually ±10%, and for lower-grade devices it is ±20%. Because CRDs are available in relatively low current values (from 35 to 50 mA), they can be easily paralleled for increased current operation, as shown in Figure 6.26.

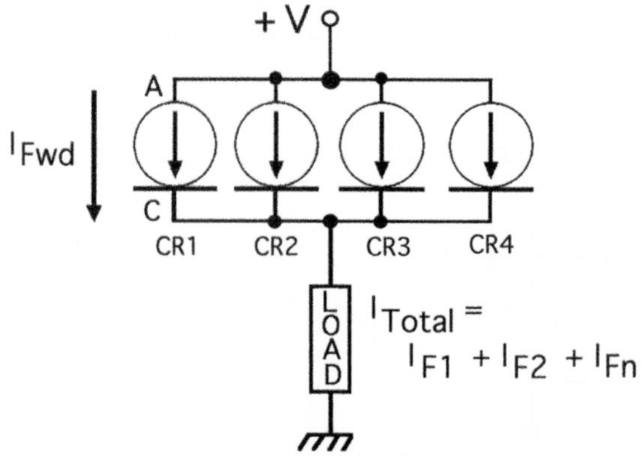

Figure 6.26. Combining CRDs in parallel for higher current operation.

At the other end of the current scale is the *limiting current* (I_L). This is the 80% point of the *minimum* forward current (I_{Fwd}), required to sustain the limiting voltage (V_L). Both the I_L and V_L points are shown in Figure 6.23 for your reference.

Another important CRD characteristic is the *knee impedance* (Z_K). This point is also shown in Figure 6.23 and indicates where the knee of the curve is located. (With a regular JFET the drain current increases linearly, then begins to be pinched off or limited at the knee of the curve. The curves that display the drain current versus the V_{DS} constitute a JFET's *output characteristics.*) With the CRD, the knee impedance should be as high as possible, to provide the most constant current for the CRD. In the diagram it is associated with "VT_6" (the knee impedance test voltage), meaning at a test voltage of 6 volts. The knee impedance is specified in MΩ, and as a *minimum* value. Low current values typically yield a high value of Z_K, while high current levels provide lower values of Z_K. This varies, however, from one CRD family to another, as shown in Table 6.4.

Series:	**Nominal Current (mA):**	**Dynamic impedance -Zd MIN: (MΩ)**	**Knee impedance -Zk: (MΩ)**
J-505	1.0-mA	0.5-MΩ	0.4-MΩ typ
IN 5297	1.0-mA	0.8-MΩ	0.205-MΩ min
CCL-1000	1.1-mA	0.65-MΩ	0.10-MΩ min
CRR-1250	1.25-mA	0.5-MΩ	0.8-MΩ typ
CR-100	1.0-mA	0.8-MΩ	0.205-MΩ min

Table 6.4 Comparison of different CRD Series: Zd and Zk values

Another impedance specified for CRDs is the *dynamic impedance* (Z_D). This point is also shown in Figure 6.23 and indicates what the minimum impedance is at full conduction and at a reasonable supply voltage. The dynamic impedance should also be as high as possible, to aid in providing the most constant current for the CRD. In the diagram it is associated with "VT_{25}" (the dynamic impedance test voltage), specified as mentioned at 25 volts. The dynamic impedance is also specified in MΩ, and again as a *minimum* value. CRDs with low current values yield a high value of Z_D (typically more than 5 MΩ), while high current devices provide lower values of Z_D (typically less than 0.5 MΩ). Again, this varies from one CRD family to another, as shown in Table 6.4.

While not usually specified on a data sheet, the CRD has a forward capacitance (C_F), which varies with the applied voltage and the particular device geometry. Its value is highest at low voltages (typically less than 20 pF at 6 volts), but falls with increasing voltage. At above 10 volts and at 1-MHz frequency, the forward capacitance is typically less than 10 pF, and it remains at this low level up to the POV limit of the device.

As mentioned previously, associated with each device's nominal current value is a defined *temperature coefficient* (TC, tempco). Some devices have a positive tempco, whereas others have a negative tempco. It is usually measured over a narrow portion of the device's total operating temperature range, such as +20°C to +50°C, of a total –55°C to +200 °C, in order to determine the slope of the TC (i.e., either positive, negative, or flat). Depending on the application and circuit arrangements, a positive or a negative tempco can be used to compensate for subsequent circuit stages in one's design. A few devices have a zero (or close to zero) TC (i.e., CR047, CR240, J502, 1N5290, CCLM0750). For many CRDs, the temperature coefficient is better than 0.15%/°C (1500 ppm).

Some manufacturers publish the TC values in their data sheets, but not all do. If the TC is important in your application, you are recommended to contact technical support at the manufacturer in question. Remember also that with higher current devices, the tempco will be larger because of higher power dissipation and self-heating. If you are unable to determine the tempco from the data sheet, you should carefully measure the current output over the temperature range of interest. You should also test the performance of the circuit by changing the value of V_{DS} and noting any change in output current. You will need to graph the results. You can determine what the circuit's tempco is in ppm/°C or as a percentage by using equations 6.23 and 6.25 shown earlier in this chapter.

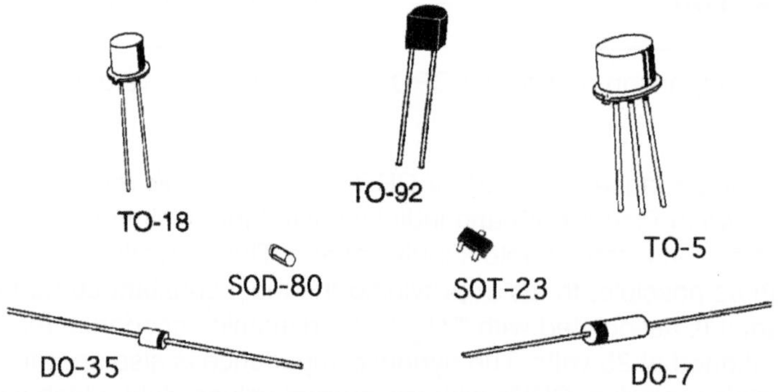

Figure 6.27. Current regulator diodes are available in various popular packages.

Generally, CRDs are available in various glass, metal, plastic, and surface-mount packages (see Figure 6.27). In any design scenario, you should consult the manufacturer or visit its Web site for up-to-date availability on the various package options. Product lines change and grow, and so does a manufacturer's packaging capability. While a couple of years ago a manufacturer may have only offered a few different package options, today it probably has most of the product line available in various packages. In one case, Crystalonics makes a high-current family of CRD devices (CIL-350 series) with currents up to 50 mA, and which are provided in a TO-5 metal can. Although some devices are proprietary to a manufacturer, many are JEDEC-registered and are available from multiple sources.

Whatever the package type of the CRD of your choice, there will be wattage and *thermal characteristics* associated with it. As with every other type of semiconductor device, the CRD's particular type of package will have a wattage rating and a derating factor. For example, an 1N5283 is provided in a glass DO-7 diode-type package. This is rated at a maximum 600 mW and has a derating factor of 4.8 mW/°C, above 75°C lead temperature. This particular package usually has its temperature measured and referenced 3/8 inch (9.525 mm), from the glass envelope. The cathode lead of the device (the end with the colored band) is where most of the heat is conducted away from the junction. Other popular package types, such as the TO-92 plastic transistor-type package, will have a different wattage rating. In this case, it is typically 350 mW, with a derating factor of 2.8 mW/°C above 25°C, room temperature. You should ensure that the package type(s) you are using in your design is operating well within the manufacturer's recommended thermal rating to ensure a long and reliable life.

6.5.2 A design guide

First establish the load, supply voltage, and constant current requirements for your circuit, along with the ±% tolerances for each. This will help determine the necessary wattage required for the CRD. Next, choose a suitable device from Table 6.5 or from a product catalog. Remember that Table 6.5 is only a summary and does *not* include the temperature coefficients that may be necessary for your design. You should make a list of suitable devices. Obtain the device(s), and then build and evaluate the circuit.

Test the performance of the circuit by varying the value of the supply voltage and load impedance. Note and record any change in output current (I_D). You may decide that over the intended temperature range, the circuit requires a change in its current/temperature coefficient to compensate. If you have to implement any changes, such as using a different CRD with a different current value and associated tempco, you should test and graph the circuit again, over the intended temperature range.

Build another circuit using the same modifications you made previously. Recalculate the new tempco, and see how that has changed from the original design. You may have to use a CRD with a current value that is an average of some of your previous tests, for a mass-production situation. Lastly, evaluate your design again, so that it is ready for preproduction stage. You may have to average your results to be more easily mass-produced. If you are involved in a design for a mass-production job (i.e., thousands of devices), you should contact the manufacturer directly and have a certain current value selected for you. To the best of the author's knowledge, the CRD manufacturers listed in this chapter will be happy to work with you in that type of situation.

6.5.3 An overview of various popular CRD families

The following is a brief summary of most of the popular CRD families in the United States. Table 6.5 briefly summarizes these for you with relevant design information.

However, you are advised to obtain more complete product literature from the manufacturer or its Web site before getting too far along in your design.

- *The CIL series* are a high current family, made exclusively by Crystalonics. They include 16 different values, subdivided into two ranges. The CIL-250 to 257 group ranges from 5.1 mA to 10 mA. The CIL-350/A to 366/A group ranges from 11 mA to 51 mA. Both groups have a guaranteed ±10% tolerance on any particular current value. The minimum peak operating voltage is 50 volts DC for CIL-250 to 257 and more than 27 volts for CIL-350/A to 366/A. Both groups operate over an extended temperature range of –55°C to +150°C. The CIL-250 to 257 series are available in a DO-7 glass package and are rated at 600 mW. The CIL-350/A to 366/A series are available in a TO-5 metal transistor-type package, rated at 1.5 Watts.
- *The 1N 5283 series* are prime-grade devices, which include 32 different values ranging from 0.22 mA to 4.7 mA, with a guaranteed ±10% tolerance on the specific current value. The peak operating voltage is 100 volts DC, and the series is typically available in a DO-7 glass diode-type package; however, Central Semiconductor makes this in a newer, more efficient DO-35 package. Either package type is rated at 600 mW. The series is specified for operation over an extended temperature range of –55°C to +200°C. Crystalonics also makes this available in military-grade JAN, JTX, and JTXV devices.
- *The CR series* are prime-grade devices, which include 32 different values ranging from 0.22 mA to 5.3 mA, with a guaranteed ±10% tolerance on the specific current value. They are available in a two-lead modified TO-18 metal-can package and operate over an extended temperature range of –55°C to +150°C. The CR series is rated at 1.25 Watts. Originally introduced by Siliconix, the range is now made by Linear Integrated Systems. Vishay-Siliconix makes CR-160 and above.
- *The CRR series* are prime-grade devices, which include eight different values ranging from 0.24 mA to 4.3 mA, with a guaranteed ±25% tolerance on the specific current value. They are available in a two-lead modified TO-18 metal-can package and operate over an extended temperature range of –55°C to +150°C. The CRR series is rated at 1.25 Watts. Originally introduced by Siliconix, the range is now made exclusively by Linear Integrated Systems.
- *The J500 series* are a lower-cost family, which include 12 different values, ranging from 0.24 mA to 4.5 mA. These devices have a guaranteed ±20% tolerance on any particular current value. The minimum peak operating voltage is 50 volts DC, and they are rated at 360 mW. The J500 series are available in a two-lead modified TO-92 plastic package (TO-226AA) and operate over an extended temperature range of –55°C to +150°C.

- *The SST500 series* are available in 12 different values, ranging from 0.24 mA to 4.7 mA. These devices have a guaranteed ±20% tolerance on any particular current value. They are basically the popular J-500 series, but in an SOT-23 surface-mount package. The SST500 series have a minimum peak operating voltage of 50 volts DC and are rated at 360 mW. They operate over an extended temperature range of –55°C to +150°C.
- *The J552 series* are a low-cost family, which include six different values, ranging from 0.25 mA to 4.5 mA. These devices have a guaranteed ±20% tolerance on any particular current value. The minimum peak operating voltage is 50 volts DC, and they are rated at more than 350 mW. The J552 series are available in a two-lead modified TO-92 plastic package (TO-226AA) and operate over an extended temperature range of –55°C to +135°C.
- *The U553 series* are available in six different values, ranging from 0.50 mA to 4.5 mA. These devices are now exclusive to InterFET and have a guaranteed ±20% tolerance on any particular current value. The minimum peak operating voltage is 50 volts DC, and they are rated at 360 mW. The U553 series are available in a modified TO-18 metal can and operate over an extended temperature range of –65°C to +150°C.
- *The CCL0035 series* is a low-cost family exclusively from Central Semiconductor, which includes 12 different values, ranging from 35 μA to 5.75 mA. These devices have a guaranteed ±16% tolerance on any particular current value. The minimum peak operating voltage is 100 volts DC, and they are rated at 600 mW. The series is available in a small but rugged DO-35 diode package. It operates over an extended temperature range of –65°C to +200°C.
- *The CCLM0035 series* is another low-cost family exclusively from Central Semiconductor, which includes 12 different values, ranging from 35 μA to 5.75 mA. These devices have a guaranteed ±16% tolerance on any particular current value. The minimum peak operating voltage is 100 volts DC, and they are rated at 800 mW. The series is available in a small but rugged SOD-80 surface-mount package. It operates over an extended temperature range of –65°C to +200°C.
- *The CCLH080 series* is a low-cost family exclusively from Central Semiconductor, which includes four different values, ranging from 8.2 mA to 15 mA. These devices have a guaranteed ±20% tolerance on each particular current value. The minimum peak operating voltage is 50 volts DC, and they are rated at 600 mW. The series is available in a small but rugged DO-35 diode package. It operates over an extended temperature range of –65°C to +200°C.
- *The CCLHM080 series* is a low-cost family exclusively from Central Semiconductor, which includes four different values, ranging from 8.2 mA to 15 mA. These devices have a guaranteed ±20% tolerance on each

particular current value. The minimum peak operating voltage is 50 volts DC, and they are rated at 800 mW. The series is available in a small but rugged SOD-80 surface-mount package that operates over an extended temperature range of –65°C to +200°C.

In summary, Linear Integrated Systems and Central Semiconductor offer the most products, followed by Crystalonics. Central Semiconductor offers the lowest-current devices (CCL-0035 and CCLM-0035 families), with current levels down to 35 μA, while Crystalonics offers the highest-current devices (CIL-250 and CIL-350 families), with currents up to 51 mA. The most second-sourced family of CRDs is the J-500 family, which is available from four different manufacturers, and in some cases is also available in surface-mount versions. Another is the 1N5283 family, which is an older device, but also available from four different manufacturers. At the time of this writing, the manufacturer with the most devices available in surface-mount packages was Central Semiconductor. Most manufacturers have now included surface-mount packages in their product offerings, but you should check to see whether package options other than those listed in Table 6.5 are available.

Some circuit examples using CRDs are shown in Figure 6.28.

Circuit A shows a typical NPN differential amplifier. Replacing the normal emitter bias resistor with a CRD significantly improves the amplifier's common-mode rejection ratio (CMRR). The CRD can boost the normal common-emitter impedance (less than 1 KΩ), anywhere from 20 to 60 dB, and also improves the temperature stability of the circuit.

Circuit B shows how to provide a stable, low-voltage reference (0.25 volt) by combining a CRD with a precision resistor. The voltage level is set by applying Ohm's law. Normally, Zener diodes and monolithic voltage references cannot provide voltages this low.

Circuit C combines a Crystalonics CIL-254 (medium-power CRD) with a Central Semiconductor 1N829A (temperature-compensated) selected zener diode, to create a two-component voltage reference. This circuit will provide a stable output for an input voltage range of approximately 8 volts to the POV rating of the CRD (in this case, 60 volts), and over a temperature range of –25°C to more than 100°C. Using a 1N5290 CRD with a (hard-to-find) 1N4569 TC zener instead provides a much lower-power alternative, with an even wider V_{in}.

Table 6.5 US Manufacturers of Current Regulator Diodes

In Circuit D, replacing a BJT's normal emitter load resistor with a CRD increases the gain of the amplifier stage, as well as increasing its input impedence.

Mfr:	Series:	Current Range (mA):	POV max (volts):	Limiting Voltage range:	Temp. range: °C	Wattage rating max	Package:
Calogic:	J-500 - J-511 (12-devices)	0.24 thru' 4.7 ± 20%	50 V	0.4 - 2.1 typ.	- 55° to + 150°	350 mW	TO-92
	SST-500 - 511 (12-devices)	0.24 thru' 4.7 ± 20%	50 V	0.4 - 2.1 typ.	- 55° to + 150°	360 mW	SOT-23
Central Semiconductor	IN 5283 - 5314 + JAN (32-devices)	0.22 thru' 4.7 ± 10%	100 V	1 - 2.9 max	- 65° to + 200°	600 mW	DO-35
	CCL-0035 thru 5750 (12-devices)	0.035 thru 5.75 ± 16%	100 V	0.4 - 4.5 max	- 65° to + 200°	600 mW	DO-35
	CCLM-0035 thru 5750 (12-devices)	0.035 thru 5.75 ± 16%	100 V	0.4 - 4.5 max	- 65° to + 200°	800 mW	SOD-80
	CCLH-080 thru 150 (4-devices)	8.2 thru' 15 ± 20%	50 V	3.1 - 4.3 max	- 65° to + 200°	600 mW	DO-35
	CCLHM-080 thru 150 (4-devices)	8.2 thru' 15 ± 20%	50 V	3.1 - 4.3 max	- 65° to + 200°	800 mW	SOD-80
Crystalonics	IN 5283 - 5314 + JAN (32-devices)	0.22 thru' 4.7 ± 10%	100 V	1 - 2.9 max	- 55° to + 200°	600 mW	DO-7
	CIL-250 to 257 (8-devices)	5.1 thru' 10 ± 10%	> 50 V	3.67 -7.2 max	- 55° to + 150°	600 mW	DO-7
	CIL-350/A to 366/A (8-devices)	11 thru' 51 ± 10%	> 27 V	6.0 max	- 55° to + 150°	1.5W	TO-5
InterFET:	J-500 - J-511 (12-devices)	0.24 thru' 4.7 ± 20%	50 V	0.4 - 2.1 typ.	- 55° to + 150°	350 mW	TO-92
	J-553 - 557 (5-devices)	0.5 thru' 4.5 ± 20%	50 V	0.75 - 1.3 typ	- 55° to + 150°	360 mW	TO-92
	U-553 - 557 (5-devices)	0.5 thru' 4.5 ± 20%	50 V	0.75 - 1.5 typ.	- 65° to + 150°	360 mW	TO-18
Linear Integrated	J-500 - J-511 (12-devices)	0.24 thru' 4.7 ± 20%	50 V	0.4 - 2.1 typ.	- 55° to + 150°	350 mW	TO-92
	SST-500 - 511 (12-devices)	0.24 thru' 4.7 ± 20%	50 V	0.4 - 2.1 typ.	- 55° to + 150°	360 mW	SOT-23
	J-552 - 557 (6-devices)	0.5 thru' 4.5 ± 20%	> 50 V	1 - 2.1 typ.	- 55° to + 135°	350 mW	TO-92
	LSCR022 - 470 (32-devices)	0.243 thru 5.83 ± 10%	100 V	0.4 -1.4 typ	- 55° to + 150°	1.25W	TO-18
	LSCRR0240 - 4300	3 thru 5.4 ± 25%	100 V	0.5 - 1.45 typ.	- 55° to + 150°	1.25W	TO-18
Microsemi	IN 5283 - 5314 + JAN (32-devices)	0.22 thru' 4.7 ± 10%	100 V	1 - 2.9 max	- 65° to + 200°	600 mW	DO-7
Solitron Devices	IN 5283 - 5314 + JAN (32-devices)	0.22 thru' 4.7 ± 10%	100 V	1 - 2.9 max	- 65° to + 200°	600 mW	DO-7/TO-92 /TO-18
Vishay - Siliconix	CR-160 - 470 (12-devices)	1.6 thru' 4.7 ± 10%	100 V	0.7 - 2.9 typ.	- 55° to + 200°	300 mW	TO-18
	J-500 - J-511 (12-devices)	0.24 thru' 4.7 ± 20%	50 V	0.4 - 2.1 typ.	- 55° to + 150°	350 mW	TO-92

NOTE: Check with manufacturer for latest device package availablity.

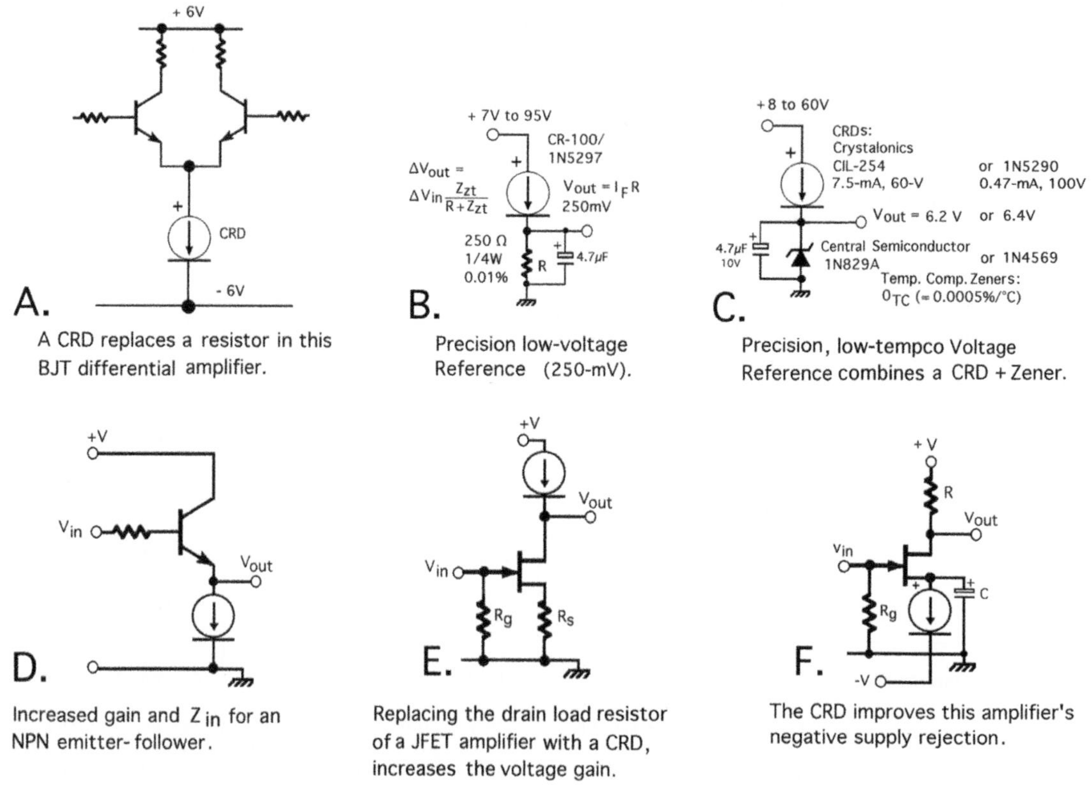

Figure 6.28. Some typical CRD applications.

Circuit E shows how using a CRD instead of the normal drain resistor can increase the voltage gain for this JFET amplifier. The CRD's high output impedance is the difference here.

In Circuit F, a CRD replaces the normal source resistor, thereby increasing the JFET amplifier's negative supply rejection.

6.6 Using JFETs to create ultra-low-leakage diodes

Dedicated low-leakage diodes typically have leakage currents in the low nano-amp range, which in some situations can actually degrade a precision measurement circuit. For example, the very-low-input bias-currents of some precision op amps are in the pico-amp range. Using a typical low-leakage diode with this type of op amp can significantly degrade the measurement capabilities of the circuit. As mentioned elsewhere in this book, JFETs can be used to create *ultra-low-leakage* diodes, which can also be of use when creating current sources or voltage references. Normally, when the JFET has its drain and source shorted together (see Figure 6.29), it provides an ultra-low-leakage diode (known as a *pico-amp diode*, PAD), with extraordinary low-leakage currents. In some conditions, such diode-connected JFETs can attain leakage-current levels that are below 200 femto-amps (2×10^{-13} A) at V_{DG} at 20 volts or less than 100 fA at V_{DG} at 5 volts—and almost immeasurable. Particular JFETs, such as these listed as follows, can attain less than 10 pA at 25°C.

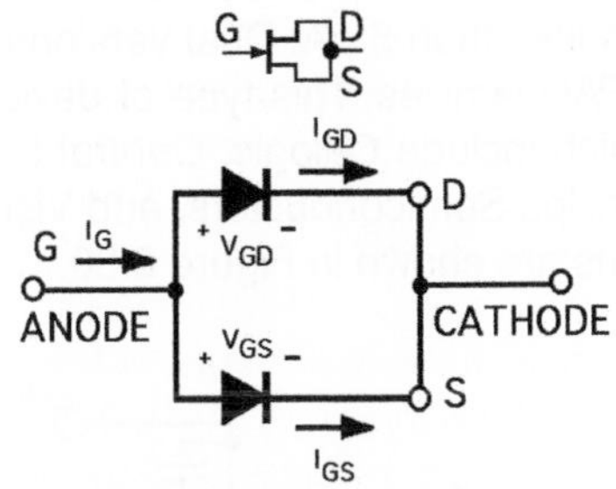

The N-channel JFET equivalent when used as a very low-leakage diode.

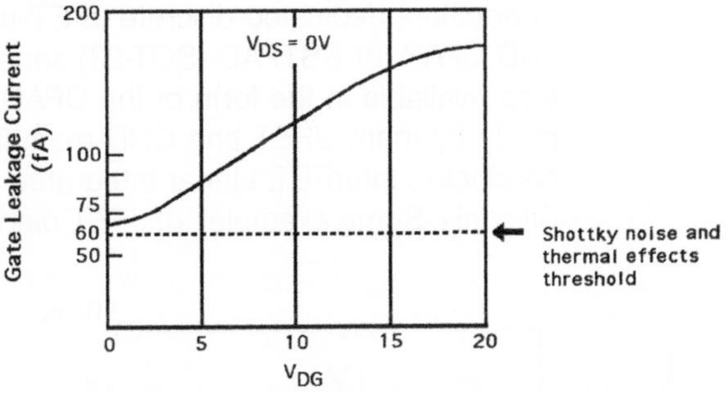

Notice the ultra-low leakage currents at low drain voltages for 2N4117A.

Figure 6.29. *JFET Pico Amp diodes.*

TO-72, four-pin metal package:

2N4117	2N4117A
2N4118	2N4118A
2N4119	2N4119A

TO-92, three-pin plastic package:

PN4117	PN4117A
PN4118	PN4118A
PN4119	PN4119A

SOT-23, three-pin surface-mount package:

SST4117

SST4118

SST4119

In addition, dedicated discrete JFET ultra-low-leakage pico-amp diodes, such as the PAD, JPAD, or SSTPAD (SOT-23) series, can attain less than 5 pA. Dual versions are also available in the form of the DPAD and SSTDPAD families. This type of device is made by many JFET and CRD manufacturers, which include Calogic, Central Semiconductor, InterFET, Linear Integrated Systems, Philips Semiconductors, and Vishay-Siliconix. Some examples of JFET diode applications are shown in Figure 6.30.

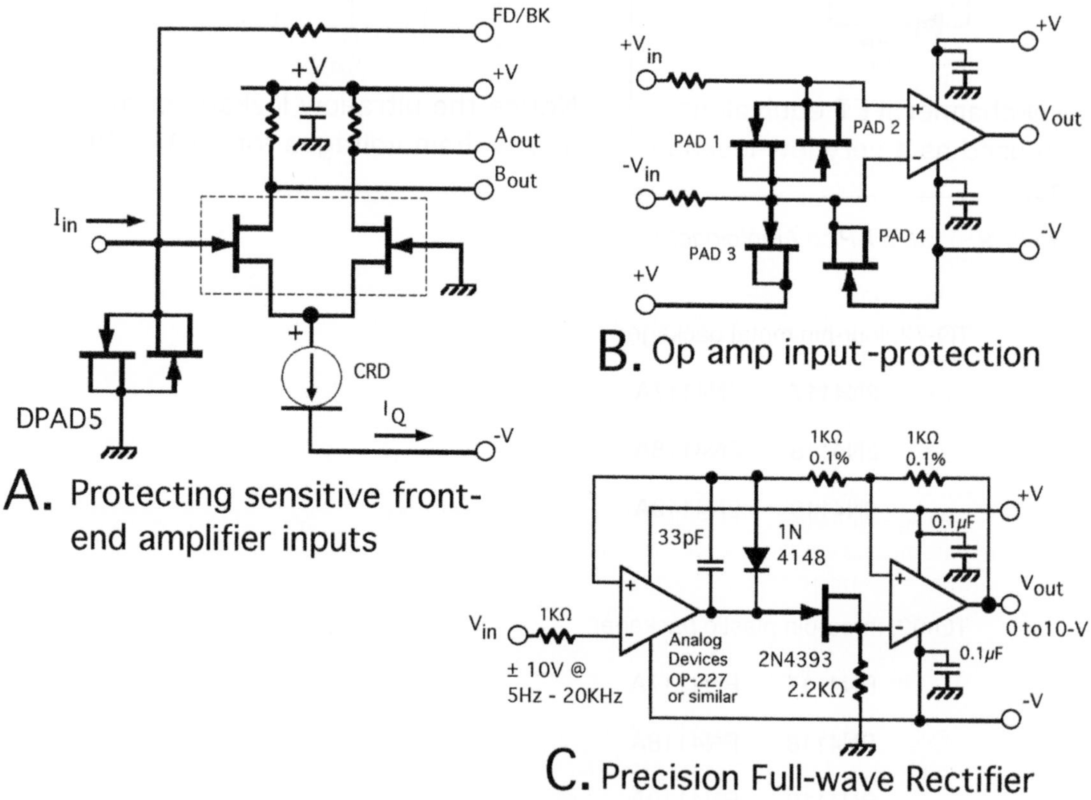

Figure 6.30. JFET/pico-amp diode (PAD) applications.

In Circuits A and B, JFET diodes protect the inputs to very-low-level amplifiers. Normally, the JFET diodes have reverse leakage currents in the low pA range, which is less than the very-low-input bias currents of most amplifiers, either discrete or monolitihic. The JFET diodes will clamp the input to approximately 0.7 volt in case a potential dangerous and costly voltage spike occurs. In Circuit B, the op amp's input differential voltage is limited to typically 0.8 volt by the JFET diodes, PADs 1 and 2, while the common-mode input voltage is limited by PADs 3 and 4, to the amount of the supply rails. Using dual PADs (DPADs) would save space.

In Circuit C, a JFET diode/PAD is used in a precision rectifier design. For best results, the op amp should have a very high CMRR value, a high gain, and a low TC V_{OS}, in which case errors will be extremely small (i.e., less than 10 ppm). Anyhow, it's worth remembering the next time you need an ultra-low-leakage diode, that the JFET can do that too.

When you consider what a simple JFET can do, it has to be one of the most useful semiconductor devices ever invented, and certainly an essential building block in analog electronics. It has come a long way since Dr. Lilienfeld first theorized on a FET-like device and Dr. Shockley subsequently invented it in the early 1950s. No doubt the JFET will be a part of our design arsenal for decades still to come.

Chapter 7

Creating Medium-Power Current Sources with DMOS FETs

After Dr. Shockley invented the basic junction FET at Bell Labs in 1952, several years elapsed during which time the epitaxial process was also developed there. In 1958, Bell Labs' researchers invented photolithography, which allowed more complex devices to be created and eventually ICs. Another important milestone was reached in 1960, when yet another Bell Labs' researcher, Dr. John Atalla, developed the first metal-gate MOSFET. However, it was more of a lab curiosity and not too stable. Just before this in 1959, Dr. Jean Hoerni, a co-founder of Fairchild Semiconductor, invented the planar process, which led him to design the first planar epitaxial JFETs. This opened the door for further development of the JFET and research of other FET types.

Development of the FET in these early years was constantly delayed by the intense industry-wide focus on the BJT, where higher frequency and higher power were the main goals. The FET was considered all but impossible to make by those who tried, because it was so unstable. Then in about 1962, Fairchild Semiconductor made a huge discovery: that sodium impurities caused instability in silicon during processing, a problem that plagued the entire semiconductor industry at the time. Soon after, two young RCA researchers, Steve Hofstein and Fred Heiman, discovered the cause of and cured another old, nagging industry-wide problem—surface states. This problem had first been uncovered and documented by Nobel Prize winner Professor John Bardeen at Bell Labs in about 1947. Surface states was a serious problem in which electron traps on the surface of the silicon dioxide limited a transistor's conduction. Their discovery in 1963 led the two RCA researchers to make some of the first stable and working MOSFETs. Both RCA and Fairchild Semiconductor unofficially shared their findings, and as a result both companies made some of the first working MOSFETs. Soon after, both of these key discoveries were shared industry-wide, much to everyone's delight. Even so, Fairchild at the time was more interested in bipolars than FETs, so FET development was given a low priority. By the mid-60s, though, RCA had some of the first small, metal-can MOSFETs in production and in their own consumer products—FM radios, TVs, and amplifiers.

By the end of the 1960s, there were several dozen competitors in the United States and globally, each of which contributed important research to the FET family. Four other stand-out American companies that helped pioneer MOS technology (and MOS-based ICs) at that time were General Micro-Electronics, General Instrument, Mostek (the first companies to have any commercial success with MOS products), and Intel—the most successful of all. One of the most potentially important and successful FET processes was developed in Japan. It was dubbed DMOS for *double-diffused MOS* and was introduced in the United States in the early 1970s. When electron-beam technology was invented in 1972, this opened the door for even more complex MOS structures, including vertical types. Since the mid-70s, continued R&D in all aspects of silicon MOSFET technology has been taking place around the globe—in the semiconductor industry, in academia, and in government research labs. These researchers have been focusing their efforts on various fronts—advanced UHF, analog switching, super-fast logic, and high-power switching devices.

In those early years, besides Bell Labs, Fairchild Semiconductor, and RCA, other notable U.S. MOSFET pioneers included Siliconix, International Rectifier, Supertex, Solitron Devices, Intel, Westinghouse Research Labs, National Semiconductor, Motorola Semiconductors, Signetics, Mostek, General Micro-Electronics, Harris Semiconductor, and General Instrument. Many significant contributions also came from Philips Semiconductors and SGS-Thomson in Europe, as well as from Hitachi and others in Japan. Without the extraordinarily bold efforts and sustained research by engineers in these companies throughout the 1960s, '70s, and '80s, we would probably not have CMOS, DMOS, or silicon power MOSFETs, all of which are now indispensable in today's portable and wireless world.

7.1 Depletion-mode DMOS FETs

Like its cousin the JFET, the double-diffused MOS (DMOS) FET can also be used for excellent current regulation. Over the last decade, DMOS FETs have become increasingly popular for many applications, particularly N-channel enhancement-mode, vertical power devices. Actually the DMOS family consists of four types, these being either *depletion-mode* or *enhancement mode*, and either *lateral* or *vertical* in their structure. Although DMOS FETs can theoretically be made in either polarity, P-channel devices are *not* available in the depletion mode. N-channel DMOS FETs are available in all four types though. The DMOS family tree and basic operation is shown in Figure 7.1. The different types and their merits with regard to functioning as current sources are reviewed in this chapter and the following one.

Being depletion-mode devices, these *normally-on* DMOS FETs operate similarly to JFETs that we looked at in the previous chapter. In fact, one could consider them as being even more powerful JFETs. They have a similar operating frequency range, although in many cases the DMOS FET is faster (more than 500 MHz). They also have ultra-low capacitance (less than 10 pF), as well as very-low-leakage currents (less than 50 nA) and very-high-input impedance (more than 100 MΩ). DMOS FETs offer much higher voltage operation (more than 200 volts) and a higher current capability (more than 100 mA) than JFETs. They also have a similar way of operating and share similar characteristics as well as most of the same terminology as the JFET, such as I_{DSS}, I_{GSS}, $V_{(BR)DSS}$, $V_{GS(off)}$, and g_{fs}. This means that many of the same

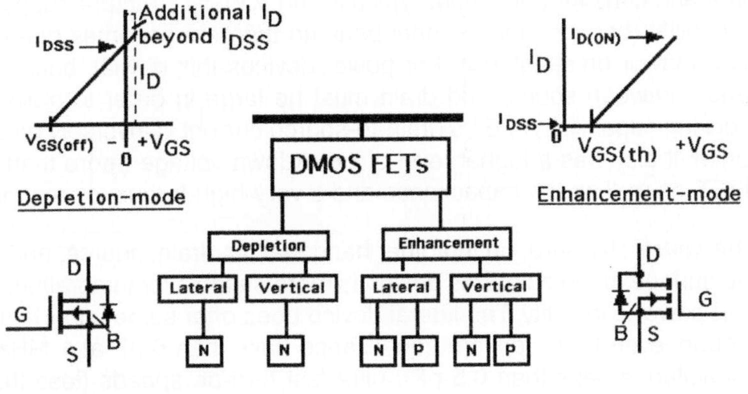

Figure 7.1. The DMOS family tree.

basic equations and models covering the N-channel JFET can be used for the depletion-mode DMOS FET (but remember, it's N-channel only).

MOSFETs have the same terminals as a JFET (gate, source, and drain), but also have an extra terminal called the *body* (B) or *substrate* (Sub). For most practical purposes, this can be considered as permanently connected to the source; otherwise, if it is available separately, connect it to the most negative point in your circuit. Figure 7.2 shows the symbol and large-signal model for the depletion-mode DMOS FET.

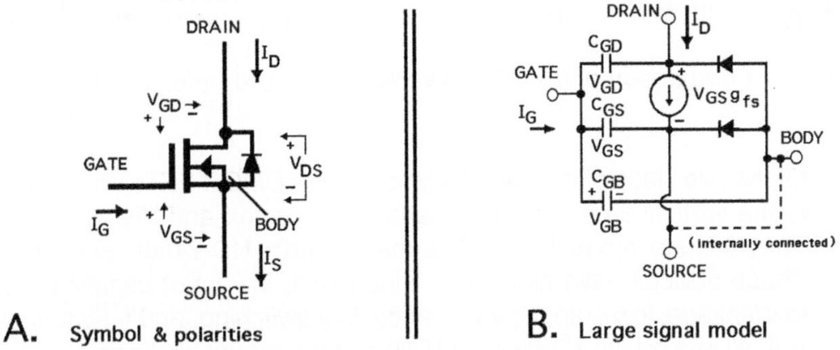

Figure 7.2. *Showing the symbol for the N-channel, depletion-mode DMOS FET, as well as its various voltages, currents, and polarities. Note this is an N-channel device only, and that the body terminal is normally tied internally to the source.*

The depletion-mode DMOS FET is constructed somewhat differently from the JFET and is actually more like its sibling, the enhancement-mode MOSFET. The structure of the device will dictate whether the current flows through it vertically (as in Figure 7.3A) or laterally (Figure 7.3B). This subject is beyond the scope of this book, but there are pros and cons for each type. Typically, the *vertical* structure supports a higher current capability (because the channel beneath the gate becomes deeper), and it therefore has a lower on-resistance. For power devices this is vital, because the physical distance between source and drain must be large in order to maintain a high voltage-blocking capability. A FET's drain-to-source current is inversely proportional to this distance. It also has a higher level of breakdown voltage (more than 200 volts) than the JFET, as well as low capacitance and a very-high-frequency response.

The *lateral* structure, on the other hand, has its drain, source, and gate located on the top surface of the chip. This design is more suitable for integration, but it does not support power capability. The lateral device does offer some important benefits of its own though: even lower forward capacitance (less than 5 pF at 1 MHz), very low reverse capacitance (less than 0.5 pF), ultra-fast turn-on speeds (less than 1 nS), and high operating frequency (more than 400 MHz). For these reasons, it is commonly used for video and high-speed analog switching, except that being harder to build, there are fewer vendors. Interestingly, all of the original MOS structures were lateral, and the very latest state-of-the-art high-speed DMOS devices are also lateral.

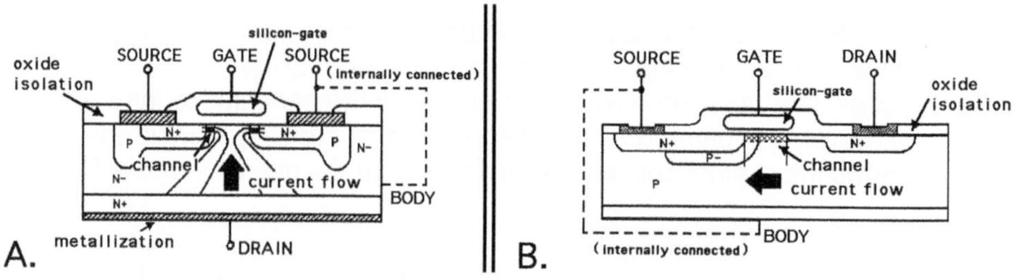

Figure 7.3. Comparison of the DMOS vertical structure (left), with the lateral structure (right).

Of the two kinds of discrete depletion-mode DMOS FETs, the most commonly available is the vertical type, which is made by Supertex and Vishay-Siliconix, but few others. They include products such as the Siliconix ND-2406L and the Supertex DN-3545. These devices have high input impedance, low input capacitance, low on-resistance, low leakage, low output conductance, fast switching, and I_{DSS} ratings between 100–200 mA, all in a plastic TO-92 or SOT-89 surface-mount package.

7.2 The importance of silicon-gate

Another notable difference between the JFET and the MOSFET is the DMOS FET's gate. While it functions similarly to the N-channel JFET, it does *not* have a reverse-biased P-N junction, but relies instead on an isolated, capacitive gate, made possible by either a *metal-gate* or a *silicon-gate*. Very simple models of this are shown in Figure 7.3. Biasing the gate to 0 volts (V_{GS} = 0V), just like with the JFET, induces full conduction through the isolated channel that runs beneath the gate. An increasingly negative bias at the gate will reduce conduction, until finally $V_{GS\,(off)}$ is reached, where conduction will cease. Such action is typical of any depletion-mode transistor.

In the late 1960s, Italian-born Dr. Federico Faggin, a Fairchild researcher at their R&D Labs in Palo Alto, California, created and developed a new technology that he called silicon-gate. Dr. Faggin made it repeatable and got it into production. Today, silicon-gate technology is used in many other semiconductor products, including analog and digital ICs, as well as certain discretes like MOS transistors. Without this key technology, the world would likely be trailing several years behind where we are today. While the earliest DMOS products (and other MOS products) used a type of metal-gate (a metal overlay deposited on top of an insulating oxide base), Dr. Faggin's device had a silicon-gate (aka polysilicon-gate or polycrystaline-silicon gate), which had an insulating oxide all around it. Typically, N-channel MOSFETs are built with an N^- type epitaxial layer grown on top of an N^+ substrate. Next, P^- regions are diffused into this epitaxial layer, before N^- regions are then diffused into the P^- regions. In fact, this is where the term *double-diffused* comes from. (The following steps originate from Dr. Faggin's design.)

The next stage has a very thin layer of silicon-dioxide (SiO_2) grown on top, which forms the gate isolation. This is followed by depositing a layer of polycrystaline-silicon doped with phosphorus, which forms the isolated silicon-gate. Once these stages are complete, the gate and source metallization is formed on the top surface of the chip, while the drain metallization is formed on the bottom surface. Although the gate is isolated from the rest of the chip by a very thin oxide layer, it is still vulnerable to ESD or voltage transients that exceed the usual ±20-volt limiting gate voltage. Should this occur, there is a good chance that the oxide layer will be damaged, by a form of punch-through, and the device rendered useless. For this reason, early models had built-in zener diode protection, although today's technology has improved the gate's vulnerability to ESD considerably.

Silicon-gate technology provides many bonuses for both the semiconductor maker and user alike, including the following:

- It is much easier to connect cells together, which could now be done by masking and diffusion, instead of the older method of metallization and bonding. With metal-gate technology, a mask is used in the final stages of processing that defines the size, shape, and position of the gate. With silicon-gate, the silicon is deposited before the gate is even created. Later the drain and source areas are doped, while the region between them is doped differently, to form a channel.
- In manufacturing, the use of a self-aligning silicon gate lends itself to much greater precision, stability, and repeatability, which leads to increased circuit density.
- It provides a cheaper, easier, and more efficient way of manufacturing these and other MOS devices (particularly CMOS). Lower manufacturing

costs helps create lower-priced products, with enhanced repeatability and reliability.

- Very importantly, it allows the same wafer fabrication process to be used for both analog and digital, something that was impossible with the older bipolar technology.
- Silicon-gate enables chips to be smaller, and thereby it reduces capacitance. This leads to increased switching speed and frequency response.
- Overall, it uses significantly less power.

Based on this technology, each of today's MOS manufacturers has subsequently added its own unique way of creating silicon-gate products, resulting in improved performance and better specifications than could ever be achieved with the original metal-gate or early silicon-gate technologies. Actually, most of the manufacturers mentioned throughout this book today use this technology. Supertex is one such stand-out, having created proprietary silicon-gate processes in which very low gate-voltages can be achieved simultaneously with high-voltage capability. This has led them to create many standard DMOS FETs, as well as custom products, such as DMOS drivers for EL display panels. Another company, which is discussed in a subsequent chapter on CMOS, is Advanced Linear Devices. They are specialists in very-low-voltage, silicon-gate analog ICs. Interestingly, both manufacturers are located in Sunnyvale, California, in the heart of Silicon Valley.

7.3 Characteristics of depletion-mode DMOS FETs

A totally unique feature of depletion-mode DMOS devices is that they can also be made to work in the enhancement mode. This is achieved by making the gate-to-source voltage (V_{GS}) slightly positive by a volt or two, which allows increased current levels beyond the normal I_{DSS} point, as seen in Figure 7.4. Although a JFET's gate can also be positively biased (N-channel) or negatively biased (P-channel), this is limited to only about 0.5 volt, otherwise the gate channel diode becomes forward-biased, and depending on the drain-to-gate voltage, the gate current can rise exponentially. If not current-limited, this can destroy the JFET. However, because the depletion-mode DMOS FET relies on an isolated capacitive gate, and not a gate channel diode, this is not a factor. Typically the absolute maximum rating for V_{GS} is usually limited to ±20 volts.

Because there is virtually no gate current, the DMOS FET has an extremely high input impedance (Z_{in}), typically more than 100 MΩ (more than 1×10^8). The output impedance of the DMOS FET depends on its gain, frequency, internal source resistance, and the value of the external source load resistor. It can range from a few ohms to more than 10 MΩ. DMOS FETs have a much lower drain-to-source resistance ($R_{DS(on)}$) than JFETs, a lower internal source resistance, and so a low output impedance. Because DMOS FETs are voltage-controlled, they have lower equivalent noise

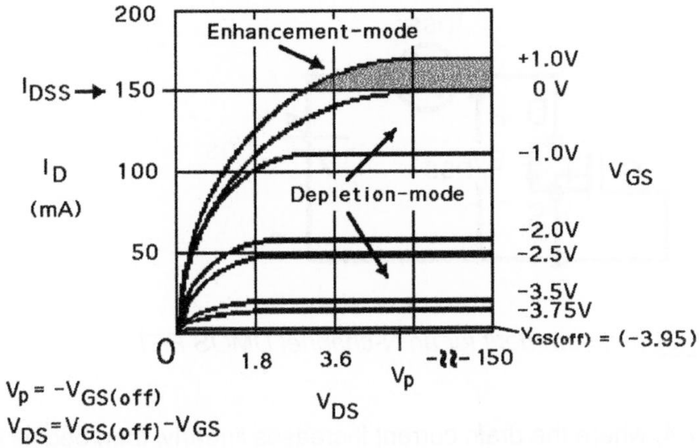

Figure 7.4. *Depletion-mode DMOS FETs provide ultra-stable I_D output levels in both depletion and enhancement modes, at higher levels than the JFET.*

characteristics (nV√Hz) than do most bipolars, which explains their popularity in VHF amplifiers. However, they do not usually have as low a noise characteristic as JFETs.

An important DMOS FET characteristic is the *drain-to-source breakdown voltage* ($V_{(BR)DSS}$), which is measured (as shown in Figure 7.5) with the FET's gate shorted to the source, at a specified gate current. This parameter is invariably specified at the beginning of an FET's data sheet. On some product data sheets, a manufacturer may refer to this as BV_{DSX}, whereas others refer to it as BV_{DSV}, but both terms mean the same as $V_{(BR)DSS}$. DMOS FETs are manufactured according to a particular process, much of which dictates the breakdown voltage of the device. As a result, DMOS FETs normally use the *same value for both* $V_{(BR)DSS}$ and $V_{(BR)DGS}$, although the latter is not usually specified other than in the absolute maximum ratings for V_{DG}. When considering using JFETs versus depletion-mode DMOS devices, the latter have much higher breakdown voltage ratings, making them eminently suitable in high-voltage circuits.

While the DMOS FET is controlled by a voltage at its gate, there is actually a tiny leakage current involved called the *gate-to-source/body leakage current* (I_{GSS}), typically in the range of 10 to 200 nA, at room temperature. Remember that with most commercially available depletion-mode (and enhancement-mode) MOSFETs, their source and the body regions are internally connected. I_{GSS} is a very small leakage current, usually measured in nano-amps, that applies at a specified V_{GS}. This leakage current doubles for every 10°C rise in temperature, which in some applications may be important.

As with the JFET, the maximum limiting current that can flow between the DMOS FET's drain and source is known as the *drain saturation current* (I_{DSS}). This occurs at a particular drain-to-source voltage (V_{DS}), when the V_{GS} is at zero. It can be seen in

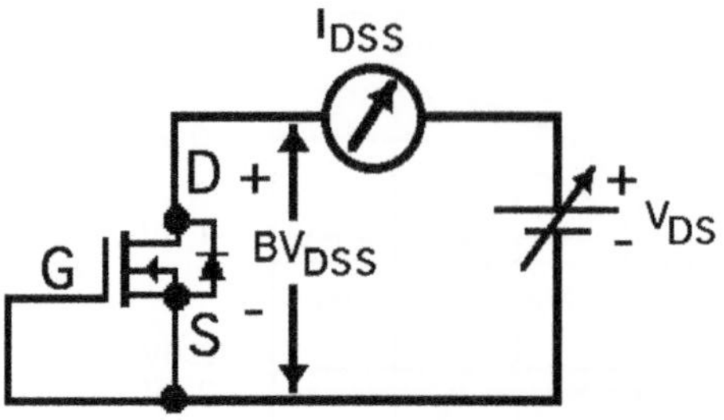

Figure 7.5. *BV_{DSS} measurement for an N-channel DMOS FET.*

Figure 7.4, where the drain current increases linearly, then begins to be pinched off or limited at the knee of the curve. This particular curve is depicted in most FET data sheets and can be easily measured, using the circuit shown in Figure 7.6. With a depletion-mode DMOS FET, this current is typically between 150 to 400 mA. It is important to remember that I_{DSS} may typically range over 3:1 for similar devices, and one should consider both the minimum and maximum ratings/curves in the range. I_{DSS} is also temperature sensitive and has a negative temperature coefficient of approximately –0.5%/°C.

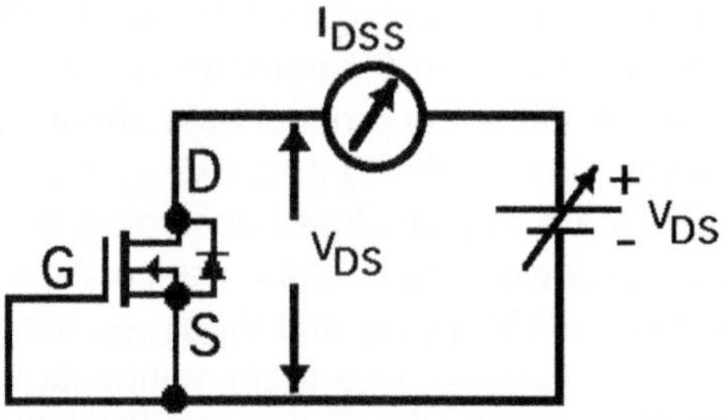

Figure 7.6. *I_{DSS} measurement for an N-channel DMOS FET.*

Some manufacturers specify $I_{D(off)}$ on their DMOS FET product data sheets. $I_{D(off)}$ is the amount of *leakage* between the drain and source. It is usually specified at a particular V_{DS} and V_{GS}. $I_{D(off)}$ is usually shown on a data sheet in micro-amps (μA), measured at room temperature, an elevated temperature (such as +125°C), and at the maximum drain-to-source rating of the device.

Another characteristic of the DMOS FET is the static *on-state resistance,* $R_{DS\ (on)}$. For small DMOS FETs it is often less than 50 Ω, whereas for a power device it can reach milli-ohm levels. It is measured as the V_{DS}, divided by a particular drain current, at a particular V_{GS}, at both 25°C and an elevated temperature. For a DMOS FET it has a positive temperature coefficient of approximately +1%/°C. Because of the device self-heating, $R_{DS\ (on)}$ is usually measured in a pulsed manner at room temperature, because heating the device decreases carrier mobility, thereby reducing the drain current for a given voltage.

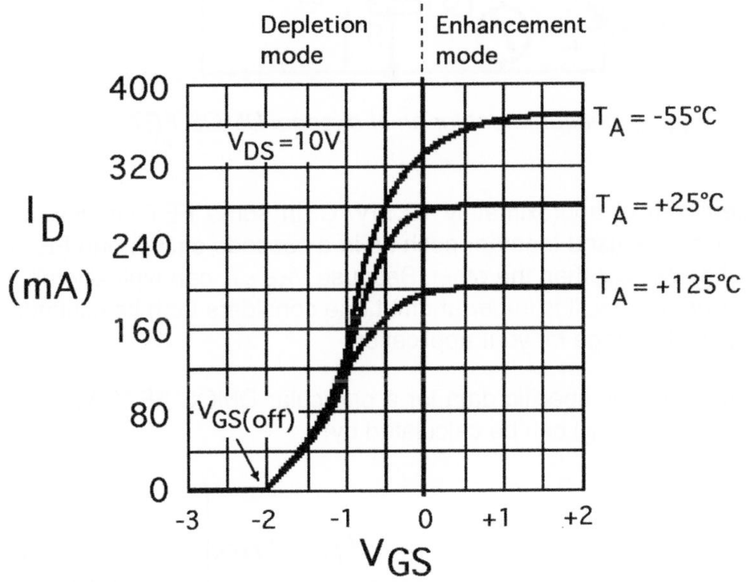

Figure 7.7. *Transfer characteristics for a typical DMOS depletion-mode FET.*

Just as with the JFET, a key characteristic is $V_{GS\ (off)}$, the *gate-to-source cutoff voltage* necessary to turn the FET off. The V_{GS} for an N-channel DMOS FET ranges from 0 volt for full conduction to some negative amount of typically several volts to turn the device off. This is shown in Figures 7.4 and 7.7. Discrete DMOS FETs usually have quite a low $V_{GS\ (off)}$ value so that they are logic-level compatible with both TTL and CMOS. For example, a Vishay-Siliconix ND 2410L has a $V_{GS\ (off)}$ of –0.5 to –2.5 volts, while a Supertex DN3135 has a $V_{GS\ (off)}$ of –1.5 to –3.5 volts. Conduction ceases when the V_{DS} is equal to the *pinch-off voltage* (V_p), so that only V_{GS} controls conduction through the channel. While V_p is a relatively low positive value, the value of $V_{GS\ (off)}$ will be the same value, but of opposite polarity. Figure 7.7 clearly shows the drain current curves versus V_{GS}, from $V_{GS} = 0$ through $V_{GS\ (off)}$, for a typical depletion-mode DMOS FET. The $V_{GS\ (off)}$ transfer curve is depicted in most DMOS FET data sheets and can be easily measured, using the circuit shown in Figure 7.8.

However, the data sheet for your chosen device may not necessarily portray the appropriate curves data in exactly the same manner, and you might have to resort to measuring devices and plotting your own curves. As you can see in Figure 7.7, I_{DSS} and $V_{GS\ (off)}$ both shift with temperature. Actually, $V_{GS\ (off)}$ has a negative temperature

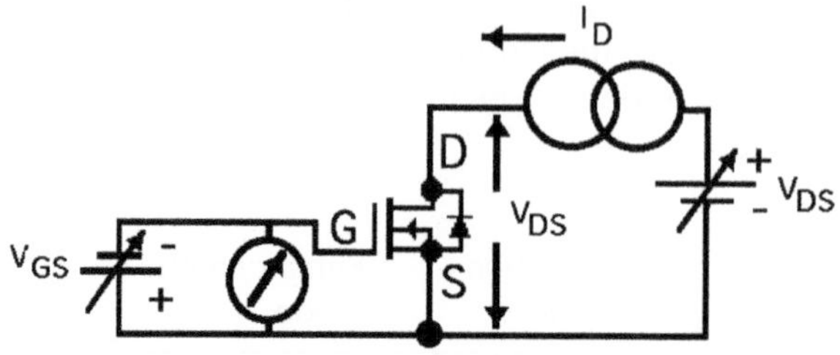

Figure 7.8. *$V_{GS(off)}$ measurement for an N-channel DMOS FET.*

coefficient of approximately –2 mV/°C. In some FET circuits, particularly when two devices are used together such as in a cascode circuit, one FET may need to have a higher $V_{GS\ (off)}$ than the other. Because $V_{GS\ (off)}$ can typically range over 3:1 or 4:1 for similar devices, it is important that one considers both its minimum and maximum ratings in the range for your application.

Once one has specific data for a particular DMOS FET's $V_{GS\ (off)}$, V_{GS}, and I_{DSS}, its drain current (I_D) can be calculated by:

$$I_D = I_{DSS}\left(1 - \frac{V_{GS}}{V_{GS(off)}}\right)^2 \qquad \text{(Eq.7.1)}$$

Another important characteristic of the DMOS FET is its *transconductance/forward conductance* (g_{fs}), or small-signal common-source forward transconductance (aka g_m and Y_{fs}). G_{fs} measures the effect of a change in drain current (I_D), for a specific change in gate voltage (V_{GS}), referenced to common-source mode.

$$g_{fs} = \frac{\Delta I_D}{\Delta V_{gs}} \mid V_{DS} = \text{constant} \qquad \text{(Eq.7.2)}$$

In practical terms, transconductance is a measure of the FET's gain, or ability to amplify, and is a figure of merit. It is usually referred to from a practical standpoint as so many mA per volt. On a graph it appears as a steep or shallow slope and is measured in milli-Siemens (mS). It is usually specified on the data sheet for most FETs with a minimum, typical, or maximum value of g_{fs}. The g_{fs} value given in some manufacturers' data sheets is measured at full conduction (I_{DSS}), where g_{fs} is at its maximum. Other manufacturers specify g_{fs} at a particular I_D, V_{DS}, and frequency, usually at 1 KHz or 100 MHz.

Yet another important characteristic of the DMOS depletion-mode FET is its *output impedance* (Z_{out}), which can range from a few ohms to more than a Meg ohm. As mentioned previously, this depends on the FET's gain (g_{fs}), input frequency, internal source resistance, and the value of the external source load resistor. This often just approximates to the value of its internal source resistance (r_S), compared with the value of the external source resistor (R_S), which is usually much larger. This internal source resistance forms a voltage divider with any external source load, such that:

$$r_s = \frac{1}{g_{fs}} \qquad (Eq.7.3)$$

The output impedance is given by:

$$Z_{out} = \frac{r_s R_s}{r_s + R_s} \qquad \text{(Eq.7.4)}$$

where r_s is the FET's internal source resistance; g_{fs} is the transconductance or gain; Z_{out} is the output impedance in ohms; and R_S is the value of the external source resistor. The output impedance is fairly close to the value of the internal source resistance, which in turn depends on the transconductance.

7.4 Depletion-mode DMOS current sources

Excellent current sources can be built using depletion-mode DMOS devices in a similar configuration and manner to that previously described for single JFETs. Only simple biasing is required. They typically offer practical low-to-medium current levels (5–50 mA), over a wide range of drain-to-source voltages ($+V_{DS}$) and gate-to-source voltages ($-V_{GS}$). Unlike the more common N-channel enhancement-mode MOSFET, whose drain current (I_D) falls to zero when V_{GS} equals 0, the depletion-mode device can be considered as working in reverse to that and has close to maximum current at zero gate voltage. In fact, its drain-to-source resistance ($R_{DS\,(on)}$) is very low when V_{GS} equals 0 volt. Normally, the V_{GS} is some value ranging between 0 volt (full conduction) and –5 volts ($V_{GS\,(off)}$). Depletion-mode devices also have a reasonably low output conductance (g_{os}) of less than 100 μs, which is close in value to many good JFET current sources.

Using the depletion-mode DMOS device in creating a constant current source is theoretically possible in either the depletion or enhancement mode, depending on the circuit's available voltage levels and polarities. In most cases, however, in order to optimize regulation, the device would be operated only in the depletion mode, at a low current level. As a result, the V_{GS} would be negative ($-V_{GS}$), just as with the JFET. Remember that the lower the current (I_{OUT}), compared with I_{DSS}, the better the regulation. Using the device in the enhancement mode forces it to run at full throttle, with current levels beyond I_{DSS}, and as a result regulation is significantly degraded.

Although DMOS FETs have higher voltage and current capabilities than JFETs by virtue of their processing, they can still operate at low voltages. For example, a Vishay-Siliconix ND-2020L with a V_{GS} of –1.2 volt and a V_{DS} of between 2 to 200 volts can provide a constant current (I_{OUT}) of approximately 10 mA. Making the drain current (I_D) quite low (e.g., less than 5% of I_{DSS}) will ensure that regulation is fully optimized and held to better than 1%. Smaller percentages of I_{DSS} will yield even tighter regulation, but this will require low currents. For example, a Supertex DN3125 is rated for a minimum I_{DSS} of 300 mA. This would mean that with a constant current level of below 15 mA (5% of $I_{DSS\,(min)}$), regulation could be expected to approach 0.5%.

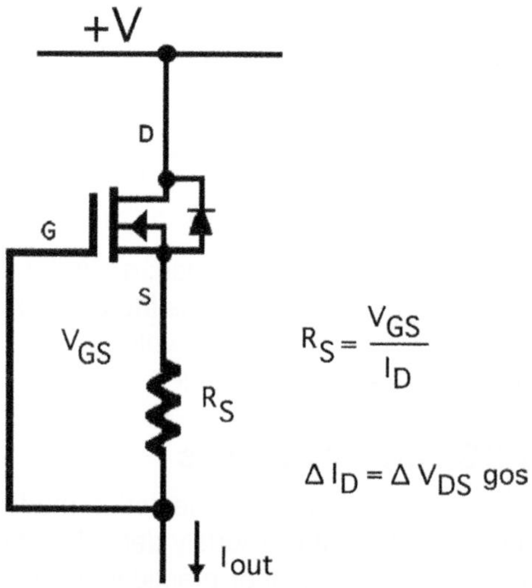

Figure 7.9. N-channel depletion-mode DMOS current source.

As shown in Figure 7.9, the depletion-mode DMOS FET may be self-biased with a resistor inserted between its gate and source. It will operate as a constant current source at any level below I_{DSS}. However, the lower the drain current (I_D), the tighter the regulation will be. The bias resistor (R_S) required between the gate and source can be determined by:

$$R_S = \frac{V_{GS}}{I_D} \qquad \text{(Eq.7.5)}$$

In most cases, it is easier to use a variable potentiometer to set the desired current level. As an example, if we required a constant current of I_D of 10 mA, and having a DMOS FET with an I_{DSS} of 200 mA and a V_{GS} of –5.6 volts, then the value of R_s would equal 560 Ω. (Use a 1-KΩ potentiometer.) A change in either the supply voltage or the load impedance will cause only a very small change in current as a result of the output conductance (g_{os}), as shown by:

$$\Delta I_D = \Delta V_{DS} g_{os} \qquad \text{(Eq.7.6)}$$

In the same example, if the particular DMOS FET device had a g_{os} value of 100 μS, a constant current of I_D of 10 mA, and a ΔV_{DS} of 500 mV change occurred, then the output current would change by only 50 μA (0.005%), which is pretty good.

Table 7.1 lists applicable depletion-mode vertical DMOS FETs from Supertex and Siliconix (Vishay). These are mostly low-cost devices and are available in a wide range of packages, including metal-can, plastic, and surface-mount, as well as in various I_{DSS}, V_{DS}, and $V_{GS(off)}$ ratings.

Part No.:	Manufacture	Package:	V	Rds(on)	Vgs(off)
ND2012L	Vishay-Siliconi	TO-92	200 V	12	-1.5 to -4
ND2020L	Vishay-Siliconi	TO-92	200 V	20	-0.5 to -2.5
ND2406L	Vishay-Siliconi	TO-92	240 V	6	-1.5 to -4.5
ND2410L	Vishay-Siliconi	TO-92	240 V	10	-0.5 to -2.5
BSS129	Vishay-Siliconi	TO-18	230 V	20	-0.7 min
DN2530	Supertex	TO-92/SOT89	300 V	12	-1 to -5
DN2535	Supertex	* See Note	350 V	25	-1.5 to -3.5
DN2540	Supertex	* See Note	400 V	25	-1.5 to -3.5
DN2620	Supertex	TO-92	200 V	4	-1 to -3
DN2624	Supertex	TO-92	240 V	4	-1 to -3
DN2640	Supertex	TO-92	400 V	6	-1 to -5
DN3125	Supertex	SOT-23/89	250V	20	-1.5 to -3.5
DN3135	Supertex	SOT-23/89	350V	35	-1.5 to -3.5
DN3145	Supertex	SOT-89	450	60	-1.5 to -3.5
DN3525	Supertex	SOT-89	250	6	-1.5 to -3.5
DN3535	Supertex	SOT-89	350	10	-1.5 to -3.5
DN3545	Supertex	SOT-89	450	20	-1.5 to -3.5

* **NOTE:** Available in TO-39, TO-92, TO-220, and SOT-89 packages. Contact the manufacturer for more details.

Table 7.1 List of depletion-mode vertical DMOS FETs

7.5 The cascode DMOS current source

As previously shown with JFETs, much improved regulation can be achieved by using *two* N-channel DMOS FETs in a *cascode* configuration, as shown in Figure 7.10. However, remembering that Q_2, the cascode FET, needs to have higher $V_{GS\ (off)}$ and I_{DSS} values than Q_1 makes it difficult to specify suitable plug-in devices without first preselecting them for $V_{GS\ (off)}$. This is because available devices from Supertex and Vishay-Siliconix have three basic categories to choose from: (1) the low $V_{GS\ (off)}$ value runs from −0.5 to −2.5 volts; (2) the midlevel runs from −1.5 to −3.5 volts; and (3) the highest $V_{GS\ (off)}$ value runs from −1.5 to −4/4.5 volts. This means that while using a device from the lowest category, one may have to hand-select a device for the cascode, which should have a higher $V_{GS\ (off)}$ value, preferably greater than −3 volts. For low-voltage circuits around, say, 5 or 6 volts (or higher), one could probably use this cascode, while below that, around say 3.5 volts, a single DMOS current source can provide a good solution.

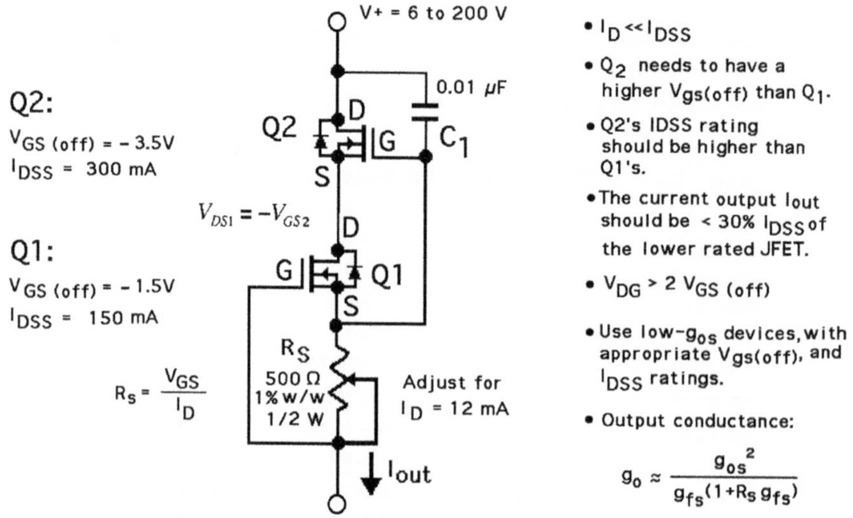
V+ = 6 to 200 V
0.01 µF
C1
Q2:
VGS (off) = - 3.5V
IDSS = 300 mA
Q2
D
G
S
VDS1 = -VGS2
Q1:
VGS (off) = - 1.5V
IDSS = 150 mA
D
G
Q1
S
RS
500 Ω
1% w/w
1/2 W
RS = VGS / ID
Adjust for
ID = 12 mA
Iout
• ID << IDSS
• Q2 needs to have a higher Vgs(off) than Q1.
• Q2's IDSS rating should be higher than Q1's.
• The current output Iout should be < 30% IDSS of the lower rated JFET.
• VDG > 2 VGS (off)
• Use low-gos devices, with appropriate Vgs(off), and IDSS ratings.
• Output conductance:
go ≈ gos² / gfs(1+Rs gfs)

Figure 7.10. *An ideal depletion-mode DMOS cascode current-source.*

Typically, cascoding improves high-frequency operation, provides even greater output impedance (Z_{out}), increases high-voltage operation or compliance, and significantly lowers the circuit's output conductance. In the circuit of Figure 7.10, the Q_2 cascode buffers the current source Q_1 from the load. So that while the source voltage of Q_2 remains constant, variations in the load voltage are accommodated by Q_2's drain. This results in the voltage drop across the current source remaining constant.

Here the drain current (I_D) is regulated by Q_1 and R_S, so that V_{DS1} equals $-V_{GS2}$. As mentioned previously, both DMOS FETs must be operated with adequate drain-to-gate voltage (V_{DG}), such that V_{DG} is preferably more than 2 $V_{GS\ (off)}$, or else the circuit's output conductance will increase significantly. FET Q_2 needs to have a higher $V_{GS\ (off)}$ than FET Q_1. A low g_{os} per I_D is achieved, because of the degenerative feedback formed by the pair. The circuit's output conductance (g_o), also caused by the forward transconductance (g_{fs}), is much less than the g_{os} of a single DMOS FET—in fact, at least 10 times lower. You can also expect the circuit to have a very high output impedance, and possibly reach GΩ levels. R_S sets the desired current, but in early development it could be substituted by a good-quality multiturn trimmer potentiometer. This would allow easy adjustment of the desired current level, which could range from a few milliamps to a few tens of milliamps. It can be used equally well in circuits of more than 200 volts or below 10 volts, depending on one's choice of $V_{GS\,(off)}$ and other device specifications. In the cascode shown, it provides an internal circuit impedance of more than 10 MΩ at a current level of 10 mA. Capacitor C_1 is included for circuit stability, but its voltage rating should be chosen according to the maximum circuit voltage. This simple and inexpensive circuit can be built using either throughhole or tiny surface-mount devices. It can be used as either a precision current source or sink, with the load connected to either terminal.

7.6 The JFET-DMOS cascode current source

Although it is possible to substitute a suitable N-channel JFET for Q_1, the circuit as shown in Figure 7.11 will produce a lower current, but with a lower output conductance (g_o), because the JFET's g_{os} is lower compared to that of most DMOS FETs. Because of the DMOS FET's high $V_{(BR)DSS}$, the circuit will operate over a much higher voltage range than an all-JFET cascode. That is the main reason for this circuit, which otherwise falls between the performance of the JFET cascode and the DMOS cascode. The simple circuit shown uses TO-92 plastic devices, but surface-mount devices could easily be substituted. Capacitor C_1's voltage rating should be chosen according to the maximum circuit voltage, which could be in excess of 400 volts.

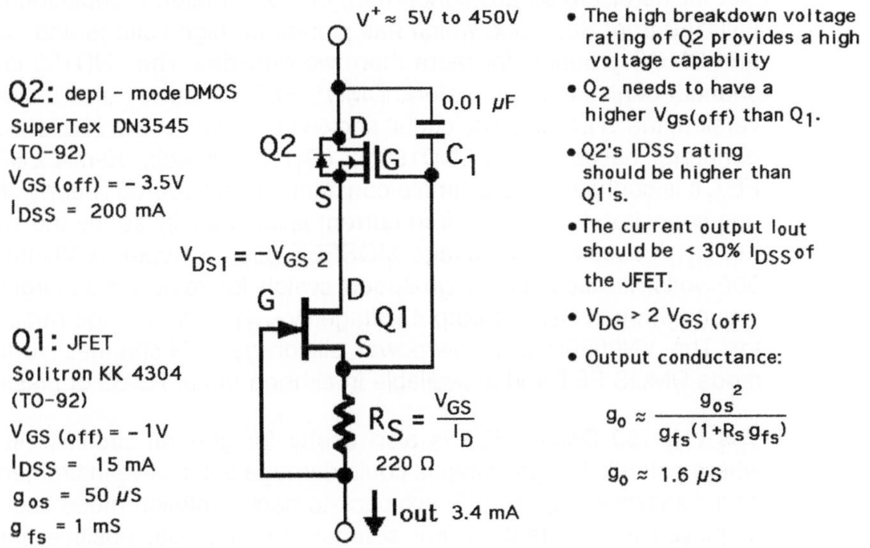

Figure 7.11. *Combining an N-channel JFET with a depletion-mode DMOS FET creates a practical, high-voltage, cascode current source.*

7.7 Lateral depletion-mode DMOS FETs

Changing the DMOS FET's geometry and processing can help enhance or exploit other features, such as switching speed. Although more difficult to manufacture, and the least available member of the FET family, the high-speed, low-voltage, lateral depletion-mode DMOS FET (LDMOS) features low distortion, a wide dynamic range, and ultra-high-frequency operation with turn-on speeds of less than 1 nS. This device can be considered a bridge between high-speed JFETs and gallium-arsenide (GaAs) FETs. The device's main advantage is the very low capacitance (less than 10 pF), which makes it eminently suitable as a high-speed switch. It is potentially suitable for biasing and current source applications in high-speed (50 volt/μS) designs, and where an improved circuit slew rate may be required.

Perhaps future technology will make this particular member of the DMOS FET family easier to manufacture and with even further improvements in performance. Since the beginning of the 21st century, lateral DMOS has already begun to gain on gallium-arsenide and is already ahead of silicon germanium (SiGe) as the main technology used to make the latest RF transmitter and receiver modules in cellular phone base stations. The lateral structure offers higher efficiency and better linearity in the 1- and 2-GHz region than competing bipolar devices can presently provide. Motorola and Philips Semiconductors are just two of several leading developers in this area, creating ICs based on advanced LDMOS technology.

As far as discretes are concerned, one such lateral, depletion-mode device is the LND150 from Supertex, which has pioneered high-voltage and low-threshold DMOS and CMOS products for more than two decades. The LND150 is a high-voltage, N-channel depletion-mode, lateral DMOS FET. (Supertex also makes a surface-mount version, the LND250.) The circuit shown in Figure 7.12 is of a low-cost 400 VDC ramp generator based on the LND150N3 (Q_1), a 500-volt, 30-mA ESD-protected DMOS FET. It is configured as a simple constant current source linearly charging a high-voltage capacitor, C_1. The desired current level is easily set by the variable potentiometer, R_1. A second high-voltage MOSFET (Q_2), a Supertex VN0550N3, also rated at 500 volts, is used as a high-speed switch for rapidly discharging the capacitor to ground through R_2. The output voltage (V_{OUT}) is the voltage ramp across the capacitor. The VN0550N is a low-power silicon-gate, N-channel, vertical enhancement-mode DMOS FET and is available in either a metal TO-39 or plastic TO-92 package.

The LND150 DMOS FET is also useful for general circuit protection applications, where potentially high voltages could damage a sensitive instrument's front end. In the circuit shown in Figure 7.13, two back-to-back depletion-mode lateral DMOS FETs are employed as constant current sources, to limit both positive and negative voltage excursions to safe values, at low current. The resistor, R_1, safely limits the current to 1 mA in either direction, but its wattage rating should be considered. The power dissipation in each LND 150 should never exceed 600 mW (max. input voltage = 500 × 1 mA = 500 mW). A nice bonus is that the LND150 DMOS FET also has built-in gate-to-source ESD protection, making it an exceptionally rugged device.

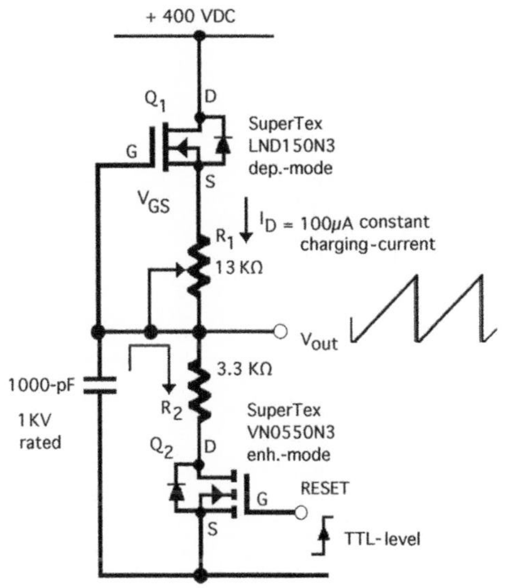

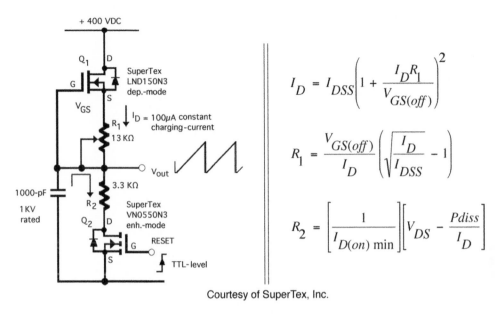

$$I_D = I_{DSS}\left(1 + \frac{I_D R_1}{V_{GS(off)}}\right)^2$$

$$R_1 = \frac{V_{GS(off)}}{I_D}\left(\sqrt{\frac{I_D}{I_{DSS}}} - 1\right)$$

$$R_2 = \left[\frac{1}{I_{D(on)\,\min}}\right]\left[V_{DS} - \frac{Pdiss}{I_D}\right]$$

Courtesy of SuperTex, Inc.

Figure 7.12. A high-voltage ramp generator using the SuperTex LND150 lateral depletion-mode DMOS FET as a constant current source to charge a timing capacitor. The RESET function is controlled by enhancement-mode MOS_{FET}, Q_2.

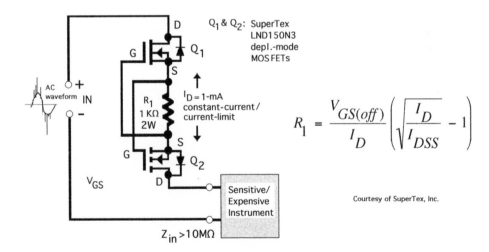

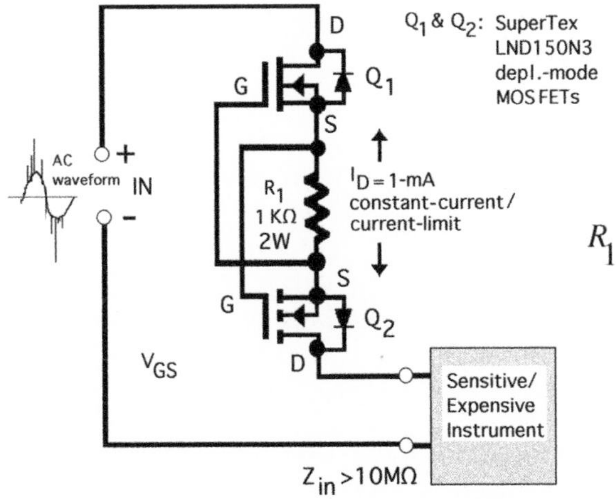

$$R_1 = \frac{V_{GS(off)}}{I_D}\left(\sqrt{\frac{I_D}{I_{DSS}}} - 1\right)$$

Courtesy of SuperTex, Inc.

Figure 7.13. High voltages could damage a sensitive instrument's front-end input. By using two Supertex lateral, depletion-mode LND150 DMOS FETs as back-to-back current sources, full protection is provided against damaging high-voltage spikes. The resistor sets the current level at 1 mA for voltages up to ± 500V-peak.

Chapter 8

Creating Current Sources with Power MOSFETs

Although the first metal-gate MOSFET had been developed at Bell Labs in 1960, the FET was considered all but impossible to make by those who tried, because it was so unstable. So it was not until 1963 when two young RCA researchers, Steve Hofstein and Fred Heiman, made the first stable and working devices. By the mid-1960s, RCA had some of the first small, metal-can MOSFETs in production and in its own consumer products. In the early 1970s, DMOS, which had been developed in Japan, became available in the United States. It was to become the foundation for subsequent power MOSFET development for the next three decades.

Then in 1976, Siliconix announced MOSPOWER®, the world's first commercially available vertical power MOSFETs with their distinctive V-groove, which was etched into the chip's structure. Up until then, MOSFETs had a lateral structure and were limited to power levels of less than 1 Watt. In comparison, with this new type, current flowed through the device vertically. It was capable of faster switching and at power levels of several amps. Unfortunately, the V-groove design was found to create a high-energy field at its point, which was hard to control, and it created stability problems. This in turn led to a truncated structure (without a sharp tip at the bottom of the V-shape), which improved operation somewhat. Although these devices were difficult and expensive to build, as well as requiring an ultra-clean processing environment, they revolutionized power MOSFET technology from then on. The same year, Japan's Hitachi introduced the world's first complementary power MOSFETs on the same chip, using a proprietary planar, lateral structure. Some of these were used in Hitachi's own consumer audio amplifiers. These breakthroughs were followed in 1979 by International Rectifier, which introduced a competing MOS structure that it dubbed the HEXFET®. This design used thousands of paralleled, hexagonally shaped *cells* on a single FET chip. This dramatically reduced the on-resistance, $R_{DS(on)}$, and enabled switching even higher currents.

Since those early days, continued R&D of silicon power MOSFET technology has been taking place around the world, and *enhancement-mode* devices have become increasingly popular. Common applications that have driven further improvements in power MOSFET technology have included switch-mode power supplies (SMPS), as well as power switches, disconnect safety switches in battery chargers, and amplifiers. The evolution of the personal computer (both desktop and laptop models) has created an enormous market for SMPS, which in turn rely heavily on magnetics and power MOSFET technology. Now, nearly 25 years after their introduction, MOS devices are made around the world by many leading companies and consist of both low-power and high-power types. The investment in semiconductor processing equipment, however, is a major consideration when building power MOSFETs. Two basic types have evolved: the more common *planar* type and the newer, more advanced *trench* type, which requires even greater investment. MOSFET manufacturers include International-Rectifier, Vishay-Siliconix, Fairchild Semiconductor, ON Semiconductor, Calogic, Philips Semiconductors, IXYS, Infineon, Powerex, ST Microelectronics, Solitron Devices, Supertex, and others.

Of the two kinds of enhancement-mode MOSFETs, the most popular and commonly available is the *vertical* type. This group is typically split into either low-power switches or high-power devices. Of all types of discrete semiconductors, vertical-type MOSFETs are part of one of the most aggressive market sectors, with intense worldwide competition. While also being the most popular of any type of FET, the technology surrounding vertical types is still evolving in the early years of the 21st century. Most of the companies mentioned here have developed proprietary MOS manufacturing processes to support their marketing areas or developed market niches based on their proprietary technologies. In some cases, manufacturers have developed several different types of proprietary process, or several generations of a particular process have evolved over time. Some examples include Vishay-Siliconix SiMOS 2.5™, LITTLE FOOT™, LITEFOOT™, IXYS HiPERFET™, and MegaMOS™; and Philips Semiconductors' TOPFET™ and μTrenchMOS™. These all encompass different features, such as improved high-voltage operation, reduced low-threshold gate voltage, avalanche voltage protection, logic-level compatibility, ultra low on-resistance, increased ESD protection, high-speed operation, fast reverse recovery, improved dv/dt performance, packages that are smaller or more efficient, changes in die size and geometry, and even higher on-chip cell densities.

8.1 Characteristics of enhancement-mode MOSFETs

Unlike depletion-mode FETs, enhancement-mode MOSFETs are *normally-off* devices and are available as either P-channel or N-channel types. In the case of N-channel types, a positive voltage between the gate and source ($+V_{GS}$) is required to turn them on. For the P-channel device, a negative gate voltage with respect to the source is required to turn it on ($-V_{GS}$). With either polarity of MOSFET, the drain current (I_D) falls to zero, when V_{GS} equals 0. This is directly opposite to depletion-mode FETs, where a negative V_{GS} turns off an N-channel device and a positive V_{GS} turns off a P-channel device. Remember that enhancement-mode MOSFETs include both low-power types (such as the popular 2N7000 switch, in a TO-92 plastic package) and the bigger high-power types (in metal, plastic, and surface-mount packages).

The low-power devices use smaller chips, have lower capacitances, and as a result are typically faster (10-nSec turn-on times), while having lower current and voltage capabilities. Power devices have much lower $R_{DS(on)}$ thus a higher current rating, but have a higher junction capacitance. This results in a slower switching speed (typically greater than 200-nSec turn-on time), when compared to low-power devices. Even so, when compared to bipolar power transistors, the enhancement-mode power MOSFET is generally superior. One major characteristic is that the MOSFET is a *majority-carrier* device, which means that it does not suffer from the minority-carrier storage time effect like bipolars do, thereby switching faster. It requires infinitely less gate drive current than a bipolar's average base current, and it has a significantly faster switching time—at least five times faster. The separate driver transistor needed to drive a large bipolar power transistor is not usually needed with power MOSFETs, unless level-shifting, or driving highly capacitive loads, or very fast switching is required. It also has a more rugged and rectangular safe operating area (SOA) curve, enabling it to be used at up to the rated breakdown voltage and maximum drain current (under pulsed conditions). Bipolars are much less forgiving.

The MOSFET also has the ability to be easily paralleled, as well as requiring a smaller heatsink, because of the very low on-resistance values now available (less than 10 milli-ohms). The resulting lower voltage drop, $V_{DS(on)}$, across the FET translates into much less heat needing to be dissipated from the chip. Last but not least, the power MOSFET's $R_{DS(on)}$ has a positive temperature coefficient, meaning that as the MOSFET heats up, its resistance also increases, thereby helping to limit the current through it. This action reduces the possibility of *thermal runaway* as the chip's junction temperature increases—something that the bipolar power transistor simply cannot match.

The symbols, polarities, and large-signal models for both P- and N-channel MOSFETs are shown in Figure 8.1, which is almost the same as the depletion-mode MOS device (the gate is depicted differently). The P-channel MOSFET device shown in Figure 8.1C and D has all of its currents and polarities reversed. Power MOS devices have a similar (although opposite) way of operating and share some of the same characteristics and terminology as the depletion-mode DMOS FET that we looked at in the previous chapter (i.e., $V_{(BR)DSS}$, g_{fs}, $I_{D(off)}$, and I_{GSS}). There the similarity ends, because only some of the same equations and models covering the N-channel depletion-mode DMOS FET can be used for enhancement-mode devices.

While a typical power MOSFET transistor is a single chip, its architecture is actually composed of thousands of densely packed, hexagonally shaped *cells* (or other geometric shapes, like squares or circles), whose sources are paralleled together. This is known as the *source geometry*. Be aware that because differing processes and technologies are involved, there are two basic types of vertical power MOSFET: the planar type or the more advanced trench type, which provides a much higher cell density. Separate metallization effectively parallels all of the gates, sources, and drains together. The source geometry and the space between sources is an important factor in determining the cell density. The number of *cells* can vary from many hundreds of thousands to tens of million *cells* per square inch (cell density) and varies according to a chip's dimensions. Current trench technology exceeds 30 million *cells* per square inch. Scaling techniques similar to Very Large Scale Integration (VLSI), which are

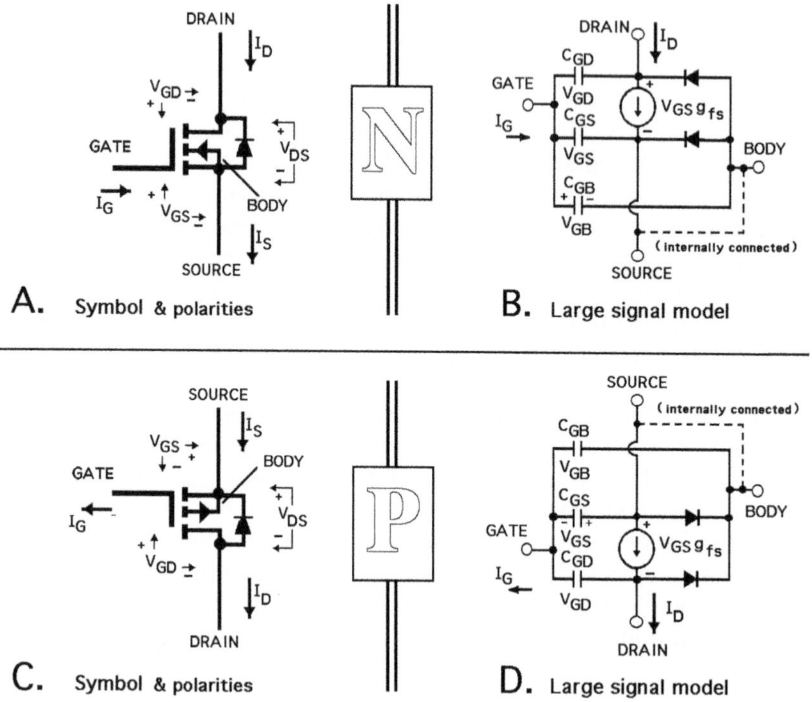

Figure 8.1. *Showing the symbols for both P- and N-channel, enhancement-mode MOSFETs as well as various voltages, currents, polarities, and models. Note that the body teminal is normally tied internally to the source.*

used in creating complex ICs like microprocessors and dense memory chips, are used in fabricating today's power MOS devices. To package this many individual cells together depends on the FET's structure, precision lithography and processing techniques, as well as a lot of manufacturing experience. This very high cell density is responsible for the MOSFET's very low $R_{DS(on)}$. A cross-sectional view of a simplified MOSFET cell is shown in Figure 8.2.

As mentioned previously, enhancement-mode MOSFETs are normally-off devices, requiring either a positive voltage (N-channel types) between the gate and source to turn them on or a negative gate voltage (P-channel types) to turn them on. For either type, biasing the gate to zero volts (V_{GS} = 0V) will reduce conduction, until finally a threshold point is reached where conduction ceases. This is a major difference from the depletion-mode device, in that the enhancement-mode device works in reverse. Like the depletion-mode DMOS FET, though, it relies on an isolated, capacitive gate, made with either a metal- or a silicon-gate. An N-channel MOS device normally has a very thin oxide layer that isolates the gate from the P-region below. Because the gate is isolated from the rest of the device, it usually has ±20-volt limiting gate voltage. By

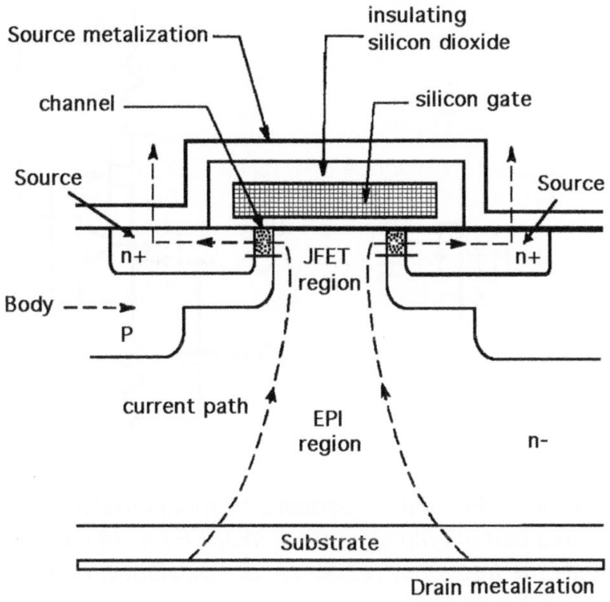

Figure 8.2. Cross-sectional view of a typical vertical DMOS FET cell.

applying a positive drain-to-source voltage of, say, 20 volts, and biasing the gate to, say, 10 volts also in the positive direction (1) repels the holes away from the surface of the P-region near the gate. This action means that the electrons are now the majority carriers by default, which (2) causes an *inversion* of the channel (aka surface inversion). (3) This allows electrons to flow in the channel, which induces full conduction between the drain and source. The current flow is vertical as it flows from the the drain, then turns abruptly horizontal through the channel. Figure 8.2 shows this sequence clearly. These actions are representative of any N-channel enhancement-mode FET.

P-channel devices work in reverse, so that by applying a negative drain-to-source voltage of several volts and biasing the gate to several volts also in the negative direction repels the electrons from the N-region near the gate, which allows holes (the majority carriers) to flow in the isolated channel that forms. Again, this induces full conduction between the drain and source. Because P-channel devices are less efficient, they require greater chip area to match the characteristics of their N-channel counterparts, and as a result are usually more expensive.

Although one might assume that the typical power MOSFET is a relatively simple device, it is not, as is depicted in a simplified form Figure 8.3. It contains several inherent *parasitic elements*, which include a bipolar transistor, a JFET, substrate diodes, and various inductances, resistances, and several important capacitances. The vari-

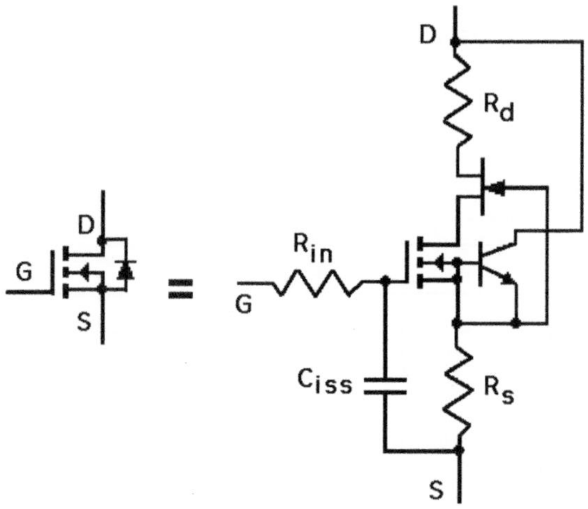

Figure 8.3. *Showing some of the intrinsic parasitic elements that are present in a typical N-channel enhancement-mode, vertical MOSFET's structure. These elements interact to dictate the frequency response, on-resistance, current capability, voltage range, and switching speed.*

ous proprietary processes mentioned previously attempt to skillfully reduce these unwelcome parasitic characteristics, while exploiting others. The end product is a compromise between many conflicting characteristics (e.g., high voltage with low current capability, versus low voltage and a high current capability).

An important power MOSFET characteristic regarding its maximum operating voltage is the *drain-to-source breakdown voltage*, $V_{(BR)DSS}$. This represents the lower limit of the device's voltage blocking capability and is measured (as shown in Figure 8.4) with the FET's gate shorted to the source, at a specified low gate current. This parameter is invariably specified at the beginning of a FET's data sheet. MOSFETs are manufactured according to a particular process, much of which dictates the breakdown voltage of the device. As a result, MOSFETs normally use the same value for both $V_{(BR)DSS}$ and $V_{(BR)DGS}$ (the drain-to-gate voltage), although the latter is not usually specified other than in the absolute maximum ratings for V_{DG}. $V_{(BR)DSS}$ has a positive temperature coefficient of 0.1%/°C.

While the MOSFET is controlled by a voltage at its isolated gate, there is actually a very small leakage current involved, called the gate-to-source/body leakage current (I_{GSS}). It is typically in the range of 10 to 200 nano-amps, when measured at room temperature, and at a specified V_{GS}. This is measured with the drain terminal shorted to the source. Remember that with most commercially available enhancement-mode MOSFETs, their source and the body regions are internally connected together. This

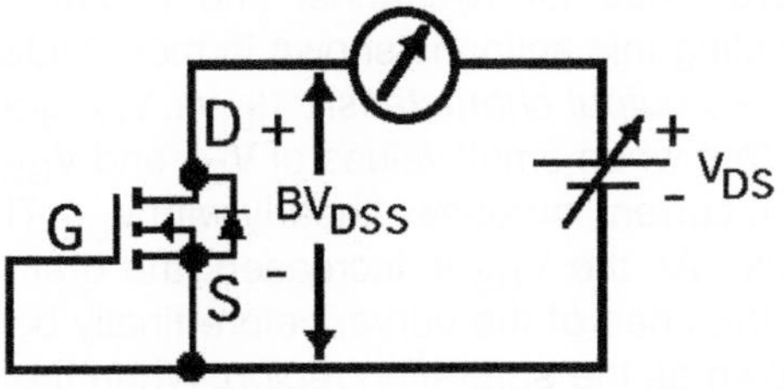

Figure 8.4. *BV_{DSS} measurement for an N-channel power MOSFET.*

leakage current doubles for every 10°C rise in temperature, which in some applications may be important.

Another *major difference* between depletion-mode devices and enhancement-mode devices is the characteristic I_{DSS}. With depletion-mode devices, I_{DSS} refers to the maximum current that will flow (full conduction) between drain and source, when the gate is shorted to the source. With the enhancement-mode device, it means something totally different; in fact, it is referring to a small *leakage* current. Remember that the two types function in reverse of one another, so that here I_{DSS} refers to the current that flows between drain and source when the gate is shorted to the source (i.e., V_{GS} = 0), in other words when the device is off. The measurement circuit is shown in Figure 8.5A. The I_{DSS} leakage current is actually that of a reverse-biased diode (see Figure 8.5B) and doubles for every 10°C rise in temperature.

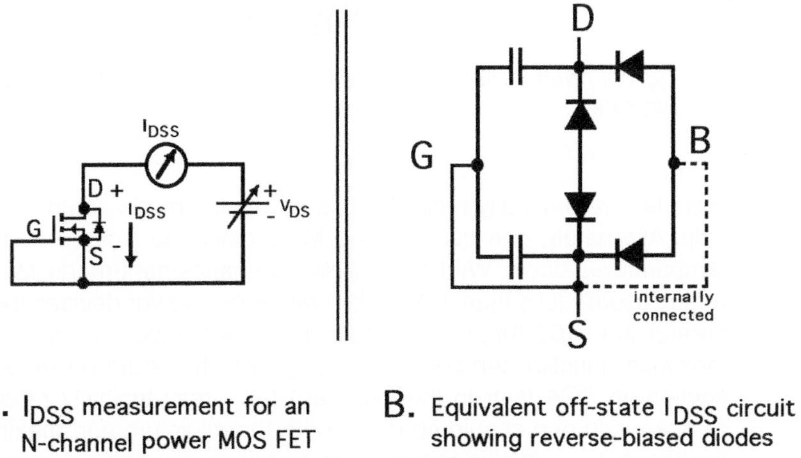

Figure 8.5. *With enhancement-mode MOSFETs, I_{DSS} is a small leakage current.*

The maximum limiting current that can flow between the MOSFET's drain and source is known as $I_{D(on)}$ or the *on-state drain current*. This occurs at a particular V_{DS} and at a particular V_{GS} (positive value for N-channel and negative value for P-channel devices). The graph depicting this action is shown in most MOSFET data sheets and is referred to as the device's *output characteristic* (I_D vs. V_{DS} graph). You can see from the curves in Figure 8.6 that when small values of V_{DS} and V_{GS} voltages are simultaneously applied, the drain current increases linearly with V_{DS}. This region is known as the *ohmic* or *linear region.* As the V_{DS} is increased, the drain current begins to be pinched off or limited at the knee of the curve, before finally becoming saturated and flattening out. This is known as the *saturation region.* When the drain current reaches saturation, it becomes proportional to the square of the applied V_{GS} and is then only slightly dependent on V_{DS}; this is known as $I_{D(on)}$. It can be easily measured using the circuit shown in Figure 8.7.

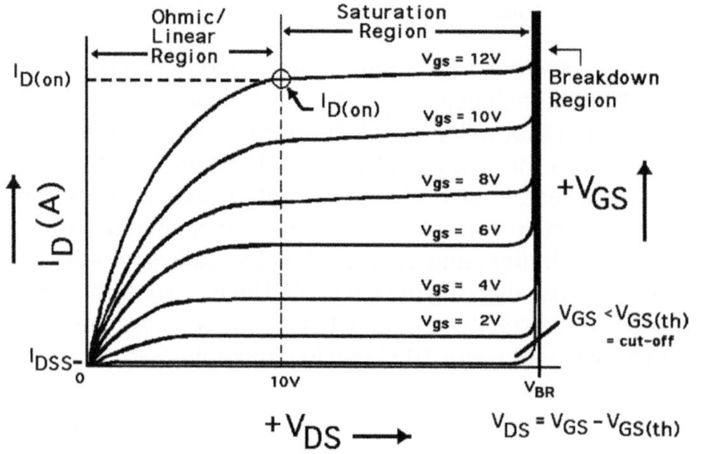

Figure 8.6. *Typical output characteristics for an N-channel enhancement-mode power MOSFET.*

Manufacturers usually measure this in a pulse mode setup, to reduce heating of the chip. Alternately, one can bolt the MOS device to a large heatsink to keep junction temperatures down. With a low-power enhancement-mode MOSFET, this $I_{D(on)}$ current is usually less than 1 Amp, but with many power devices the drain current can be greater than 100 Amps. It is important to remember that $I_{D(on)}$ is also limited by the maximum junction temperature, $T_{j(max)}$, which should never be exceeded. (See the section on SOA later in this chapter.) $I_{D(on)}$ may typically range over 3:1 for similar devices, and one should consider both the minimum and maximum ratings/curves in the range. $I_{D(on)}$ is also temperature sensitive and has a negative temperature coefficient of approximately –0.5%/°C.

Another important characteristic of the power MOSFET is its static *on-state resistance*, $R_{DS\ (on)}$. For small MOSFETs it is often less than 50 Ω, whereas for power devices it can reach milli-ohm levels. The lower the on-resistance, the higher the cur-

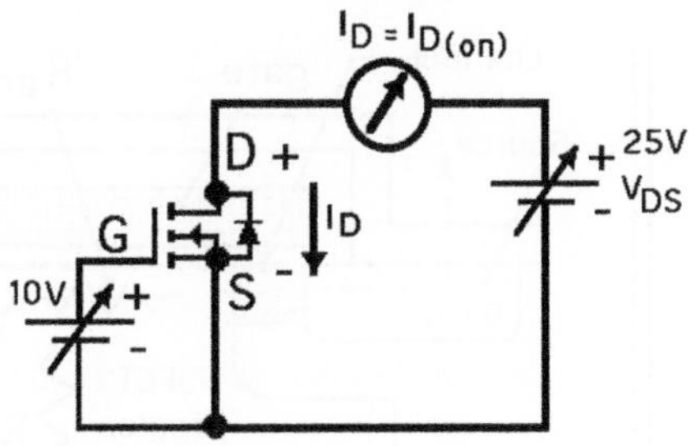

Figure 8.7. *$I_{D(on)}$ measurement for an N-channel power MOSFET.*

rent that the device can switch. $R_{DS\ (on)}$ is measured as the V_{DS}, divided by a particular drain current, at a particular V_{GS}, at 25°C and at an elevated temperature. It can be expressed as:

$$R_{DS(on)} = \frac{V_{DS}}{I_D} \qquad \text{(Eq. 8.1)}$$

Because of device self-heating, $R_{DS\ (on)}$ is usually measured in a pulsed manner, at room temperature, because heating the device decreases carrier mobility, thereby reducing the drain current for a given voltage. Figure 8.8A shows the equivalent circuit for $R_{DS\ (on)}$. It is actually composed of several smaller resistances (seen in Figure 8.8B), which include the channel resistance (R_{ch}), the JFET region's resistance (R_{jfet}), the drift region's resistance ($R_{epi\text{-}drift}$), the substrate resistance (R_{sub}), and so on. At low voltages, $R_{DS\ (on)}$ is governed mainly by the channel resistance, whereas at high voltages the (central parts) epi and JFET regions are most significant. At high voltages beyond about 200 volts, $R_{DS\ (on)}$ increases. It is interesting to note that the $R_{DS\ (on)}$ for P-channel devices is typically two to three times higher than for similar N-channel devices. As a result, in order to make complementary pairs with similar on-resistances, the P-channel device needs to be a physically bigger chip. This in turn alters its capacitance values, and because it is bigger, it costs more. For a MOSFET, $R_{DS\ (on)}$ has a positive temperature coefficient of approximately +1%/°C. Actually, this is a beneficial characteristic, because as the device heats up and the on-resistance increases, the drain current automatically reduces.

Another significant difference between depletion-mode devices and enhancement-mode devices is the characteristic $V_{GS(th)}$. This applies only to enhancement-mode devices and is the *gate-to-source threshold voltage* necessary to just turn the MOSFET on, at a very low current level. $V_{GS(th)}$ is actually the voltage required to cause the surface inversion of the channel, which allows any forward current to flow. The V_{GS} for an N-channel DMOS FET ranges from 0 volt, where the device is fully cut off, to some positive amount of several volts (typically between 2 and 4 volts), to turn the device on. This is shown in Figure 8.9. Most MOSFETs have a fairly low $V_{GS(th)}$ value, so that

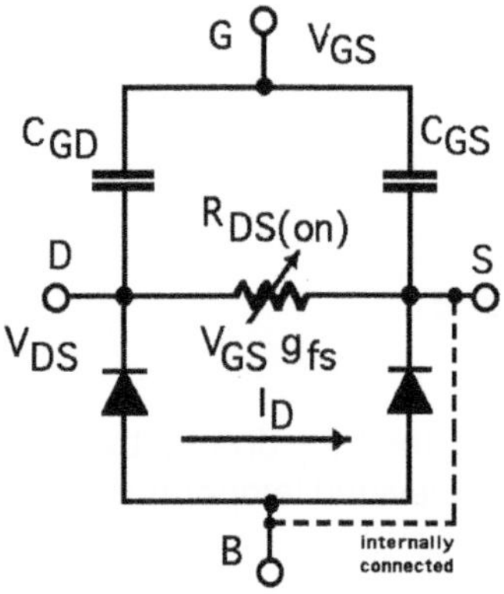

A.
The MOS FET's variable on-resistance, $R_{DS(on)}$ depends on both V_{GS} and V_{DS}.

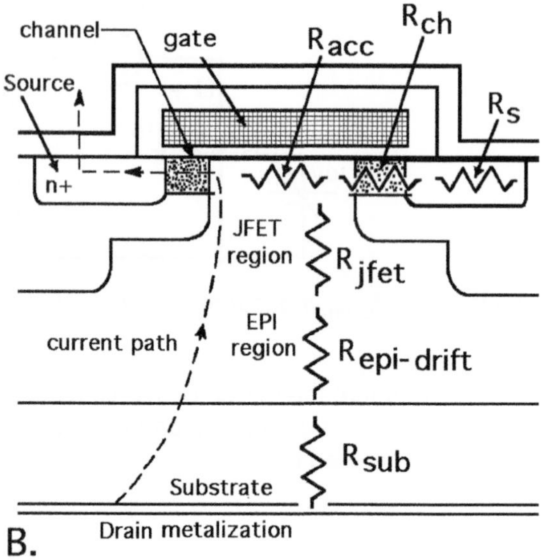

B.
Cross-sectional view of the various elements that make up a MOS FET's $R_{DS(on)}$.

Figure 8.8. On-state resistance.

they are logic-level compatible. Several manufacturers, such as SuperTex, Fairchild Semiconductor, and ON Semiconductor make very-low-threshold devices ($V_{GS(th)}$ less than 2 volts). Such devices provide overall faster switching, because less charge current is needed to charge the parasitic input capacitances. (We will look at the subject of gate charge later in this chapter.) However, they are (at least theoretically) more susceptible to voltage transients, which can cause spurious turn-on of the device. Typically, a MOSFET's isolated gate has ±20-volt limiting gate voltage. Beyond about ±16 volts, increases in V_{GS} tend to increase switching time, as well as making the gate more vulnerable to voltage transients. If this is an issue, it is common practice to insert a zener diode (16 to 18 volts; ±10%; 400 mW or higher) between the gate and source, to protect the sensitive gate.

Conduction ceases when the V_{GS} drops to less than the threshold voltage; otherwise, the value of V_{GS} mostly controls conduction through the channel. To ensure that the device is fully conducting, a V_{GS} of between 10 and 12 volts is recommended. Figure 8.9 shows the drain current curves versus V_{GS} for a typical enhancement-mode power MOSFET. This I_D versus V_{GS} transfer characteristic is depicted in most MOSFET data sheets and can be easily measured, using the circuit shown in Figure 8.10. Most manufacturers use a standard drain current value of 1 mA or 250 μA to determine the $V_{GS(th)}$ value. Unlike the circuit shown in Figure 8.10, manufacturers typically tie the

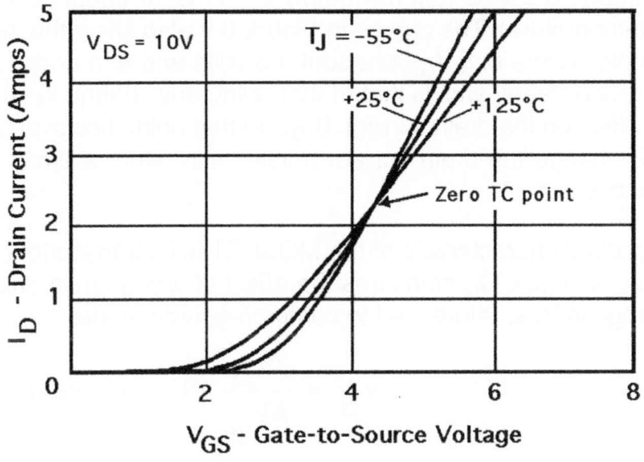

Figure 8.9. *The I_D versus V_{GS} transfer characteristics of a typical N-channel enhancement-mode MOS FET.*

drain and gate together to determine the gate threshold voltage, which is easier to test in a production environment.

The data sheet for your chosen device may not necessarily portray the appropriate curve data in exactly the same manner, and you might have to resort to measuring devices and plotting your own curves. This is because most manufacturers' curves are created on a small-scale graph, and this may not be sufficiently large and accurate for some purposes, such as in creating low-/medium-level precision current sources. Because $V_{GS(th)}$ can typically range over 3:1 for similar devices, it is important that you consider both its minimum and maximum ratings in the range for your application. As you can see from the transfer characteristics, shown in Figure 8.9, both $I_{D(on)}$ and $V_{GS(th)}$ shift with temperature. Actually, $V_{GS(th)}$ has a negative temperature coefficient

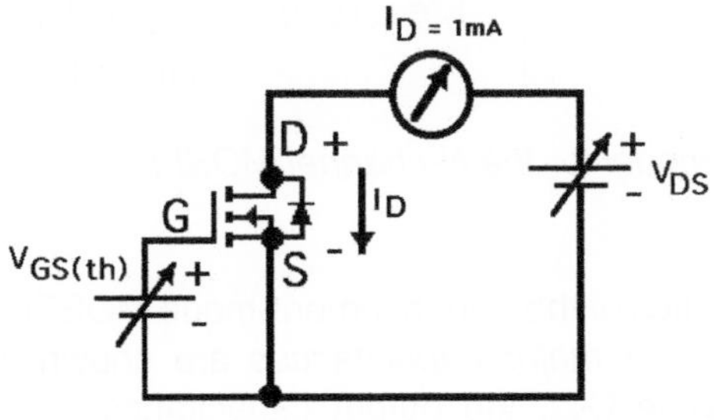

Figure 8.10. *$V_{GS(th)}$ measurement for an N-channel power MOSFET.*

of approximately –0.2%/°C (approximately 5 mV/°C, or about 10% for each 45°C rise in junction temperature). The curves in Figure 8.9 also show the zero TC point, which in this example occurs at a V_{GS} of about 4.3 volts and at a drain current of about 2.3 Amps. The graph clearly shows that at that point, any change in temperature will have virtually no effect on the drain current. Beyond that point, however, any change in temperature will change the drain current a little, with either a small positive or a small negative tempco.

Another important characteristic of the MOSFET is its transconductance/forward conductance (g_{fs}, aka g_m). G_{fs} measures the effect of a change in drain current (I_D) for a specific change in V_{GS}, referenced to common-source mode.

$$g_{fs} = \frac{\Delta I_D}{\Delta V_{gs}} \,\Big|\, V_{DS} = \text{constant} \qquad \text{(Eq. 8.2)}$$

In practical terms, transconductance is a measure of the FET's gain and is a figure of merit. It is usually referred to from a practical standpoint as amps or milliamps per volt (mA/V). On a graph it appears as a steep or shallow slope and is measured and referenced in milli-Siemens (mS). For example, 10 mA/V is a more shallow slope than 100 mA/V. G_{fs} is usually specified on the data sheet for most MOSFETs with a minimum, typical, or maximum value. The g_{fs} value given in some manufacturers' data sheets is measured at full conduction ($I_{D(on)}$), where g_{fs} is at its maximum. Other manufacturers specify g_{fs} at a particular I_D and V_{DS}.

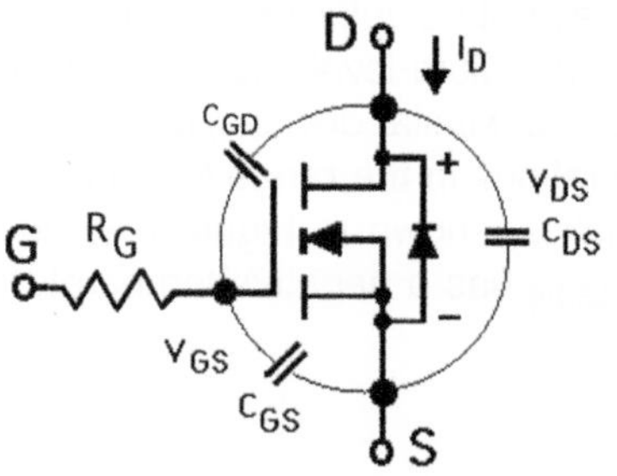

Capacitances:

INPUT: $C_{iss} = C_{GS} + C_{GD}$, with C_{DS} shorted

OUTPUT: $C_{oss} = C_{DS} + C_{GD}$

REVERSE TRANSFER: $C_{rss} = C_{GD}$ Most important.
= Miller capacitance.

R_G = Gate resistance

Freq. resp. = $R_G \times C_{iss}$
(upper frequency limit)

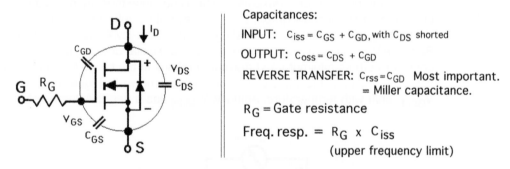

Figure 8.11. A simple capacitance model for the N-channel MOSFET.

Other important characteristics of the enhancement-mode MOSFET include its capacitances and gate charge. The *major capacitances* are shown in Figure 8.11 and include the input capacitance C_{iss}, the output capacitance C_{oss}, and the reverse-transfer capacitance C_{rss}. Of the three, C_{rss} is the most dominating, because it is part of the feedback loop between the device's output and its input. It is also known as the Miller capacitance. The frequency response of the MOSFET is governed by the charging and discharging of C_{iss}, which is composed of the gate-to-source capacitance (C_{GS}), the gate-to-drain capacitance (C_{GD}), and the resistance of the gate overlay structure (not to be confused with the very large input resistance of more than 100 MΩ). This is typically around 20 ohms for silicon-gate and about 10 ohms for metal-gate FETs.

Typically, the upper frequency limit of a silicon-gate power MOSFET is in the range of 1 to 10 MHz and about twice that for metal-gate devices. Small-power MOSFETs have a higher frequency response, because the size of the chips is smaller, thereby reducing the value of C_{iss}. Because C_{iss} is unaffected by temperature effects, neither is the MOSFET's switching speed. The capacitance values for low-power MOSFETs and power devices is significant, most of it being the result of differences in chip size. For example, a popular 2N7000/7002 device (60-volt, 2.1-volt threshold, 0.2 A, 7-nS switching, TO-92 case) has typical data sheet capacitance values of C_{iss} = 22 pF; C_{oss} = 11 pF; C_{rss} = 2 pF. In comparison, a popular power device such as an ON Semiconductor MTB75N05HD (50-volt, 4-volt threshold, 75 A, 170-nS switching, D^2PAK package) has typical capacitance values from its data sheet of C_{iss} = 2600 pF; C_{oss} = 1000 pF; C_{rss} = 230 pF (all about 100 times higher!).

A characteristic directly related to a MOSFET's capacitance and switching is the *gate charge.* While capacitance values are helpful in many design situations, differences in chip size, operating voltage, and transconductance between different devices, and from different vendors, make accurate comparisons difficult. Even though these different characteristics are given on most data sheets, it is more common to simply use the gate charge, rather than capacitance values, in order to determine the gate drive requirements.

Remember that a MOSFET has an isolated capacitive gate, and that being essentially a capacitor it takes time to charge or discharge that capacitance, as well as to support a certain amount of charge. It works in the following way, slowed down here so that we can follow along: Assuming that an N-channel MOSFET is connected with a 10-volt positive supply voltage at the drain, a source that is grounded, and the gate temporarily at ground, if the gate is now connected to the positive supply voltage, the V_{GS} starts to increase. Soon after it will reach the $V_{GS(th)}$ at, say, 3 volts, at which point a small drain current will start to flow, and the C_{GS} will begin to charge. Once C_{GS} has fully charged, the gate voltage becomes constant and begins to charge the C_{GD} (also known as the Miller capacitance). This takes longer than when charging C_{GS}, because it is a larger capacitance. Once C_{GD} has finished charging, the V_{GS} starts increasing again until it finally reaches the 10-volt supply voltage. When it reaches this point, that is the total time (referred to in a manufacturer's data sheet as "Q_T") needed to turn the device full-on. It can be expressed by the following equations:

$$Q = It \quad \text{; in coulombs} \qquad \text{(Eq. 8.3)}$$

$$I = \frac{Q}{t} \quad \text{; in Amps} \qquad \text{(Eq. 8.4)}$$

$$Q = CV \quad \text{; in coulombs} \qquad \text{(Eq. 8.5)}$$

$$C = \frac{Q}{V} \quad \text{; in Farads} \qquad \text{(Eq. 8.6)}$$

$$I = C\frac{\partial V}{\partial T} \quad \text{; in Amps} \qquad \text{(Eq. 8.7)}$$

Most MOSFET manufacturers specify the gate charge in the switching characteristics section of their product's data sheet, and it is usually expressed in *nano-coulombs* (nC $= 1 \times 10^{-9}$). Using the gate charge makes it easier for the circuit designer to calculate the amount of charging current needed to switch the device on. For example, if a device had a specified gate charge of 50 nC, it could be switched on in 50 microseconds at a gate current of 1 milliamp, or alternatively it can be switched on in 50 nanoseconds at 1 Amp.

An important consideration for using any power MOSFET is its *maximum power dissipation* (P_D). While MOSFETs have much better temperature characteristics than bipolars (in that the on-resistance is positive, thus reducing the current flow), and the *Safe Operating Area* (SOA) is more rectangular, where a bipolar's is more limited, it will still need to dissipate power, in the form of heat, away from the chip. For a MOSFET, it is limited by its breakdown voltage, current rating, on-resistance, power dissipation, and maximum junction temperature. Exceeding any one of these, particularly its maximum $V_{(BR)DSS}$ rating, could be fatal to the device and the circuit.

Typically, the manufacturer's SOA curve also provides information that trades off voltage and current levels, versus DC or pulsed operation. One should always keep within the limits set by the SOA's boundaries for reliable operation. Operating at low V_{DS} values (less than 60 volts) but high current levels (I_D more than 6A) may be limited by the $R_{DS\ (on)}$ of the device. Figure 8.12 clearly illustrates this point. In the example, the SOA's boundaries show that the maximum continuous drain current, I_D, of 23 Amps at 4.25 V_{DS}, through I_D of 3 Amps at 30 V_{DS}, down to I_D of 1 Amp at 95 V_{DS} (the device here has a 100-volt breakdown rating). As you can see, in pulsed mode, higher drain currents are possible. Remember, though, that in case your application gets too close to one of the SOA's boundaries, it is always easy to parallel MOSFETs and achieve a high, but safer current that way.

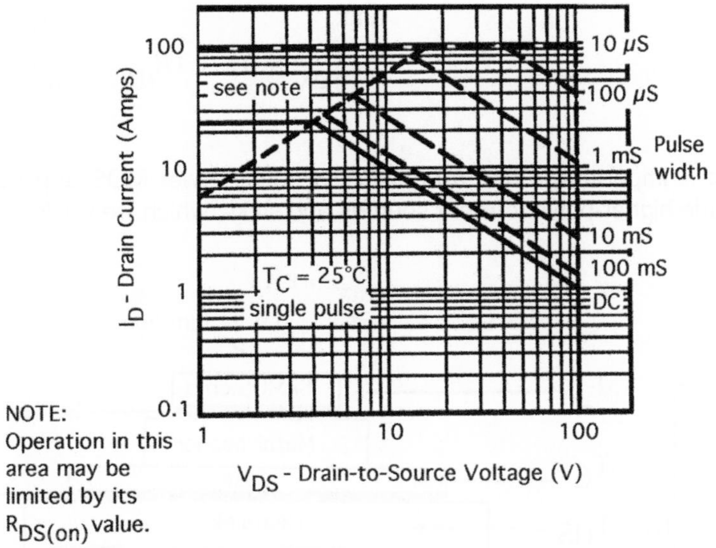

Figure 8.12. *The Safe Operating Area (SOA) curves for a typical power MOSFET with a 100V breakdown voltage, and 80-amp current limit, a 100-m$\Omega_{DS(on)}$, and a 100-W maximum power rating.*

A manufacturer will typically rate the device with a maximum allowable junction temperature T_{jmax} (usually specified in the data sheet at 150°C), as well as specifying the junction-to-case thermal resistance ($R\theta_{JC}$), according to the package type. Here it is assumed that the external case temperature is held at a constant 25°C, by securely bolting the device to a large finned heatsink. In many cases, the manufacturer also specifies the junction-to-ambient thermal resistance ($R\theta_{JA}$), for applications requiring no heatsink, where the package is dissipating heat into free air (see Figure 8.13). For example, a typical power MOSFET in a standard TO-220 package and with a 100-Watt rating can only dissipate about 2 Watts into free air, without using a heatsink. Generally, the power dissipation (P_D), expressed in Watts, can be calculated from the following:

$$P_d = \frac{Tj\max - 25°C}{R_{\theta JC}} \qquad \text{(Eq. 8.8)}$$

also:

$$P_d = I_D^2 \times R_{DS(on)} \qquad \text{(Eq. 8.9)}$$

and:

$$T_{j\,max} = T_C + (R_{\theta JC} \times P_d) \qquad \text{(Eq. 8.10)}$$

So it is important to use a heatsink for most power MOS applications, and this will include higher-power current sources and sinks, which we will look at next.

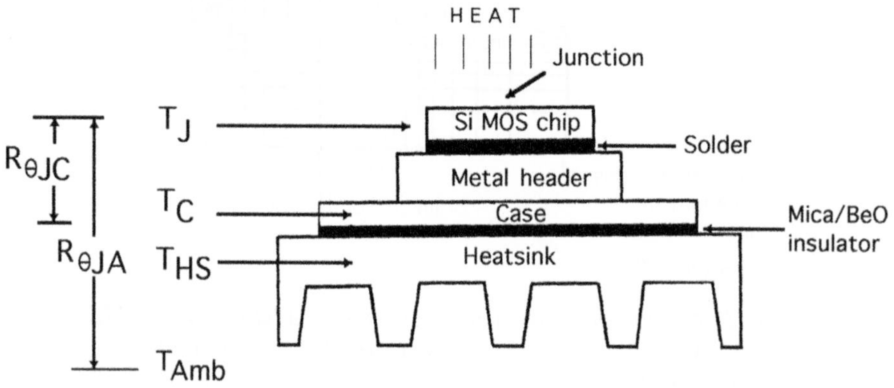

Figure 8.13. Power dissipation in a MOSFET.

8.2 Using the enhancement-mode MOSFET as a current source

The simplest current source using an N-channel power MOSFET can be made with a zener diode and a bias resistor on the gate, as shown in Figure 8.14. Simply changing the FET's drain connection enables one to create either a current source or a current sink. In this circuit, the zener's voltage determines the MOSFET's gate voltage, and thereby its drain current (typically ranging from hundreds of milliamps to several amps). Because the FET has a very-high-input impedance, its gate requires virtually no input current from the bias resistor, hence the resistor establishes only the zener's current. One can determine the output current and necessary V_{GS} using the chosen MOSFET's transfer and output characteristic curves from the data sheet.

Although fairly accurate, the simple circuit is not very temperature stable, because of the zener's temperature coefficient (tempco). Below 3 volts the zener diode's temperature coefficient is negative, while above 8 volts it is positive, but between these two levels the tempcos can be either positive or negative. The overall tempco will be determined according to the quality and tempco of the resistor, R_1, and the particular zener diode, its voltage, current, and the operating temperature range. For example, if the voltage chosen gave a positive tempco, this can be compensated for by adding one or more silicon diodes (which have a negative tempco) and reducing the zener's voltage value accordingly (for the diode's forward voltage drop of approximately 0.65 volt). A small heatsink may well be required, depending on the desired current level (I_D), the voltage drop across the FET (V_{DS}), the chosen MOSFET's $R_{DS(on)}$, and its package type.

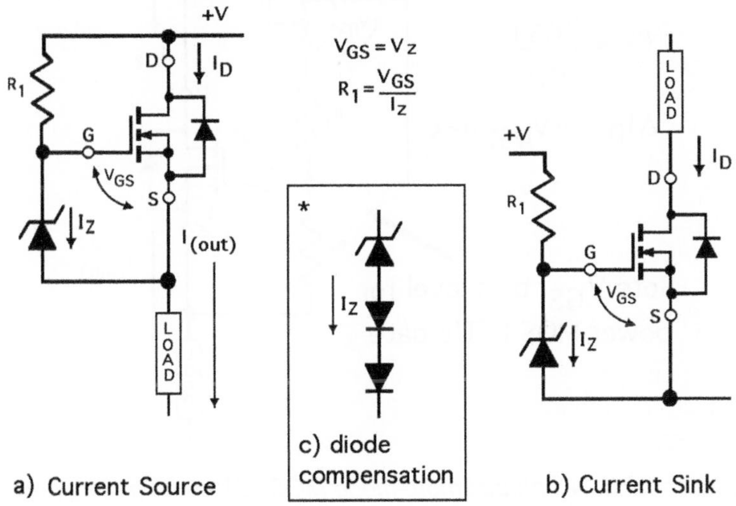

Figure 8.14. *The N-channel, enhancement-mode vertical power MOSFET can be used as an effective current source (a) or current sink (b), in this simple Zener-based circuit. Improved temperature stability and zener tempco can be attained by adding silicon diodes, as in (c).*

One can also build a power current source using a depletion-mode DMOS FET in conjunction with an enhancement-mode power MOSFET, in a combinational circuit (see Figure 8.15). The depletion-mode device is set up in the manner described previously (see Chapter 7), so that in this application its drain current, together with the gate-to-source resistor (R_2), biases the power MOSFET's gate at a particular voltage, and thereby provides the desired output current. This circuit is useful at medium current levels (100 mA to 3 Amps) and because of its four-item component count. Again, a small heatsink will likely be required for the power MOSFET, depending on the desired current level, operating voltage, and package type. One can decide on the output current and necessary V_{GS} using the transfer and output characteristic curves from the MOSFET's data sheet. Knowing the required V_{GS} will allow one to determine the necessary resistor value for R_1. This design may be useful in higher voltage (50- to 200-volt) circuits, where the FET's $V_{(BR)DSS}$ rating will establish the high voltage limit. It can also be used at levels down to about 6 volts.

An alternative to these simple circuits uses the combination of a JFET current regulator diode (CRD), an NPN transistor, a capacitor, and a resistor to set the power

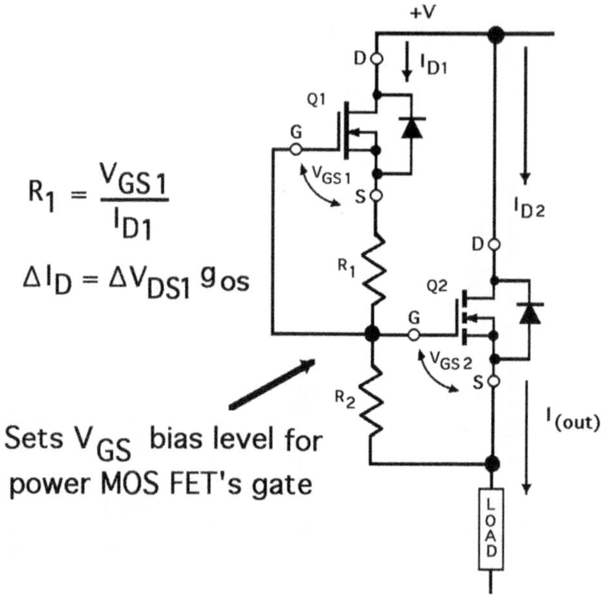

Figure 8.15. *A depletion-mode + enhancement-mode MOSFET combination, provides a simple current-source or sink, at medium current levels. Q_1 sets the voltage level at Q_2's gate, thereby setting the desired durrent level.*

MOSFET's drain current, as shown in Figure 8.16. In this circuit, the JFET CRD provides the operating bias for the NPN transistor, Q_2, and ensures that its emitter current and hence its V_{BE} remain relatively constant, irrespective of power supply fluctuations. The transistor's V_{BE} (approximately 0.65 volt) is used as a reference voltage across the power FET's source resistor, R_S. This forces the FET to pass constant drain current irrespective of its V_{DS}. The source resistor should be a precision, low-tempco, wirewound type, with its wattage dependent on the maximum current through it (+30% margin).

In practice, though, both Q_2's V_{BE} and R_S drift with temperature. The transistor's V_{BE} has a negative tempco of –2.2 mV/°C. This may be improved a little by thermally connecting both components so they are at the same temperature. One could also add a silicon diode of appropriate wattage in series with the resistor, as shown as an option here (and in the equation). This will help compensate for the transistor's V_{BE}. Another improvement would be to replace the CRD with a cascode JFET combination, as shown in Figure 8.16B, which can provide tighter current regulation.

Building on the previous design, one could substitute a Widlar bandgap voltage reference for the simple V_{BE} reference, as shown in Figure 8.17. This would greatly improve regulation, and particularly the temperature characteristics of the previous cir-

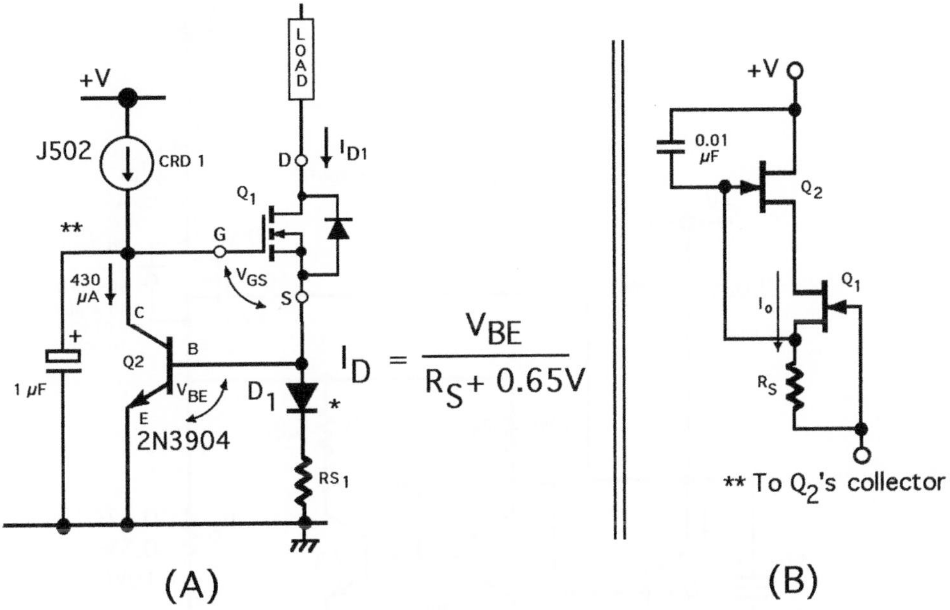

Figure 8.16. *The circuit in (A) shows a simple current-sink using a CRD, a BJT, and a power MOS FET. The diode D_1 is optional, but compensates for the BJT's V_{BE}. The CRD may be replaced by a JFET cascode, shown in (B), which will provide improved regulation.*

cuit. The bandgap circuit inherently provides both an ultra-stable temperature coefficient, which in this design occurs at approximately 25°C, and a precise 1.22-volt voltage reference. The three small-signal NPN transistors shown, which make up the bandgap voltage reference, should be part of a monolithic array (i.e., CA3045, CA3046, MMPQ3904, MMPQ2222A), so they share the same temperature characteristics. The particular JFET current regulator diode (CR1) was chosen because at 430 μA it has a very flat temperature characteristic yielding virtually 0 ppm/°C, and it provides a very stable operating bias for the NPN transistor, Q_1. This in turn provides a stable reference for the bandgap reference circuit, whose output is impressed across the power resistor, R_S. This should be a precision, metal-film, high-wattage, very-low-value power resistor, and it should have a low tempco. As a result, this design provides a medium-power (2.6-Amp) current sink. In this circuit, the power MOSFET uses a TO-220 package, which should be mounted on a good-quality heatsink.

An alternative to all of the preceding designs is to generate the MOSFET's gate voltage from a low-power, monolithic voltage reference, as shown in Figure 8.18. While a little more expensive than other solutions, because of the adjustable voltage reference, several benefits make up for that. The desired output current can be accurately set between approximately 100 mA and several amps, by varying the

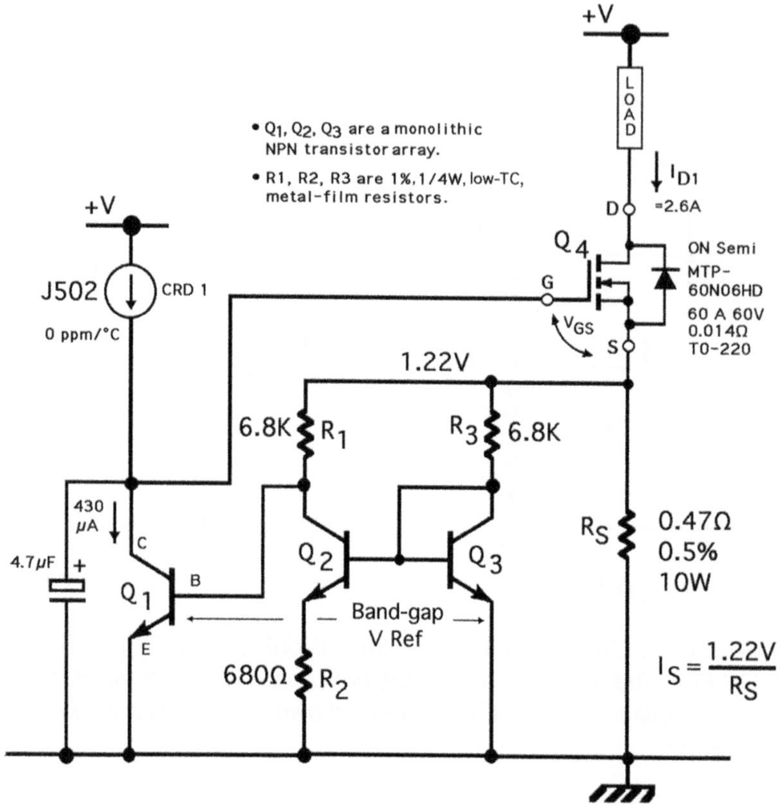

Figure 8.17. *This improved current-sink uses a Widlar bandgap voltage reference to improve temperature stability and current regulation.*

reference IC's output voltage. Once set, the gate voltage is stable and will be even more temperature stable because of the high accuracy of the voltage reference (and its internal tempco technique). Although most IC voltage references are factory-set and fixed at a particular value, adjustable versions are available. Two such devices are the Maxim MAX6160 and the National Semiconductor LM4041-CIZ-ADJ. Both devices are rated at 100 ppm/°C maximum (with typical values of less than 20 ppm/°C) and are available in surface-mount packages and others. Both can be varied over the range of 1.25 volts to more than 10 volts. The high end of the range is of little interest, because this will be where the MOSFET's drain current is at its maximum, but the low end of the range will be of more practical use. In any case, the MOSFET's output conductance (G_{os}) and transconductance (G_m) will both be lower at low current levels, which is more desirable for a good current source (see Chapter 15, adjustable series bandgaps).

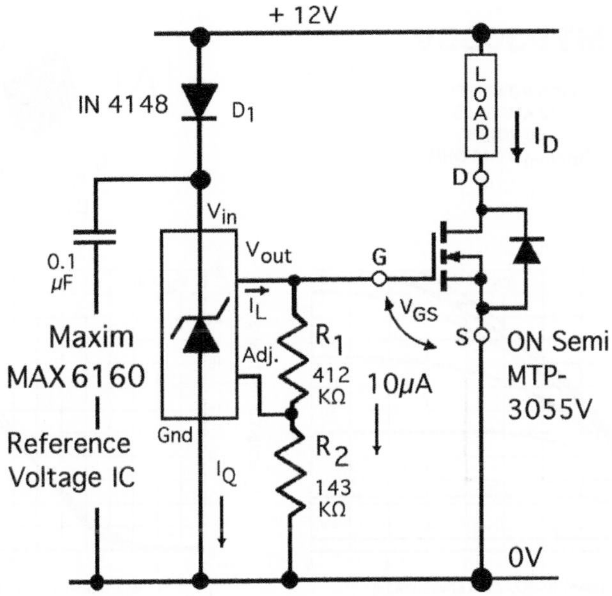

Figure 8.18. *Using a CMOS surface-mount, adjustable voltage-reference, and a power MOSFET, in order to create a stable current-source/sink.*

We will use Figure 8.18 as a design example, because the combination of a monolithic voltage reference together with a power MOSFET would be useful in many applications. In this example of a power current sink, one should first determine the voltage supplies and the current output required. Let's suppose that we have 12 volts (±5%) available and that the current required through the power FET is 4 Amps. The power MOSFET required needs to have a $V_{(BR)DSS}$ of at least 18 volts and a current rating (at 100°C) of at least 6 Amps. We should check the FET's SOA curve, to determine that the device can safely sustain a continuous DC rating of 4 Amps.

A suitable device in a TO-220 power package is the popular ON Semiconductor MTP-3055V. This device has a 60-volt, 12-Amp, 48-Watt rating and a maximum $R_{DS(on)}$ of 150 milli-ohms. It has a gate threshold voltage of between 2 and 4 volts, with a typical value of 2.7 volts. The SOA curve (Figure 8.19E) shows us that the device can sustain a continuous DC current of 4 Amps, with a V_{DS} range of between 1 and 8 volts. From the device transfer characteristics (Figure 8.19C), we can see that for a continuous DC current of 4 Amps, we need a V_{GS} of approx 4.75 volts. The I_D versus V_{DS} curves, referred to as the on-region characteristics (Figure 8.19B), confirm this and that the

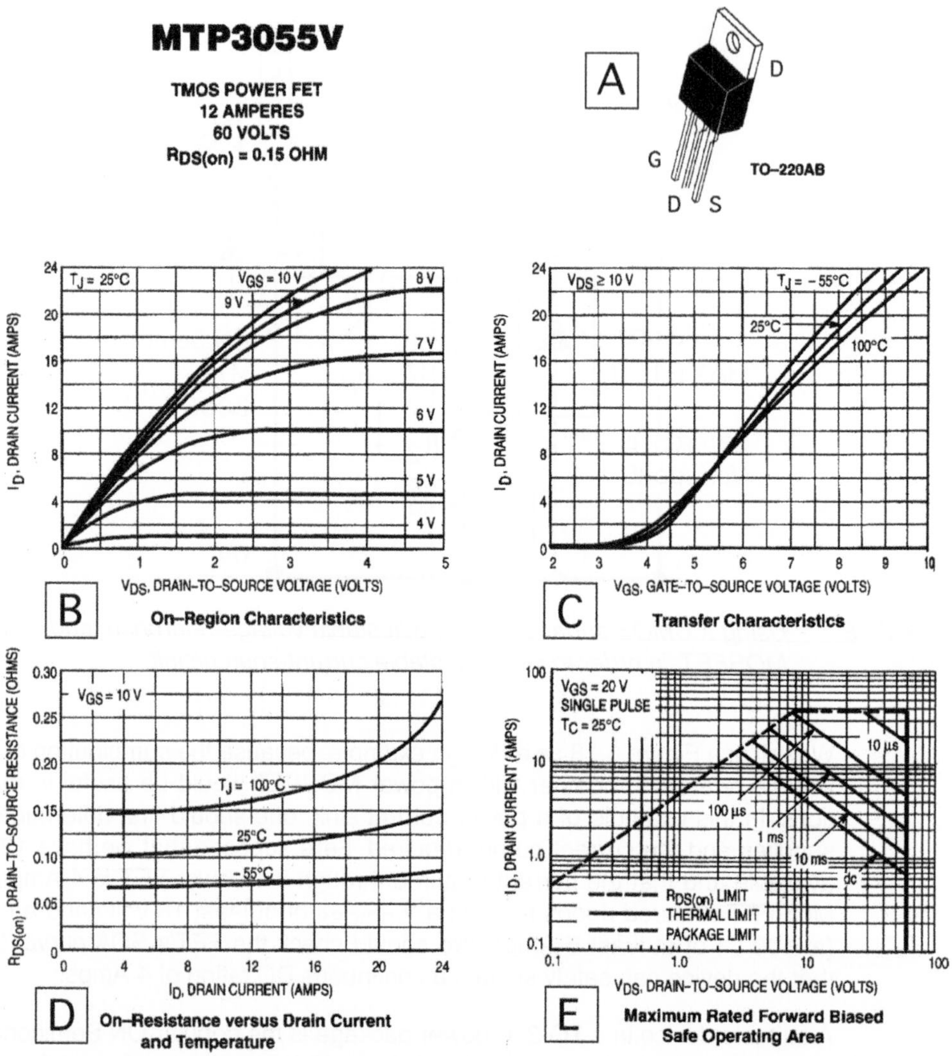

Figure 8.19. Important design considerations using ON Semiconductor MTP-3055V Power MOSFET. (Courtesy ON Semiconductor)

minimum V_{DS} that can sustain this will be about 2 volts. We can determine the maximum power dissipation from the following:

$$P_d = I_D^2 \times R_{DS(on)}$$

$$16A \times 0.2\Omega = 3.2 \text{ Watts} \qquad \text{Eq.8.11}$$

This will necessitate using a small heatsink, because the TO-220-AB package (Figure 8.19A) cannot by itself dissipate more than 2 Watts into free air. Also, like many devices in this particular package, the drain is the center terminal as well as the metal tab. This means that these points should both be isolated for safety purposes, because the heatsink will most probably be grounded (0V) to a metal chassis. A mica or beryllium oxide (BeO) insulator *must* be used between the TO-220 package and the heatsink, and an insulator plug should be used when mounting a metal nut, screw, and washers via the hole in the package's metal tab.

Turning now to the adjustable voltage reference, in this example we will use a Maxim MAX6160, because the device is easy to use. Maxim describes the CMOS device as a low-dropout, micro-power voltage reference. The output voltage range of the adjustable version is from 1.23 volts to 12.4 volts, with a total output current capability of up to 2 milliamps. The operating temperature range of the MAX6160 spans –40°C to +80°C. In this design, the voltage reference tolerance for the device is ±1%, with an output tempco of typically 15 ppm/°C, which is more than adequate. Remember that 100 ppm translates into better than 12-bit accuracy, but Maxim's device with its typical 15 ppm equates to around 16-bit accuracy.

From our previous calculations we know that we need approximately 4.75 volts (V_{GS}) to provide the 4 Amps constant drain current through the power MOSFET. Therefore, the output of the voltage reference should be nominally calculated for this level. We will assume that the 12-volt power supply (±5%) is a low-noise, well-regulated type providing between 11.4 and 12.6 volts. Because the 12.6-volt level is close to the absolute maximum rating of the MAX 6160, a diode has been inserted into the positive supply, to drop the voltage arriving at the VIN terminal (by 0.65 volt) to 11.95 volts, a safer level. The design equations that we will use (from Maxim's MAX6160 data sheet) are shown in Figure 8.20.

First, the known values are the supply voltage $+V_S$ equals 12 volts; the output voltage V_{OUT} equals 4.75 volts (which is the MOSFET's desired gate voltage V_{GS}); the total load current we will assume as I_L equals 1 mA; and the MAX6160's own quiescent current we will assume as I_Q equals 100μA.

Maxim's data sheet shows a resistive divider composed of R_1 and R_2, which together with the ADJ. pin set the output voltage (V_{OUT}) to the desired level. Now we will calculate the resistor value for R_1 and R_2, using equations 1 and 2 in the diagram. The internal voltage reference V_{ADJ} is set at 1.23 volts, and the 106K scaling factor used in equation 1 compensates for the MAX-6160's change in ADJ input-current, over the temperature range. Using equations 1 and 2, the values calculated result in R_1 equals 409 KΩ and R_2 equals 143 KΩ. The nearest standard value for R1 is 412 KΩ, while R_2 is already a standard value. Both resistors should be a quality metal-film type, with a tolerance of 1% and with a 1/8th watt rating. For optimum performance, the MAX6160 should be bypassed close to its input with a 0.1-μF ceramic disk capacitor. No output bypass capacitor is necessary in this application. The amount of current flowing in the

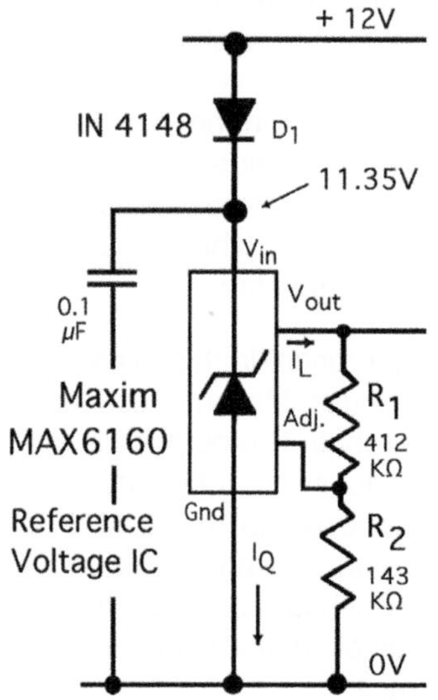

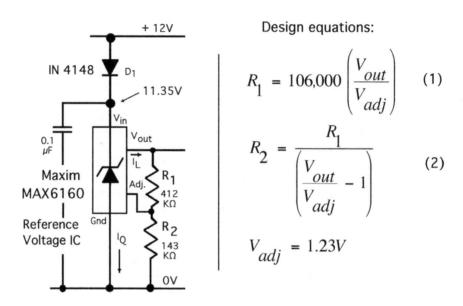

Design equations:

$$R_1 = 106{,}000\left(\frac{V_{out}}{V_{adj}}\right) \quad (1)$$

$$R_2 = \frac{R_1}{\left(\frac{V_{out}}{V_{adj}} - 1\right)} \quad (2)$$

$$V_{adj} = 1.23V$$

Figure 8.20. Calculating resistor values for the MAXIM MAX6160 reference.

resistive divider is approximately 10 µA, meaning that approximately 1 mA is available to charge the MOSFET's gate. The total gate-charge value (Q_T) for the ON Semiconductor MTP-3055V FET is less than 20 nC, meaning that the device would take approximately 20 µS to turn on, once the reference voltage has stabilized at V_{OUT}.

If only a fixed output voltage reference is available (e.g., 5 volts), then use a variable potentiometer (in series with a fixed resistor), which will allow selection of different output voltages. The fixed resistor will prevent the reference IC's output being accidentally shorted to ground when trimming the potentiometer. Choose a voltage reference IC that can adequately drive the lowest value of resistor load involved.

While the technique of using a monolithic voltage reference works well, it is more usual to use an op amp to buffer the reference IC's output to the gate of the power MOSFET. This may also provide better stability when even higher current levels are involved. Such an arrangement is shown in Figure 8.21, where a precision voltage reference is used (see Part 2 of this book). Part of the reference's output is picked off via the multiturn potentiometer and used as a reference level for the op amp's noninverting input. The op amp should be a quality, single-supply, precision type with a low-input offset voltage (less than 1 mV). Some examples include National Semiconductor LMC6081, Analog Devices OP184, Linear Technology LT1077, and Maxim MAX495. A precision, metal-film, high-wattage, very-low-value power resistor is used to sample the power FET's source current, and it should have a low temperature coefficient.

Depending on the exact voltage level across the power resistor, this difference will cause the op amp to drive the power FET's gate by feedback, so that a state of equilibrium is reached. The output current is set by the voltage across the power resistor, divided by its ohmic value (Ohm's law). One premium source of precision power resistors is Caddock Electronics, whose MP 8xx Kool-Pak series are available in a standard TO-220 style and other heatsinked packages. Other manufacturers can be found in the listing in Appendix D, Precision Passive Manufacturers.

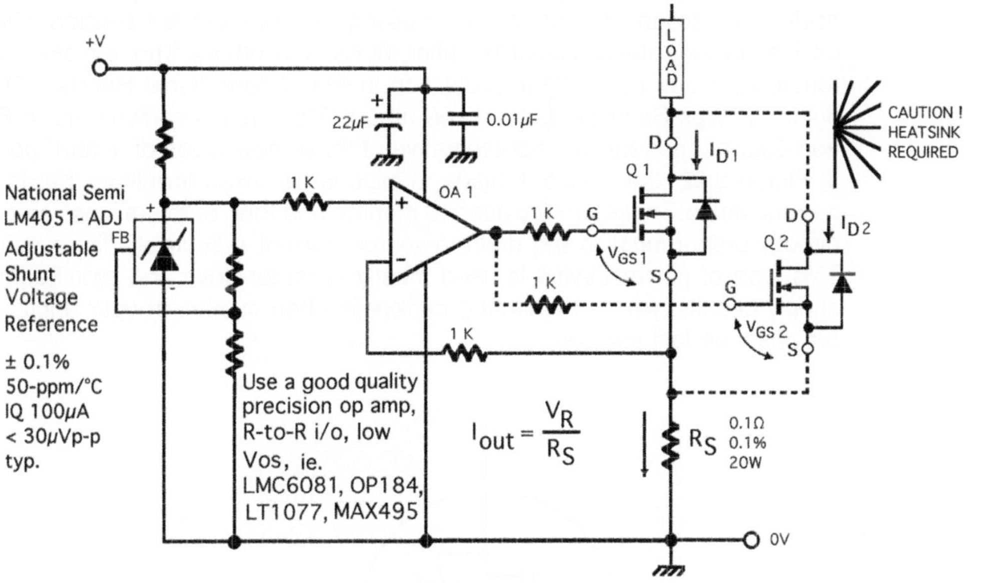

Figure 8.21. Creating a high-power (>10-amp) Current-sink using a precision voltage reference, an op amp, and power MOS FETs. The op amp's input common-mode range should include ground. Paralleling MOSFETs can provide even higher current levels, but a good efficient heatsink is essential.

Even higher currents are easily obtained simply by paralleling the DMOS power FET with another or others—a unique property that power DMOS FETs have over bipolar power transistors. (See the dotted lines in the circuit.) Initially, the FET with the highest transconductance (g_{fs}) will draw the most drain current, and as a result will have the highest power dissipation. However, the resulting increase in chip temperature will also increase its $R_{DS(on)}$, with its positive tempco, which will help limit the drain current. As a result, the total drain current will automatically balance itself through all of the other power FETs in this circuit. A separate gate resistor (1 KΩ) should be used between the op amp's output and the gate of each FET and located as close as possible to the gate. All resistors should be close tolerance, high-quality metal-film types. A good efficient heatsink will be required for the power device(s). In this way it would be possible to easily build a high-quality 20-Amp (or higher) power current sink.

8.3 Using "smart" power MOSFETS

In creating power current sources, there are some additional power MOS devices to consider using. It is worth noting that some manufacturers have introduced DMOS power devices that have integrated some of the techniques shown in the proceeding circuits on-chip, for the purposes of either current-monitoring or current-limiting and device protection. Manufacturers making on-chip current-sensing DMOS power devices include International Rectifier, IXYS, and others. Those making on-chip current-limiting and overvoltage protection include International Rectifier, ON Semiconductor, Philips Semiconductors, and others. The circuits shown here in Figures 8.22 and 8.23 are generically representative of these new types of "smart" power devices. In Figure 8.22, the device is made to include two extra terminals: Kelvin source and current sense. These are provided to monitor and then output a small current, which is directly proportional to the drain-to-source current (effectively like a current mirror). This type of power device is used mostly in motor-drive and switched-mode power supply circuits, where monitoring current is often crucial. In your application, it may provide a perfect solution.

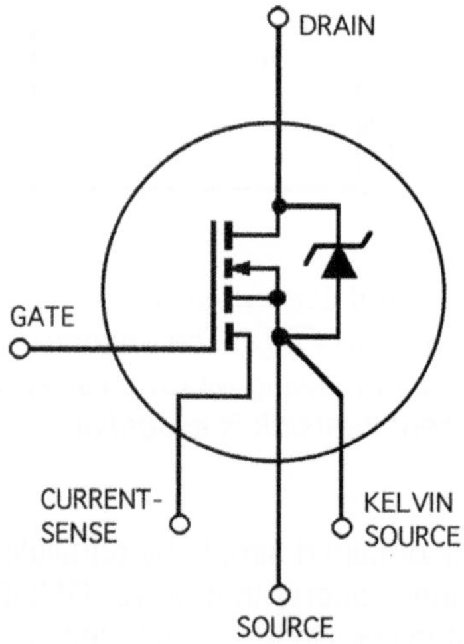

Figure 8.22. A "smart" enhancement-mode, vertical power MOSFET, with "on-chip" current sensing.

In Figure 8.23, the power MOSFET is equipped with additional protection features that include built-in short-circuit and current-limiting, as well as protecting it against ESD and avalanche voltage conditions. This type of power device is useful when switching inductive loads and where large voltage transients are common. Notice that both the gate and drain are protected against ESD (typically 2 KV or more) and avalanche voltages, and the drain-to-source current is monitored by the bipolar transistor. In the event of a short-circuit or overcurrent condition, the bipolar transistor effectively turns the power device off. This type of power DMOS FET is a close cousin of the current sink circuit previously shown in Figure 8.16. It is particularly useful in motor-drive and switched-mode power supply circuits. Some people might think of it as being a bullet-proof MOSFET.

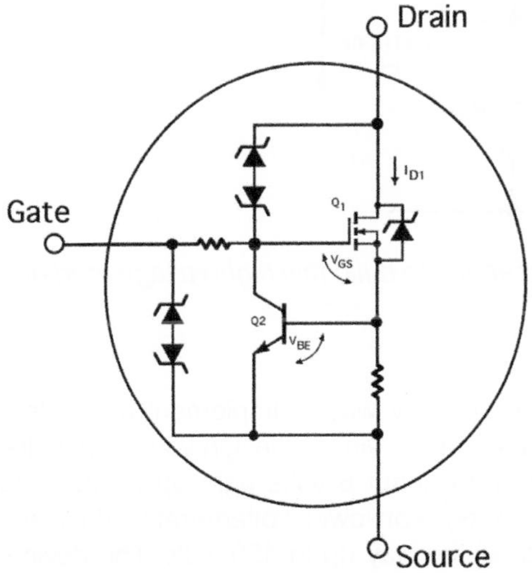

Figure 8.23. *This "smart" MOS device protects itself against short-circuit, overcurrent, and overvoltage conditions that could otherwise destroy it.*

8.4 IXYS power MOS current sources

One of the market leaders in power MOSFETs is IXYS Corporation (Santa Clara, CA), which offers various power devices, including power MOSFETs, SCRs, power diodes, FREDs, bridges, IGBTs, custom devices, and dedicated MOS current regulators (their IXC series). These devices, which are based on a proprietary IXYS high-voltage process, provide excellent temperature stability that yields a tempco of typically 50 ppm/°K. Devices are provided in popular three-lead, power transistor packages including TO-220 and D-PAK. One simply chooses the appropriate device from a wide range of available currents, just like with a JFET current-regulator diode.

High-voltage operation is possible up to 900 volts in some of these IXC devices, at fixed currents ranging typically from between 2 and 110 mA. The exact current level is factory-adjusted during manufacture. Devices may be used as stand-alone for DC circuits (as shown here in Figure 8.24) or connected in series to form high-voltage AC regulators. In addition, while most models are fixed and nonswitchable, some models are

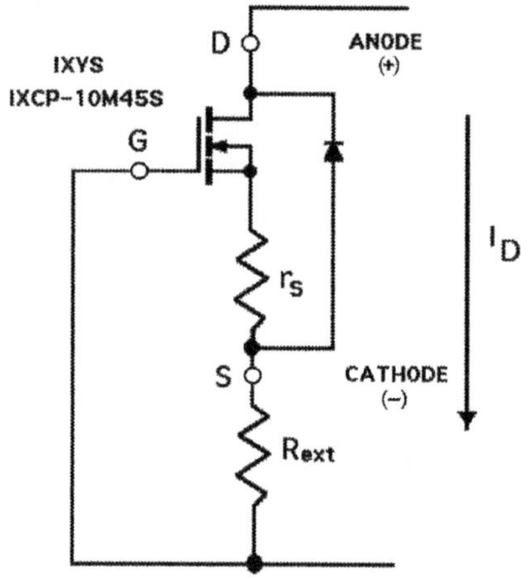

Figure 8.24. *Only two parts are needed to build this high-voltage medium-power current source from IXYS Corp.*

switchable. They provide an easy way to implement a simple, low-cost, high-voltage power current source rather than trying to design your own. Voltage operation for a typical device could range from about 6 volts right up to the breakdown voltage of the device (more than 800 volts). For lower voltage-rated devices, this typically extends down to about 3.5 volts, all the way up to 450 volts. The devices typically offer a very high output impedance, as well as a high bandwidth of several hundred MHz.

IXYS also offers a large range of rugged N- and P-channel, power MOSFETs based on its own technologies (HiPerFET™ and MegaMOS™), in metal, plastic power, and surface-mount packages. Some devices are rated for breakdown voltages in excess of 1 KV. Others are for lower voltage operation, higher current, with ultra-low on-resistances.

8.5 Lateral enhancement-mode MOSFETs

The last member of the MOS family is the *lateral* enhancement-mode FET. These products are made by Vishay-Siliconix, Supertex, Calogic, Fairchild Semiconductor, Philips Semiconductors, Linear Integrated Systems, and others. They include products such as the Siliconix SD-210/211/5000/5400, the Philips BSD12 and BSD 22, and the Supertex low-threshold P-channel LP0701/0801 and TD9944. The very-high-speed, low-voltage, lateral, enhancement-mode FET features a wide dynamic range, low distortion, and high frequency operation with ultra-fast switching, with turn-on speeds of less than 1 nSec. Such devices can be considered a bridge between high-speed JFETs and gallium-arsenide (GaAs) FETs and are primarily used as high-speed analog signal switches, having high input impedance (Z_{in}) and very low input capacitance (C_{in}). Although current sources can be built using these devices, other types of MOSFETs are really more suitable.

Chapter 9

Using Analog CMOS Arrays to Create Current Sources

9.1 RCA pioneered CMOS

Once the first stable MOSFETs had been created in the early 1960s by RCA researchers Steve Hofstein and Fred Heiman, and by Dr. Frank Wanlass at Fairchild Semiconductor, it opened the door to researching and developing other types of MOS transistors. This included small-signal *complementary-MOS* (CMOS) devices, which would theoretically have both N-channel and P-channel transistors coexisting on the same chip together. Although other researchers around the world had tried, no one had yet succeeded in doing this. It was an uncharted area, considered by many to be a lost cause. However, everyone agreed that in theory it was a wonderful concept, particularly for switching functions such as in digital logic gates. The N- and P-channel devices would consume minute amounts of power only during switching (when compared with TTL logic gates), but how to get both types on the same chip was what eluded everyone.

In 1963, Dr. Frank Wanlass at Fairchild Semiconductor discovered that the cause of instability in making MOS devices was trace amounts of sodium. Once the sodium was eliminated, his MOS transistors were perfectly stable. Soon after this discovery, he focused his attention on CMOS, for which he saw a great future. Unfortunately, he could not get his process to work properly because of processing problems. This was a huge technical challenge at the time, but despite not being able to get the device to work properly, he wrote a patent for it anyway. In 1963, Fairchild Semiconductor was granted the first patent for metal-gate CMOS by the U.S. Patent Office. At about the same time, the two RCA researchers, Hofstein and Heiman, developed the first working multitransistor MOS array IC (but not complementary). This in turn led to several government contracts for RCA, for custom MOS devices, although none were made available commercially at the time.

Since the beginning of the 1960s, another small group of researchers at RCA's labs in Princeton, New Jersey, had been researching small-signal MOS devices. They were being pushed by RCA's management to made a breakthrough in this area because it was hoped that MOS would become a viable replacement for some of the vacuum tubes used in its TVs. RCA was primarily a TV set maker. The group, which was managed by Jerry Herzog and led by Israel Kalish, included Al Medwin and Art Lipschutz. Together they developed the first low-power CMOS chip technology and incredibly made the process work, when no one else could. They started by creating simple gates, then decided on a complete digital logic family. At first the devices were speci-

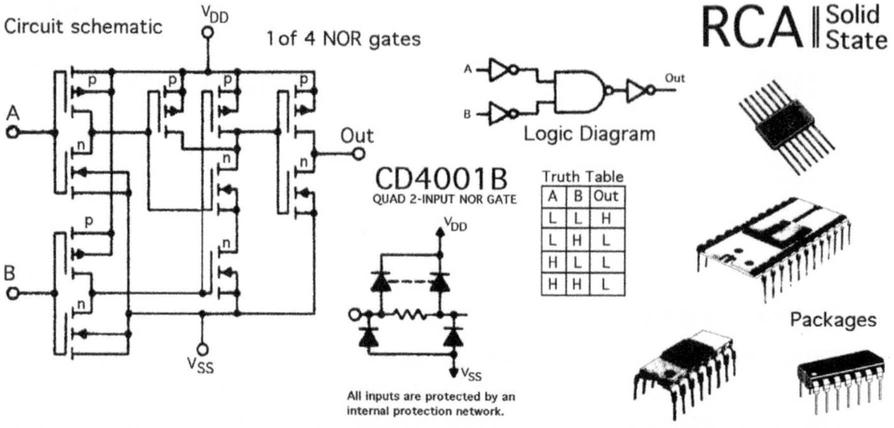

Figure 9.1. *When RCA introduced the first commercially available CMOS devices in 1968 it created a whole new facet to the semiconductor industry and was a major milestone in the industry's history. The first devices were simple gates in ceramic flat-packs, then followed ceramic D-I-L packages as shown here.*

fied at up to 18-volt operation, but this slowly came down, as their expertise grew. Finally, RCA managed to produce parts that would run on 5-volt (TTL) supplies. When it was ready for the marketplace, RCA called the new technology *complementary symmetry metal-oxide* silicon (abbreviated COS/MOS®) and trademarked it.

So in 1968, RCA unveiled its first CMOS logic family, the CD 4000 series (Figure 9.1). Initially it was only available in ceramic packages (like the flat-pack, which the military, NASA, and contractors loved). This introduction was a significant milestone in the marketplace. Up until that time, the logic family of choice had been the bipolar TTL 5400/7400 series, pioneered by Texas Instruments. (TI did not get into CMOS for a long time after it was introduced by National, Motorola, and Fairchild, which quickly shared the new fledgling marketplace with RCA.) Later, others throughout the industry referred to it by the generic name CMOS, which has stuck to this day.

RCA surprisingly had production problems. It was Al Medwin again who found that because most of RCA's production staff were people who had formerly worked with vacuum tubes, they had no concept of handling MOS devices properly or of packaging them. For that reason, Medwin cleverly designed the CMOS family to include built-in input and output protection. It primarily guarded against static *ESD charges* and brief voltage spikes. One major source of problems had been in the packing and shipping area. Now, however, by eliminating static materials from the work area, by much more careful handling and using antistatic packaging, it cured the problem. This little remedy trickled out to the rest of the industry, as it began to license and make CMOS products. It's a procedure that still applies today: always handle CMOS products *care-*

fully, preferably when you are wearing a grounded (via a 1-MΩ resistor) antistatic wristband, on a grounded, static-free work surface.

By the late 1960s, the problems that had previously delayed progress with MOS transistors were all but over. By the early 1970s, the same group of RCA researchers had created more complex products, including the "1801"—the first two-chip CMOS microprocessor developed by Bob Winder. By 1974, another member of the group, Eddie Dingwell, had created the first single-chip CMOS microprocessor, the "1802." In 1976, Japan's Hitachi introduced the world's first complementary power MOSFETs on the same chip, using a proprietary planar, lateral structure. Hitachi followed this just two years later by developing the world's first fast CMOS devices, in the form of a 4K static CMOS RAM memory chip. It matched the speed of the industry-standard NMOS design (Intel) at the time, but drew only about a tenth of its power. The researcher credited with heading that development was Toshiaki Masuhara, who had been educated at Kyoto University, then later at the University of California at Berkeley. Today he is President of Hitachi Microsystems in Japan.

From the mid-60s through the mid-80s, RCA Solid State, as the division had been named, was a major technology leader in advanced semiconductor devices. Besides the 4000-series logic chips, this included both bipolar discretes (like the famous 2N3055 power transistor), CMOS microprocessor families, CMOS A-to-D converters, the first bipolar-MOS op amps, CMOS A/D "flash" converters, and MOS RF parts. Within a few years, RCA's CMOS technology was licensed and used by many microprocessor and digital IC makers worldwide. By the early 1970s, RCA's CMOS devices were being designed into consumer products such as pocket calculators and watches—a huge global market at the time. With this experience in mass-producing CMOS came other technologies and products along the way. This included silicon-gate CMOS, which provided even lower power consumption, smaller chips, and helped open the door for many of today's battery-powered consumer products.

RCA was eventually absorbed back into General Electric in 1986, and most of its semiconductor portfolio was sold off to Harris Semiconductor. This was part of an even wider reorganization by GE, which later decided to get out of the semiconductor business altogether. Without RCA's efforts, we probably would not have CMOS as we know it today, where it now dominates various market segments. In the digital marketplace, for example, we have CMOS microprocessors (like Intel's Pentium™ and Xeon™) and the IBM/Motorola Power PC™ G5 used in today's Apple Macs, glue-logic, graphics controller chips, and CMOS RAM, to name but a few. In today's analog marketplace, CMOS op amps, analog switches, A-to-D converters, and power supply controllers are commonplace.

Although RCA had introduced CMOS commercially—first in digital products, then later with analog parts - other companies soon followed. Three U.S. companies that have exclusively pioneered analog CMOS have been Intersil, Maxim Integrated Products, and Advanced Linear Devices. Intersil came first (in 1969) and challenged RCA's leadership with several cutting-edge products that included analog switches and op amps. They introduced the first CMOS dual-slope A/D converter (ICL7106), as well as the first CMOS voltage inverter (ICL7660). Both became very popular design-ins and helped further establish analog CMOS products in the marketplace. Intersil's enormous success however resulted in their being acquired by General Electric in 1980. However, this action resulted in the key people who had set up and run Intersil to leave and co-found another CMOS pioneer—Maxim Integrated Products (Sunnyvale, CA), in 1983. Maxim has subsequently established itself as a world leader in analog CMOS with a wide range of products that include op amps, precision voltage references, A/D converters, charge pumps, digital pots, and SMPS products, to name but a few.

When General Electric decided to pull out of the semiconductor business in 1986, it sold off parts of RCA Solid State and Intersil to Harris Semiconductor. Later Harris Corporation decided to spin off its semiconductor operation. Thus Intersil Corporation was reborn in 1999, combining parts of Harris Semiconductor, and GE Solid State. In the ensuing years, it has subsequently grown, and acquired other companies including Elantec Semiconductor, and Xicor Corporation. Today Intersil is ranked in the NASDAQ-100® top companies (as is Maxim), and is once again a global leader in the semiconductor industry. Like Maxim, Intersil also makes a wide range of analog products, and some of the world's most advanced voltage references, that we will read about in Part 2 of this book.

The third analog CMOS pioneer is Advanced Linear Devices (Sunnyvale, CA). Founded in 1985, ALD is probably best known for redesigning the popular "555" timer as a low-voltage, silicon-gate, CMOS device (ALD-555-1), which runs on a miniscule 1-volt supply voltage. ALD also has an exciting range of low-power CMOS products that include op amps, timers, comparators, A/D converters, and electrically programmable analog devices (EPADs® and ETRIMs™), that can be used to build current sources and voltage references. Other major U.S. semiconductor companies with significant analog CMOS product lines today include Analog Devices, National Semiconductor, and Texas Instruments.

The company that has more experience than any other at mass-producing CMOS is National Semiconductor. It was one of the first to begin shippping 74C and 4000-series digital CMOS. For more than 30 years, National has added many other products in CMOS, both digital and analog; some have been simple like a NOR gate, others have been extremely complex, like entire microprocessor families. In addition National has pioneered several A/D and D/A families, as well as CMOS op amps and CMOS voltage references. Today, National Semiconductor is also a world leader in advanced CMOS optical arrays, which are used in some of the the world's most advanced digital cameras, such as those from Hasselblad and Sigma. In a joint venture with Foveon Inc., National co-designs and manufactures CMOS image sensors that are cutting-edge, super-high-resolution technology (more than 16 megapixels), capable of the highest color quality measurable. Yes—National knows all about CMOS too!

9.2 Characteristics of CMOS FETs

CMOS FETs share most of the same characteristics of the low-power enhancement-mode devices DMOS FETs that are described in Chapter 8. However, CMOS designs combine both N-channel and P-channel transistors on the same substrate, whether digital gates or analog op amps. Typically, the N-channel MOS device is used as the driver, while the P-channel transistor is used as its active load. The sensitive inputs and outputs of most CMOS devices employ protective diodes to guard against dangerous static ESD voltage transients. Without this added protection, the ultra-thin oxides used in fabricating CMOS devices can be easily ruptured (see Figure 9.2).

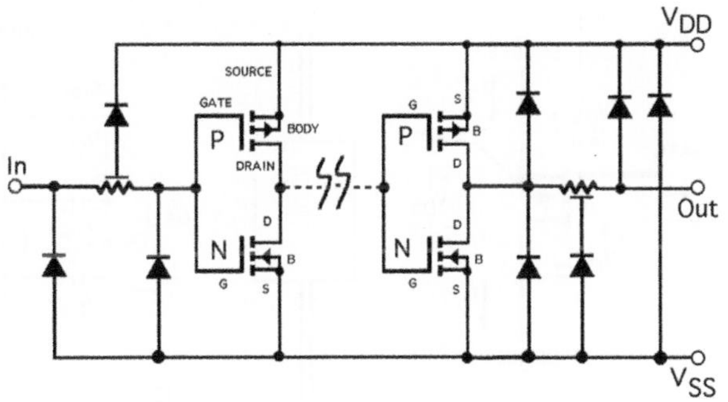

Figure 9.2. *CMOS input and output protection networks.*

Being low-power enhancement-mode MOSFETs, they are *normally-off* devices and are created as both P-channel or N -channel types on the same substrate. They also have ultra-low current and voltage (low-power) capabilities. In the case of N-channel types, a positive voltage between the gate and source ($+V_{GS}$) is required to turn them on. For the P-channel device, a negative gate-voltage with respect to the source is required to turn it on ($-V_{GS}$). With either polarity of MOSFET, the drain current (I_D) falls to zero when V_{GS} equals 0. CMOS devices use small chips with very low capacitances, resulting in fast (10-nSec) turn-on times.

One major characteristic is that the CMOS FET is a majority-carrier device, which means that it does not suffer from the minority-carrier storage time effect like bipolars do, thereby switching faster. It requires infinitely less gate drive current than a bipolar's average base current or a large MOSFET, and it has a significantly faster switching time—at least five times faster. The CMOS FET also has the ability to be easily paralleled, as well as having a low on-resistance. The resulting lower voltage drop, $V_{DS(on)}$, across the FET translates into much less heat needing to be dissipated from the chip. Last but not least, the CMOS FET's on-resistance, $R_{DS(on)}$, has a positive temperature coefficient. This means that if the FET heats up, its resistance also increases, thereby helping to limit the current through it (the drain current has a negative tempco). This action reduces the possibility of thermal runaway, as the chip's junction temperature increases.

The symbols, polarities, and large-signal models for both P- and N-channel MOSFETs are shown in Figure 9.3, which is virtually the same as the regular enhancement-mode MOS device. The P-channel MOSFET device shown in Figure 9.3A and B has all of its currents and polarities reversed. CMOS devices have a similar way of operating and share most of the same characteristics and terminology (i.e., $V_{(BR)DSS}$, g_{fs}, $I_{D(off)}$, and I_{GSS}) as the enhancement-mode DMOS FET that we looked at in the previous chapter.

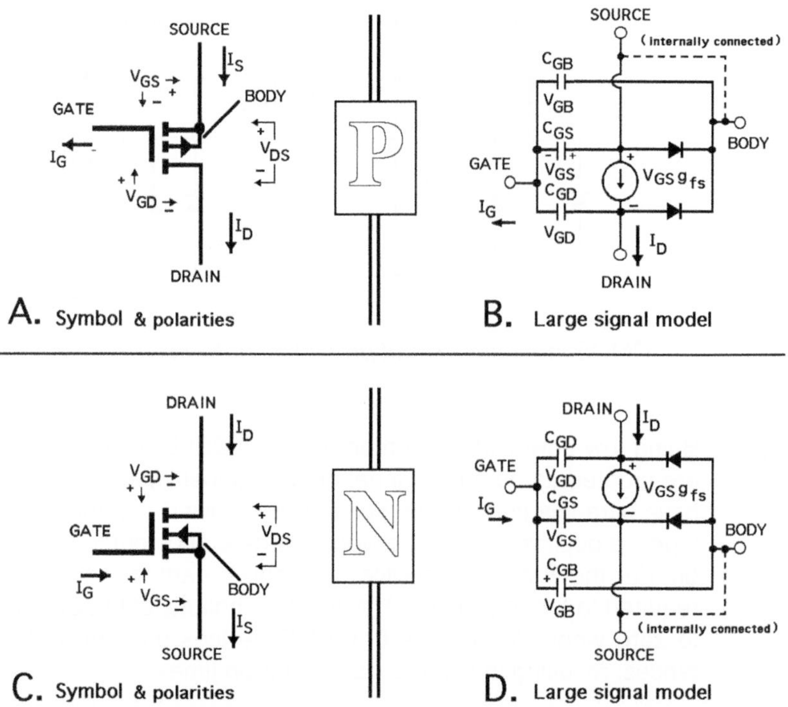

Figure 9.3. *Showing the symbols for both P- and N-channel CMOS FETs, as well as various voltages, currents, polarities, and models. Note that for most devices the body terminal is normally tied internally to the source, but not all.*

As mentioned previously, enhancement-mode MOSFETs are normally-off devices, requiring either a positive voltage (N-channel types) between the gate and source to turn them on or a negative gate voltage (P-channel types) to turn them on. For either type, biasing the gate to zero volts (V_{GS} = 0V) will reduce conduction, until finally a threshold point is reached where conduction ceases. It relies on an isolated, capacitive gate, made with either a metal- or a silicon-gate. An N-channel CMOS device normally has a very thin oxide layer that isolates the gate from the P-region below. Because the gate is isolated from the rest of the device, it usually has ±15-volt limiting gate voltage (some are considerably lower at less than ±5 volts). By applying a positive drain-to-source voltage of 10 volts and biasing the gate to say 5 volts also in the positive direction, (1) repels the holes away from the surface of the P-region near the gate. This action means that the electrons are now the majority carriers by default, which (2) causes an inversion of the channel (aka surface inversion). (3) This allows electrons to flow in the channel, which induces full conduction between the drain and source. These actions are representative of any N-channel CMOS FET.

P-channel devices work in reverse, by applying a negative drain-to-source voltage (or a more negative voltage) of several volts and biasing the gate to several volts in the negative direction. This first repels the electrons from the N-region near the gate, which allows holes (the majority carriers) to flow in the isolated channel that forms. Again, this induces full conduction between the drain and source. Because P-channel devices are less efficient, they usually require greater chip area to match the characteristics of their N-channel counterparts.

Although one might assume that the typical CMOS FET is a relatively simple device, it actually contains several inherent parasitic elements, which include a bipolar transistor, a JFET, substrate diodes, and various inductances, resistances, and several important capacitances. Each manufacturer uses various proprietary processes to skillfully reduce these unwelcome parasitic characteristics, while exploiting others. The end product is often a compromise between many conflicting characteristics.

An important CMOS FET characteristic regarding its maximum operating voltage is the drain-to-source supply voltage ($V_{(BR)DSS}$). This represents the upper limit of the device's voltage blocking capability and is invariably specified at the beginning of a device's data sheet. Typically, the FET is actually connected to the V_{DD} rail, to which this applies. Although some CMOS devices can work at below 1.5 volts, V_{DD} is more typically between 3 and +12 volts. CMOS devices are manufactured according to a particular process, much of which dictates the breakdown voltage of the device. As a result, CMOS FETs normally use the same value for both $V_{(BR)DSS}$ and $V_{(BR)GSS}$ (the gate-to-source voltage), although the latter is not usually specified other than in the absolute maximum ratings for V_{GS}. Many analog CMOS devices can run on split power supplies (+V and –V) or between 0V and say –10V, so long as $V_{(BR)DSS}$ is never exceeded. As with any type of enhancement-mode device, $V_{(BR)DSS}$ has a positive temperature coefficient of 0.1%/°C.

Although the CMOS FET is controlled by a voltage at its isolated gate, there is actually a very small leakage current involved, called the gate-to-source/body leakage current (I_{GSS}). It is typically in the range of 1 to 50 pico-amps, when measured at room temperature and at a specified gate-to-source voltage. (At elevated temperatures, the leakage current can rise to around 10 nA.) This is measured with the drain terminal shorted to the source and with 10 volts applied between gate and source. The measurements and levels are virtually identical for both N-channel and P-channel devices. Remember that with most commercially available CMOS FETs, their source and the body regions are internally connected together. This I_{GSS} leakage current doubles for every 10°C rise in temperature, which in some applications may be important. Even so, at 125°C it is almost certainly still below 50 nano-amps, which is quite small.

Another important characteristic is $I_{DS(off)}$, which refers to the small leakage current that flows between drain and source when the gate is shorted to the source (i.e., $V_{GS} = 0$); in other words, when the device is off. This is the same for either N-channel or P-channel devices. At room temperature it is typically a very low value, usually several hundred pico-amps. At elevated temperatures (such as 125°C), it can reach a few microamps. The $I_{DS(off)}$ leakage current is actually that of a reverse-biased diode, and it doubles for every 10°C rise in temperature.

The maximum limiting current that can flow between the CMOS FET's drain and source is known as $I_{DS(on)}$, or the on-state drain current. This occurs at a particular drain-to-source voltage, V_{DS}, and at a particular gate-to-source voltage, V_{GS} (positive value for N-channel and negative value for P-channel devices). The graph depicting this action is shown in the data sheets of all enhancement-mode MOS devices, and as with discretes, it is referred to as the device's *output characteristic* ($I_{DS(on)}$ vs. V_{DS} graph) (see Figure 9.4). When small values of V_{DS} and V_{GS} are simultaneously applied, the drain current increases linearly with V_{DS}. This region is known as the *linear region*. As the V_{DS} is increased, the drain current begins to be pinched off or limited at the knee of the curve, before finally becoming saturated and flattening out. This is known as the *saturation region*. When the drain current reaches saturation, it becomes proportional to the square of the applied V_{GS}, and is then only slightly dependent on V_{DS}; this is known as $I_{DS(on)}$. Manufacturers usually measure this in a pulse-mode setup, to reduce heating of the chip. With a CMOS FET, this $I_{DS(on)}$ current is usually less than 25 milliamps. It is important to remember that $I_{DS(on)}$ is also limited by the maximum junction temperature, $T_{j(max)}$, which should never be exceeded. Generally, the N-channel device will always have a higher current rating than the P-channel device, and so $I_{DS(on)}$ can typically range 2:1 for complementary devices in the same package. For matched-pairs, however, this is virtually identical. $I_{DS(on)}$ is temperature sensitive and has a negative temperature coefficient of approximately –0.5%/°C.

Another relevant characteristic of the CMOS FET is its static on-state resistance, $R_{DS\,(on)}$. For many CMOS FETs, it is often less than 500 Ω. The lower the $R_{DS\,(on)}$, the higher the current that the device can switch. $R_{DS\,(on)}$ is measured as the V_{DS}, divided by a particular drain current, at a particular V_{GS}, at 25°C or at an elevated temperature. It can be expressed as:

$$R_{DS(on)} = \frac{V_{DS}}{I_D} \qquad \text{(Eq. 9.1)}$$

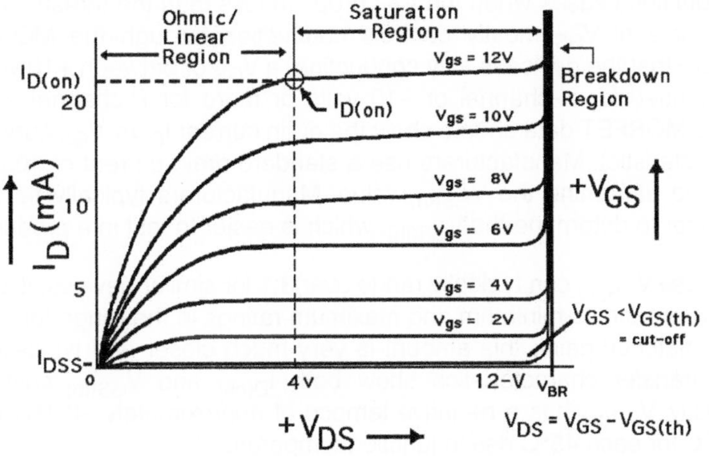

Figure 9.4. *Typical output characteristics for an N-channel enhancement-mode CMOS FET.*

Because of device self-heating, $R_{DS\ (on)}$ is usually measured in a pulsed manner, at room temperature, because heating the device decreases carrier mobility, thereby reducing the drain current for a given voltage. At low voltages, $R_{DS\ (on)}$ is governed mainly by the channel resistance, whereas at higher voltages the epi and JFET regions are most significant. It is interesting to note that the $R_{DS\ (on)}$ for P-channel devices is typically three to four times higher than for similar N-channel devices on the same substrate. As a result, in order to make complementary pairs with similar on-resistances, the P-channel device needs to be a physically bigger area. With CMOS matched-pairs, a typical figure for $R_{DS\ (on)}$ mismatching is less than 0.5%. For an N-channel device with a typical $R_{DS\ (on)}$ of 50 ohms, this would equate to a mismatch of just 250 milli-ohms. For a CMOS FET, $R_{DS\ (on)}$ has a positive tempco of approximately +1%/°C. Actually, this is a beneficial characteristic, because as the device heats up and the on-resistance increases, the drain current automatically reduces.

Another significant characteristic is the *gate threshold voltage*, $V_{GS(th)}$. This is the gate-to-source threshold voltage necessary to just turn the CMOS FET on, at a very low current level. $V_{GS(th)}$ is actually the voltage required to cause the surface inversion of the channel that allows any forward current to flow. The V_{GS} for an N-channel MOS-FET ranges from 0 volt, where the device is fully cut off, to some positive amount of several volts (typically between 2 and 4 volts) to turn the device fully on. Most CMOS FETs have a very low $V_{GS(th)}$ value, typically ranging between 0.4 and 1 volt. Such devices provide overall faster switching, because less charge current is needed to charge the parasitic input capacitances. However, they are somewhat susceptible to voltage transients, which can cause spurious turn-on of the device. Typically, an analog CMOS FET's isolated gate has ±8 to ±15-volt limiting gate voltage, depending on the device.

Conduction ceases when the V_{GS} drops to less than the threshold voltage; otherwise, the value of V_{GS} mostly controls conduction through the MOSFET's channel. To ensure that the device is fully conducting, a V_{GS} of between +10 volts or higher is recommended for N-channel or –10 volts or more for P-channel. Most enhancement-mode MOSFET data sheets show the drain current I_D vs. V_{GS} curves (aka the transfer characteristic). Manufacturers use a standard drain current of 10 μA (and with $V_{GS} = V_{DS}$) to determine the $V_{GS(th)}$ value. Manufacturers typically tie the drain and gate together to determine the $V_{GS(th)}$, which is easier to test in a production environment.

Because $V_{GS(th)}$ can typically range over 3:1 for similar devices, it is important that you consider both its minimum and maximum ratings in the range for your application. For dual matched-pairs, this amount is very much closer and typically less than 10 mV. The transfer characteristics show both $I_{D(on)}$ and $V_{GS(th)}$ shift with temperature. Actually, $V_{GS(th)}$ has a negative tempco of approximately –0.1%/°C (approximately 2 mV/°C for each 45°C rise in junction temperature).

Another characteristic of the CMOS FET is its transconductance/forward conductance (G_{fs}). G_{fs} measures the effect of a change in drain current (I_D), for a specific change in gate voltage (V_{GS}), referenced to common-source mode.

$$g_{fs} = \frac{\Delta I_D}{\Delta V_{gs}} \bigg|_{V_{DS}} = \text{constant} \qquad \text{(Eq.9.2)}$$

In practical terms, transconductance is a measure of the FET's gain and is a figure of merit. It is usually referred to from a practical standpoint as milliamps per volt (mA/V). On a graph it appears as a steep or shallow slope and is measured and referenced in milli-Siemens (mS). For example, 1 mA/V is a lot more shallow slope than 10 mA/V. G_{fs} is usually specified on the data sheet for some CMOS FETs with a minimum or typical value. The G_{fs} value given in a manufacturer's data sheet is measured at a particular I_D and V_{DS}. Typical values for analog CMOS transistors are between 5 and 10 μS for N-channel FETs and between 2 and 4 μS for P-channel devices.

Another important characteristic of the CMOS FET includes its capacitances. This includes the input capacitance C_{iss}, the output capacitance C_{oss}, and the reverse-transfer capacitance C_{rss}. Of the three, C_{rss} is the most dominating, because it is part of the feedback loop between the device's output and its input. It is also known as the Miller capacitance. The frequency response of the MOSFET is governed by the charging and discharging of C_{iss}, which is composed of the gate-to-source capacitance (C_{GS}), the gate-to-drain capacitance (C_{GD}), and the resistance of the gate overlay structure (not to be confused with the very large input resistance of more than 100 MΩ). This is typically around 20 Ω for silicon-gate and about 10 Ω for metal-gate FETs. Typically, the upper frequency limit of a silicon-gate CMOS FET is in the range of 10 to 50 MHz and about twice that for metal-gate devices. CMOS FETs have a higher frequency response, because the size of the chips is smaller, thereby reducing the value of C_{iss}. Because C_{iss} is unaffected by temperature effects, neither is the CMOS FET's switching speed.

Remember that a MOSFET has an isolated capacitive gate and that being essentially a capacitor, it takes time to charge or discharge that capacitance, as well as to support a certain amount of charge. It works in the following way: Assuming that an N-channel MOSFET is connected with a 10-volt positive supply voltage at the drain, a source that is grounded, and the gate temporarily at ground, if the gate is now connected to the positive supply voltage, the V_{GS} starts to increase. Soon after it will reach the $V_{GS(th)}$ at say 1.5 volts, at which point a small drain current I_D will start to flow, and the C_{GS} will begin to charge. Once C_{GS} has fully charged, the gate voltage becomes constant and begins to charge the C_{GD} (also known as the Miller capacitance). This takes longer than when charging C_{GS}, because it is a larger capacitance. Once C_{GD} has finished charging, the V_{GS} starts increasing again until it finally reaches the 10-volt supply voltage. When it reaches this point, that is the total time needed to turn the device full-on.

As with all other types of FET, an important characteristic of the CMOS FET is its output conductance (G_{os}). When the FET is applied as a current source, the quality of its current regulation is strongly dependent on its output conductance, which is in turn closely related to its drain current (I_D). The lower the drain current, the lower the conductance, and the better the regulation will be. Remembering that conductance is the reciprocal of resistance, very low conductance translates into high resistance (i.e., a 0.2-mS $g_{os} \approx 5\ K\Omega$). In practical terms, output conductance is measured and referenced in milli-Siemens (mS). G_{os} is usually specified on the data sheet with a typical value, measured at a particular I_D and V_{DS}.

An important consideration for using any MOSFET is its maximum power dissipation (P_D). While MOSFETs have much better temperature characteristics than bipolars (in that the on-resistance is positive, thus reducing the current flow) and the Safe Operating Area (SOA) is more rectangular, where a bipolar's is more limited, it may still need to dissipate power, in the form of heat, away from the chip. For a CMOS FET, it is limited by its breakdown voltage, current rating, on-resistance, power dissipation, and maximum junction temperature. Exceeding any one of these, particularly its maximum $V_{(BR)DSS}$ rating, could be fatal to the device and the circuit. For CMOS matched-pairs and quads, this specification will probably be around 500 mW, with an operating temperature of 0°C to +70°C. Some devices have a military operating temperature range of –55°C to +125°C.

9.3 Using CMOS linear arrays to create current sources

If you asked most electronic designers how they would create current sources using analog CMOS devices, they would probably tell you that no such devices exist to use in their designs. Wrong! Actually they do exist in the form of various matched-pairs, matched-quads, and matched complementary pairs from Advanced Linear Devices Inc. This Sunnyvale, California–based company specializes in analog CMOS and has an exciting and growing product line. (Most engineers have heard of ALD's super low-power CMOS "555" timer chip, which runs on an incredible 1 volt.) Besides timer chips, ALD also makes CMOS op amps, A-D converters, comparators, custom ASICs products, and electrically programmable analog devices (EPADs®). The ALD products that are particularly suitable for use as current sources are shown in Table 9.1. They are available in various packages, including surface-mount and throughhole, and in either the commercial or military temperature ranges.

ALD's MOSFET devices are enhancement-mode (normally-off) FETs, manufactured with their proprietary state-of-the-art silicon-gate CMOS process. This process results in creating small, high-speed, very-low-power chips. They are available in several offset voltage (V_{OS}) grades of between 2 mV and 10 mV. This is the maximum difference in gate-to-source voltage (ΔV_{GS}) between individual transistors on the chip. They offer an extremely high input-impedance, low $R_{DS(on)}$, a fairly low g_{os}, and very low V_{DS}

ALD Part #:	Description:	Package options:
1101	Dual N-channel matched MOS FET Pair	8-pin TO-99, DIP 8-pin, SO-8
1102	Dual P-channel matched MOS FET Pair	8-pin TO-99, DIP 8-pin, SO-8
1103	Dual N-channel and P-channel matched MOS FET Pair	DIP 14-pin, SO-14
1106	Quad N-channel matched MOS FET Pair	DIP 14-pin, SO-14
1107	Quad P-channel matched MOS FET Pair	DIP 14-pin, SO-14
1108E	Quad N-ch. programmable EPAD™ Precision matched Pair	CERDIP-16, DIP 16-pin, SOIC-16
1110E	Dual N-ch. programmable EPAD™ Precision matched Pair	CERDIP-8, DIP 8-pin, SOIC-8
1116	Dual N-channel matched MOS FET Pair	DIP 8-pin, SO-8
1117	Dual P-channel matched MOS FET Pair	DIP 8-pin, SO-8

Table 9.1 ALD's matched CMOS transistors

operation. They also have low input capacitance (C_{iss}), therefore fast switching, and have a negative temperature coefficient. These devices all have a guaranteed very low threshold voltage (V_{th}) of 1 volt maximum, for both N-channel *and* P-channel products. Because they are low-power devices, current sources designed with them will be used in designs typically requiring less than 2.5 milliamps. ALD's EPADs® (which we cover later this chapter) are low-voltage, electrically progammable devices that can also be used as precision current sources.

Comparing these devices to JFETs and DMOS FETs that we have looked at previously, these CMOS devices have some major advantages for the designer. First, these devices are matched monolithic pairs or quads, versus unmatched single discrete devices, which one would likely have to try and match in some applications. They definitely provide easier handling and savings, particularly in terms of inventory management and incoming Q&A, not to mention the tedious burn-in, testing, and device matching procedures needed with some discretes. As a result of the monolithic construction, there is excellent thermal tracking between devices and close matching of

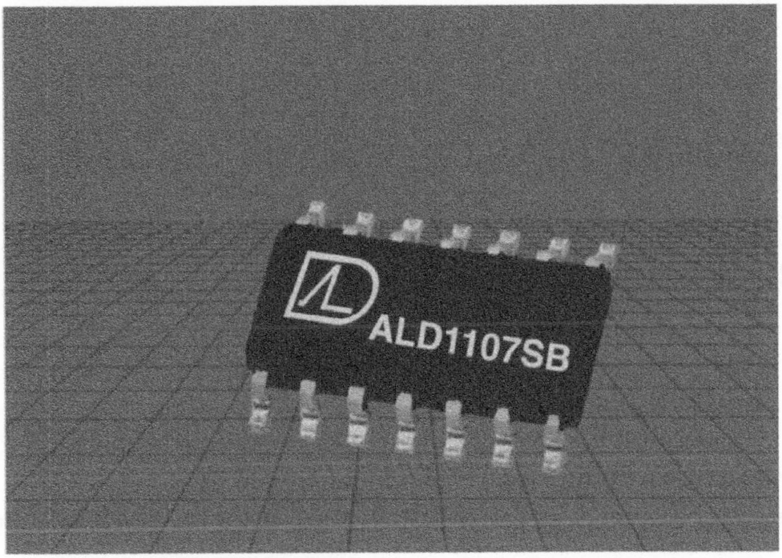

Photo 9.1. *Ultra-low threshold CMOS transistor arrays from Advanced Linear Devices, Inc. (Photo courtesy of ALD, Inc.)*

other characteristics. They can provide constant currents over a range of drain-to-source voltages from about 2 volts to 12.5 volts.

One major advantage is that being monolithic duals or quads, they can be made into current mirrors far more easily than discrete FETs. Some practical circuit examples of this are shown in Figures 9.5 to 9.11. As mentioned before, current regulation strongly depends on output conductance (g_{os}), which in turn is closely related to the drain current (I_D). Although in theory a MOSFET may be biased to operate as a current source at any level below its maximum drain current, $I_{D(on)}$, the lower the drain current, the tighter the regulation will be. Best performance is obtained when using a MOSFET that is biased well below its maximum drain current. Because the minimum drain current for ALD's N-channel devices is 30 mA, and for the P-channel devices it is about 11 mA, one should ideally try to make the current source's desired output level 10% or less of either of those values, in order to maximize regulation. While the typical value for output conductance is about 200 μs for the N-channel FETs and about 500 μs for P-channel FETs, a circuit's real output conductance (g_o) will be significantly lower (more than 10 times) than the data sheet value for an individual CMOS transistor (because of a low value of drain current). Cascoding devices (which we will look at shortly) can result in reducing the circuit's output conductance even further (100 times).

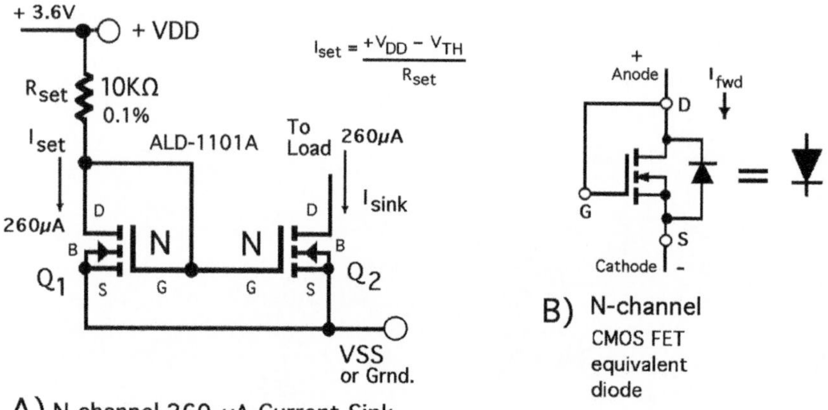

A) N-channel 260-μA Current Sink

Using a CMOS N-channel dual matched-pair to create a mirrored current sink. Note the equivalent diode connection.

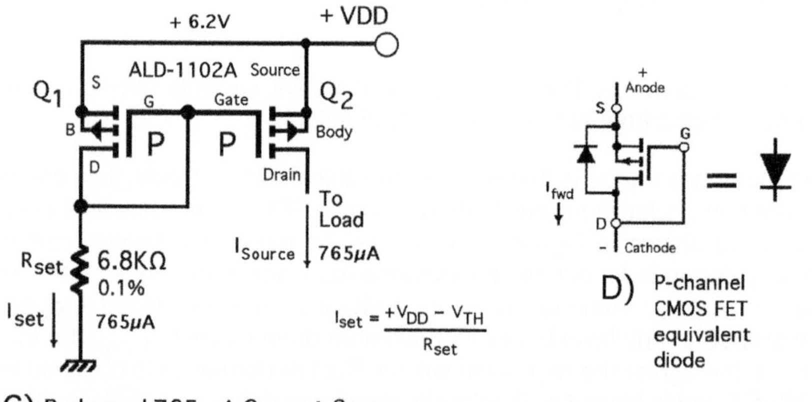

C) P-channel 765-μA Current Source

Using a CMOS P-channel dual matched-pair to create a mirrored current source. Note the equivalent diode connection.

Figure 9.5. *Simple current mirrors using CMOS matched pairs.*

Looking at the circuit in Figure 9.5A, this shows a simple current sink using a matched N-channel pair (such as an ALD-1101A). In this circuit transistor Q_1 is *diode-connected* (see Figure 9.5B), and as a result the V_{GS} and the transistor's forward transconductance (g_{fs}) control the drain current. This can be shown by:

$$I_D = V_{GS} \bullet g_{fs} \qquad \text{(Eq.9.3)}$$

If a second transistor (Q_2) is now connected so that its gate and source are in parallel with Q_1's gate and source, Q_2's drain current will *mirror* Q_1's. In this way the mirror current I_{sink} will equal the set current I_{set}. The exact value of current can be determined by the following equation:

$$I_{SET} = \left(\frac{V_{DD} - V_{(TH)}}{R_{SET}} \right) \qquad \text{(Eq.9.4)}$$

where the gate threshold voltage $V_{(th)}$ is guaranteed to be 1 volt max and V_{DD} is the positive supply. It is assumed that the voltage supply is well-regulated and decoupled with a 0.1-μF disk ceramic capacitor, located close to the current source. In the example shown in Figure 9.5A, a 260-μA current sink is created from a regulated 3.6-volt supply. The resistor R_{SET} should be a good-quality metal-film, 1/4-watt type with a 0.1% tolerance or better and with a low tempco. It is interesting to note that the best performance and regulation occurs when using the MOSFET biased well below its maximum drain current (in this case it is 30 mA minimum). I mentioned previously that one should try to make the current source's desired output level 10% or less of its $I_{DS(on)}$ drain current. In this case, the 260-μA current source equates to just 0.0087% of the minimum $I_{DS(on)}$ value, which would have superb regulation (better than 0.001%).

The circuit in Figure 9.5C shows an ALD-1102A, a matched P-channel pair being used as a simple current source. This circuit works in exactly the same way, except the polarities are reversed, because it uses P-channel FETs. The source current is found by the same formula as in equation 9.4. The equivalent diode-connection for the P-channel connection is shown in Figure 9.5D.

9.3.1 CMOS cascode current sources

Although performance is good using an FET matched-pair, much improved regulation can be achieved by using a *cascode* configuration, as shown in Figure 9.6. Cascoding improves high-frequency operation, provides even greater output impedance (Z_{out}), increases voltage operation and compliance, and reduces output conductance even further. This circuit is a P-channel cascode version of the mirror-pair shown previously in Figure 9.5B and provides even sharper regulation. The operation is essentially the same, except that in this circuit one should remember that *two* gate threshold voltages exist (Q_1 and Q_3, and Q_2 and Q_4), hence the equation is modified to:

$$I_{SOURCE} = \left(\frac{V_{DD} - 2V_{(TH)}}{R_{SET}}\right) \quad \text{(Eq.9.5)}$$

The circuit shown uses the ALD-1107, a monolithic matched P-channel quad transistor array, and as a result good performance can be achieved. Remember too that because the devices all share the same substrate, there will be excellent thermal matching. This quad transistor device is available in either a 14-pin DIP or a 14-pin SO surface-mount package.

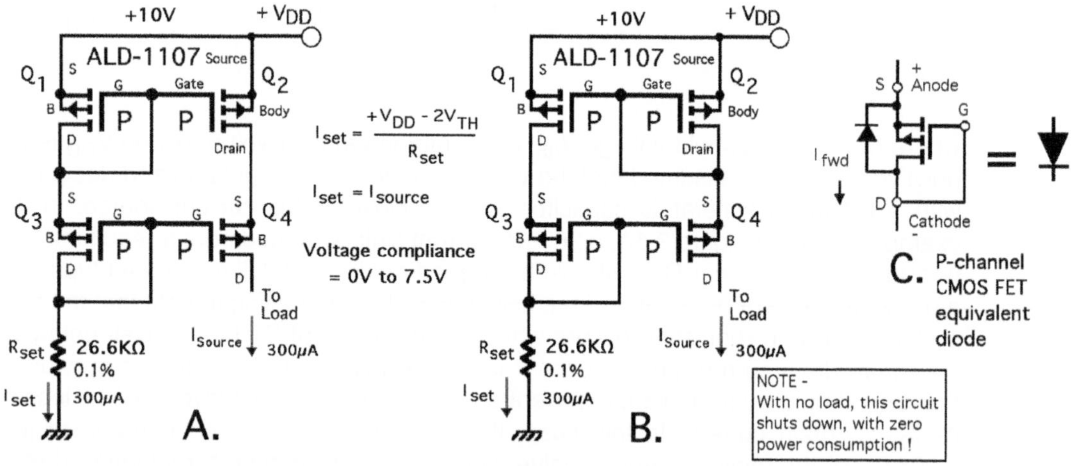

A 300-µA true cascode current source using a Quad P-channel ALD-1107.

A 300-µA quasi-cascode current source implemented as a Full Wilson in CMOS.

Figure 9.6. Different types of cascode current sources.

The circuits in Figure 9.6 show two different cascode current sources using two matched P-channel pairs (Duals: ALD-1102, 1117; or Quad: ALD-1107). In circuit A, (a true cascode), both Q_1 and Q_3 are diode-connected (Figure 9.6C), and as a result their V_{GS} and their g_{fs} control the drain current. If a third and fourth transistor (Q_2 and Q_4) are now connected so that their gates and sources are in parallel with Q_1 and Q_2's gates and sources, Q_2's drain current will mirror Q_1's. In this way the mirror current I_{source} will equal the set current I_{set}. In this circuit, cascoding buffers the current source from the load, so that variations in the load voltage are accommodated by Q_1's drain-to-source. As a result the voltage drop across the current source remains constant. Here the drain current (I_D) is regulated by Q_3 and R_{SET}, so that it is mirrored by Q_2 and Q_4. Both FETs must be operated with adequate V_{DS}, or else the circuit's output conductance will increase significantly. A very low g_{os} value is achieved by either cascode, because of degenerative feedback and the circuit's lower output conductance (which in turn is caused by the combined forward transconductance). Here the circuit's output conductance (g_o) is much less than the g_{os} of a single CMOS transistor—about 100 times lower. You can also expect the cascode circuit's minimum output impedance to be more than 100 MΩ, at current levels less than 1 mA, and for best linearity compliance will range from 0 to 7.5 volts. Regulation is at least 10 times better with the cascode circuit.

Figure 9.6B shows a Full Wilson 300-μA source provided from a 10-volt supply. Here transistors Q_2 and Q_3 are diode-connected, but operation is otherwise similar. One should be aware though that in this circuit, changes in the load (going to high impedance or open-circuit) alter the bias conditions, forcing the circuit to switch *off.* This could be a significant benefit in some applications, where a minimum power dissipation is required. So it is important to remember that this configuration is load-dependent. Again, R_{SET} should be a good-quality metal-film, 1/4-watt type with a 0.1% tolerance or better and with a low tempco. In this case the 300-μA current source equates to just 0.0272 % of the $I_{DS(on)}$ value, which again would provide superb regulation (better than 0.001%). One could easily use these cascodes in low-voltage circuits of around 2.5 to 5 volts. Ideally, one would like more than 2 volts across each FET, but one can probably get down to a 3-volt supply, because the transistors are ALD's proprietary low-voltage, silicon-gate CMOS devices. These current sources can be built using surface-mount components, taking up minimal circuit board space.

The circuits shown in Figure 9.7 use the ALD-1106, a monolithic matched N-channel quad array. In circuit A, Q_1 and R_{set} determine the 222-μA current level. Transistors Q_2 through Q_4 mirror this, to provide three slave 222-μA current sinks. In circuit B, Q_1 and R_{set} again determine the current level, which is set for 750 μA. Transistors Q_2 through Q_4 mirror the I_{set} current, which is effectively multiplied by the integer (whole number) of additional transistors used in order to create a total I_{sink} current of 2.25 mA. Again, because of the monolithic construction and ALD's close matching, excellent performance in either circuit can be achieved. Notice also that the three CMOS FET transistors are easily paralleled, without regard to base current mismatches or power-hogging, as can occur with BJTs.

The circuit shown in Figure 9.8 uses the ALD-1107, a monolithic matched P-channel quad array. In this circuit, Q_1 and R_{set} determine the 500-μA I_{set} current level. This is found by the formula previously shown in equation 9.4 for determining I_{set}. Transistors Q_2 through Q_4 each mirror the I_{set} current, to provide three separate 500-μA current sources. This popular configuration (two or three P-channel sources) provides optimum performance because of its monolithic construction. It provides excellent thermal matching, as well as close matching of some of its key (amplifier/switch) characteristics (maximum V_{os} equals 2 mV; ΔG_{fs} = 0.5% max.; $\Delta R_{DS(on)}$ = 0.5% max. etc).

This circuit may be easily cascoded, as shown in Figure 9.9. This will provide higher output impedance, higher frequency operation, increased voltage operation, reduced output conductance, and improved regulation. In this circuit, Q_5 and R_{set} determine

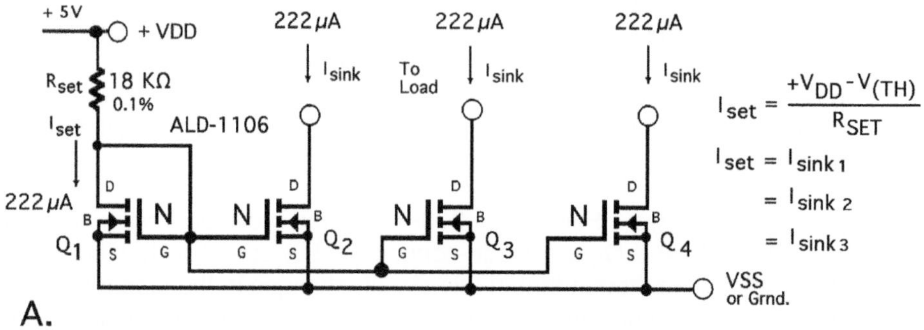

A.

A CMOS multiple current sink using an ALD-1106 Quad N-channel array. In this example I set is mirrored by the other transistors in the array.

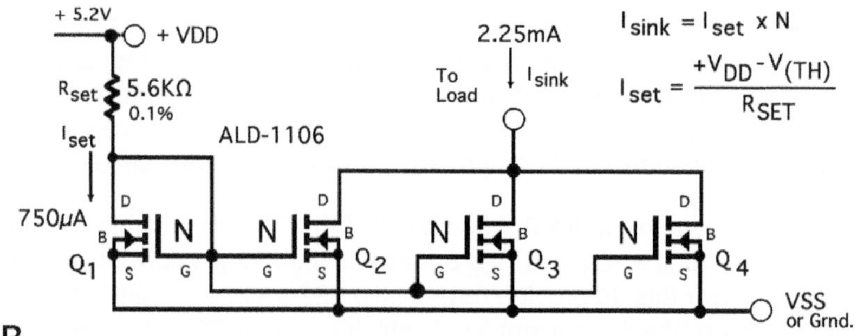

B.

A CMOS current sink multiplier using an ALD-1106 Quad N-channel array. Here I_{set} is multiplied by the number of mirror-transistors used to determine the value of I_{sink}.

Figure 9.7. *Current sinks can be easily implemented with ALD's CMOS arrays.*

the 120-μA I_{set} current level. This is found by the formula previously shown in equation 9.5 for determining I_{set}. Transistors Q_6 through Q_8 each mirror the I_{set} current, to provide three separate 120-μA current sources. Operation is essentially the same as shown previously, except that here again remember that *two* gate threshold voltages exist per vertical pair (i.e., Q_1 and Q_5), as per equation 9.5. The cascode circuit shown here uses *two* ALD-1107s, monolithic matched P-channel quad transistor arrays, which are available in either 14-pin DIP or SO surface-mount packages.

Using essentially the same cascoded circuit, one can multiply the separate outputs, as shown in Figure 9.10. In this circuit, Q_5 and R_{set} again determine the 120-μA I_{set} current level. Transistors Q_6 through Q_8 each mirror the I_{set} current, but here the multiplier

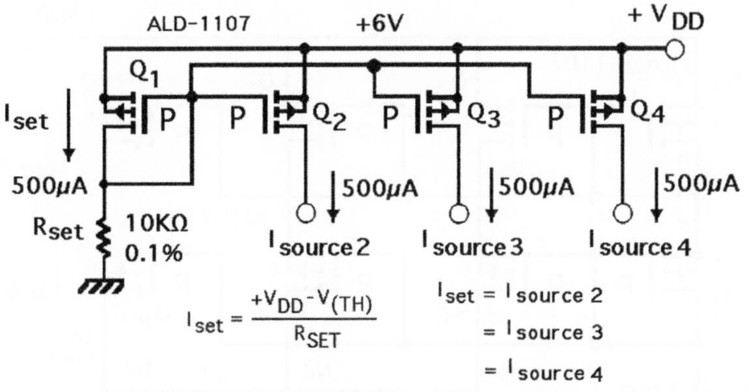

Figure 9.8. *A CMOS multiple current source using an ALD-1107 Quad P-channel array.*

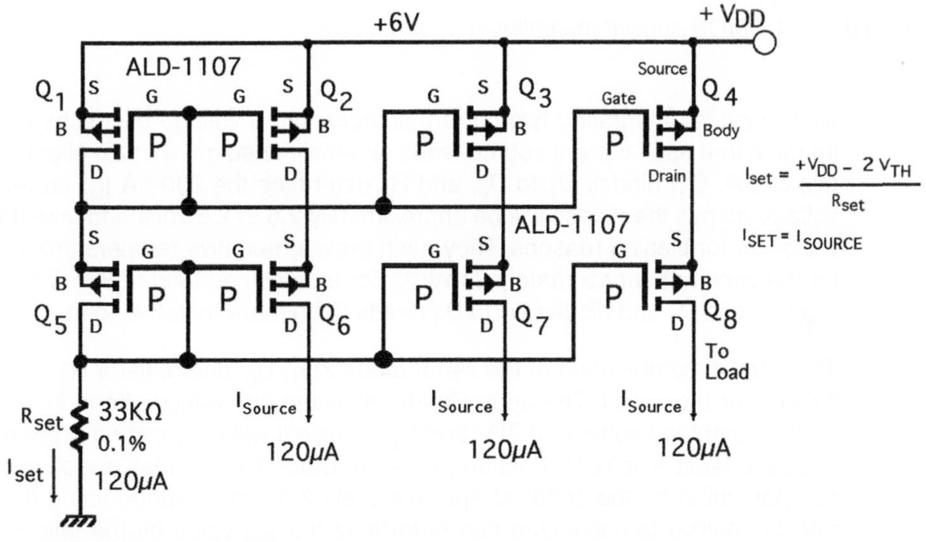

Figure 9.9. *Cascoding this multiple current mirror boosts its frequency response, Z_{out}, and regulation.*

is three times, which results in a single output current providing 360 μA. Operation is otherwise the same, including the *two* gate threshold voltages that exist for each vertical pair (i.e., Q_1 and Q_5), as per equation 9.5. The cascode circuit shown here uses *two* ALD-1107s, monolithic matched P-channel quad transistor arrays.

There are times when a conventional current-source circuit just has to be modified a little to meet the needs of a design application. The circuit shown in Figure 9.11 does

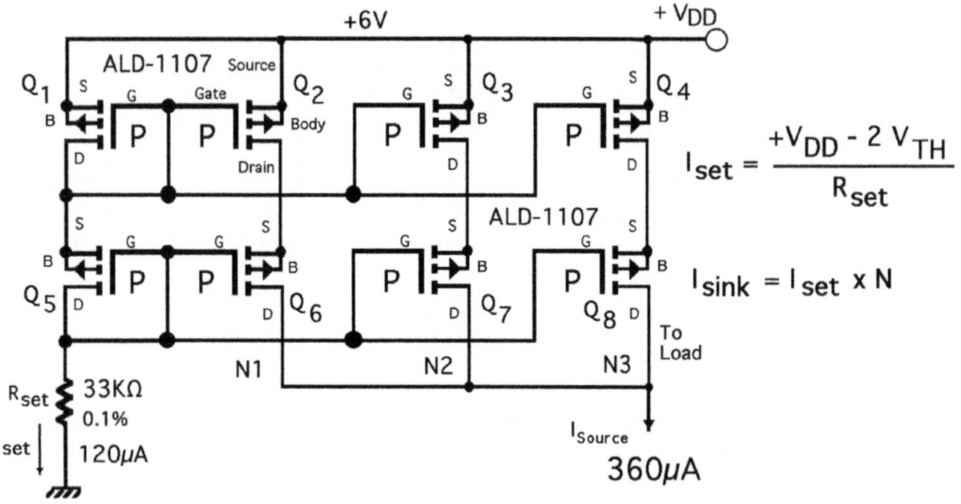

Figure 9.10. *A CMOS current multiplier.*

such a job by integrating two current sources into a voltage bias scheme. One could imagine that each current source feeds an amplifier stage, with excellent regulation. In this circuit, Q_1, diodes D_1 to D_4, and R_1 determine the 200-μA I_{set} current level. The voltage across the diodes will be approximately 2.6 to 2.8 volts and take the place of a transistor for biasing reasons. They each provide negative temperature compensation for the circuit at approximately 2 mV/°C, for a combined total of 8 mV/°C. The voltage drop across Q_1 and diodes D_1 to D_4 needs to be more than 3.8 volts.

The parallel combination of the zener diode ZD_1, D_5, and resistor R_2 set up the main biasing for the circuit. The approximate minimum operating voltage for the circuit will be the combined voltage of ZD_1 and D_5, or else it will not function properly. (In fact, it needs at least 8 volts to function properly). Diode D_5 provides negative temperature compensation for the zener at approximately 2 mV/°C, counteracting the zener's +5 mV/°C positive tempco. One can determine the I_{set} value by the following modified formula:

$$I_{SOURCE} = \left(\frac{V_{DD} - V_{(TH)} - 2.8V}{R_{SET}}\right) \qquad \text{(Eq.9.6)}$$

Transistors Q_2 through Q_5 each mirror the I_{set} current, to provide two separate 200-μA current sources. Operation is essentially the same as shown previously. This cascode circuit could use both quad and dual P-channel monolithic matched arrays, which could be either in DIP or SO surface-mount packages.

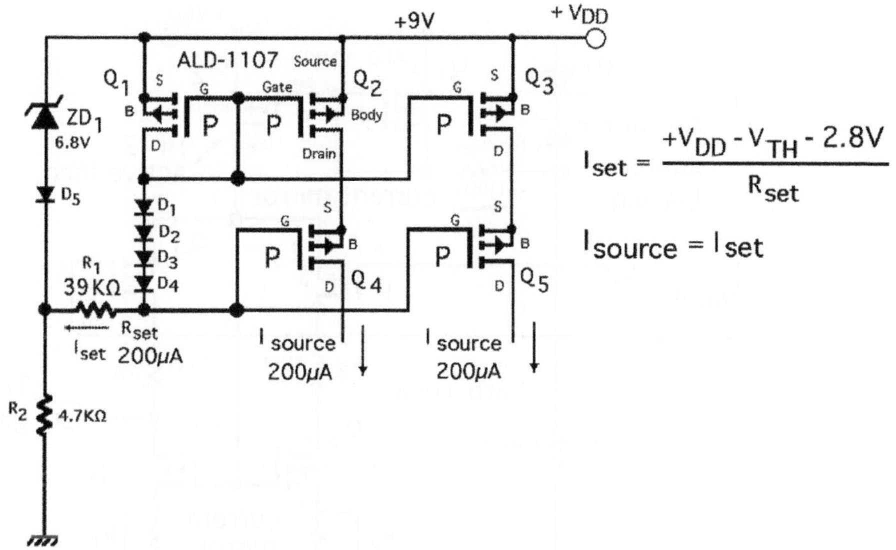

Figure 9.11. Part of the current and voltage biasing circuitry for a typical CMOS op amp. Notice the modified P-channel cascodes, which supply 200 µA to the separate amplifier stages.

You can see another example of how a similarly modified biasing scheme is employed for a typical input stage of a voltage-feedback CMOS op amp in Figure 9.12. It shows how some of ALD's matched CMOS transistor arrays could theoretically be used in order to create one's own op amp. The actual input amplifier stage is created by using a pair of P-channel MOSFETs (in a quad transistor package), while their drain load consists of a pair of N-channel MOSFETs (consisting of the second half of the matched complimentary quad), functioning as the current mirror-sink. The zener diode is part of the bias network, working in conjunction with the upper P-channel current mirror-source (in a dual transistor package). This establishes a constant current for the differential amplifier. The drain loads for the differential pair consists of R_3 R_4, and the mirror sink consisting of Q_5 and Q_6. The amplifier's offset-voltage (V_{os}) can be adjusted if necessary by connecting a 10-KΩ to 1-MΩ potentiometer across the offset-null terminals. The single-ended output for the following gain stage is provided from the drain of Q_6. Building such a circuit can provide a great deal of insight into how the front end of a real op amp functions. It gives one an appreciation of many of the op amp's features, such as its input offset-voltage (V_{os}), input offset-voltage drift (TCV_{os}), low input bias currents, input noise voltage (e_n), slew rate (S_R), input and output voltage range, large signal voltage gain, CMRR, PSRR, power dissipation, supply current, and so on. Luckily for us, ALD makes some excellent CMOS op amps for different applications (high speed, precision, low-voltage operation, and even programmable EPAD® op amps), so you don't really need to build your own.

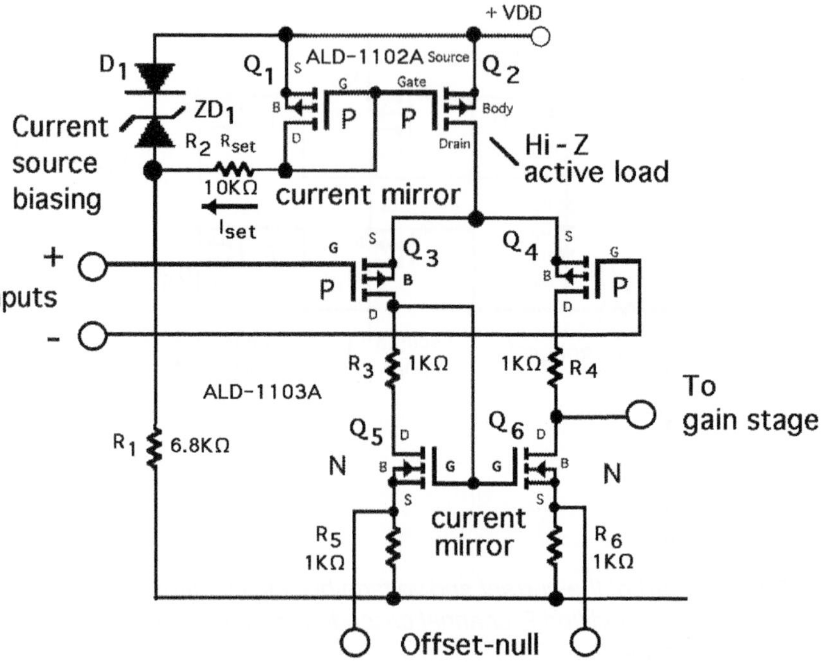

Figure 9.12. Creating a simple op amp front-end amplifier using two pairs of current mirrors to set bias levels.

9.4 Using ALD's programmable EPADs® to create precision current sources

The EPAD® is a kind of analog version of the digital EPROM. Advanced Linear Devices pioneered the electrically programmable analog device (EPAD®) in the late 1990s. Initially, these were matched-pairs and quad transistor arrays, but subsequently they have introduced a whole range of op amps, A/D converters, and other devices. The ALD 1108E (quad) and ALD1110E (dual) precision matched-pairs are examples of this exciting new technology. These CMOS transistor arrays are designed to operate over a 2- to 10-volt supply range and have ultra-low power consumption. They also have a unique electrically programmable gate threshold (V_{th}) feature, which can be easily set by the user. This gate threshold voltage can be set with great precison over a range of between 1 and 3 volts in 100-μV steps. The initial threshold voltage for new unprogrammed devices is 1.000 volts (±1%). Once set, the device will retain this precise setting for more than 10 years (like a nonvolatile RAM or EPROM), with a drift typically less than 2-mV per 10 years. Such an array, whether quad or dual, can either be programmed in-circuit or with the use of an ALD EPAD® programmer connected to one's PC (see Figure 9.13).

In an OEM environment, it can be programmed on-site before assembly or after assembly as part of a systemwide calibration. With additional circuitry, it can be pro-

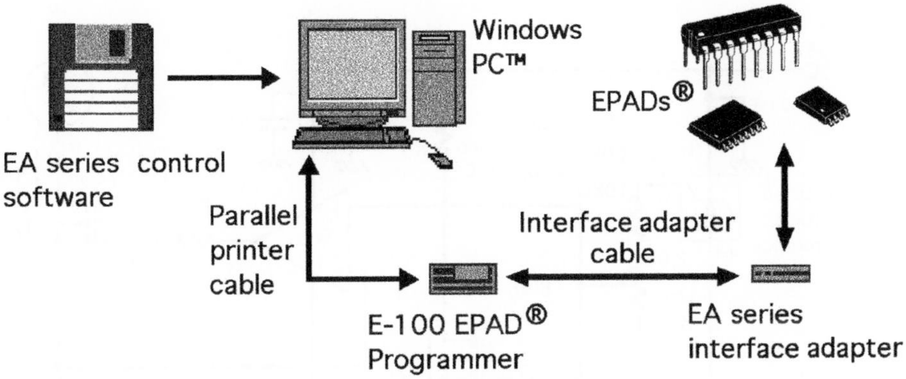

Figure 9.13. A typical EPAD® programming system.

grammed remotely via a network or even over the Internet. This type of device is perfect for manufacturers of potted/sealed assemblies, where trimming or fine calibration is required, or it can be programmed remotely for applications involving hazardous environments or remote locations. The gate threshold voltage can be trimmed, set, and left in a one-time calibration type of application, or it can be increased several times (see Figure 9.14). Once set, the gate threshold setting cannot be reduced or cleared. However, bidirectional adjustments can be made simply by using two devices together with an op amp, where one EPAD® can be made to increase the threshold voltage, while the other uses the op amp to invert the level.

Being N-channel MOS devices, the 1108E and 1110E have very low input currents, and as a result a very high input impedance (10^{12} Ω). Because the gate voltage controls the on-resistance and drain current, either of these characteristics can be effectively trimmed and set as required. Thus, in terms of current sources, they can be programmed to provide a precise constant current over a 100-nA to 3-mA range, and with either a positive, negative, or zero tempco. The devices have a zero tempco current of 68 μA over a range of threshold voltages, as seen in Figure 9.15.

Once programmed and set, the devices function like a very-high-quality current sink or current mirror. An example of a current mirror-source using an ALD-1110E EPAD® and an ALD-1102A P-channel MOSFET pair is shown in Figure 9.16.

9.5 ALD breaks the gate-threshold barrier

Advanced Linear Devices introduced some exciting new products in early 2005, which include three new families of precision-matched monolithic pairs and quads. These remarkable ETRIM™ products are based on ALD's well-proven EPAD® technology, but in this case are preprogrammed at the factory in various voltages. What makes these products so attractive is that they have ultra-low gate threshold voltages, typically down to 0.2 volt, and they can run at very low drain-to-source voltages as well. All

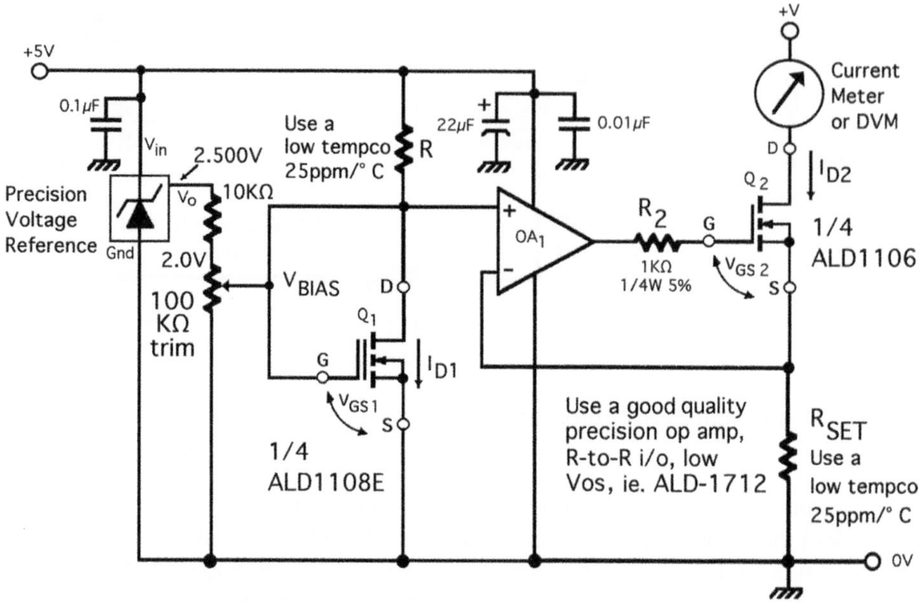

Q_1 is the device being programmed, while OA_1 and Q_2 mirror its current, displaying this on the meter. Once you have determined the combination of R and V_{BIAS}, and noted these, you are ready to start programming.

Figure 9.14. Setting up an EPAD® precision matched transistor.

of these devices are perfect for battery-powered and portable instrumentation applications. The operating temperature for all of these products is specified for 0° to +70°C. Devices are presently all N-channel transistors and available in either eight-pin DIP or SOIC packages for duals or in either 16-pin DIP or SOIC packages for quads. Because these products are all MOSFET devices, they have a very high input impedance (1×10^{14} Ω), and can provide a very large current gain (1×10^{8}; 100 M) in low-frequency applications. Their maximum gate input leakage current at 25°C is specified at 100 pA, or 1 nA at 125°C, which is also impressive.

One ETRIM™ family of N-channel precision matched-pairs (ALD110900/A) and quads (ALD110800/A) has a zero-threshold voltage, which eliminates input to output level shifts. These devices have unique characteristics that make them *both* depletion and enhancement types simultaneously. Using these small-signal devices, it has been possible for the first time to build an amplifier input stage that operates from a tiny supply of just 0.2 volt. Another family, the ALD1108xx (quad) and ALD1109xx (dual) precision matched-pairs, are more examples of this exciting new technology. These CMOS enhancement-mode (normally-off) transistor arrays are designed to operate over a 0.2- to 10-volt supply range and have ultra-low $V_{GS(th)}$ voltages. Depending on the

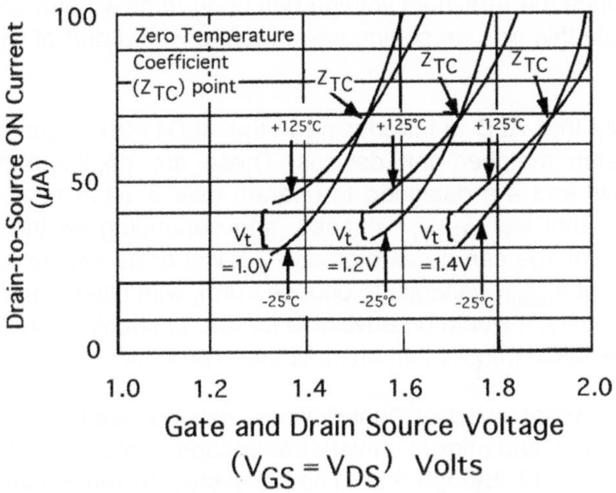

Figure 9.15. *The ALD-1108E/1010E EPAD's drain-to-source ON current, bias current versus ambient temperature.*

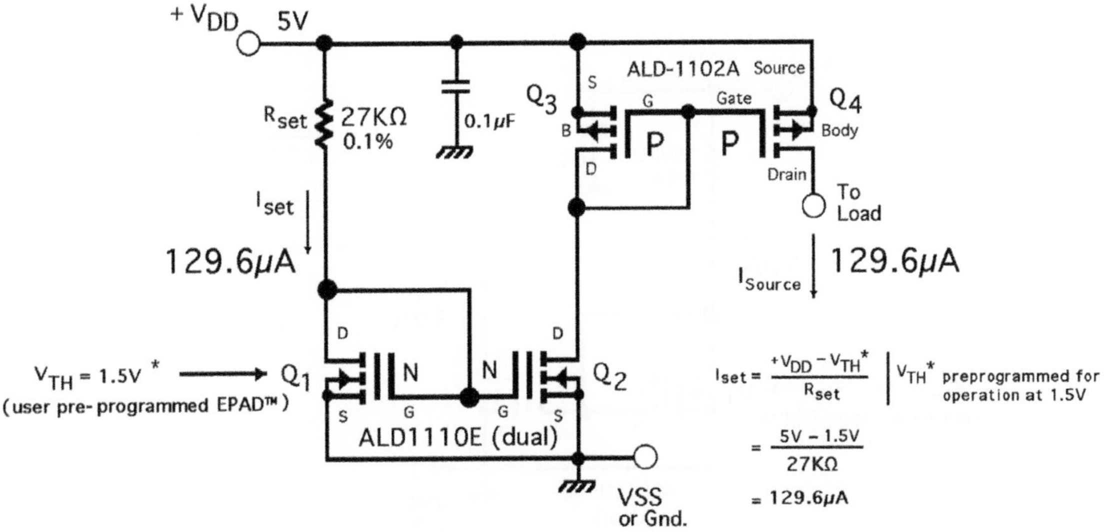

Figure 9.16. *A mirror source using an EPAD™ precision-matched pair together with a P-channel MOSFET pair. Note the EPAD™ here has been user programmed for operation at 1.5V_{TH}. Z_{out} is > 10MΩ.*

grade chosen, the gate threshold voltage can be as high as 1.42 volt or as low as 0.18 volt (there are four different $V_{GS(th)}$ voltages to choose from), and with different package options. Once the threshold voltage has been trimmed and set at the factory, the device will retain this precise setting permanently, with a drift of typically less than ±2 mV/°C.

The third family that ALD is introducing is the ALD1148xx (quads) and ALD1149xx (duals), precision matched-pair devices. These are depletion-mode (normally-on) transistor arrays and are designed to operate over a ±0.2- to ±5-volt supply range. They too have ultra-low $V_{GS(th)}$ voltages, and depending on the grade chosen, the gate threshold voltage can be as high as –3.5 volt or as low as –0.43 volt (there are also four different $V_{GS(th)}$ voltages to choose from), with two different package options. Being new products, it would be advisable for one to check ALD's Web site (*www.ald-inc.com*) for the latest information on all of these products.

Such arrays, whether quad or dual devices, can be used to create accurate, low-power current sinks and mirrors, in very-low-voltage applications. Some examples are shown in Figures 9.17 through 9.21. The necessary current-setting design equations are shown in each example. The circuit of Figure 9.17 shows a precision-matched N-channel pair (ALD110902), which is used to create a simple 5-μA current sink, in a circuit running from a minuscule 1-volt supply. With a BJT or JFET current sink, this would not be possible, because in most cases those devices need more operating headroom than the supply voltage here allows.

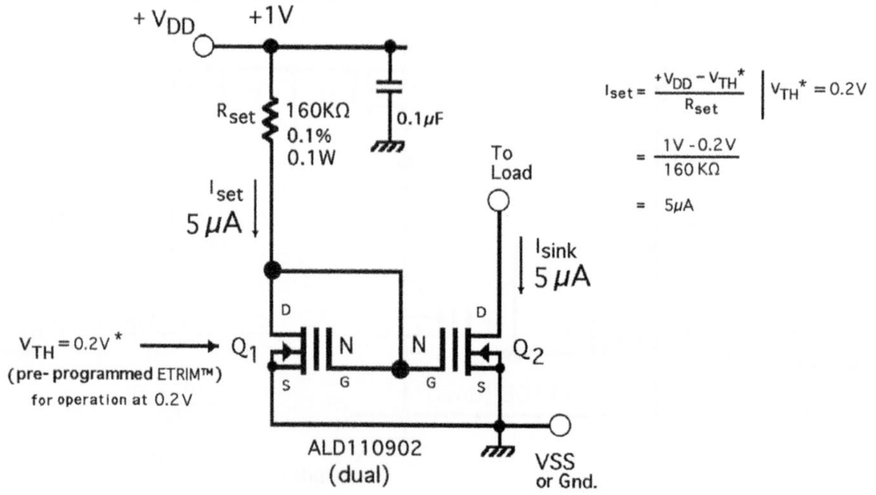

Figure 9.17. *A mirror current sink using an ALD ETRIM™ precision-matched N-channel pair. The device has been programmed for operation at a very low gate threshold voltage of 0.2V.*

The circuit shown in Figure 9.18 shows an N-channel precision-matched quad (ALD110802), which is used to create a multiple current mirror sink, again with a very low supply voltage (1.2 volt), along with an ultra-low 0.2-volt gate threshold voltage.

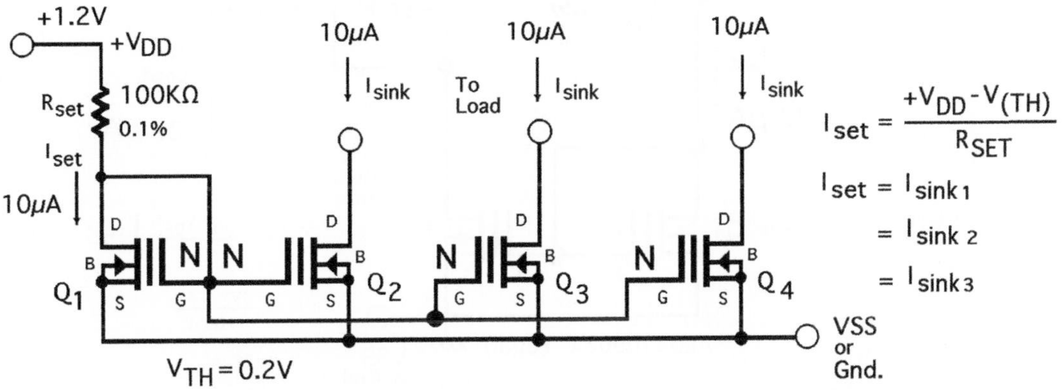

Figure 9.18. A CMOS multiple current sink using an ALD 110802 ETRIM™ ultra-low-threshold, precision-matched, quad N-channel array. In this example I_{set} is mirrored by the other transistors in the array. The supply voltage is a single cell battery.

In Figure 9.19, an e-trimmed N-channel pair (ALD110902A) is combined with a regular ALD-1102 (a matched P-channel dual) to provide a simple current source. The N-channel pair sets the current level, while the P-channel devices mirror this current to the load. The ALD110902A devices in this example have a gate threshold voltage of only 0.2 volt.

Figure 9.20 shows an N-channel matched quad (ALD110802), which is used to create a current sink multiplier, running on an ultra-low supply voltage (0.75 volt). In this circuit, the set current set by Q_1 and R_{set} is mirrored by transistors Q_2, Q_3, and Q_4, but with their drains in parallel. This effectively multiplies the I_{set} current by the number of mirror transistors. This circuit illustrates the importance of having a very low gate threshold voltage, which enables lower voltage operation than is possible with most BJTs, JFETs, or other MOS devices.

The circuit in Figure 9.21 shows two N-channel ETRIM™ transistors using matched array pairs (ALD114904A and ALD110900A), which are used to create a simple current sink. The ALD110900A is a zero-threshold device, while the ALD114904A is a very low threshold depletion-mode (normally-on) device. As a result, this design can run from a very low supply voltage (1 volt). When connected as shown here, the devices can provide a constant current sink of between 5 to 100 μA, with a near-zero tempco.

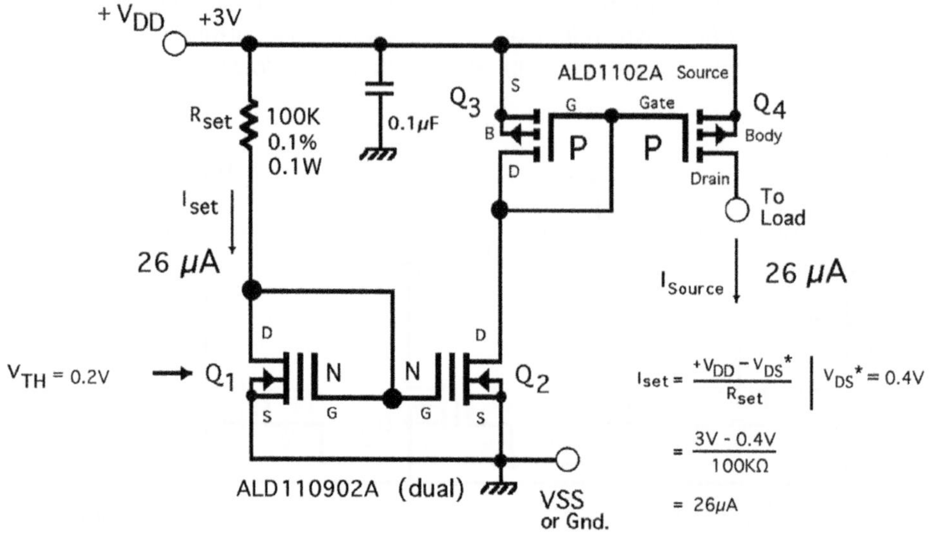

Figure 9.19. *A mirror current source using an ALD ETRIM™ precision-matched N-channel pair, with a precision P-channel MOSFT pair. The ETRIM™ has been programmed for operation at a very low 0.20V_{TH}.*

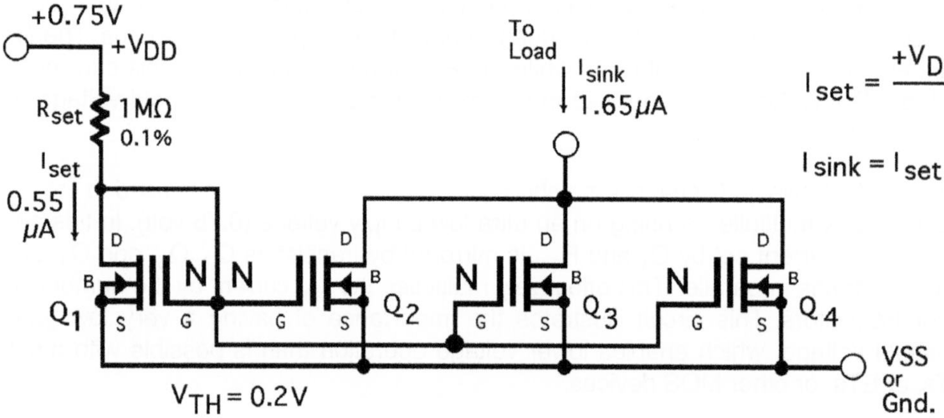

Figure 9.20. *A current sink multiplier using an ALD110802 ETRIM™ ultra-low-threshold, quad N-channel array. In this example I_{set} is multiplied by the other transistors. The supply voltage is 0.75V (a typical diode's V_{Fwd}), and the FET's gate V_{TH} is only 0.2V.*

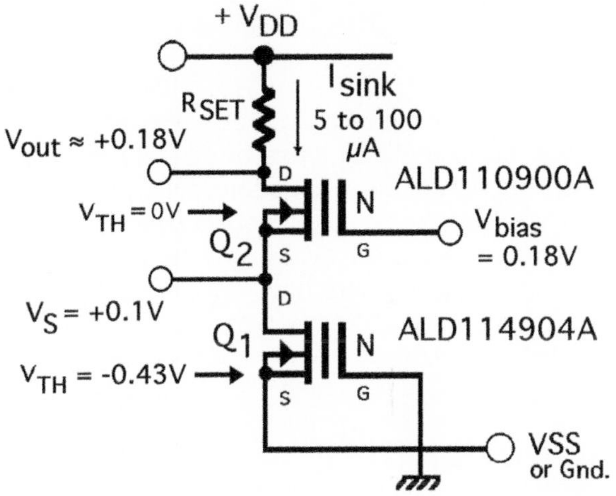

Figure 9.21. *A current sink using two ETRIM™ precision-matched N-channel pairs. The devices uniquely provide V_{out}, V_S, and a low current, with a near-zero TC when biased as shown here.*

Chapter 10

Using Monolithic IC Current Sources and Mirrors

While much of this book focuses on how to build current sources using different kinds of discretes, this chapter looks at some dedicated, monolithic IC current source circuits. These devices can be used in many diverse applications requiring microamps to several milliamps and are readily available in either throughhole or surface-mount packages. Actually, two dedicated monolithic current source ICs exist in the U.S. analog marketplace. One of them is National Semiconductor's LM134 series, and the other is TI/Burr-Brown's newer REF-200 (which we will review later in this chapter).

10.1 National's LM134—a monolithic IC current source

Of all the thousands of analog semiconductor products that have been introduced and then later discontinued over the past two decades, one that has survived the test of time and changing market needs is National Semiconductor's venerable LM134. This is a dedicated three-terminal adjustable bipolar/JFET current source IC. The LM134 was originally designed by Carl Nelson at National Semiconductor and introduced in the late 1970s. (Carl Nelson is now with Linear Technology.) Over the past 25 years, it has been at the cornerstone of thousands of electronic designs around the world, and because of its popularity, it is likely to be in National's product lineup for a long time to come.

The LM134 family has various options (134/234/334), which are shown in Table 10.1, and the range is second-sourced by Linear Technology Inc. From the outset the LM134 was designed for ease of use and to play a dual role: to be both an accurate current source and a temperature sensor (for reasons that will become clear later). Actually, the LM134 also helped spawn many dedicated temperature-sensing ICs, which have evolved over the past two decades, not only from National but from others too. (*Note*: The list shown in Table 10.1 was originally longer, but several temperature sensor products have been deleted over the past decade, as newer, cheaper products have become available. Additionally, some options that were originally available in the costly TO-46 metal package have also been made obsolete by both National and Linear Technology. Table 10.1 shows many of the devices that were available as of early 2005. Check the manufacturer's website for availability.)

The various current source devices in the series are adjustable over a practical range of 1 μA to 5 mA. They can be used in applications above 5 mA (to 10 mA maximum),

Mfr.	Part #:	Package options:	Temp.range °C	Maximum Voltage:	Cs or Ts
NS	LM 134 H	TO-46, 3-pin metal can	-55° to +125°	+40V	CS
NS	LM 234 H	TO-46, 3-pin metal can	-25° to +100°	+40V	CS
NS	LM 234 Z3	TO-92, 3-pin plastic	-25° to +100°	+30V	TS
LTC	LM 234 Z3	TO-92, 3-pin plastic	-25° to +100°	+30V	TS
NS	LM 234 Z6	TO-92, 3-pin plastic	-25° to +100°	+30V	TS
LTC	LM 234 Z6	TO-92, 3-pin plastic	-25° to +100°	+30V	TS
NS	LM 334 H	TO-46, 3-pin metal can	0° to +70°	+40V	CS
NS	LM 334 Z	TO-92, 3-pin plastic	0° to +70°	+40V	CS
LTC	LM 334 Z	TO-92, 3-pin plastic	0° to +70°	+40V	CS
NS	LM 334 M	SO-8, plastic SMD	0° to +70°	+40V	CS
NS	LM 334 SM	SO-8, plastic SMD **	0° to +70°	+40V	CS
LTC	LM 334 S8	SO-8, plastic SMD **	0° to +70°	+40V	CS
NS	LM 334 MX	SO-8, plastic SMD **	0° to +70°	+40V	CS
NS	LM 334 SMX	SO-8, plastic SMD **	0° to +70°	+40V	CS
NS = National Semiconductor LTC = Linear Technology ** = alternative pinout			CS = current source TS = temperature sensor ($1\mu A/°K$)		

Table 10.1 LM134 family options

although this is *not* recommended because of the effects of self-heating, which seriously impairs regulation. Most of the LM134 family has an operating voltage range of between 1 to 40 volts, but temperature sensors have a 30-volt maximum. All of the devices can withstand being reverse-biased by up to 20 volts. This means that the devices can be safely used in low-voltage AC applications as well. The current source products offer excellent current regulation and are true "floating" current sources. This means that while they require no separate power supply connections in order to function, they can be used as either current sources or current sinks. Initial untrimmed current accuracy is a maximum of ±3% for LM134 and LM234, for currents of between 10 μA and 1 mA. The LM334 has a maximum of ±6%, as well as a lower operating temperature range of 0° to +70°C.

Figure 10.1 shows a simplified view of the LM134's internal circuitry, which incorporates both NPN and PNP bipolar transistors, as well as some very-low-voltage P-channel JFETs and an integrated capacitor. The JFETs are used for start-up biasing and enable the BJT current mirrors. In very-low-current applications (less than 3 μA), this may limit the power-on response time to about 1 mS because of the very low internal currents involved and the charge time of the integrated capacitor.

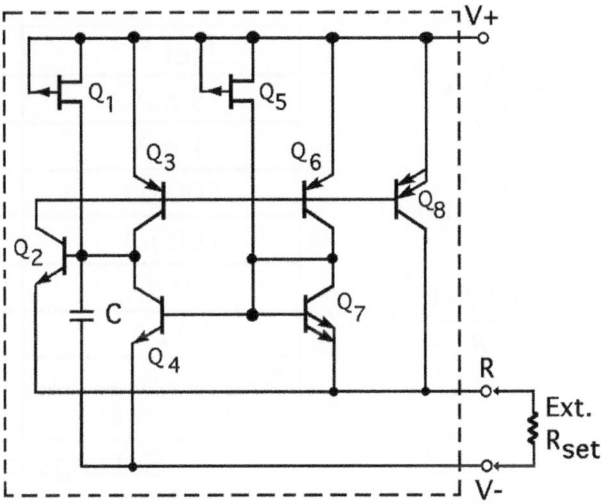

Figure 10.1. LM134 simplified internal circuit.

The current source's desired current level (I_{set}) can be set simply with just one external (precision) resistor (R_{set}), between its "R" and "V-" terminals, as shown in Figure 10.2. I_{set} consists of two currents: the current through the set resistor (I_R) and the IC's own bias current (I_{bias}). The current through the set resistor depends on the voltage across it (V_R), which because of feedback always equates to 67.7 mV. This voltage is directly proportional to absolute temperature (V_{PTAT}) and is the equivalent reference voltage at 298°K (25°C), while at 0°K (absolute zero), it extrapolates to 0 mV. On a graph this produces a straight line with a positive slope of +227 μV/°C. The current flowing out of the LM134's V- terminal (I_{bias}) is always 1/18th of the current flowing into the V+ terminal. It changes at a rate of 0.336%/°C, and on a graph has a maximum slope error of ±4%. This makes the LM134 an accurate, linear, and repeatable device, whether it is used as a current source or as a temperature sensor (1 μA/°K). However, because I_{set} is directly proportional to absolute temperature (I_{PTAT}), this +0.336%/°C tempco may not be suitable for every application.

As a current source, the best area for operation of the LM134 lies between 5 μA and 1 mA. In fact, beyond 100 μA internal heating of the die can have a significant effect on overall current regulation. For example, with I_{set} at the 1-mA level, each 1-volt increase across the device increases the junction temperature approximately 0.4°C. This together with the LM134's tempco of +0.336%/°C will yield a change in I_{set} of 0.132%, a significant amount. To minimize this, first try to minimize the voltage across the device, and if possible provide a small heatsink. If a zero temperature coefficient ($T_{C(0)}$= less than 50 ppm) is needed, this can be made by adding a second (precision) resistor and a good-quality, low-leakage silicon diode, as shown in the circuit of Figure

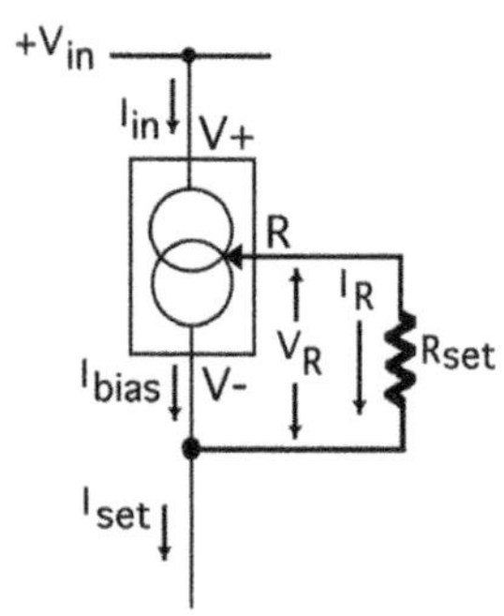

$$I_{set} = \frac{V_R}{R_{set}} + I_{bias}$$

or

$$I_{set} = \frac{V_R}{R_{set}}(1.059)$$

or

$$I_{set} = \frac{67.7mV}{R_{set}} \text{ @ } 25°C$$

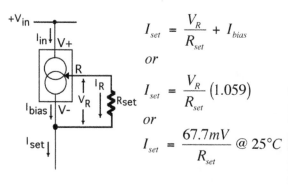

I_{set} *	R_{set} (Ω)
5 mA	14 Ω
1 mA	68 Ω
500 μA	135 Ω
100 μA	680 Ω
50 μA	1.3 KΩ
10 μA	6.8 KΩ
*Assumes 1.2V minimum across the LM134	Use 1% metal-film resistors

Settings chart

Figure 10.2. The LM134 used as a basic two-terminal current source.

10.3A. Here the positive tempco of the LM134 (+227 μV/°C) is offset by the forward-biased silicon diode's negative tempco (approximately –2.5 mV/°C). In this circuit, I_{set} comprises several combined currents, such that:

$$I_{set} = I_{bias} + I_1 + I_2 \quad \text{(Eq.10.1)}$$

$$= \frac{V_R}{R_1} + \frac{V_R + V_D}{R_1} \quad \text{(Eq.10.2)}$$

$$= \frac{67.7mV}{R_1} + \frac{67.7mV + 0.6V}{10R_1} \quad \text{(Eq.10.3)}$$

$$= \frac{0.134V}{R_1} \quad \text{(Eq.10.4)}$$

For best results, the circuit shown in Figure 10.3A should first be built and tested, and then its tempco should be carefully measured over the required temperature range (i.e., –10°C to +50°C or +14°F to +122°F), in a tightly controlled temperature environment. R_2's value is made to be 10 times the value of R_1. In this example, the diode's forward voltage drop (V_D) is assumed to be 600 mV, and its tempco should either be obtained from the manufacturer's data sheet or measured and plotted in advance. National's LM134 data sheet shows a Fairchild Semiconductor 1N457, a silicon high-conductance, low-leakage diode in a DO-35 case. This was originally chosen because its tempco is centered at 11 times the tempco of the LM134, which allows R_2 to be 10 times R_1. Some close alternatives made on the same die/process include FDH3595 or

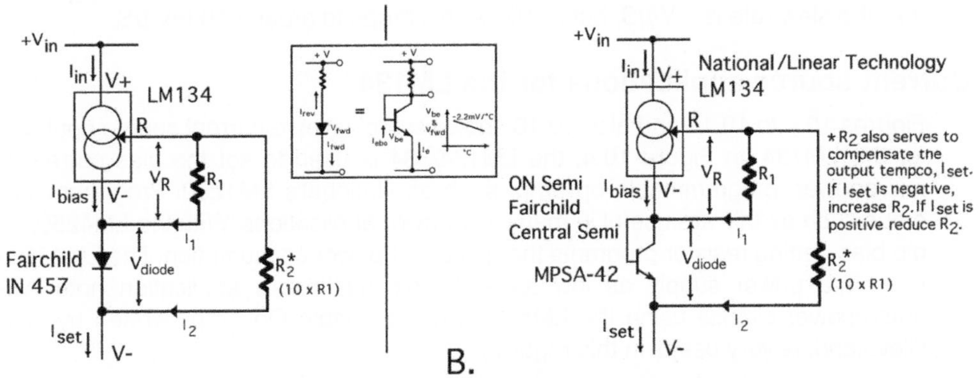

A.
Adding a low-leakage diode to the LM134, helps create a current source with a near-zero TC.

B.
Substituting a wide-base NPN transistor can provide even more improvements.

Figure 10.3. *Methods for improving the LM134's tempco.*

FDH300 (in DO-35 diode packages); FDLL457 or FDLL300 (LL-34 SMD package); and MMBD1501 (SOT-23 SMD package). Ideally, one should measure and accurately plot the diode's tempco before designing it into a circuit.

An alternative to using a diode is to use a diode-connected BJT, as shown in Figure 10.3B, because a transistor's double-diffused structure is closer to ideal than a diode's, and particularly a wide-base NPN transistor is linear with temperature. A good choice is the ever-popular MPSA-42, a TO-92, high-voltage 300 $V_{(BR)CEO}$ max, 500-mA I_c max device (made by ON Semiconductor, Fairchild Semiconductor, Central Semiconductor, and others). It is important to use this particular device and not cross-reference it with a near-equivalent, which will probably have different temperature characteristics. In any design situation, be sure to locate the diode (or diode-connected BJT), the R_{set} resistor, and the LM134 as close together as possible, so they share the same temperature. Ideally, the resistor used for R_{set} should be a good-quality metal-film or wirewound type, with 0.01% tolerance or better, and less than 30-ppm/°C tempco. The resistor should also have good long-term stability, to reduce circuit drift errors.

If as a result of your temperature tests the circuit's I_{set} tempco is positive rather than zero, then R_2 (in Figure 10.3) should be reduced; if negative, then R_2's value should be increased. The minimum voltage across the LM134 depends on I_{set}, but would typically be around 1 volt at between 1 mA to 5 mA I_{set}. At between 6 μA to 1 mA, the minimum voltage would typically be about 0.9 volt, and between 2 μA to 60 μA expect it to be as low as 0.75 volt. The slew rate, which may be important in some applications, is naturally affected at very low currents. National states that at an I_{set} level of 1 mA, the slew rate is 1 V/μS, but at 10 μA this drops to around 10 mV/μS.

10.2 Current source applications for the LM134

Figures 10.4 to 10.11 and also 10.18 show several diverse current source applications for the LM134. In Figure 10.4, the LM134/334 is used to set the bias current for a micropower, programmable op amp, such as National's LM4250. Programmable op amps such as this are useful in many low-power applications. With the LM4250, a single bias-setting resistor programs the quiescent power consumption. The LM4250 can run off a power supply as low as ±1.5 volt. (National's application note AN-71, "Micropower circuits using the LM4250 Programmable Op Amp," written by George Cleveland, is very useful in this regard.)

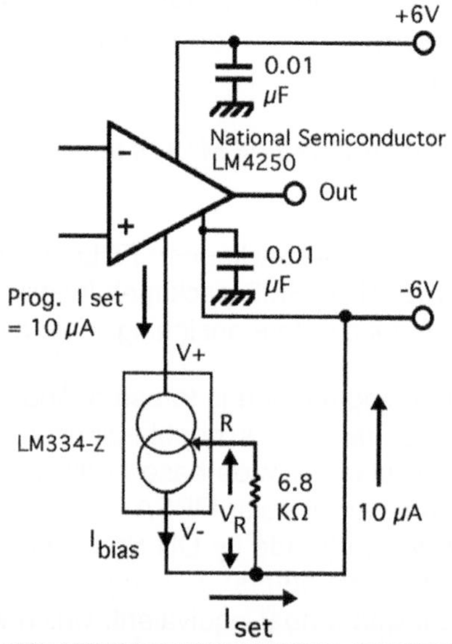

Figure 10.4. *Using the LM334 to set the bias current for a micropower, programming op amp.*

The circuit in Figure 10.5 shows a remarkable 10-nano-amp current source using several Linear Technology devices, including the second-sourced LM134 series. The additional devices include an LT1008 precision op amp (an upgrade of the legendary LM108) and an LT1004 micropower 1.2-volt bandgap voltage reference. The circuit's output impedance is approximately 1 GΩ. (*Note*: This is one of the *lowest* level current sources in the book!)

The circuit in Figure 10.6 shows the LM134/334 being used to set the operating current for two micropower bandgap voltage references, in a series-connected stack.

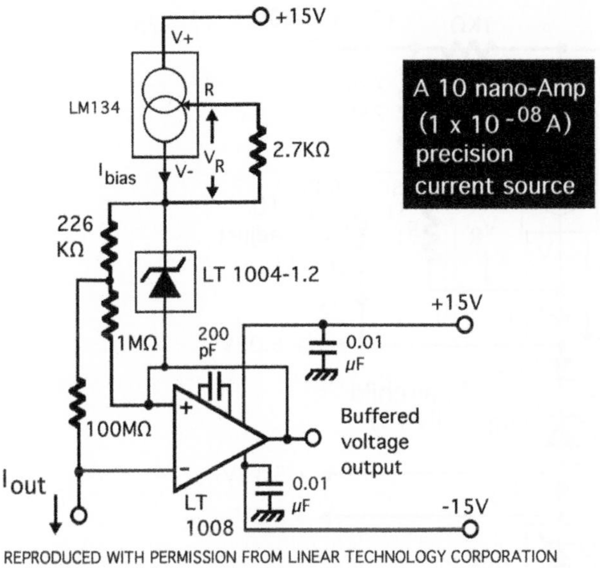

Figure 10.5. Using the LM134/334, together with a Linear Technology LT1004 voltage reference and an LT1008 op amp, to create a precision 10-nA current source.

The LM385 reference IC can operate on as little as 20 μA, but here operates at 100 μA. Once the circuit's performance has been measured and plotted over the required temperature range, its tempco can be adjusted by increasing or decreasing the value of R_2.

In some applications, the 15 pF effective shunt capacitance of the LM134 can be reduced to less than 5 pF by cascoding the IC with a low g_{os} JFET (see the circuits in Figures 10.7 and 10.8). This can also increase the output impedance and boost regulation performance, while not altering the circuit's other DC characteristics. The JFET should be selected such that there is at least 1 volt across the LM134, as defined by:

$$V_{GS(off)}\left(1 - \frac{I_{set}}{I_{DSS}}\right) \geq 1.2V \qquad \text{(Eq.10.5)}$$

In Figure 10.7A, an N-channel general-purpose amplifier, a J204, is combined with the LM134 in a simple, single-FET cascode. Adding the JFET helps reduce the shunt capacitance of the LM134 from its normal 15 pF to less than half that, so that to start with, high-frequency operation is improved. In some cases, cascoding can also allow increased operating voltage range and compliance (the LM134 has a maximum operating voltage of 40 volts, but here so does the J204), when the FET has a higher $V_{(BR)GSS}$ and $V_{(BR)GDS}$ than 40 volts. Cascoding can also increase the output imped-

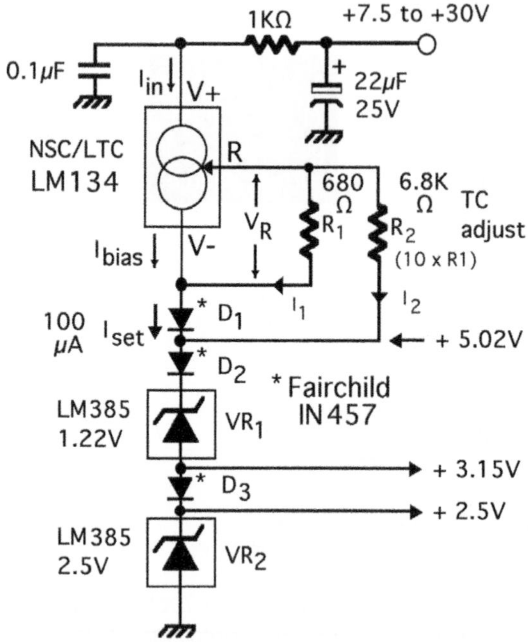

Figure 10.6. *The LM134 sets the 100-µA operating current for this micropower voltage reference stack. R_2 allows one to adjust the circuit's tempco.*

ance (Z_{out}), so that in this circuit it can provide an internal circuit impedance of more than 10 MΩ at current levels less than 1 mA. Overall, the cascode provides a sharper, better quality of current regulation.

Here the FET's drain current (I_D) is regulated by the LM134 and R_{set}. R_{set} still determines the I_{set} current, as previously described, which in turn sets the JFET's drain current, I_D. A constant voltage now appears across the LM134, which produces an even more stable constant current. A level of equilibrium is maintained voltage-wise, between the JFET's gate and source, to maintain this constant voltage and produce this constant current. In circuit A, cascoding buffers the current source from the load, by using Q_1. This means that while the source voltage of Q_1 remains constant, variations in the load voltage are accommodated by Q_1's drain. As a result the voltage drop across the current source remains constant.

The JFET should be operated with a low drain current (preferably less than 10% of I_{DSS}) and with adequate drain-to-gate voltage (V_{DG}), so that V_{DG} is ideally more than two times $V_{GS\ (off)}$, or else its output conductance will increase significantly. (The JFET's output conductance, g_{os}, in this cascode configuration is much less than its typical data sheet value—about 10 times lower.) Based on this knowledge, the minimum operating voltage using the particular JFETs shown in Figure 10.7 would then be

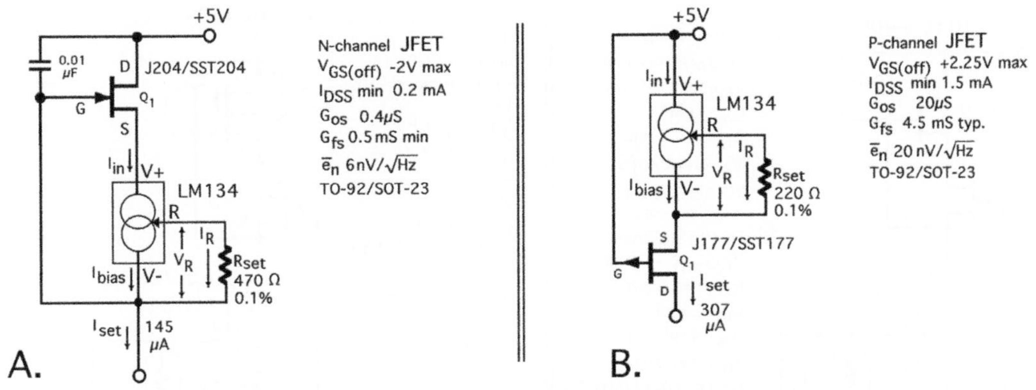

Figure 10.7. *Combining the LM134 with a JFET in a simple cascode configuration increases the output impedance (Z_{out}), lowers the LM134's shunt capacitance, and improves the quality of regulation.*

around 5.25 volts, but this is based on the maximum $V_{GS\ (off)}$ value of –2 volts. Because the $V_{GS\ (off)}$ will almost certainly be somewhat lower, the minimum operating voltage would then be around 4.75 volts. With different choices of JFET, and with lower values of $V_{GS(off)}$, this minimum operating voltage may be reduced down to about 3 volts. (Remember, for volume applications, JFET manufacturers will select devices to meet your requirements.)

It is important to remember to allow at least 1.25 volts across the LM134 in any of these circuits. The circuits shown in Figure 10.7 can be used equally well in designs of up to 40 volts or below 6 volts, depending on one's choice of $V_{GS(off)}$ for the JFET(s). The desired current level could range from more than 1 μA to less than 5 mA. Capacitor C_1 is included just for circuit stability. This inexpensive circuit (four components) can be used as either a precision current source or sink, because the load can be connected to either terminal. This results in an improved level of regulation, compared with the single LM134 on its own. The circuit in Figure 10.7B shows a J177 JFET, a P-channel analog switch used in exactly the same manner, but with its polarities reversed. Operation is otherwise the same, as are the previously mentioned considerations. (For more on using JFET current sources/cascodes, please refer to Chapter 6.)

Figure 10.8A shows a precision 500-μA current source, using two JFETs together in cascode (a J109 and a 2N4393, both N-channel analog switches), further combined with an LM134 into a *super-cascode* configuration. Regulation will be further improved and output impedance increased by this configuration. The JFETs could each have a lower value of $V_{GS(off)}$ than with the parts shown, and (allowing at least 1.25 volts across the LM134) could support a minimum operating voltage of about 7 volts (the circuit shown has a higher minimum operating voltage of 10 volts). The LM134 circuit

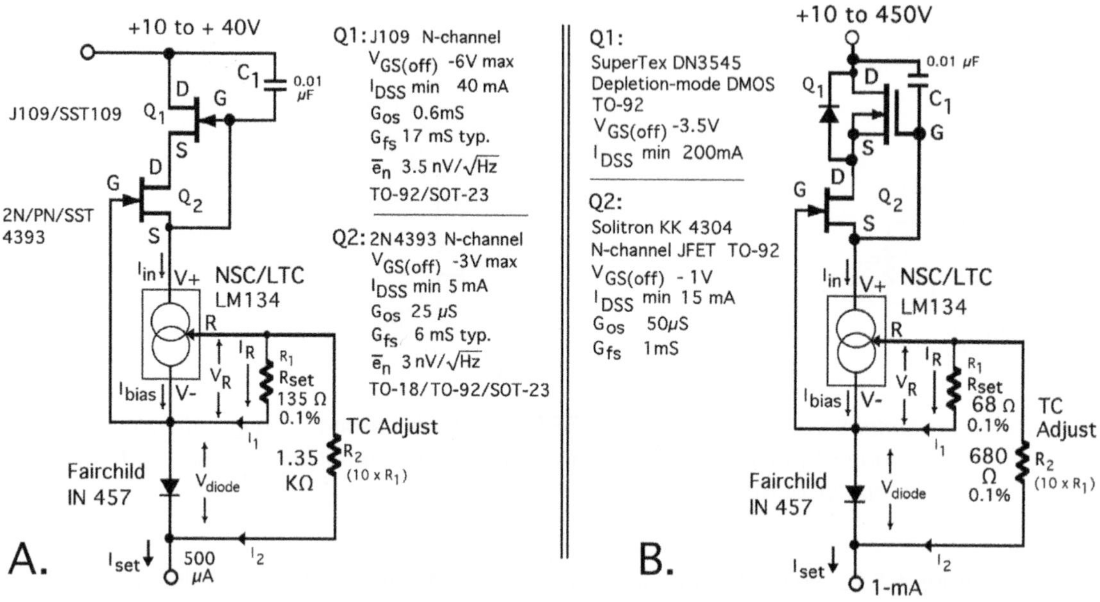

Figure 10.8. *Super-cascodes! Combining the LM134 with either an all-JFET cascode or a JFET-DMOS combination can provide excellent current regulation and a high output impedance. The FETs should be chosen for the desired $V_{GS(off)}$ values while allowing at least 1.2 volts across the LM134. Both circuits provide adjustment of the temperature coefficient, via R_2. Circuit A is low-noise, while circuit B provides high-voltage.*

also includes the compensating diode and the tempco trim using R_2 (as previously described and shown in Figure 10.3). Although this is a more expensive circuit because of the number of parts, it can be expected to provide an output impedance of more than 100 MΩ, as well as the best possible regulation.

The circuits shown in Figure 10.8 could both be considered to be super-cascodes. In Figure 10.8A, two N-channel JFETs in a cascode configuration are combined with the LM134. This circuit, while providing high output impedance, an adjustable tempco, and lower shunt capacitance, also provides a very low noise feature, by choice of the JFETs. In Figure 10.8B, a Supertex DN3545 N-channel DMOS FET and a very low $V_{GS(off)}$ −1.0-volt max, N-channel JFET (a Solitron KK4304 general-purpose amplifier) are combined with the LM134 in an unconventional cascode to provide a precision 1-mA current source. While retaining all of the major characteristics of the circuit in A, the most unique feature is its high-voltage capability, which can reach 450 volts because of the special processing by SuperTex of all their depletion-mode DMOS FETs. (See Chapter 7 for more information on DMOS.)

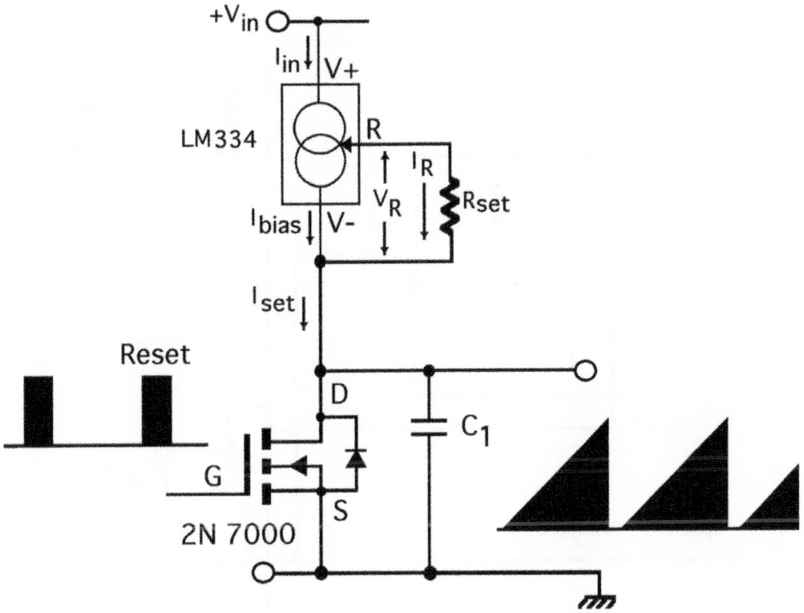

Figure 10.9. *The LM334 is used here as a precision ramp-generator, where the current source linearly charges a capacitor. The DMOS FET is used as a fast reset switch.*

The circuit in Figure 10.9 shows the LM134/334 being used as a precision ramp generator, where its constant current is used to linearly charge a timing capacitor, C_1. The 2N7000 small-signal DMOS FET is used as a fast reset switch.

In Figure 10.10, the LM334 is being used as a stable reference for the level settings of a quad comparator IC, in a classic window comparator circuit. Compensating the LM334 for a near-zero temperature coefficient helps improve the overall accuracy of the window comparator.

10.3 Using the LM134 as a temperature sensor

In addition to being a current source, the device also doubles as an accurate, remote temperature sensor, as shown in Figure 10.11. Because of its current-mode operation, it does not lose accuracy as a result of long cable runs. This is because the internal sense voltage (64 mV) used to establish the operating current is directly proportional to absolute temperature, in degrees Kelvin. The temperature dependence of the operating current is approximately +0.33%/°C. So for a device set up at room temperature to generate an operating current of 298 μA, it will have a temperature coefficient of 1 μA/°C (298°K = 25°C). Devices in three-lead transistor packages that are specified as temperature sensors are useful over the industrial (–25°C to +100°C) temperature

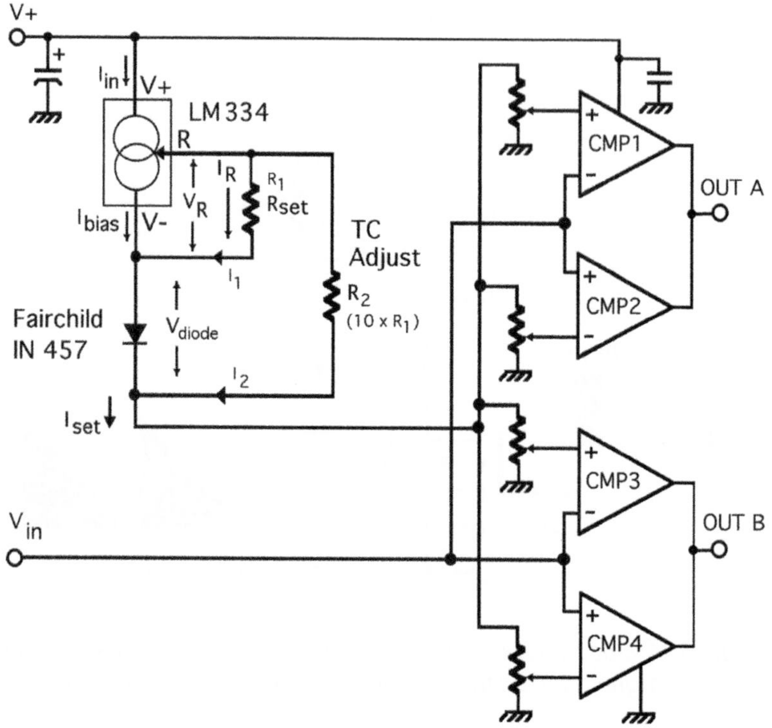

Figure 10.10. *The LM334 provides a zero tempco current source that provides a very stable reference for a window comparator.*

range. They have an equivalent temperature error of either ±3% (Z-3 devices) or ±6% (Z-6 devices).

Because of this predictable temperature dependence, the LM134 family is particularly useful in applications where remote temperature sensing or long cable runs are necessary. As a result, two devices in the family are specified as temperature sensors over the 100-μA to 1-mA range: the LM234-Z3 (±3%, –25°C to +100°C) and LM234-Z6 (±6%, –25°C to +100°C). These are shown in Table 10.1 and are available from either National Semiconductor or Linear Technology. As mentioned previously, the LM134 has spearheaded the development of the dedicated temperature-sensing ICs over the past two decades. Today's devices, while based on the same fundamental technology, are in many cases smaller, even easier to use, and have more practical outputs (i.e., 10 mV/°C or 10 mV/°F). Although some devices are relatively simple, others have digital interfaces and are programmable by and compatible with the computer. The enormous growth of the PC over the past decade has spawned many temperature-sensing applications. These range from monitoring the temperature inside the computer and safeguarding expensive CPUs to remote telemetry. Five U.S.

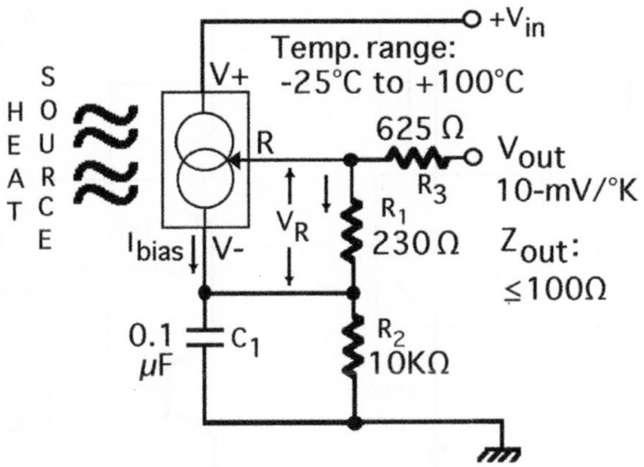

Figure 10.11. *A simple Kelvin thermometer, with a two-wire interface, can be built using an LM234-Z3.*

semiconductor makers of these types of product include National Semiconductor, Linear Technology Corp., Analog Devices, Maxim Integrated Products, and Fairchild Semiconductor.

In the words of National's staff scientist, technical guru, and popular columnist Bob Pease, the LM134 covers a wide range and is "so versatile." That it is, and probably underscores the reason why after some 25 years, the LM134 series is a tried-and-trusted workhorse and still a very popular item.

10.4 TI/Burr-Brown's REF-200 monolithic current source

TI/Burr-Brown's REF-200 is only the second of two standard catalog ICs that exists in the U.S. analog marketplace that are dedicated monolithic current sources. Burr-Brown was one of the electronics industry's oldest and most respected companies, having been started in the mid-50s. It had a strong tradition of providing innovative and state-of-the-art analog products. Burr-Brown was acquired by Texas Instruments in 2000 to complement TI's analog products portfolio. It is now a part of Texas Instruments' analog products group.

Unlike National Semiconductor's LM134 single, adjustable current source, the REF-200 has two fixed 100-μA current sources, as well as a precision current mirror, all packaged in an eight-pin surface-mount SOIC package. The REF-200 was introduced in 1990 and uses a proprietary Burr-Brown dielectrically isolated process (Difet®), which completely isolates the three circuits, making them independent of one another. Each of the 100-μA current sources uses a precision bandgap cell to provide a zero temperature coefficient, while the current mirror uses the reliable Full Wilson architecture. Each of the circuits is subject to laser-trimming at the wafer level, to provide the

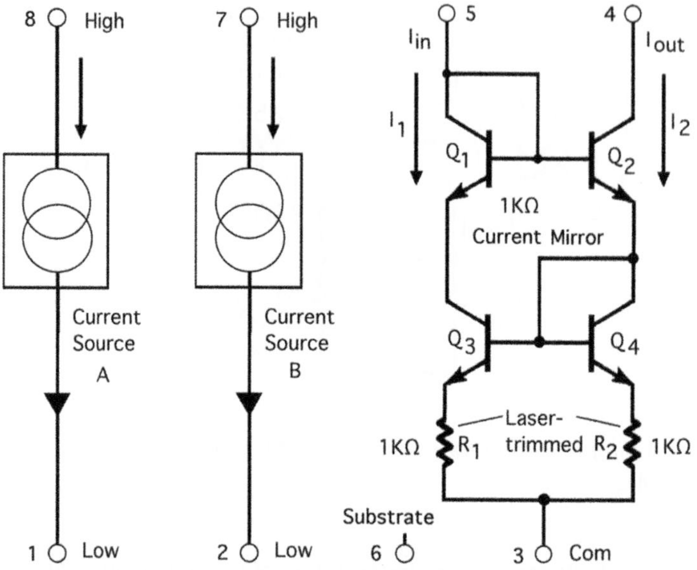

REPRODUCED WITH PERMISSION FROM TEXAS INSTRUMENTS, INC.,

Figure 10.12. TI/Burr-Brown's REF-200 houses two (zero-TC) current sources and a current mirror, in either a plastic SOIC-8 or eight-pin DIP package.

highest accuracy. A simplified diagram of the REF-200 is shown in Figure 10.12, which shows the three circuits.

The minimum operating voltage for the REF-200's current sources is approximately 2.5 volts, up to an absolute maximum rating of 40 volts. For the current mirror, the minimum operating voltage is lower at about 1.4 volts, also up to the maximum rating of 40 volts. In the case of the current sources, the minimum output impedance is specified at more than 20 MΩ, but is typically more than 100 MΩ. The output impedance of the current mirror is specified at more than 40 MΩ minimum and typically more than 100 MΩ. Both types of circuit have a typical temperature drift of +25 ppm/°C, which is very accurate (0.0025%). The REF-200 has an operating temperature range of between –25°C and +85°C.

To safeguard the REF-200 against the possibility of reverse polarity, one should use either a parallel-connected, reverse-biased diode, as shown in Figure 10.13A, or a forward-biased diode in series with the REF-200. For best AC performance, the manufacturer advises leaving pin 6 (substrate) open and any unused sections unconnected. An AC application is shown in Figure 10.13B, where a 200-μA bidirectional current source is shown. The REF-200 can be pin-strapped to provide currents of 50 μA, 100 μA, 200 μA, 300 μA, or 400 μA. By adding external circuitry, one can create virtually any current, either smaller or greater than this (some examples are shown later).

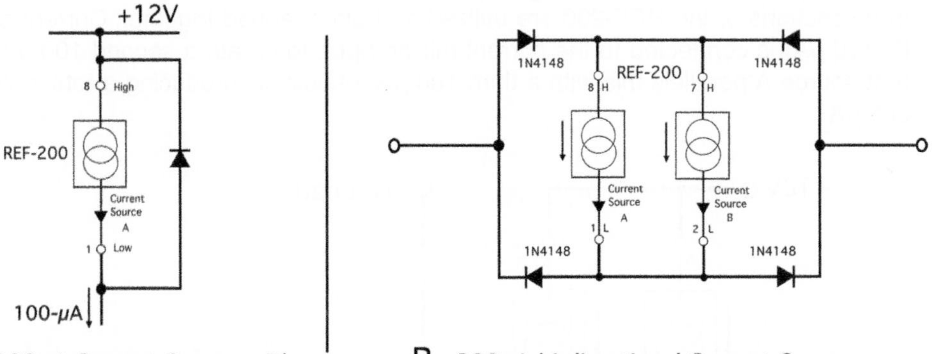

A. 100μA Current Source with reverse-bias protection.

B. 200μA bi-directional Current Source.

Figure 10.13. REF-200 reverse polarity protection.

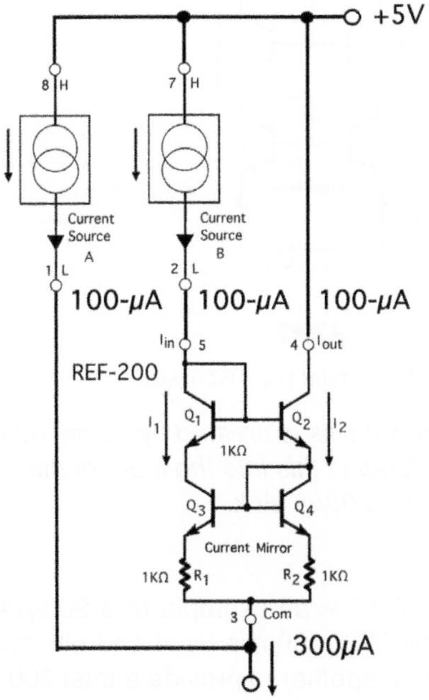

Figure 10.14. 300μA floating current source using all sections of the REF-200.

An example of a 300-μA floating current source is shown in Figure 10.14. Here all three sections of the REF-200 are utilized and pin-strapped together. Current source B (100 μA) is connected to the current mirror input, to create a second 100 μA. Current source A parallels this with a third 100 μA, effectively producing a total output of 300 μA.

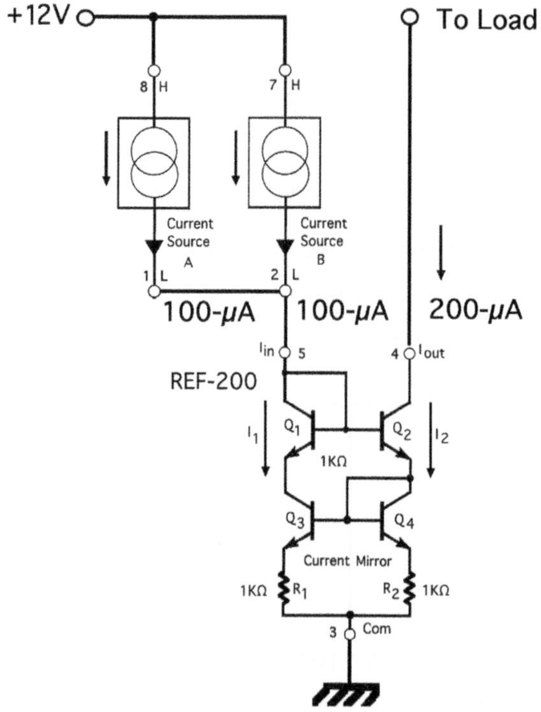

Figure 10.15. This 200μA precision current sink is created by summing the currents from the two current sources and feeding this into the input of the current mirror. The resulting mirror current at I_{out} is a 200μA sink.

The circuit shown in Figure 10.15 is an example of a 200-μA precision current sink. Again all three sections of the REF-200 are used and pin-strapped together. Current sources A and B are paralleled together to provide a total 200 μA, which is then fed to the current mirror input. This current is faithfully mirrored, to create a precise 200-μA sink.

Borrowing from some of the techniques regarding cascoding, as previously described in Chapter 6 and subsequent FET chapters, Figure 10.16 shows how combining a JFET with the REF-200 can make a superior current sink or source.

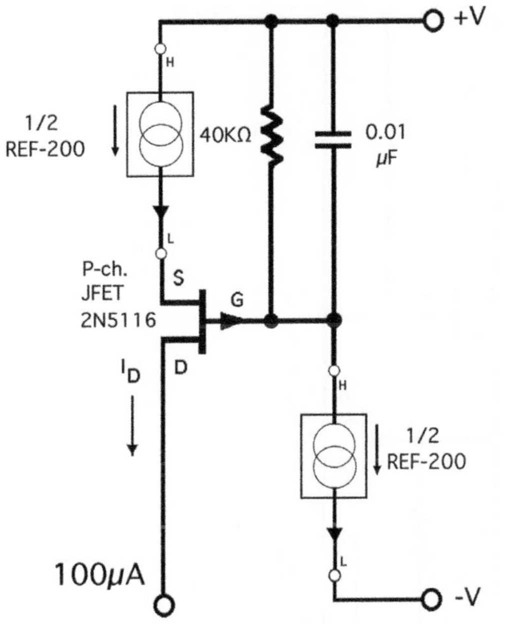

Figure 10.16. *Combining a P-channel JFET with the REF200 results in a high Z_{out} (>1GΩ), well-regulated 100μA* current source.

Here a 2N5116 P-channel, low-noise JFET ($V_{BR(GSS)}$ = 30 volts; $V_{GS(off)}$ = 4 volts max; I_{DSS} min = 5 mA; G_{os} = 20 μS typically) is combined with the REF-200's two current sources to create a cascode circuit, with an output impedance of more than 1 GΩ. The JFET's necessary gate voltage is met by biasing its gate from the REF-200's second current source.

The circuit in Figure 10.17 shows a current source with a very high output impedance. It is made by combining both of the REF-200's current sources, together with a pair of low-output conductance discrete JFETs. (More information on each JFET can be found in Table 6.2 in Chapter 6.) The current source is made by combining the upper current source with the 2N5116 P-channel JFET, together with a cascoded current sink using the lower current source and the 2N4340, an N-channel JFET. This combination provides exceptionally good regulation (0.00005% per volt), along with a high output impedance of more than 5 GΩ. Compliance in this circuit is bounded at the high end by the lower $V_{BR(GSS)}$ JFET rating (which here is 30 volts) and at the low end by the combination of the two V_{GS} values (approximately 8 volts).

An unusual example is shown in Figure 10.18 of a circuit that combines *both* the REF-200 and the LM234 current sources in a super-cascode 229-μA current source. Add-

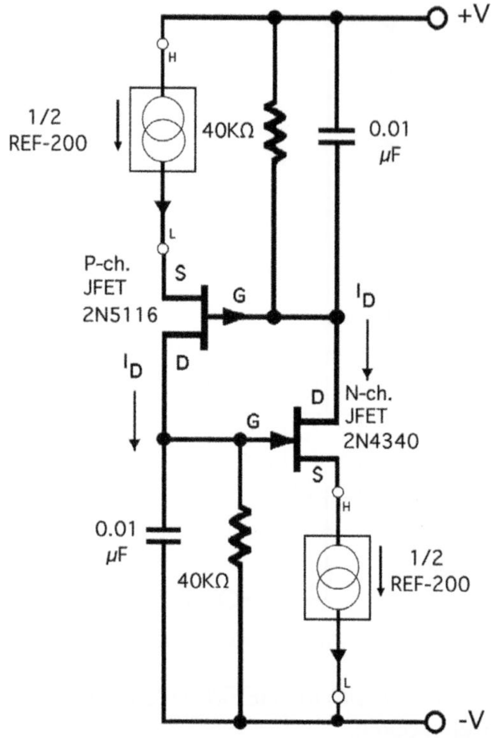

Figure 10.17. A 200μA cascoded current source using JFETs and the REF200.

ing a pair of JFETs helps boost the output impedance and improves overall regulation. Two hundred microamps are supplied from the REF-200's two current sources, which each have a zero TC. The LM234 supplies 14.4 μA, also with a zero TC. This current is fed to the input of the REF-200's current mirror, which outputs the same value. JFET Q_2's source feeds the output of the current mirror, the input of the LM234, and both of the REF-200's current mirrors. The total current flowing out of the circuit is 228.8 μA, with a zero TC, and with a very high output impedance of more than 1 GΩ.

The circuit could be built using just the LM234, but with a lower output impedance and not quite as low a tempco. It could also be built using just the REF-200, although additional circuitry (including an op amp) would be needed to create the 229 μA. Alternately, it could be built using just the two cascoded JFETs, but the tempco would have to be determined by time-consuming measurement and adjustment over the intended temperature range. Combining both the LM234, the REF-200, and the JFETs provides a superior solution in this particular circuit, because it uses the best features of them all.

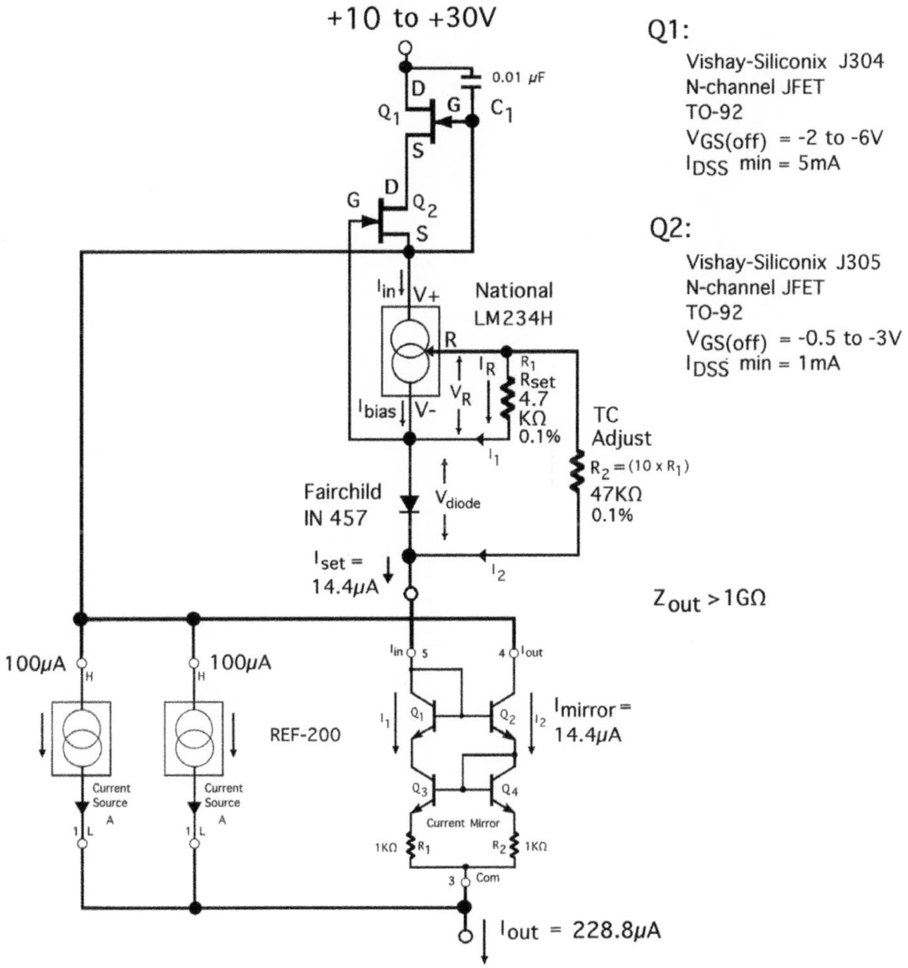

Figure 10.18. Combining two N-channel JFETs with the LM234 and the REF-200 monolithic current sources all in one design. It results in a super-cascoded 229μA current source with ultra-high output impedance, and with a near-zero TC.

In summary, using a dedicated monolithic current source IC such as the REF-200 or the LM134 can often save the designer much time and effort. Although the LM134 is directly adjustable from about 3 μA to 1 mA, the REF-200's fixed 100-μA output can be pin-strapped to create multiples of this. A zero tempco is achievable with the LM134, with the addition of a low-leakage diode and passive components, while the REF-200 has a laser-trimmed zero tempco. The two devices complement one another, though, and will continue to find many precision reference applications, which can now be built using surface-mount components. Combining an op amp with either of these two devices can provide even more flexibility, as will be seen in Chapter 11.

Chapter 11

Creating Precision Current Sources with Op Amps and Voltage References

11.1 How op amps evolved

Operational amplifiers (op amps) can be used to create (1) precision current sources and sinks, which range from submicroamps to milliamps, and (2) current regulators, which range from milliamps to amps. Equipment and instrumentation designers have used the op amp for these purposes for many years past, in fact ever since the third or fourth generation of op amps was introduced. Those products brought with them greater stability and were much more affordable than their predecessors. In this chapter we will look at how to use the op amp in both current source and current regulator applications, but first we will take a brief look at how today's op amp evolved.

An op amp circuit was first created during World War II for the U.S. military by an American scientist Loebe Julie. Following World War II, a few small companies emerged and started looking at analog computer applications. The best-known pioneer was Philbrick Researches, whose founder, another American engineer, George A. Philbrick, was the pioneer of both the op amp and of analog computing. In the late 1940s and early '50s, this original op amp circuit was constructed using several vacuum tubes, to form a multistage amplifier block. Julie's best design, known as the K2W, was adopted by Philbrick, which successfully commercialized it. Eager buyers include instrumentation and analog computer makers.

Initially, Philbrick made op amp circuits with vacuum tubes, but once the transistor became available, the company switched to making circuits with germanium transistors. The transistors of that era were not as good or as stable as today's silicon devices, so it was a challenging and time-consuming task—drift was a constant enemy. The transistor circuits Philbrick created were hand-assembled and then trimmed to meet the specs. The keys to success were that by using transistors they offered a product that was significantly smaller. Being smaller, and running on lower voltages and currents, meant that the transistor circuit was less power hungry. It also offered higher precision, because the circuits were trimmed and calibrated. Offering the product line as a group of separate function modules was the best part. In fact, Philbrick offered modules that included many mathematical functions, such as square root and log converters, integrators, differentiators, multipliers, and so on. With these functional building blocks, design engineers could now begin to create many different analog sensing and control circuits. One of Philbrick's main competitors in the 1950s was Burr-Brown Corporation (now a part of Texas Instruments). In the 1960s, Philbrick was bought by Teledyne, along with another company, Nexus, and combined into a new company, Philbrick-Nexus.

Another of Philbrick's competitors was Fairchild Semiconductor, which for a time also made these kinds of transistorized function modules (this company made the industry-leading epitaxial planar transistors). So Fairchild introduced the first *monolithic* op amp, in 1964. This was called the μA702 op amp and had been designed by a new, young designer named Bob Widlar. The μA702 was introduced in a metal can and contained nine transistors and eleven resistors. For an op amp it had a limited input common-mode range and low gain, among other problems. In fact, Widlar didn't want to put the device into production, because he didn't think it was good enough. Eventually, Fairchild's management decided to go ahead anyway and put it into production. Unfortunately, the μA702 was difficult to make, and it had a bad yield. It was the first of its type, and there was little or no experience to draw on. Widlar was determined to create a better device, though, and he followed up with an improved version about a year later. At that time the revolutionary IC op amp cost roughly twice as much as its transistorized, modular counterpart.

Widlar's follow-up op amp was called the μA709, which Fairchild introduced in November 1965. It quickly became the industry standard and helped establish Fairchild as the undisputed analog IC leader for several years to come. It also helped spawn a whole new part of the semiconductor industry, known today as the analog IC market. The μA709 became an industry standard even though it cost $50 each and needed a resistor and two capacitors for compensation. (In addition, it sometimes suffered from a phenomenon known as input latch-up and was prone to occasional oscillation. It was also vulnerable to having its output short-circuited. Even a momentary excursion would destroy the device, so one treated it *very* carefully.)

For Fairchild, although the device was hard to manufacture and had a poor yield, the company could not keep up with demand for this revolutionary product and had back orders for several years after it was introduced. The person who Fairchild's management teamed up with Widlar, and who marketed these early op amps so successfully, was Jack Gifford. He first introduced the dual-in-line (DIL) IC package to the market, and then later as Fairchild's Linear Circuits Product Manager, he laid the foundations for all of today's analog IC business. (Jack Gifford is now the CEO of Maxim Integrated Products, yet another successful analog IC company.)

Later, at National Semiconductor, Widlar continued with his op amp designs, which included the LM101, the first *uncompensated*, general-purpose op amp (actually it needed just a 30-pF capacitor for compensation), which fixed all of the problems of his previous designs from when he was at Fairchild. (The LM101A is still in production today at National.) Probably Widlar's most remarkable op amp design was introduced by National Semiconductor in the late '70s and is also still in production today, more than 25 years later. It is the LM10, combining one monolithic chip, a very-low-voltage op amp, and a low-drift bandgap voltage reference. The LM10 is an audio-frequency op amp with good specifications. Although it may not be as good as many of today's newer, faster, and differently processed op amps, it is unique in that (1) it can run off a total voltage supply of between 1.2 volts and 45 volts, and (2) it contains a useful 0.2-

Photo 11.1. *Robert J. Widlar, inventor of the IC op amp, the bandgap voltage reference, and the IC voltage regulator. This picture shows him looking at the layout of his LM10 low-voltage op amp design when he was at National Semiconductor in the late 1970s. (Photo courtesy of National Semiconductor Corporation.)*

volt voltage reference, all in the same eight-pin package. It has extremely low drift characteristics, a very low input bias current, low-noise input, and low power. How far ahead was Bob Widlar in 1978? Well, let's just say it took the industry more than a decade for other very-low-voltage analog products to appear in the marketplace.

Widlar was a design phenomenon, but Fairchild had several other talented IC designers too. One was Dave Fullagar, who developed a metal-oxide capacitor that improved the stability of the μA709 op amp, and thus boosted production yields. He also developed its successor, which was to become the first monolithic bipolar, *precision* op amp, the μA725 (this was later finished by George Erdi). Fullagar also designed the legendary general-purpose op amp—the μA741. It was reliable, which its predecessors were not, and it required *no* frequency compensation. Better still, its output was short-circuit proof. This op amp's phenomenal success and acceptance further established Fairchild Semiconductor as the analog IC leader at the time. The μA741 became the worldwide industry standard and the best-selling op amp for more than two decades, and is still in production today. (Dave Fullagar is now a co-founder, vice president, and staff scientist at Maxim Integrated Products Inc., in Sunnyvale, California.)

Another Fairchild designer was George Erdi, who helped design the first monolithic bipolar 10-bit digital-to-analog converter (DAC), which became the μA722. Next he completed the design that Fullagar had started on the μA725, and which Fairchild introduced in 1969. Erdi is also credited with the research work he did at Fairchild Semiconductor in the late 1960s on monolithic ICs. One example was his technique of incorporating thermal symmetry into the design and layout of monolithic analog ICs and power chips.

(*Reference*: G. Erdi, "Minimizing Offset Voltage Drift with Temperature in Monolithic Operational Amplifiers," from the *Proceedings of the National Electronic Conference*, Vol. 25, 1969.)

After recently finishing development and preproduction work on the μA725, Erdi had no doubt gained a lot of insight into the various strengths, weaknesses, and pitfalls of that design. The next design would be unique in that it would be all his own design, from beginning to end. After co-founding Precision Monolithics Inc., Erdi designed the legendary high-performance op amp, the OP07. This was the first *real* precision op amp. It coupled his op amp design with a technique he developed of adjusting the op amp's input offset voltage (V_{OS}) to a very low level. This technique became known as *zener zapping* and has been used industry-wide for many years to adjust op amps and other analog products. Erdi's ultra-stable OP07 has a very low-input offset voltage of 25 μV maximum, with a maximum drift of 0.6 μV/°C. It also has a low input noise characteristic of about 10 nV/√Hz. (Although it is now more than 30 years old, the OP07 is a good example of a precision op amp and is still an excellent choice in many current source applications.) The OP07 brought with it such a major improvement in precision that it became the industry standard and is still a steady production item to this day, with Analog Devices (which later bought PMI). George Erdi later left PMI and in 1981 co-founded Linear Technology Inc.

A fellow designer of Widlar's when he was at National was Bob Dobkin. He designed the first high-speed op amp, the LM118 (typical small-signal BW = 15 MHz, and typical slew rate at 70 V/μS), as well as others. He also invented the feed-forward technique that enables the op amp to achieve a higher slew rate and bandwidth.

(*Reference:* Robert C. Dobkin, "Feedforward Compensation Speeds Op Amp," National Semiconductor, Linear Brief LB-2, March 1969.)

Like his friend Bob Widlar, Dobkin can make his transistors do some amazing tricks. Dobkin also designed the first three-terminal, adjustable voltage regulator (as well as the buried-zener voltage reference). After being with National for more than a decade in various capacities that included analog IC designer and then later Director of Advanced Circuit Development, Dobkin, together with Bob Swanson, George Erdi, and Widlar, co-founded Linear Technology, Inc.

Besides Fairchild, National, and Precision Monolithics, other companies joined the fledgling analog IC marketplace throughout this time. They included Texas Instruments, Analog Devices, Burr-Brown, Signetics, Union Carbide, Transitron, Crystalonics, AMD, Sprague, Raytheon, Motorola Semiconductor, Linear Technology, Silicon General, Teledyne Semiconductor, Unitrode, RCA, and Intersil (and others such as Philips, SGS, Siemens, and Plessey in Europe). The main products they all offered were bipolar op amps and comparators. All the technology of the day was based on BJT transistor technology (or bipolar FET), because many of these manufacturers were also making transistors and TTL logic as well, which ruled the digital market at that time. Not all were though: for example, Analog Devices was originally a precision analog/data-converter module manufacturer like Philbrick and Burr-Brown.

All that changed in the mid-70s when RCA Solid State introduced the first analog CMOS op amps. Later, another analog CMOS pioneer, Intersil, introduced op amps and other ground-breaking analog products. As many of those companies mentioned here slowly got into CMOS, they too started offering CMOS op amps. They were joined on the way by some exceptionally talented new companies (which specialized in CMOS), such as Maxim and Advanced Linear Devices. While some discovered how to make high-speed, precision op amps, others discovered how to make CMOS op amps that would run off a single AAA battery. The key to that was using silicon-gate technology, which we read about in Chapter 7 and which has also evolved over time.

By the early 1970s, Analog Devices started making analog ICs, and a decade later it was making op amps in various processes that included bipolar and bipolar MOS. At that time it and others were trying to find ways that would help improve speed and bandwidths, because the transistor-based technology of that era had serious limitations. NPNs had much higher f_Ts than their PNP counterparts, and this PNP bottleneck limited the overall bandwidth to just a few MHz. Analog Devices eventually introduced a working process that became known as *complementary bipolar* (CB). This process uses *ion implantation,* and for the first time was able to include high-speed (f_T = 700 MHz) PNP transistors alongside high-speed NPN devices. The CB process was a major advancement in the overall evolution of the op amp, because it provided symmetrical operation internally, lower power with higher output currents, and effectively improved the bandwidth by at least an order of magnitude. Others like National Semiconductor with its Vertically Integrated PNP (VIP™) process, and Texas Instruments with its Excalibur™ process, have also contributed enormously to this important high-speed technology.

Begun in 1964 by Fairchild Semiconductor with the introduction of the first *monolithic* op amp, today the number of U.S. dedicated op amp manufacturers is around 12, but the analog IC business that they and others from around the world generate is worth $ billions. In the United States, only a few of the original pioneers remain or are still making op amps. The insatiable global PC market, together with the cell phone, digital camera, music Pod, DVD player, and high-definition TV markets, will help fuel the analog industry for many years to come.

11.2 Some op amp characteristics

Since their introduction, op amps have evolved across many generations. Today they are available in various distinct categories, such as general-purpose, precision, high-speed, single-supply, and so on, and are manufactured with any one of several different processes (e.g., bipolar, complementary bipolar, silicon-gate CMOS). Although it is a little more complex than is depicted here, because of the many different processes and categories of op amps, the relevant part of the op amp family tree is shown in Figure 11.1. (This part is relevant to the types of op amps needed to create good-quality current sources.) In many cases there is a trade-off between operating voltage and speed. Some processes enhance high speed, whereas others provide high precision or very low power consumption.

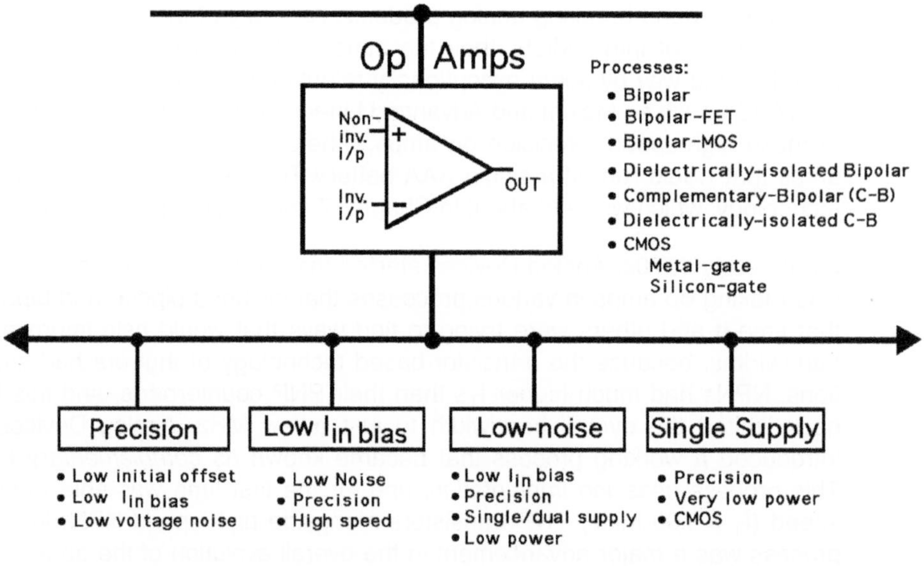

Figure 11.1. Part of the op amp family tree.

Just like its predecessors, the op amp consists of several stages (see Figure 11.2). These usually include a differential amplifier at the front end, some gain stages, and a (low-impedance) buffered output stage. The differential input stage is commonly used because it offers high-input impedance, as well as sharply rejecting common-mode input noise. Early op amps like the μA709 and 741 typically had only two gain stages, while later products incorporated three stages or more. In the two-stage amplifier, the first stage acts as a voltage-to-current converter, while the second stage is a current-to-voltage converter. By rolling off the frequency response of the second stage at about 6 dB per octave so that the overall gain is unity at the design frequency, the device can then be stabilized for virtually any feedback conditions. This is achieved by incorporating a tiny metal-oxide capacitor (typically between 12 pF and 47 pF) in the chip's design. This created two types of op amps: the *internally compensated* type and the *uncompensated* type, which lets the designer customize the necessary frequency response. Today, most op amps are internally compensated.

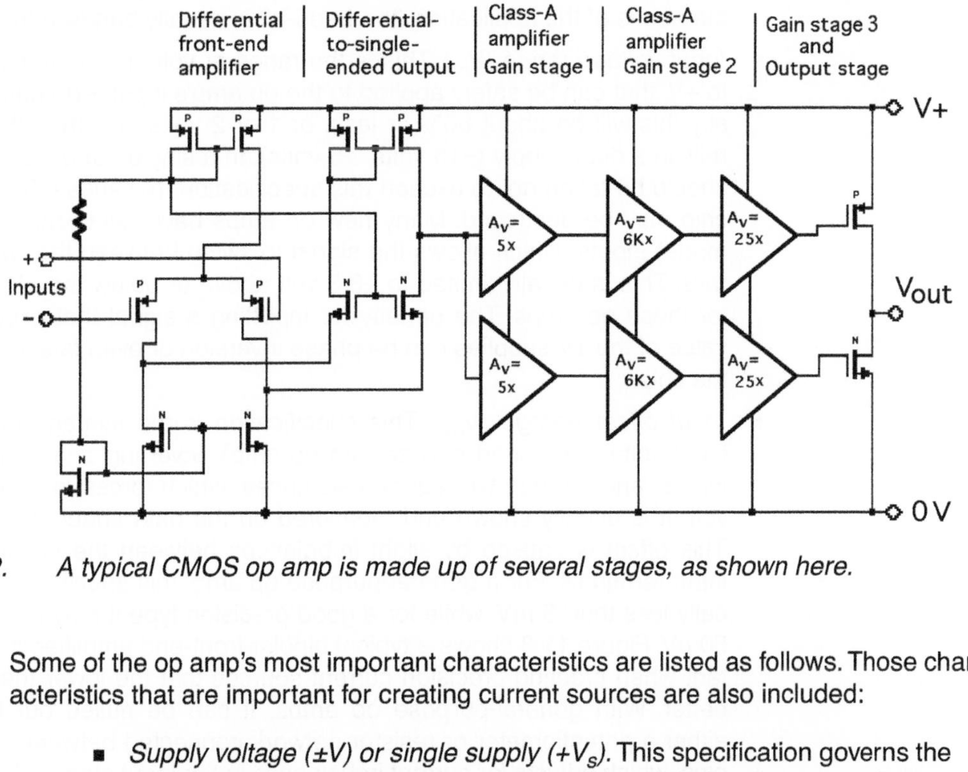

Figure 11.2. *A typical CMOS op amp is made up of several stages, as shown here.*

Some of the op amp's most important characteristics are listed as follows. Those characteristics that are important for creating current sources are also included:

- *Supply voltage (±V) or single supply (+V_S).* This specification governs the range of supply voltages that can be safely applied to the op amp's supply terminals (+V and -V, or +V and 0V). The maximum voltage supply should never be exceeded, because it may reverse-bias/overvoltage some junctions, which may degrade some of the characteristics or cause fatal damage to the circuit. The minimum voltage supply should always be observed too, so that the lowest foreseeable voltage (i.e., a battery's end-of-life) still supports normal operation. Some op amps, including some precision types, run best on a dual supply (i.e., ±12 volts), whereas others, including many CMOS op amps, run on single and low-voltage supplies (i.e., less than +5 volts). There is often a trade-off in using very-low-voltage devices, such as a lower bandwidth and lower precision.

- *Supply current (I_{SS}).* This specification provides the typical and maximum operating current(s) according to the range of supply voltages applied to the op amp's supply terminals (+V and -V dual supplies, or +V and 0V single supply). The maximum supply current(s) should be noted and budgeted in supporting normal operation. It is usually specified on the data sheet in either milliamps or microamps. Although CMOS op amps usually have some of the lowest supply currents, many bipolar devices also have low supply currents. Several bipolar *micropower* products also have low supply currents and allow you to adjust their supply currents to meet the needs of the application (the trade-off is usually bandwidth).
- *Input voltage range (V_{in}).* This is the range of voltages, either ±V, or 0V to +V, that can be safely applied to the op amp's input terminals. Generally this will be about 80% or less, or 1 or 2 volts less than the supply rails in a dual supply (±15 volts, ±5 volts). In a single-supply circuit, care should be taken not to exceed this specification, because otherwise the chip may be destroyed. Many new op amps have rail-to-rail (common-mode) inputs, which allows the signal to swing between the two supply rails. This is usually limited to ±0.3 volt above or below the ±V supplies for those op amps. The penalty for inputting a signal that exceeds the value of the ±V supplies can be phase inversion or even fatal damage to the chip.
- *Input offset voltage (V_{OS}).* This specification is the average differential input voltage required between an op amp's inverting and noninverting inputs, and through two equal resistances, which forces the output to 0 volt. It is usually shown and measured on the data sheet in mV or μV. This offset is caused by slight imbalances between the op amp's two input terminals. For a general-purpose op amp, this specification is typically less than 5 mV, while for a good precision type it may be less than 50 μV. Figure 11.3 shows a typical bipolar front-end amplifier. It is important when creating precision current sources that the lower the V_{OS}, the better. With general-purpose op amps, it can be nulled out by using either a potentiometer or resistor network connected between the offset pins, which adjusts for current imbalances in the input stage. With a high-precision op amp, one should use caution in adding a potentiometer or resistor network, because the input circuit is normally laser-trimmed precisely at the factory. Adding external resistance for additional offset adjustment can actually cause the V_{OS} drift to increase, rather than the intended decrease.
- *Input offset voltage drift (TCV_{OS} in $\mu V/°C$).* This specification is the average ratio of change in input offset voltage (ΔV_{OS}) to a measured change

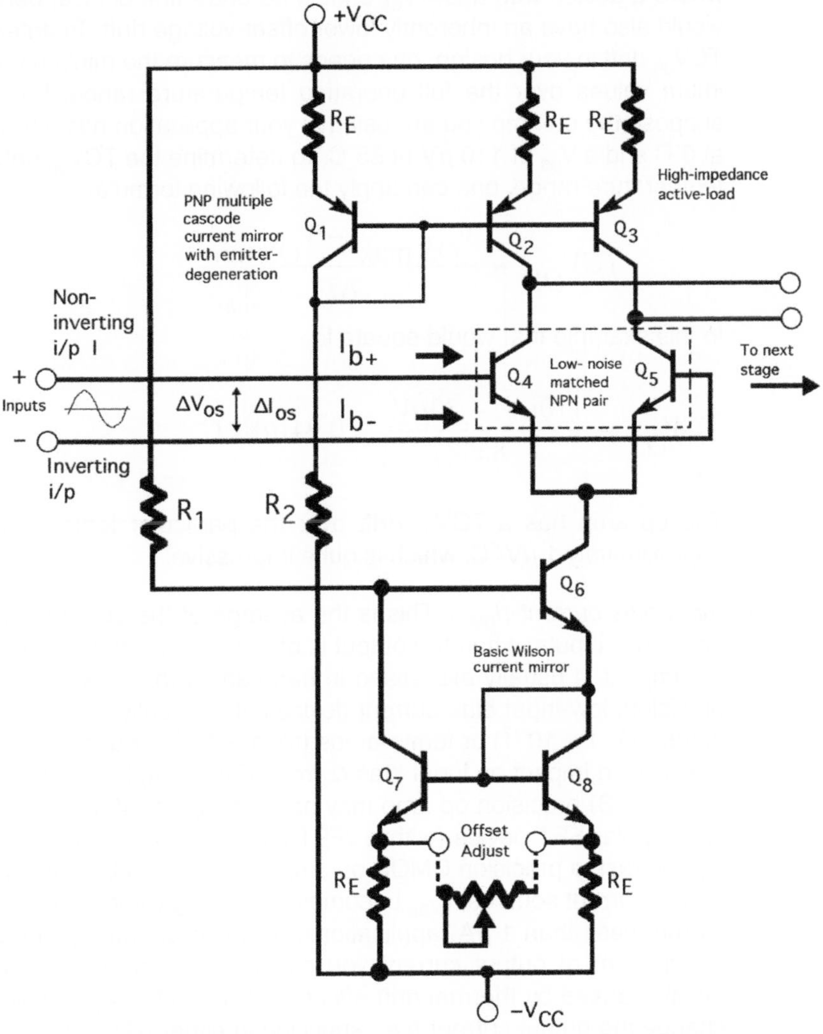

Figure 11.3. Showing a bipolar op amp's typical front-end amplifier and biasing circuits.

in temperature (Δ°C). It is specified in μV/°C for precision op amps and not usually for other types. Typically, TCV_{OS} ranges from less than 1 μV/°C to more than 50 μV/°C. A good precision op amp may have a specification of less than 5 μV/°C and a long-term drift of less than 10 μV/month. (As noted in V_{OS} previously, V_{OS} drift may actually increase in a precision op amp simply by adding an external resistor network connected between its offset pins.) Usually, op amps with large initial V_{OS} will have a higher TCV_{OS}. It is also important for current source applications, where a device with a low V_{os} should be one's first choice, because this would also have an inherently lower offset-voltage drift. To determine the TCV_{os} drift in your design, one needs to measure the minimum and maximum values over the full operating temperature range. For example, suppose the op amp you are using in your application has a V_{os} of 30 μV at 0°C and a V_{os} of 110 μV at 85°C. To determine the TCV_{os} drift over the temperature range, one can apply the following formula:

$$TCV_{OS} = \frac{V_{OS\,\max} - V_{OS\,\min}}{\Delta T} \qquad \text{(Eq.11.1)}$$

In this example that would equate to:

$$TCV_{OS} = \frac{110\mu V - 30\mu V}{85^{\circ}} = 0.941\mu V/^{\circ}C \qquad \text{(Eq.11.2)}$$

The op amp has a TCV_{os} drift over the particular temperature range of approximately 1 μV/°C, which is quite impressive.

- *Input bias current (I_{bias}).* This is the average of the currents into the op amp's two inputs, when the output is at 0 volt and with no load. For most op amps it is usually expressed in nano-amps (nA, 1×10^{-9}). For high-precision, low-input bias current devices, it is usually expressed in pico-amps (pA, 1×10^{-12}) or femto-amps (fA, 1×10^{-15}). Each analog process has its own impact on input bias current. For example, a complementary bipolar (CB) precision op amp may have an I_{bias} of 40 nA, while a precision bipolar FET device (with a JFET input) may have an I_{bias} of 10 pA. By contrast, a precision CMOS op amp may have an I_{bias} of less than 50 fA. For current sources, I_{bias} becomes increasingly important in very-low-current (less than 1 μA) applications, because at some point it will limit the amount of output current you can reliably sink or source. This is mostly caused by thermal drift effects with I_{bias} (TC I_{bias}), which tend to change the output current (i.e., specified in either nA/°C or pA/°C). Table 11.1 serves as a quick summary of this concept. Input bias current is usually specified at room temperature (25°C) and increases with temperature, so that it approximately doubles for every 11°C rise in temperature.
- *Input noise voltage density (e_n).* This is a measure of an op amp's input noise voltage in a 1-Hz center band, centered on a particular frequency (such as at 10 Hz, 100 Hz, or 1 KHz). A precision op amp will typically have an input noise voltage of less than 25 nV/√Hz, while a particular low-noise device may have an input noise voltage of less than 5 nV/√Hz.
- *Input impedance and resistance (Z_{in} and R_{in}).* This specification is the input impedance (expressed in Ω), looking directly into the op amp's inputs, and is usually more than 1 MΩ. A high-input impedance allows

the op amp to directly interface with a high-impedance signal source. There are actually two distinct kinds: (1) the *differential input impedance* is the total impedance when looking from one input terminal to the other, and (2) the *common-mode impedance* is the common impedance when measured to ground. Although most manufacturers do not specify the

To generate this low constant current:	The op amp needs an I_{bias} of under:	with a 1% drift of:
1-mA	1-μA	10-nA/°C
100-μA	100-nA	1-nA/°C
10-μA	10-nA	100-pA/°C
1-μA	1-nA	10-pA/°C
100-nA	100-pA	1-pA/°C
10-nA	10-pA	100-fA/°C
1-nA	1-pA	10-fA/°C
100-pA	100-fA	1-fA/°C
10-pA	10-fA	100-aA/°C
NOTE: Input bias current approx. doubles for every 11°C rise in temperature.		

Table 11.1 Input bias current required for current source.

input impedance per se, some do specify the input *resistance* in either *differential-mode* or *common-mode*. The input resistance in differential mode is defined as the ratio of small-signal change in input voltage to a change in the input current, at either input terminal, but with the other terminal grounded. An industry-standard precision op amp such as ADI's OP07 has a minimum input resistance of 20 MΩ (with ±15-volt supplies). The input resistance in common mode is defined as the ratio of input voltage range to a change in the input bias current over this range. The OP07 has a typical input resistance in this mode of 200 GΩ (with ±15-volt supplies).

- *Output impedance (Z_{out}).* While the ideal op amp would have zero output impedance/resistance, practical op amps do have a small output resistance of around 100 Ω. Having a low-output resistance enables the op amp to deliver its full output signal to a low-impedance load. This is the resistance measured at the output, with the output voltage at or near zero. Some manufacturers do not specify output impedance, but instead specify the *open-loop output resistance* (R_{out}), which is defined as the small-signal resistance of the output terminal with respect to ground, at a specified DC output voltage and current.

- *Output voltage swing (V_{out}).* This refers to the peak output voltage swing, referenced to ground that is available (without clipping or distortion) and running off a defined supply voltage(s) into a specified load (R_L). With older op amps, which had higher internal voltage drops, this voltage would generally be about 90% or less, or 1 or 2 volts less than the supply voltage(s). Many new op amps have rail-to-rail outputs, which can swing very close to the two supply rails, typically within 20 mV, but only with a high impedance load (more than 100 KΩ).
- *Output current capability (I_{out}).* This is a measure of how much current is available at the op amp's output to the load, during the peak output voltage swing (again, without clipping or distortion). If it is specified at all, then it's with a particular load resistance and running from a specified voltage supply. Because the output in most cases is a complimentary output stage and is often generating an AC signal waveform, the manufacturer might give output current information in the form of sink and source capability. These values will usually be close together. Generally speaking, low-power op amps will have a considerably lower output drive capability, because they essentially trade speed for lower power consumption. High-speed op amps, on the other hand, will usually have more output current capability, because they are designed for driving low-impedance loads (an example is ADI's venerable OP50, which can drive ±50 mA into 50-Ω loads). The typical op amp falls between these two extremes and will usually have a maximum output of between 10 and 25 milliamps. Of course, running at near maximum output current will cause unwanted heat to accumulate in the chip and may well cause serious and unpredictable thermal drift problems. Most of today's op amps have continuously rated output short-circuit protection, which trips in at a particular value and shuts down the output stage until the chip cools and/or the short is removed.

For current source designs, one has to simply consider whether the output current is going to be sufficient for the task, according to whether one is creating a current source or a current sink, and whether the op amp's output is going to be directly sourcing or sinking the load. Most times it is not. With a typical precision op amp, the current range will be capable of nano-amps up to a few milliamps. Beyond that, it is more advisable to have the op amp drive a separate device such as a power MOSFET or BJT, which can supply much more current, while keeping the op amp cool.

- *Open-loop voltage gain (A_O).* The op amp's open-loop gain is the differential voltage gain occurring between the input and output. It can be referred to either as a number or in decibels (e.g., an A_O of 100 K = +100 dB). The A_O symbol refers to gain specified as open loop. It could be considered as the theoretical maximum gain possible. Seldom though is the op amp operated in open-loop mode (i.e., no feedback), because feedback is used to reduce the gain to more manageable levels. This type of loop gain is referred to as the *closed-loop gain* and is determined by the choice of input resistance in conjunction with feedback resistance. The feedback here will be *negative feedback*, where a portion of the output signal is fed back to the op amp's inverting input. This reduces the voltage gain, while increasing the bandwidth. For op amp current source applications, a high gain helps enhance the current source circuit and provides it with high accuracy.
- *Common-mode range.* This is simply the range of voltages that can be safely applied to the op amp's input terminal inputs, without causing the output to start clipping the signal waveform or become nonlinear. Traditionally, this has been about 85% of the supply voltages, so that for an older op amp running off ±15-volt supplies, its common-mode range will be about ±12 volts. For many CMOS op amps, because of their internal design their input common-mode voltage range includes the power supply rails, hence rail-to-rail input.
- *Common-mode rejection ratio (CMRR).* This specification is the ratio of the op amp's gain with differential signals to its gain with common-mode signals. It is a measure of how much the op amp can reject common-mode signals. It can also be considered as the ratio of the change of input offset voltage to the change of input common-mode voltage producing it. Because it is a very high value, it is usually expressed in decibels (dB). For example, a good-quality op amp will probably have a CMRR of around 100 dB or higher.
- *Power supply rejection ratio (PSRR).* PSRR is specified in most op amp manufacturers' data sheets. It is expressed as a ratio, which measures the change in the op amp's input offset voltage to the change in the level of the power supply/supplies that produce it. It can be expressed either in decibels (dB) or in microvolts per volt (μV/V), although the latter is probably more widely used. To measure PSRR, the common-mode input is held constant, while the power supply/supplies are symmetrically varied over a specified range. For an op amp–based precision current source, PSRR is important. For example, the popular Analog Devices' *OP37E* (a low-noise, low-drift precision op amp) has a maximum PSRR of 10 μV/V, at room temperature, when the dual power supplies are symmetrically varied over a range of ±4 volts to ±18 volts—quite impressive. Likewise, Linear Technology's LT1007 (a low-noise, high-speed precision op amp) has a minimum PSRR of 120 dB, under identical conditions—also very

impressive. For precision current source applications, having an op amp with either a high PSRR of more than 90 dB or a low PSRR of less than 20 μV/V will be helpful in one's design, because this will buffer changes in the supply rails and the load voltage, from the constant current output to the load.

- *Gain bandwidth product (G_{BW} or f_T).* This specification refers to the frequency range from DC to the frequency at which the open-loop gain equals unity. Each op amp is designed so that its gain reduces as the frequency rises, and this falls off at a rate of about 6 dB/octave. Normally, the op amp would have a low-frequency open-loop voltage gain of, say, 100 dB up to 1 KHz, but its response from there onward will fall off as frequency rises, so that at, say, 100 MHz its gain will be unity.
- *Slew rate (S_r).* The slew rate is a measure of how much the op amp's internally limited output limits (i.e., slews) the maximum rate of change of output voltage (ΔV/T), when a large-amplitude step function is applied to its input. Slew rate is usually measured in volts per microsecond (V/μS). It can be used to make a greater bandwidth available for small output signals than for large output signals. Exceeding the specified slew rate limit can turn a sine wave into a triangle wave. Many general-purpose op amps have slew rates of around 5 V/μS, while some low-power precision types may have slew rates of 0.25 V/μS. High-speed op amps can often have slew rates in excess of 100 V/μS.

11.3 Op amp supply bypassing and input protection

An op amp's power supplies should always be well-regulated, with the local regulator(s) preferably located near the op amp circuit. The circuit board containing the op amps and local regulator may require some additional bypass capacitance in the form of an electrolytic capacitor (such as 22 μF), in parallel with a disc ceramic (0.1 μF)—both rated according to the circuit voltages and the regulator used. In addition, each op amp should be individually decoupled with a disc ceramic capacitor (0.01 μF), located as close as possible to the op amp's voltage supply pin(s) (see Figure 11.4). For very-low-bias currents and high-impedance circuits, the op amp's signal input pins may need to use guard bands around them, Teflon™ standoffs, and very short wire lengths.

When using expensive op amps, such as some low-input bias current or precision types, it might be fitting just to mention protection, even though it may not always apply to current source circuits. In Figure 11.5, JFET diodes protect the inputs to an expensive very-low-level amplifier. Dedicated low-leakage diodes typically have leakage currents in the low nano-amp range, which in some situations can actually degrade a precision measurement circuit (e.g., the very-low-input bias currents of some precision op amps are in the pico-amp range). Using a typical low-leakage diode with this type of op amp can significantly degrade the measurement capabilities of the circuit. As mentioned in Chapter 6, JFETs can be used to create ultra-low-leakage diodes.

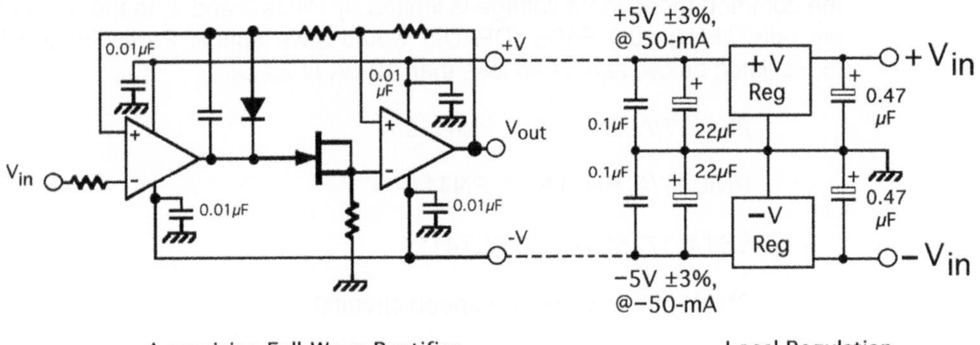

Figure 11.4. *The importance of decoupling the op amp power supplies.*

Normally, when the JFET has its drain and source shorted together, it provides an ultra-low-leakage diode (a.k.a. pico-amp diode, or PAD), with ultra-low-leakage currents. In some conditions, such diode-connected JFETs can attain leakage-current levels that are below 200 femto-amps (2×10^{-13} A) at V_{DG} equals 20 volts, or less than 100 fA at V_{DG} equals 5 volts, which is almost immeasurable.

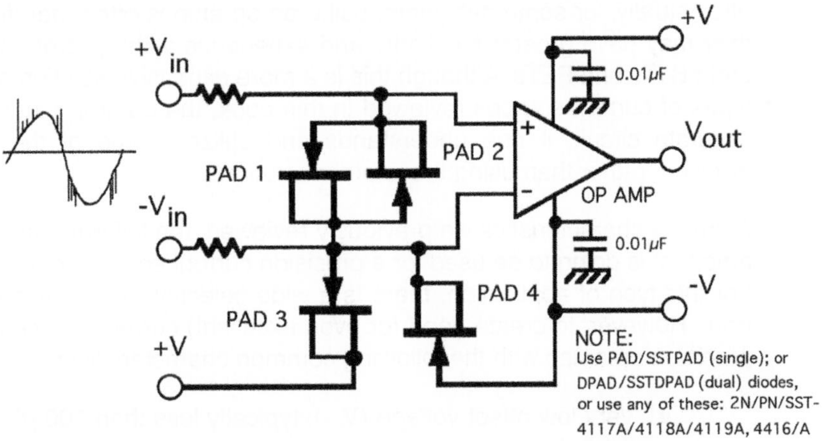

Figure 11.5. *Simple op amp input-protection using JFET pico-amp diodes.*

Normally the JFET diodes have reverse leakage currents in the low pA range, which is less than the very-low-input bias currents of most monolithic amplifiers. The JFET diodes will clamp the input to approximately 0.7 volt if a potentially dangerous and costly voltage spike occurs. In the circuit shown in Figure 11.5, the op amp's input differential voltage is limited to typically 0.8 volt by the JFET diodes, PADs 1 and 2, while the common-mode input voltage is limited by PADs 3 and 4, to the amount of the supply rails. Using dual PADs (DPADs) would save space. Particular JFETs, such as those listed below, can attain less than 10 pA at 25°C:

2N4117/A, 2N4118/A, 2N4119/A

PN4117/A, PN4118/A, PN4119/A

SST4117, SST4118, SST4119

2N4416/A (in very-high-speed circuits)

In addition, dedicated discrete pico-amp diodes such as the PAD, JPAD, or SSTPAD series can attain less than 5 pA. Dual versions are also available in the form of DPAD and SSTDPAD families. These devices are made by Calogic, Central Semiconductor, InterFet, Linear Integrated Systems, Philips Semiconductors, Vishay-Siliconix, and others.

11.4 Creating current sources with op amps

Just as BJTs and FETs can be used to create current sources, so too can op amps. The designer can readily utilize op amps in low current-source applications, or by adding either BJTs or power MOSFETs, one can provide significantly higher current levels. Actually, for some designers, using an op amp is often their first choice, because they may have greater familiarity and experience with op amps than with using discrete BJTs and FETs. Although this is a more expensive solution compared with other types of current sources reviewed in this book, the op amp can often render a more accurate circuit, if one understands and utilizes some of the critical parameters involved, rather than using these other solutions.

From the characteristics we previously reviewed, the following are important for an op amp that is going to be used for a precision current source or current sink application. For this type of application, there is a wide selection of possible op amps to choose from. However, to create good (or even excellent) current sources, one typically uses precision op amps with the following common characteristics:

- Very low offset voltage (V_{OS}): typically less than 100 μV, in fact the lower the better
- Very low offset voltage drift (TCV_{OS} in μV/°C): less than 5 μV/°C, the lower the better
- Very low input bias current (I_{bias}): less than 50 nA would be desirable
- Very low offset current drift (TCI_{OS} in pA/°C): less than 10 pA/°C, the lower the better
- A high loop gain (A_o): more than 100 dB, the higher the better
- A high CMRR: more than 100 dB, the higher the better

- A high PSRR: more than 100 dB, the higher the better
- Low noise (e_n): typically less than 10 nV/√Hz, the lower the better
- A reasonably low quiescent current (I_s): according to the application

Some examples of precision op amps with many of these desirable characteristics are shown in Table 11.2. (This is *not* a complete listing, however, and there are other good products available that may not be shown.) It is evident from the 11 U.S. manufacturers listed that the biggest range of this type of op amp clearly comes from Linear Technology Inc., with more than 100 different products, followed next by Texas Instruments (which includes some former Burr-Brown devices) and National Semiconductor. One should not overlook devices from other manufacturers simply because they have a smaller range. Some devices from Maxim, Analog Devices, and ALD, for example, are truly remarkable, cutting-edge products. After all, if a company made the very *best* device on the market, why would it need to make others?

As mentioned, a precision op amp would be the first choice. As a rule of thumb, for low-precision current source and current regulator circuits that help generate several milliamps, one can use an op amp with less stringent specs, because those small offsets and drifts have less impact on a circuit producing tens of milliamps or more. For high-precision circuits, which generate microamps, nano-amps, or even pico-amps, one must use a very-high-performance precision op amp, with stringent specs that include an ultra-low-input bias current ($I_{in\ bias}$) and a very-low-offset voltage drift (TC V_{os}). Ambient temperature changes, noise, circuit board layout, thermal gradients across active components or across the circuit board, wiring and circuit board track lengths, thermocouple effects, as well as parasitic leakage can all become potential problems, depending on the current level one needs to generate. This is because the same small offsets and drifts have a much greater impact on a current source circuit producing, say, 5 nA than they do on one generating 10 milliamps or more. Thermal drift effects in any type of low-level circuit are always a concern, so the circuit may need to be enclosed in a shielded box, with ambient temperature precisely controlled.

One should also consider whether this circuit will be configured as a current source or a current sink, what supply voltage(s) are available, the load, its tolerances, load compliance, intended operating temperature range, and so on. In some applications a high-voltage, dual-supply, precision op amp might be required for driving a high-compliance load, whereas in others it might call for a lower-voltage single-supply type. It is likely that voltage drops in the op amp's output as well as across the conversion resistor will reduce the range of voltages available to the load, which directly affects the load compliance. For those applications you might consider using a higher-voltage op amp or one that has a much lower output voltage drop (i.e., CMOS rail-to-rail output).

Manufacturer:	Products:
ADVANCED LINEAR DEVICES	ALD-1712A/B, ALD-1721/E, ALD-1722/E, ALD-2702, ALD-2711A/B, ALD-2724, ALD-4702
ANALOG DEVICES	OP-07, OP-22, OP-27, OP-37, OP-50, OP-77, OP-97, OP-113, OP-177, OP-184, OP-297, OP-497, AD-704, AD-705, AD-706, AD-707, AD-795, AD-797, AD-820, AD-822
FAIRCHILD SEMICONDUCTOR	FAN-4113, FAN-4171, FAN-4272, KM-4110, KM-4112, KM-4120, KM-4210, KM-4212
INTERSIL	HA-5127/A, HA-5135, HA-5137A, HA-5147, HA-5170, ICL7650S
LINEAR TECHNOLOGY	LT-1001/A, LT-1002A, LT-1006, LT-1007/A, LT-1008, LT-1012A, LT-1013A, LT-1014/A, LT-1022, LT-1024/A, LT-1028/A, LT-1037/A, LT-1055, LT-1056, LT-1057, LT-1058, LT-1077/A, LT-1078/A, LT-1079, LT-1097, LT-1112/A, LT-1113/A, LT-114/A, LT-1122, LT-1124/A, LT-1125/A, LT-1126/A, LT-1127/A, LT-1128/A, LT-1169, LT-1178/A, LT-1179/A, LT-1211/A, LT-1212, LT-1218, LT-1219, LT-1413A, LT-1457/A, LT-1462/A, LT-1463, LT-1464/A, LT-1465, LT-1467, LT-1468, LT-1492, , LT-1493, LT-1496, LT-1499, LT-1674, LT-1677, LT-1678, LT-1679, LT-1806, LT-1807, LT-1880, LT-1881A, LT-1882, LT-1882, LT-1884, LT-1885, LT-2078, LT-2079/A, LT-2178/A,LT-2179/A, LT-6011, LT-6012, LTC-1047, LTC-1049, LTC-1050/A, LTC-1051, LTC-1052, LTC-1053, LTC-1150, , LTC-1151, LTC-1152, LTC-1250, LTC-2050, LTC-2051/HV, LTC-2052/HV, LTC-2054
MAXIM	ICL-7650, ICL-7652, MAX-400, MAX-420, MAX-421, MAX-422, MAX-423, MAX-427, MAX-432, MAX-437, MAX-438, MAX-478, MAX-4236/A, MAX-4237/A, MAX-4238, MAX-4239, MAX-4460, MAX-4475, MXL-1007, MXL-1178
MICROCHIP/TELCOM SEMI	MCP-606, MCP-607, MCP-608, MCP-609, MCP-616, MCP-617, MCP-618, MCP-619, TC-7650, TC-7652, TC-913A, TC-913B
NATIONAL SEMICONDUCTOR	LF-411A, LM-10, LM-108A/208A/308A, LM-833, LM-837, LM-7301, LMC-6001, LMC-6061/A, LMC-6062/A, LMC-6064/A, LMC-6081/A, LMC-6082/A, LMC-6084A, LMC-6462, LMC-6464, LMC-6482A, LMC-6484, LMV-722, LMV-751, LMV-771, LMV-772, LMV-774, LMV-2011
ON SEMICONDUCTOR	MC-33077, MC-33102, MC-33272/A, MC-33274/A, MC-33282, MC-34080B
PHILIPS SEMICONDUCTORS	SA-5230, SA-5234, SE-5512, SE-5514, SE-5532/A, SE-5533/A, SE-5534/A
TEXAS INSTRUMENTS	OPA-111, OPA-124, OPA-132, OPA-177, OPA-227, OPA-228, OPA-234, OPA-241, OPA-251, OPA-27, OPA-277, OPA-334, OPA-335, OPA-336, OPA-340, OPA-344, OPA-345, OPA-350, OPA-363, OPA-364, OPA-37, OPA-380, OPA-602, OPA-606, OPA-627, OPA-637, OPA-734, OPA-735, TLC-2201/A, TLC-2652/A, TLC-2654/A, TLC-4501/A, TLC-2021A/B, TLE-2027/A, TLE-2037/A, TLE-2141A, TLE-2161B, TLV-2211, TLV-2221, TLV-2231
NOTE: This is only a partial listing, and may not be the most up-to-date. You should contact the manufacturer for the latest information. You can reach each manufacturer via the Contact information in the Appendix.	

Table 11.2 Some Precision Op Amps from U.S. Manufacturers

The accuracy and stability of the current source also depends on which grade or level of precision op amp is used, relative to the level of accuracy needed. At the top end, devices are trimmed with great precision, burned-in and tested over the full military/industrial temperature range, have low drifts, and cost more. At the other end, devices receive less attention, have more relaxed specifications, drift more, and cost less. Low-voltage, single-supply current sources may have less precision and accuracy than dual-supply types and have a much smaller supply voltage range. The choice is between which product may translate into requiring someone to hand-calibrate an instrument rather than buying a more expensive op amp for a few dollars more that eliminates that step.

Because the op amp requires a stable, low-drift input voltage in order to provide a stable output voltage, it is common to use a monolithic *voltage reference IC* at its input. The op amp and the voltage reference need to share similar specifications though. There is no point in using a very-low-noise, low-offset-voltage precision (expensive) op amp, if the (cheap) voltage reference is noisy and drifts. Likewise, it is no use having a very-low-noise, super-precision voltage reference if the op amp is a cheap general-purpose type with a high offset voltage and accompanying high drift. *Both* devices need to be low-drift, precision types. As we will discover in Part 2 of this book (Voltage References), various types of monolithic voltage references are available. The accuracy and stability of the current source also depends on the type of voltage reference IC used, together with what level of accuracy (measured in digital bits, for reference) is required, and over what temperature range. For example, in order to keep any conversion error to 1/2 LSB in a 16-bit system, over a temperature range of say 100°C, then the voltage reference will need to have a tempco of less than 0.1 ppm/°C. By contrast, in an 8-bit system, over the same temperature range, and with a 1-LSB conversion error, the voltage reference will need to have a tempco of less than 35 ppm/°C, as shown in Table 11.3.

Table 11.4 shows that Xicor's FGA™ technology leads in terms of initial percentage error and attainable bit resolution. It will provide the type of performance needed by 14- to 16-bit systems. If that level of accuracy is not required, then consider each of the three remaining types: buried-zener, bandgap, and ADI's XFET®. Generally, a three-terminal reference will have higher accuracy and lower drift, but cost somewhat more than a two-terminal bandgap type. Some of the best two-terminal bandgaps are available from Linear Technology, Maxim, and National Semiconductor.

Some of the circuits shown here just use a simple two-terminal shunt bandgap voltage reference, which may be sufficient in many applications, easier to work with, and offer a lower-cost solution. Subsequent circuits shown use a three-terminal voltage reference for higher precision, better accuracy, and a lower tempco. Most of the examples show an op amp and voltage reference, both suited for the –25°C to +85°C industrial operating temperature range.

Number of bits:	Allowable drift over 100° temp. range for 1/2 LSB error (ppm/°C)	Allowable drift over 100° temp. range for 1 LSB error (ppm/°C)
8	19.6	35
10	4.90	8.5
12	1.25	2.25
14	0.32	0.64
16	0.08	0.16
18	0.017	0.035
20	0.005	0.01

Table 11.3 Maximum allowable error for n-bit system

Feature:		Band-gap IC	Buried zener IC	ADI X-FET®	Xicor FGA™ V_{Ref}
Reference voltage	Volts	5.000	5.000	5.000	5.000
initial error %	%	0.5	0.03	0.06	0.01
Output voltage tempco	ppm/°C	30	3	8	3
Long term stability (Note 1)	ppm/1K	120	20	0.2	10
Suitable for n-bit systems	bits	<10-bit	>12-bit	<12-bit	>14-bit
Thermal hysteresis	ppm	25-ppm	20-ppm	15-ppm	50-ppm

NOTES:
1. Measured over 1000-hours @ 25°C.
2. Xicor is now part of Intersil Corp.

Table 11.4 Typical performance comparisons of different types of 5-Volt references

In case you are wondering why not simply have the voltage reference IC drive the load and omit the op amp, here is why: Although it is theoretically possible to use the output of a voltage reference IC and only a current-setting resistor to directly drive the load, this will usually introduce a large error voltage. Actually, the current will change very little, but only if the source resistance (R_S) is much larger than the load resistance (R_{Load}) and the load voltage (V_{Load}) is much smaller than the reference voltage (V_{Ref}).

This is shown by equation 11.3. In fact, this would only happen if the reference voltage was extremely large—on the order of several hundred volts.

$$I_{out} = \frac{V_{REF} - V_{Load}}{R_S} \quad \text{(Eq.11.3)}$$

So, adding an op amp to the circuit isolates the voltage reference from the load and avoids this problem completely.

Creating current sources with op amps and voltage references also depends on using a precision resistor to convert the reference voltage into a precise and stable reference current. In such a design, the op amp circuit effectively functions as a precision voltage-to-current converter. The precision resistor you use, though, needs to be a high-quality type (metal-foil, wirewound, or thin-film), with a less than 0.1% tolerance and with a low tempco of less than 20 ppm/°C, or that suits the application. Assuming that the op amp and voltage reference have minimal drift, then the circuit's overall accuracy, temperature drift, and long-term stability will be largely governed by the resistor's drift characteristics. The lower the resistor's percentage tolerance, TCR, and long-term stability, will usually provide the best performance and the lowest drift.

That works fine for fixed resistances, but in some cases one may have to *trim* the voltage across the current-setting resistor (i.e., adjust to an exact current level), which means there are two possible options: (1) use one or two additional precision resistors in series with the main resistor, which retains the accuracy and overall tempco of the resistors involved (In most cases, where high accuracy and low drift are desired, this is the best option.), or (2) use a multiturn trimmer, but in this case adding the trimmer will degrade the overall performance of the fixed resistor, particularly its low TCR drift, not to mention introducing poor matching characteristics.

It is best to try to minimize the effects of the trimmer. This is done by using a small value for the trimmer to give the desired adjustment range (i.e., less than 10% of the value of the fixed resistor) and by preferably making the critical adjustment value occur approximately in the center of the trimmer's range. Here it might be more prudent to use a precision resistor with a higher percentage tolerance (and a slightly higher ppm/°C drift), rather than a lower tolerance (see Figure 11.6). Typical cermet trimmers usually have a tolerance of ±10% and a tempco of more than ±100 ppm. A better, but more expensive solution is found by using either (1) a Bourns precision wirewound trimmer with a low ±5% tolerance and a ±50-ppm tempco, or (2) a Vishay metal-foil trimmer that also has a ±5% tolerance but with a ±10-ppm tempco. (If resistors are used in pairs, such as in gain setting, then close tracking will be important. Precision thin-film resistor arrays have excellent tracking and low drift and will usually be the best solution.) You should contact precision resistor makers such as Bourns, Caddock, or Vishay for their product information.

In summary, you need to consider the following important factors:

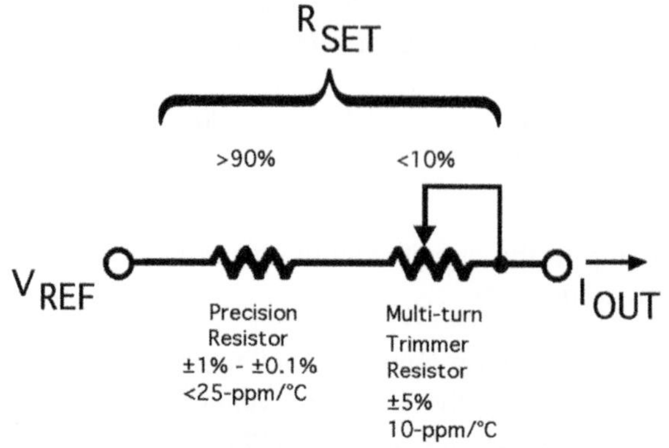

Figure 11.6. Adding a trimmer to the R_{SET} current-setting resistor changes its TCR drift.

- The type of input: whether it is a fixed voltage or an AC voltage source. This will establish whether it's an op amp plus voltage reference combination or simply an op amp circuit.
- The level of accuracy required, over the required temperature range (i.e., 0.1%, 0.01%; 50 ppm/°C, less than 5 ppm/°C; 8-bit, 12-bit, 16-bit, etc.)
- Whether the level of accuracy needed in a fixed voltage design requires a two- or a three-terminal voltage reference, because this will dictate the circuit configuration needed
- The resulting cost involved with your design, including hand-calibration involved during Q&A testing
- Can a current source/sink with similar specifications be made more inexpensively, using an alternate strategy? For example, using BJT matched pairs, or with JFETs, DMOS, etc., or monolithic current sources like the LM134/334 or the REF200 (see Chapter 10), both of which have a typical tempco of less than 30 ppm/°C.

In the following current-source and current-regulator examples, one can realize virtually any of these designs in either high or low performance, by building them accordingly with higher/lower accuracy op amps, voltage references, and resistors, that match or exceed the accuracy required. Figures 11.7 and 11.8 both show current sources created using two-terminal bandgap voltage references. It is important to remember that because of that, these circuits have at best an accuracy of between 0.05% to 0.1%, with a drift of between 20 ppm and 150 ppm. Some circuits use a single supply, whereas others require dual supplies. Also, because of the flexibility of the circuit configuration, the load can be connected to ground or any other convenient voltage, relevant to the power supplies and the output capability of the chosen op amp. The polarity of the voltage reference IC and the op amp's supplies are otherwise the only other differences between the sink and source circuits. Because the voltage reference IC is in the op amp's feedback loop, it could potentially cause a problem with frequency, because there would then be two poles in the loop. To provide stability to the circuit, the reference IC can be bypassed by capacitor C_1, which leaves the op amp in control of the loop at higher frequencies. Depending on which reference IC is used, it may require a different value than shown. Or it may not even be necessary, depending on the voltage reference and op amp combination used.

The examples in Figure 11.7 both provide a constant current for the load. The op amp isolates the voltage reference from the load, by virtue of its high PSRR value (more than 100 dB). In both circuits the two-terminal voltage reference and its current-converting resistor are bootstrapped to the load voltage, and the reference voltage is then converted to a constant current. Because the reference voltage is fixed between the op amp's inverting input and its output, feedback helps establish the same reference voltage V_{REF} between the op amp's noninverting input and the output. As a result, the reference voltage V_{REF} is made to appear across the current-setting resistor, R_S. Feedback maintains a state of equilibrium between the op amp's two inputs and provides a constant current through R_S and to the load (R_L), regardless of the load voltage. The basic formula for determining the output current in any circuit in Figures 11.7 to 11.9 is shown in equation 11.4. For low currents (less than 100 μA), use equation 11.5 instead.

$$I_{OUT} = \frac{V_{REF}}{R_S} \qquad \text{(Eq.11.4)}$$

$$I_{OUT} = \left(\frac{V_{REF} + V_{OS}}{R_S}\right) + I_{BIAS} \qquad \text{(Eq.11.5)}$$

Figure 11.7A shows a very-low-power current sink, using two National Semiconductor products: an LMC6061A op amp and an LM4040AI-2.5 bandgap reference. The LMC6061A is a silicon-gate CMOS op amp (typical values include V_{os} equals 100 μV; TCV_{os} equals 1 μV/°C; I_{in} bias equals 10 fA; A_v equals more than 130 dB; I_s equals 20 μA). The low-power op amp's supply range is from 4.5 volts to 15 volts and can be used in single or dual supplies (±2.5 volts to ±7 volts). A lower-voltage version of this circuit is practical, but only by using a lower-voltage op amp with similar specs and by changing down to a 1.23-volt voltage reference. The LM4040AI-2.5 is a micropower bandgap shunt reference (2.5 volts, tolerance equals ±0.1%; with a maximum temperature drift of 100 ppm/°C, over the –40°C to +85°C operating temperature range).

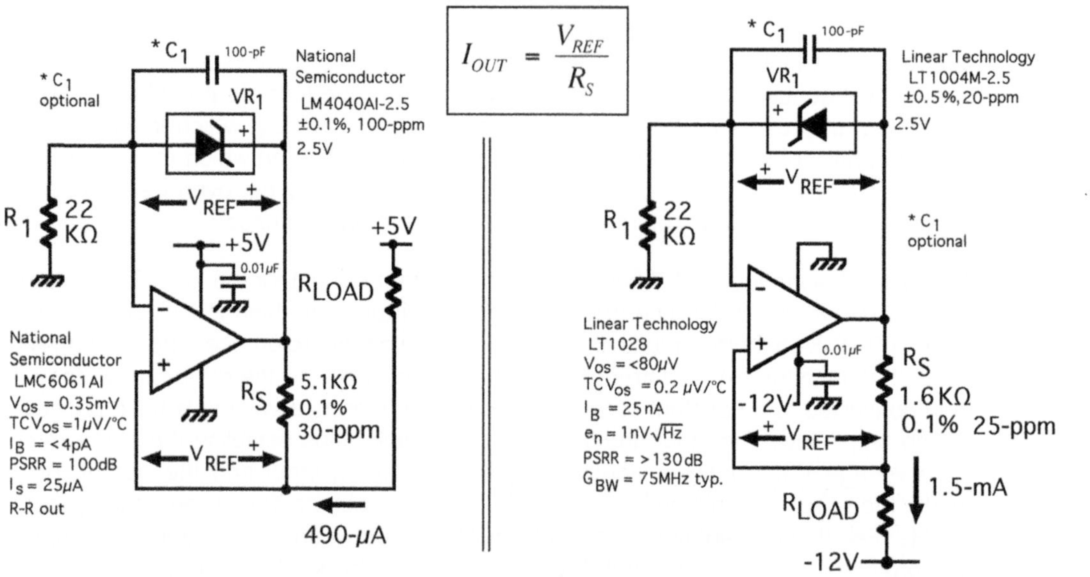

A. A micro-power current sink

B. A low-noise, high-speed current source

Figure 11.7. *Simple but reliable op amp current sources.*

In Figure 11.7A, the op amp's typical offset voltage alters the voltage across the current-setting resistor R_S by only 100 μV (0.004% or 40 ppm), with respect to the 2.5-volt reference voltage (see Table 11.5). Because its input bias current (100 pA max) is so low, this adds a negligible amount of error (less than 0.2 ppm; less than 0.00002%) to the 490-μA constant sink current value. These tiny errors can be reduced if required, by trimming the current-setting resistor R_S (make R_S a 4.7-KΩ, 0.1%, 20-ppm resistor, and R_{trim} a 500-Ω, ±5% multiturn trimmer, with a 20-ppm/°C tempco). Trimming can reduce these errors to a point where only thermal drifts are the remaining sources of error. Combining the op amp's typical TCV_{os} drift of 1 μV/°C (approximately 0.5 ppm or 0.00005%/°C), with the voltage reference's drift ($\Delta VR/\Delta T$) of approximately 0.03%/°C equates to a total output drift of less than 0.035%/°C. Precise control of room temperature will impose a practical limit on any circuit's trimmed accuracy, to approximately 0.05%. (Small air currents and long-term ambient temperature control are both important factors in any precision calibration/measurement environment.) In this 490-μA constant sink circuit, that equates to 0.24 μA. Therefore, this circuit's output has a trimmed accuracy of ±0.05%, with a maximum drift of approximately 130 ppm/°C. Output impedance for this circuit is more than 10 MΩ.

While the best attainable accuracy for any of these circuits is going to be greater than about ±0.025%, drift can be significantly reduced to single (ppm) digits, by using precision devices with improved specifications. Figure 11.7B illustrates that, and shows a

Volts	Volts	ppm	%
1.0V	2.5V	1M	100
0.10V	0.25V	100K	10
10mV	25mV	10K	1
1.0mV	2.5mV	1K	0.1
100μV	250μV	100	0.01
10μV	25μV	10	0.001
1.0μV	2.5μV	1	0.0001
100nV	250nV	0.1	0.00001
10nV	25nV	0.01	0.000001
1.0nV	2.5nV	0.001	0.0000001

Table 11.5 PPM and % converter for 1V and 2.5V references.

low-noise, high-speed current source circuit, which runs off a single negative supply of –12 volts. The circuit uses two Linear Technology devices: the op amp is an LT1028 (V_{OS} equals 50 μV typically; TCV_{OS} equals 0.2 μV/°C; e_n equals 1 nV√Hz, with a typical gain bandwidth of 75 MHz). The reference IC is an LT1004M 2.5-volt (having a voltage tolerance of ±0.5%, with a maximum temperature drift of 20 ppm/°C, over the –40°C to +85°C operating temperature range). Although the circuit shown here runs on a negative supply, it could work just the same with a +12-volt and ground supply or even dual supplies.

In this circuit the reference is isolated from the load by the op amp's excellent PSRR of more than 130 dB. Here the op amp's typical offset voltage can alter the voltage across the current-setting resistor R_S by only 25 μV (0.001% or 10 ppm), with respect to the 2.5-volt reference voltage. Because its typical input bias current of 25 nA is very low, this adds a negligible amount of error (less than 0.002%) to the 1.5-mA constant current. The circuit sources this 1.5-mA current into the load, despite changes in the load voltage, and with (trimmed) accuracy of about ±0.05% and drift of less than 50 ppm. The output impedance for this circuit configuration is more than 4 MΩ. It is a very good circuit!

The next two examples (again showing a current source and a current sink) are higher-performance versions. They use even higher precision op amps, voltage references, and resistors. However, they are both limited from even greater precision by (1) the precise control of room temperature, which imposes a practical limit on the circuit's trimmed accuracy, to approximately 0.05%, and (2) the more than 20-ppm/°C drift of the typical two-terminal bandgap reference. (Using a three-terminal reference, with higher accuracy and lower drift, requires a completely different circuit configuration, as we will soon discover.)

The examples in Figure 11.8A (a precision current sink) and 11.8B (a precision current source) both provide a constant current with a grounded load. They use a similar configuration to the previous circuits, except for the added JFET cascode, which does a better job of biasing the voltage reference than the resistor in the previous circuits. While the JFET cascode provides a constant current for the voltage reference IC, it also helps in other ways too. Cascoding generally improves high-frequency operation, provides greater output impedance (Z_{out}), increases high-voltage operation and voltage compliance, and reduces the JFET's output conductance. In this circuit, cascoding buffers the current source Q_1 from the load, by using Q_2. This means that while Q_2's source voltage remains constant, variations in the load voltage are accommodated by Q_2's drain. As a result, the voltage drop across the current source remains constant. The drain current (I_D) is regulated by Q_1 and R, so that effectively V_{DS1} equals $-V_{GS2}$. Both JFETs must be operated with adequate drain-to-gate voltage (V_{DG}), so that V_{DG} is ideally more than two times $V_{GS\ (off)}$, or else the circuit's output conductance will increase significantly. Also, JFET Q_2 needs to have both a higher $V_{GS\ (off)}$ and a higher I_{DSS} rating than Q_1. In the JFET circuit, resistor R sets the desired current (at about three times the minimum operating current of the voltage reference). This cascode has an internal circuit impedance of more than 5 MΩ at current levels less than 1 mA. Capacitor C_1 is simply included here for circuit stability. The cascode's output conductance (g_o) is about 100 times lower than the g_{os} value of either individual JFET. Thus, the current biasing for the voltage reference is significantly improved, as is the overall output resistance.

The examples in Figure 11.8A and 11.8B both provide a constant current with a grounded load and provide better overall performance than those shown previously. In both circuits, the op amp isolates the voltage reference from the load, by virtue of a high PSRR value of more than 120 dB. Figure 11.8A shows a very-low-drift current source, using Analog Devices' OP177 precision op amp and an ADR525B, a 2.5-volt bandgap reference. The circuit shown here runs on ±12-volt supplies and sources 25 μA to the load. It has a best accuracy of about ±0.1% (although the trimmed accuracy limit is about ±0.05%), with a drift of less than 40 ppm. This circuit's output impedance is more than 250 MΩ, because of the use of the JFET cascode and of the value of R_S.

Figure 11.8B shows a low-noise circuit that sinks a constant 6 μA from the load and runs off a dual supply of ±15 volts. This circuit uses two Linear Technology devices; the op amp is an LT1012 (V_{os} equals 30 μV typically; TCV_{os} equals 0.2 μV/°C; e_n equals 20 nV/√Hz, with a PSRR more than 125 dB). The reference IC is an LT1004M (a 2.5-volt reference with a tolerance of ±0.5% and a maximum drift of 20 ppm/°C, over the –40°C to +85°C temperature range). In this circuit, the op amp's very low offset voltage can alter the voltage across the current-setting resistor (R_S) by only 25 μV (0.001% or 10 ppm/°C), in respect to the 2.5-volt reference voltage. Because the input bias current (100 pA max) is also very low, this adds only a tiny amount of error

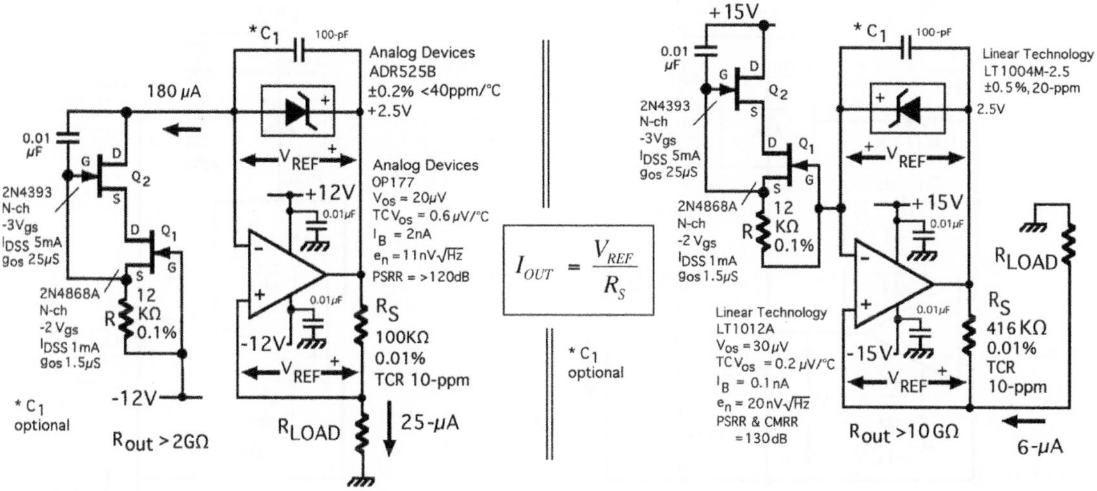

A. A low drift, precision current source

B. A low drift, low noise, precision current sink

Figure 11.8. *Precision op amp current sources with added JFET cascodes.*

(0.002%, 20 ppm) to the 6-μA current sink. These errors can be reduced to zero if required, by trimming the current-setting resistor R_S (make R_S a 390-KΩ, 0.1% resistor, and R_{trim} a 50-KΩ, 10-ppm/°C multiturn trimmer). The circuit shown here has a best accuracy of about 0.025% (although the trimmed accuracy limit is still about ±0.05%), a drift of less than 30 ppm/°C, and an output impedance of more than 10 GΩ. Long-term stability of the voltage reference used in this design is typically around ±20 ppm/1000 hours, which is also very impressive.

The next two circuits, shown in Figure 11.9, use *three-terminal* voltage references for higher precision, better accuracy, lower tempco, and improved long-term stability. Both examples show an op amp and voltage reference suited for the –25°C to +85°C industrial operating temperature range. In the example of the current sink (Figure 11.9A), the load may be connected to either ground or a positive supply (equal to or less than the circuit's main supply). With the current source (see Figure 11.9B), the load may be connected to either ground or a negative supply (equal to or less than the circuit's main negative supply). In both cases the reference voltage V_{REF} is made to appear across the current-setting resistor R_S, and provides a constant current to the load (R_L), regardless of the load voltage. The formulae for determining the output current in any circuit in Figure 11.9 is the same as shown previously in equations 11.4 and 11.5. Again, the current-setting resistor R_S can be effectively trimmed by adding a small-value, multiturn precision trimmer R_{trim}, with a very low tempco, in series with it.

Figure 11.9A shows a low-drift, low-noise current sink, using two popular Analog Devices products: an OP37E op amp and an ADR431B, a 2.5-volt XFET® voltage ref-

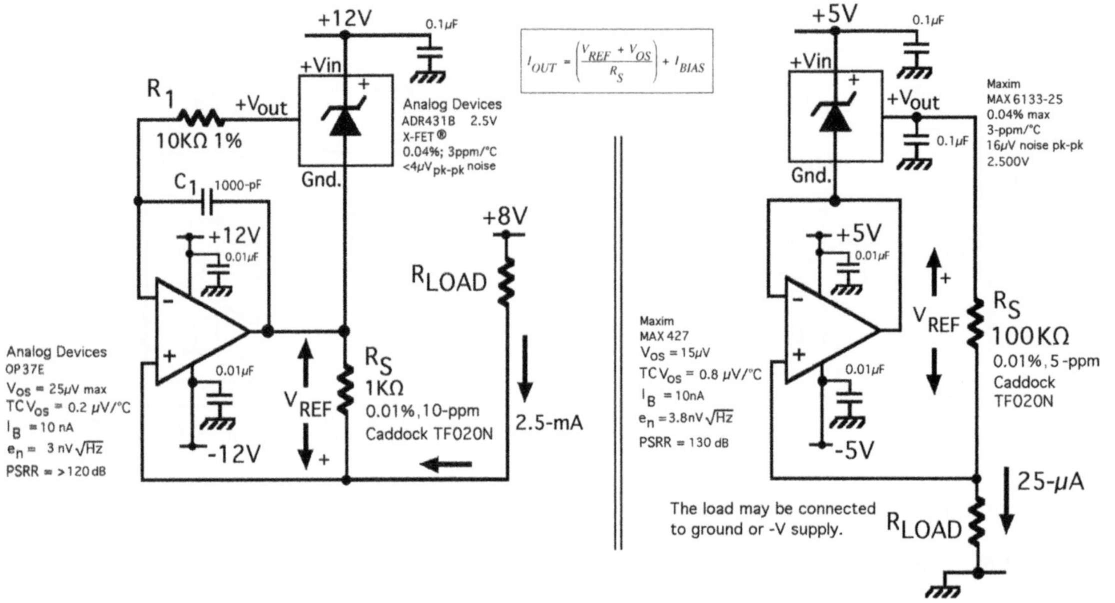

A. A low-drift, low-noise, precision current sink with an overall accuracy of <0.05%/°C.

B. An ultra-low drift, low-noise, precision current source

Figure 11.9. Ultra high-precision, low-drift op amp current sources.

erence. In this circuit the op amp is running on dual supplies of ±12 volts and sinking a load current of 2.5 mA. The op amp drives both the current-setting resistor R_S and the voltage reference IC's ground terminal. This causes the output terminal of the voltage reference to equal the load voltage. Because the op amp is configured here as an inverter, this means that –2.5 volts is impressed across the current-setting resistor R_S, regardless of the load voltage. In this way the reference voltage is converted to a constant current. The R_1-C_1 combination shown provides noise filtering, as well as loop stability.

The OP37E is a precision high-speed op amp (typical values include V_{os} equals 20-μV; TCV_{os} equals 0.2 μV/°C; long-term drift of only 0.2 μV/month; I_{in} bias equals 10 nA; PSRR more than 120 dB, and GBW of typically 40 MHz). The ADR431B, an XFET® voltage reference (2.5 volts, with a tolerance of ±0.04%, with a maximum temperature drift of only 3 ppm/°C over the –40°C to +125°C operating temperature range. This XFET® reference also features a very low noise voltage of under 4 $\mu V_{pk\text{-}pk}$). In this circuit, the op amp's maximum offset voltage alters the voltage across the current-setting resistor R_S by only 25 μV (0.001% or 10 ppm), with respect to the 2.5-volt reference voltage. Because its input bias current (40 nA max) is very low, this adds only a tiny amount of error (20 ppm; 0.002%) to the 2.5-mA sink current. These errors can be reduced as mentioned, by trimming the current-setting resistor R_S (make R_S a 850-Ω,

0.1%, 10-ppm resistor, and make R_{trim} a 250-Ω, ±5% multiturn precision trimmer with a 10-ppm tempco). The circuit has a best accuracy of about 0.08% (although the trimmed accuracy limit is still approximately ±0.05%), with a drift of less than 10 ppm, and with an output resistance of more than 5 MΩ. Because of the use of the ADR431B voltage reference, long-term stability here is typically less than 1 ppm/1000 hours, which is truly exceptional!

Figure 11.9B shows an ultra-low-drift, low-noise current source circuit that provides a constant 25 μA to the load and running off a dual supply of ±5 volts. In this configuration the CMOS op amp is connected as a voltage follower, which forces the voltage reference IC's ground terminal to equal the load voltage. This results in the voltage reference's output being impressed across the current-setting resistor R_S, regardless of the load voltage. As a result, the reference voltage output is converted to a constant current. This example uses two Maxim devices: the op amp is a MAX427 (V_{os} equals 15 μV typically; TCV_{os} equals 0.8 μV/°C; e_n equals 3 nV√Hz, and PSRR more than 130 dB). The reference IC is a new generation of bandgap, a MAX6133 2.5-volt (having a voltage tolerance of ±0.04%, with a maximum temperature drift of 3 ppm/°C, over the –40°C to +85°C operating temperature range, and a 16-μV noise voltage). The output impedance for this configuration is more than 100 MΩ. The circuit has a best accuracy of about 0.025% (trimmed accuracy limit approximately ±0.05%), and with a drift of approximately 15 ppm. The long-term stability of the voltage reference is about 40 ppm/1000 hours, which is also very good.

So far we have looked at current sources using a fixed reference voltage to create a constant current. Now we will look at some current sources that are controlled by a varying AC signal voltage instead, which are shown in Figures 11.10 and 11.11. First, you will notice that there is no voltage reference IC in these circuits, only the precision op amp and the conversion resistor. Remember that you can realize any of these designs in either high or low performance, by building them accordingly with higher/lower accuracy op amps and resistors. At whichever level of precision you have to work at, their specs should match or exceed the accuracy required. Both examples show an op amp suited for the –25°C to +85°C industrial operating temperature range, which provides a current into a grounded load. As shown previously, the current can be effectively trimmed by adding a small-value multiturn precision trimmer R_{trim}. Best performance and lowest drift will always be achieved when using just the current-setting resistor R_S only. The resistors that are used in pairs/multiples for gain setting in these circuits are best realized as precision resistor networks/arrays. These will usually provide the best solution, because they will have good ratio-matching accuracy, excellent tracking, and very low drift. Ideally, such precision resistor network specifications will include tolerance equals ±0.1%; absolute TC equals 25 ppm/°C; ratio of 1:1, 9:1, 10:1, etc.; ratio tolerance equals 0.1% to 0.01%; ratio TC equals less than 10 ppm/°C. (These are available in common DIP, SIP, and SMD packages from various U.S. manufacturers such as Bourns, Caddock Electronics, Ohmite, Vishay, and others.)

The first example (Figure 11.10) shows a simple bilateral circuit that produces a positive output current proportional to the negative input voltage (0 to –10 volts). The output current is set by R_5 and the gain setting of the precision op amp, and provides current to the load, regardless of load voltage (0V or -V). The load can be connected to ground or any other convenient voltage, relevant to the power supplies and the output capability of the op amp. (This application may be ideal for a rail-to-rail input/output precision op amp.) The formulae for determining the output current as well as resistor matching are shown in the diagram. The source resistance should be low compared with R_1 (1 MΩ), or else both the output resistance and gain will be affected. The output resistance (R_{out}) for the values shown (with 0.1% tolerance resistors) is more than 2 MΩ. The circuit has an accuracy of about 0.1%, with a drift of approximately 100 ppm/°C.

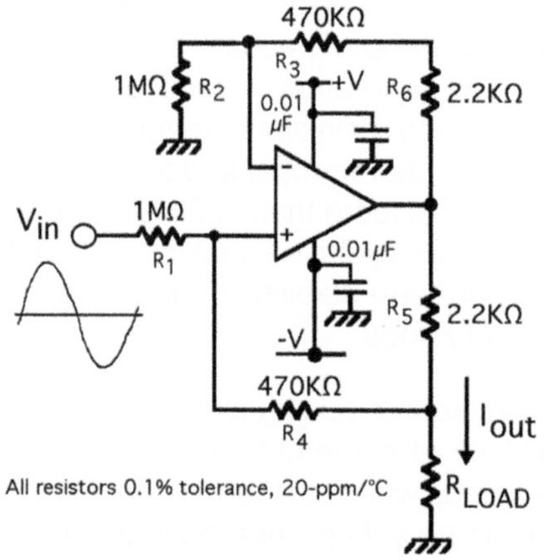

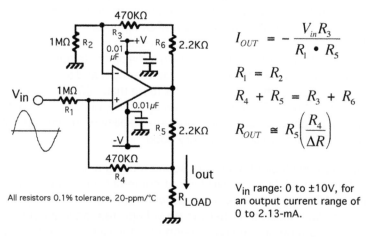

$$I_{OUT} = -\frac{V_{in} R_3}{R_1 \bullet R_5}$$

$$R_1 = R_2$$

$$R_4 + R_5 = R_3 + R_6$$

$$R_{OUT} \cong R_5\left(\frac{R_4}{\Delta R}\right)$$

V_{in} range: 0 to ±10V, for an output current range of 0 to 2.13-mA.

Figure 11.10. Single-ended, bilateral op amp current source.

The next example (Figure 11.11) shows a differential-input, bilateral current source circuit, also known as a *Howland Current Pump.* It looks similar to the previous example but has higher accuracy and lower drift. It produces a bidirectional output current, proportional to the differential input voltage (0 to ± 200mV). In this circuit a precision trimmer potentiometer (VR_1) is used to balance the feedback loop, because this also helps increase the circuit's output resistance (typically the output resistance is 1000 times R_6), which here is more than 1 MΩ. Feedback helps regulate the current through R_6 and to the load (R_{load}), regardless of the load voltage. The formulae for determining the output current as well as the resistor matching are shown in the diagram. The op amp must be a precision type, preferably with a V_{os} of less than 50 µV and an offset drift of less than 0.5 µV/°C, because this will improve circuit accuracy and linearity over the operating temperature range. The output frequency characteristics will depend on the chosen op amp's bandwidth and slew rate.

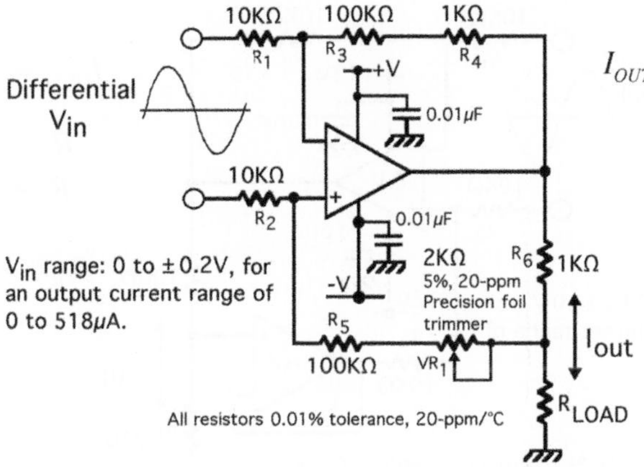

$$I_{OUT} = \frac{V_{in} \bullet A_V}{R_6} = \frac{V_{in} R_3}{R_1 \bullet R_6}$$

$$= \frac{0.2V \bullet 10.1}{1000}$$

$$= 518\mu A$$

$$R_1 = R_2$$

$$R_5 + VR_1 = R_3 + R_4$$

$$R_{OUT} \cong R_6 \left(\frac{R_5}{\Delta R} \right)$$

Figure 11.11. The Howland current pump.

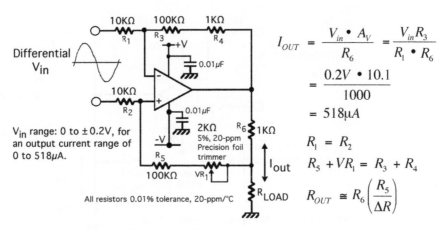

An alternative to this circuit is shown in Figure 11.12, which is a fast, differential-input, bilateral current pump. It looks similar to the previous example but uses a second op amp in place of the trimmer. It produces an output current of up to ±10 mA, proportional to the differential input voltage, which in this example is limited to ±1 volt. A higher output current is achievable if one uses an op amp with a higher-rated output current. In this circuit the second op amp A_2 is used as a buffer in the feedback loop of the first op amp. This helps increase the circuit's output resistance to more than 2.5 MΩ and maintains accuracy and linearity over the operating range. The voltage compliance in this circuit is ±8 volts (to the load), with ±12-volt supplies. The formulae for determining the output current and resistor ratios are shown in the diagram. Both op amps must be fast precision types (ideally, both part of a dual device), and all resistors should be 0.1% or better, with low (less than 25 ppm/°C) tempcos, to maintain circuit accuracy. Using a monolithic precision resistor network for the 10-KΩ resistors, as well as using a dual op amp, will enhance low-drift performance and minimize circuit board space.

11.5 Creating precision current regulators with op amps

The following circuits are some current regulators that may be useful in your application. Generally speaking, current regulators provide higher current outputs (from tens of milliamps to many amps), but with a little less precision than the current sources we have looked at up until now. This is mostly because of the higher TCRs of precision power resistors, which are typically more than 100 ppm/°C. Usually a current regulator circuit will use a *three-terminal* voltage reference along with an op amp to drive the power transistor, which is in series with the load. As with the op amp/voltage reference circuits we looked at previously, the output can be arranged to either sink or source the load current.

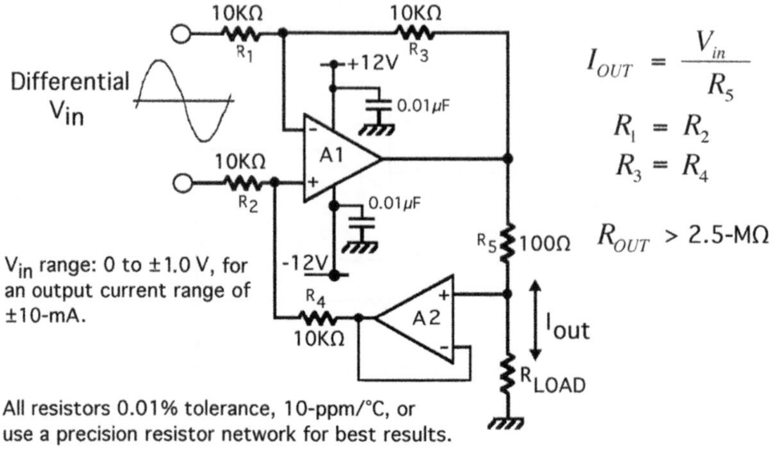

Figure 11.12. A fast, precision current pump.

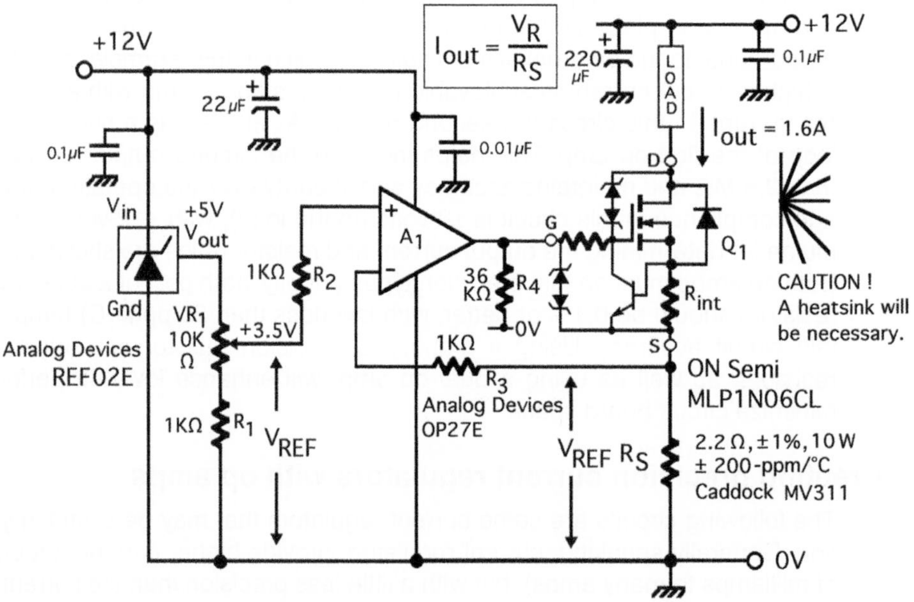

Figure 11.13. A 1.6-Amp precision current sink with a 2-Amp current-limited "smart" power MOSFET.

Figure 11.13 shows an effective current-sinking regulator using a "smart" N-channel MOSFET to drive the load. The circuit is a low-drift current regulator that provides a constant 1.6-Amp current sink for the load, and running off a single supply of +12 volts. The op amp buffers the reference IC's output and the resistive divider from the gate of the power MOSFET. This provides better stability when higher current levels are involved. In this design, part of the REF02E voltage reference's output is picked off from the multiturn potentiometer (or a resistive divider) and used as a reference level for the op amp's noninverting input. This results in the V_{REF} voltage setting being impressed across the current-setting resistor, R_S. As a result, the reference voltage output is converted to a constant current, regardless of load voltage. The op amp should be a good precision type, such as the popular (multisourced) Analog Devices OP27E shown here, with an input common-mode range that includes ground. A precision, low-value power resistor (Caddock MV311) with a fairly low tempco is used to sample the power FET's source current. The op amp drives the gate of the MOSFET by feedback, so that a state of equilibrium is reached between its input terminals. The output current is set by the voltage across the power resistor, divided by its ohmic value. (R_S times the output current, I_{OUT} is forced to equal V_{REF})

This example uses two Analog Devices products: an industry-standard op amp, the venerable OP27E (V_{os} equals 25 μV max.; TCV_{os} equals 0.2 μV/°C; e_n equals 3 nV/√Hz, and PSRR equals 10μV/V), and the reference IC, also an industry standard, a REF02E, a precision 5-volt bandgap device, having a voltage tolerance of ±0.3%, with a typical temperature drift of 3 ppm/°C, over the –55°C to +125°C operating temperature range, and a 10-μV peak-to-peak noise voltage. Both devices can be safely operated over an 8- to 20-volt supply range. The MOSFET is an ON Semiconductor MLP1N06CL, a 60-volt, 2-Amp N-channel "smart" device, with an on-resistance of less than 1 Ω and a maximum gate threshold of 2 volts. It is available in a standard TO-220 power transistor package and should be mounted securely on a small finned heatsink. The precision power resistor is a 2.2-Ω, 10-watt device with a tolerance of 1% and a tempco of 200 ppm/°C. (Such precision power resistors in various packages are available from Caddock Electronics, Ohmite, Vishay, Huntington Electric, and others.) The other resistors shown in the circuit should be 0.1% tolerance, 50-ppm precision types. R_4 is simply a protective measure to ensure that the gate of the MOSFET does not float, in the event of a circuit failure. Normally, a small-value resistor will connect the op amp's output to the gate of a MOSFET to prevent oscillation. In this case, though, the smart MOSFET contains its own gate resistor. The 10-KΩ trimmer (VR_1) could be replaced by a fixed precision resistor. This would enable the circuit to have a lower drift. As it is, the circuit has a trimmed accuracy of approximately ±0.5%, with a drift of approximately 200 ppm (mostly a result of the drift characteristics of the precision power resistor).

Using a 2.2-Ω precision power resistor with a tolerance of 1% yields a maximum resistance of 2.222 Ω and a minimum resistance of 2.178 Ω. With a fixed voltage of 3.5 volts impressed across this nominal 2.2-Ω resistor provides a current of 1.6 Amps. However, because of the resistor's ±1% tolerance, these two extreme values yield a maximum 1.607 Amps, or a minimum 1.575 Amps. In this industrial application it is necessary to provide a current limit/short-circuit for the load, so this is why the smart MOSFET is used, because it contains its own built-in current limit. The MOSFET also includes overvoltage and ESD protection to 2 KV. Under normal conditions, the MOS-

FET conducts the nominal 1.6 Amps constant current into the load. (At 25°C its current limit occurs at 2.4 Amps.) However, in the event that there is a problem with the load, the MOSFET will get hotter, and at 100°C its current limit will be 1.65 Amps. Beyond that it will shut down automatically, protecting both itself and the current source from potential damage.

Even higher currents can be obtained simply by paralleling the power MOSFET with another or others—a unique property that MOSFETs have over bipolar power transistors. Initially, the FET with the highest transconductance (G_{fs}) will draw the most drain current, and as a result will have the highest power dissipation. However, the resulting increase in chip temperature will also increase its on-resistance, $R_{DS(on)}$ (a positive tempco), which will help limit the drain current. As a result, the total drain current will automatically balance itself through all of the other power FETs in that branch of the circuit. A separate gate resistor (1 KΩ) should be used between the op amp's output and the gate of each FET (to prevent oscillation), and located as close as possible to the gate. A good, efficient heatsink will be required for high-power device(s). In this manner it is possible to easily build a reliable, low-drift, high-power current regulator, capable of more than 50 Amps. (A 10-Amp circuit is shown in Chapter 8, Figure 8.21.)

Figure 11.14 shows a precision current regulator that sources 100 mA to a load connected to the negative supply. The circuit is powered from ±10-volt dual supplies and uses a quad precision op amp (such as an Analog Devices OP470), to reduce the number of component parts. The quad op amp has specs that include V_{os} equals 0.4 mV max; TCV_{os} equals 2 µV/°C; e_n equals 5-nV/√Hz; A_o more than 100 dB, and PSRR/CMRR more than 110 dB). The reference used here is an Analog Devices REF03, a precision bandgap type (+2.5-volt, ±0.6%, and a TCV_o equals 50 ppm/°C). The output of the voltage reference is inverted by amplifier A_1, to provide –2.5 volts, which is buffered by amplifier A_2. This –2.5 volts then drives the resistive combination of VR_1 and R_3. (The VR_1 trimmer could be replaced by a discrete precision resistor, which would then be part of a resistive divider with resistor R_3.) In this example, a voltage of –2.2 volts is picked off from the multiturn potentiometer (or a resistive divider) and used as a reference level for amplifier A_3's inverting input. (This amplifier needs to have a high open-loop gain (A_o) in order to enhance the current source and with an input common-mode range that includes ground.)

The op amp drives the gate of the MOSFET by feedback, so that a state of equilibrium is reached between its input terminals. This results in the reference voltage setting being impressed across the current-setting resistor R_S, and thereby the output is converted to a constant current, regardless of the load voltage. The output current is set by the voltage across the power resistor, divided by its ohmic value. A precision, 22-Ω, ±1%, 1-watt power resistor (Caddock MV217), with a 100-ppm tempco, is used to sample the FET's drain current. The MOSFET used is a SuperTex TN0602N2, an N-channel device rated up to 20 volts, 2.5 A, and with a maximum $R_{DS(on)}$ of 0.85 Ω. This device is a vertical DMOS type, with a very low gate threshold of 1.6 volts maximum, and available in various metal and plastic packages. The circuit has a trimmed accuracy of approximately ±0.1%, with a drift of approximately 125 ppm (mostly a result of the drift characteristics of the precision 1-watt power resistor). With a lower current and/or a 1.2-volt voltage reference instead, the resistor could be reduced to a 1/2-watt type, with a lower tolerance and lower TCR. This would enable a higher accuracy design with a lower drift.

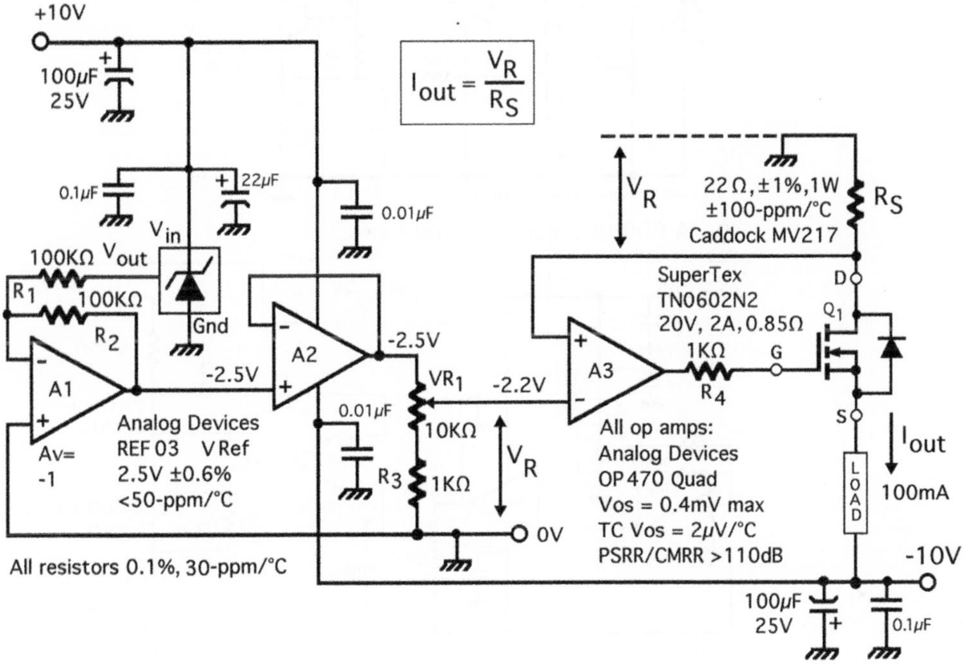

Figure 11.14. A 100-mA precision current regulator.

For smaller-value currents (i.e., below 10 mA), one can use either of the two circuits shown in Figure 11.15. In these circuits, rather than use a MOSFET, they use a combination of a small N-channel JFET and a small-signal NPN transistor (otherwise they use the same configuration as in the two previous examples). With these circuits, much higher accuracy and lower drift are both possible, because they could use precision 0.1% or 0.01% resistors throughout.

Figure 11.15A shows a precision 600-μA current sink that runs off a single supply. The op amp buffers the reference IC's output and the resistive divider from the gate of the JFET. The JFET is a low-noise amplifier (2N4339 or near-equivalent J201), with a low I_{DSS} (less than 2 mA) and is used to drive the small-signal NPN transistor (2N2222/A), which conducts virtually all of the load current. In this circuit, part of the voltage reference's output is picked off from the multiturn potentiometer (or a resistive divider) and used as a reference level for the op amp's noninverting input. The op amp drives the

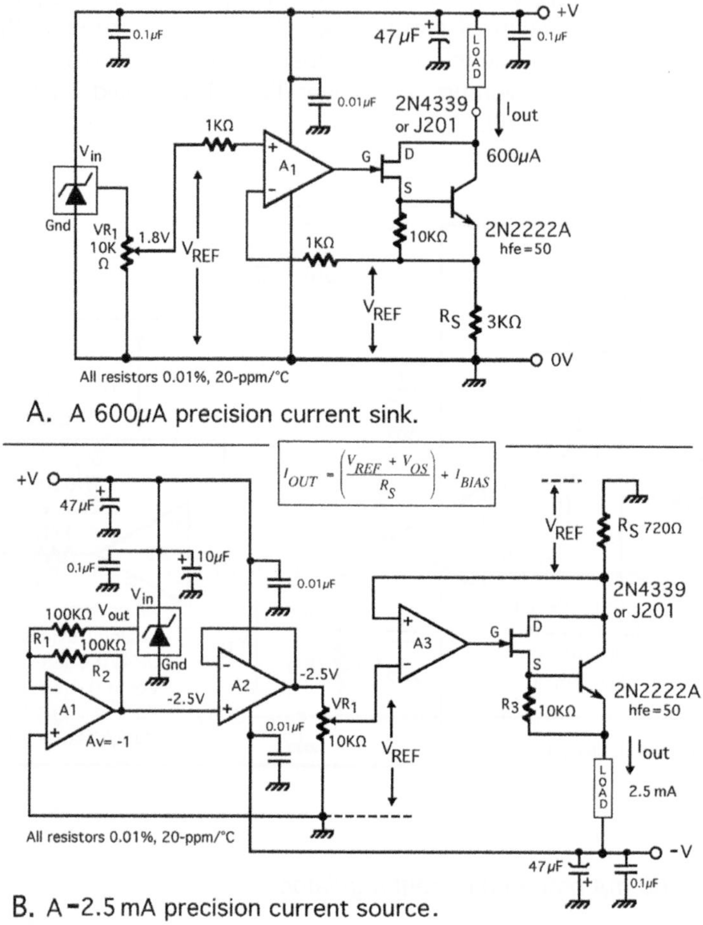

A. A 600μA precision current sink.

B. A -2.5 mA precision current source.

Figure 11.15. Low power op amp current sources using a BJT/JFET combination to drive the load.

gate of the JFET by feedback, so a state of equilibrium is reached between its input terminals. This results in the V_{REF} voltage setting being impressed across the current-setting resistor, R_S. As a result, this reference voltage is converted to a constant current, regardless of load voltage. The op amp should be a good precision type, with an input common-mode range that includes ground. The output current is set by the voltage across the current-setting resistor, R_S, divided by its ohmic value. Even though no particular precision op amp or voltage reference IC has been specified in this design, one could expect such a circuit to have a trimmed accuracy of better than ±0.5%, and with a drift of less than 100 ppm, depending on your choice of components. The design equation shown in the diagram is that of equation 11.5, which is more accurate for designing low-current circuits.

Figure 11.15B shows a precision 2.5-mA current source that runs off a single negative supply. It works similarly to the previous circuit, but with some of the polarities reversed. It too can provide good accuracy, depending on your choice of op amps, voltage reference IC, and the quality of the precision resistors used. Quad precision op amps are rare. After all, it is hard enough to make a precision single or dual device, let alone a quad. So once again, as in Figure 11.14 (showing a 100-mA current regulator), you may use an Analog Devices' OP470 or similar quad device.

In summary, some precise current sources can be built with op amps, and currents can be generated precisely and easily. The highest-precision current sources use high-performance op amps, voltage references, and precision resistors. When you consider how far today's monolithic op amp has come since Bob Widlar invented it in the mid-60s and how it has evolved since, it has to be the most useful analog products ever invented. The versatile op amp will surely be a major part of the analog designer's arsenal for a long time to come, and with it the ability to create precision current sources.

Chapter 12

An Introduction to Voltage References

12.1 Introduction and history

Semiconductor voltage references are basic electronic building blocks that are used extensively in OEM circuit board designs, as well as integrated into the designs of many monolithic A/D and D/A converters and voltage regulators. They are created by combining diodes and resistors with either transistors (BJTs and/or FETs) or op amps. Various techniques for creating and optimizing voltage references will be reviewed in this section of this book. While most forms of today's instrumentation use either voltage or current references, the former are far more readily available. Voltage references include discrete zener diodes, temperature-compensated zeners, as well as several different types of integrated circuit devices: bandgap, buried zener, Analog Devices' XFET®, and Xicor's FGA™ types. Of these different types, the zener diode is the oldest (1960s). The bandgap reference is the oldest IC type (early 1970s), as well as the most common, while Xicor's FGA reference is presently the newest. As mentioned, many A/D and D/A converter products contain their own on-board bandgap references. Figure 12.1 shows the voltage reference family tree. We will look more closely at these various types in this and following chapters.

The advantages of using voltage references is their inherent constant voltage outputs, which (ideally) are independent of changes in supply voltage, temperature, load resistance, load current, or aging over time. One could liken the voltage reference to that of a precision, low-noise, low-current voltage regulator. These advantages, when compared with using a simple fixed resistor divider, include the following:

- Greater precision
- Better repeatability
- Improved temperature stability
- Increased long-term stability
- Lower long-term drift

Applications for voltage references include biasing and stabilization, where both rely on precise and stable reference levels. For example, in a 16-bit A/D system, a preci-

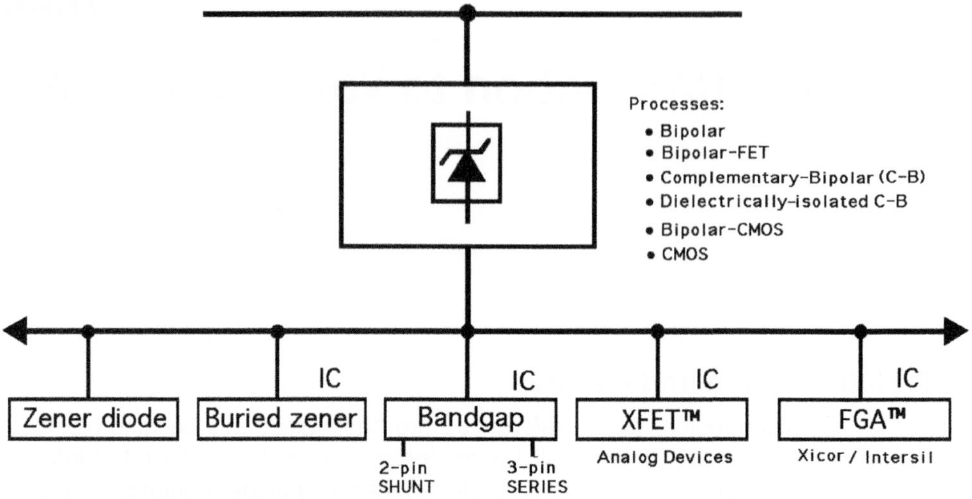

Figure 12.1. *The voltage reference family tree.*

sion voltage reference is used to set the converter's full-scale input voltage range. In a level-sensing switching circuit, the comparator has a particular voltage trip-point set by a precision voltage reference. In a high-quality bench power supply unit, a low-noise voltage reference circuit is used to provide a stable reference. A voltage reference IC is used in a digital panel meter (DPM) to set the full-scale input voltage range. In the design of an expensive color-analyzer instrument, an optical sensor and its amplifier are both biased by a precision voltage reference, so that very-low-level signals can be measured precisely. In calibrating this instrument, a comparator's voltage trip-points are set by the system processor, while a precision voltage reference is used to set the analog output voltage range for the D/A converter (as shown in Figure 12.2). As we learned in Chapter 11, designers frequently use voltage references together with op amps and precision resistors to create highly accurate current sources or sinks. There are innumerable other examples of applications for voltage references, but those mentioned here are common. General market segments that rely on high-precision references include industrial control and instrumentation, medical instrumentation, avionics, aerospace, and space flight instrumentation, as well as most lab calibration, test, and measurement equipment.

Voltage references are not a new innovation; they predate the integrated circuit by several decades. Before being used within integrated circuits, or designed as dedicated IC products, voltage references took the form of bulky and expensive laboratory standards. These include the Weston cell, the Clark cell, expensive cryogenic Josephson arrays, as well as some types of batteries. These types of laboratory standards have been used for decades. The best-known standard cell is the Clark cell (named after its inventor Latimer Clark), which provides a constant voltage of 1.434 volts at room temperature. The Clark cell has two electrical poles: the positive pole is mercury

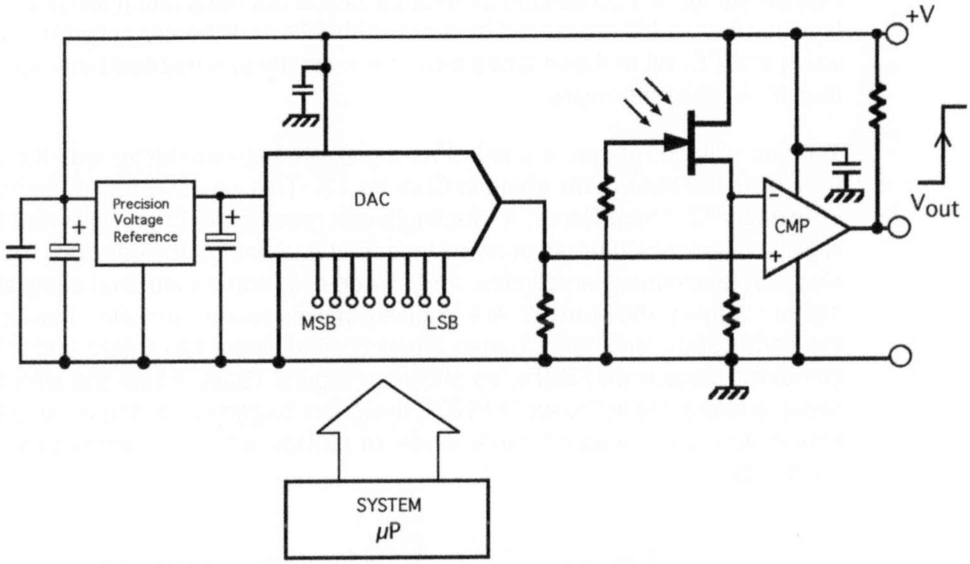

Figure 12.2. *In this circuit the voltage reference sets the DAC's output range, while the DAC provides the comparator's trip point, which is set by the system's processor. As a result, the sensitivity of the high-speed optical detector can be adjusted and set by the system.*

(Hg), in contact with mercurous sulphate (Hg_2SO_4), while the negative pole is zinc (Zn), in contact with a solution of zinc sulphate ($ZnSO_4$). Another is the Weston cell (named after its inventor Edward Weston), which is similar to the Clark cell. The Weston cell also has two poles: the positive pole is mercury (Hg), in contact with mercurous sulphate (Hg_2SO_4), while the negative pole is cadmium (Cd), in contact with a solution of cadmium sulphate ($CdSO_4$). It produces a constant voltage output of 1.019 volts, which is virtually independent of temperature change. Both of these standard cells are composed of an H-shaped glass container and are self-powered from unique chemical actions. (Electrolysis is based on the work of the renowned English scientist Michael Faraday, who pioneered the process in the early 1800s. He is best known for his work in electromagnetism and capacitance though.)

Both standard cells produce stable, well-defined voltages, accurate to a few parts per million or better. Care should be taken though not to tip or knock over the devices, or to load the output with any appreciable current, because this will temporarily cease operation of the cell, where recovery time can take weeks. Josephson arrays are expensive and require cryogenics to function. They produce an output of 1.018 volts, with stability better than 1 ppm. Mercury cells, developed during World War II, have also been used as voltage references, mostly because they are cheap and immune to the physical problems of the fragile glass standard cells. They provide an output of 1.35 volts at several milliamps but have much lower accuracy. They typically have a life measured in years, with 85% to 90% capacity retained after two years at 20°C. All of these types have been mostly superseded by modern semiconductor voltage references.

The first voltage reference created as a discrete semiconductor was the zener diode (which we will read more about in Chapter 13). This was created in the late 1950s by its inventor Clarence Zener, a Westinghouse researcher. Even today the zener diode is still a popular workhorse in many industrial and commercial designs, because of its reasonable accuracy, small size, and low cost. In some industrial applications where higher voltages and currents are involved, power zeners are often the only option. In the early 1960s, engineers began creating small linear regulators and references by combining zeners with BJTs, as shown in Figure 12.3A. When the silicon bipolar IC became affordable in the early 1970s, designers began to use the op amp to buffer the temperature-compensated zener diode, to provide a more stable reference (see Figure 12.3B).

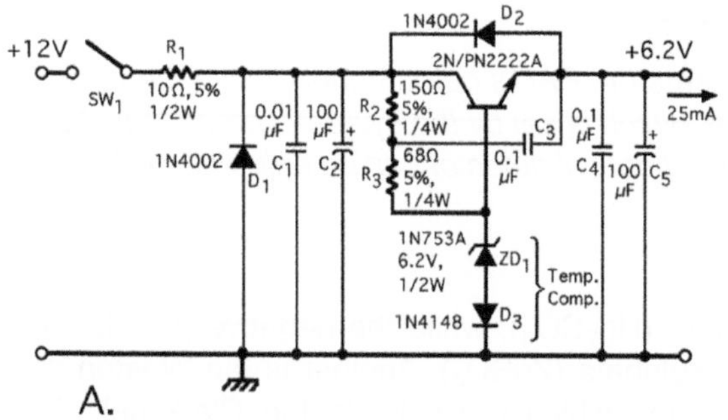

A.

Building a small linear voltage regulator in the sixties, meant combining a zener diode regulator with an NPN pass transistor, like the circuit shown here.

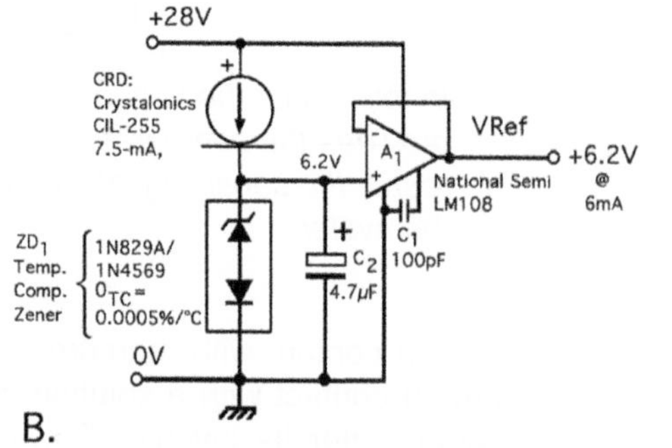

B.

A precision avionics voltage reference from the early seventies, using a temperature-compensated zener diode, a current-regulating diode, and a precision op amp.

Figure 12.3. Examples of early semiconductor references.

Then in 1969, the legendary Bob Widlar at National Semiconductor became the first analog IC designer to create an *integrated* voltage reference (based on the bandgap principle), as part of a power regulator IC design, the LM109. This was the first monolithic, 5-volt, 1-Amp linear voltage regulator (see Figure 12.4) and was a deliberate challenge for Widlar. He showed that it was possible—despite industry-wide skepticism at the time—to build a precision linear power regulator on *one* monolithic chip, despite changes in chip temperature. Then in 1971, Widlar and his friend Bob Dobkin co-designed National Semiconductor's LM113—the first dedicated, monolithic bandgap reference.

Figure 12.4. *National Semiconductor's LM109, designed by the legendary Bob Widlar, and the first 5-volt linear regulator to use a bandgap reference.*

Much development has focused on creating monolithic voltage references ever since the early '70s, and it still continues today at a fast pace. Early innovators besides National Semiconductor included Analog Devices, Texas Instruments, RCA, and Motorola Semiconductors, among many others. Some of the early pioneering work was shared on an industry-wide basis in the form of published technical papers. Technical articles also appeared in the electronics industry's leading-edge magazines of that era, as well as in manufacturers' own in-house application notes. Up until this point, these first-generation voltage reference ICs were marketed and referred to as though they were zener diodes, which were the established standard at the time. This is shown by the fact (see Figure 12.5) that they were depicted on a circuit diagram with the zener diode's symbol. The only way to tell the difference was that a zener typically had a JEDEC-registered number such as 1N4733A (a 5.1-volt, 1-Watt zener), while a voltage reference IC had a part number, such as LM285-1.2. To this day, it is still common to diagram the two-terminal bandgap voltage reference IC using the zener diode's symbol.

This all changed in the mid-70s when National Semiconductor introduced the LM199/399, the first *buried-zener* voltage reference. This legendary product, designed by Robert Dobkin, offered a 1-ppm temperature drift, which was better than anything at the time and is still better than most devices today. However, it had some significant drawbacks, including (1) its cost, (2) an output voltage of 6.9 volts (too high for many applications), and (3) drawing a high quiescent current because of powering its on-board temperature stabilizer (the heater). Another milestone was reached when Paul Brokaw at Analog Devices Inc. created the first *precision* bandgap reference (the AD580) in the late 1970s, based on what is referred to today as the *Brokaw cell.* This product was destined to become one of the most successful voltage reference ICs ever introduced and ushered in a new level of precision for the affordable voltage reference IC. Up until then, most voltage references were either 10 volts or 6.9 volts, and the remainder were mostly 1.2-volt devices. Other companies followed by introducing some other landmark products that include Precision Monolithics' (now part of Analog

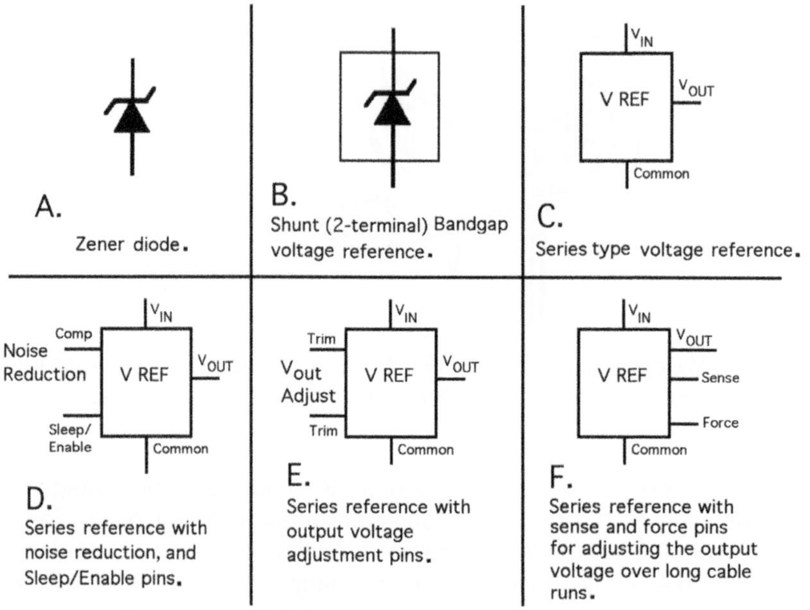

Figure 12.5. Symbols used to depict voltage references.

Devices) REF01 and REF02; Burr-Brown's (now part of Texas Instruments) 10-volt REF102; Linear Technology's amazing LTZ1000 (buried-zener, also designed by Robert Dobkin) and LT1021, designed by Carl Nelson; Maxim's innovative MAX676 with on-board temperature-correction ROM, which provided a tempco of 0.6 ppm/°C; National Semiconductor's low-cost bandgap devices LM185/285/385 (again designed by Bob Dobkin); Intersil's popular ICL8069 (a two-terminal bandgap); and others with more recent products, which we will look at in Chapters 14 and 15.

12.2 Understanding voltage reference specifications

The complete range of semiconductor voltage references includes discrete zener diodes and several different types of monolithic devices (bandgap, buried-zener, Analog Devices' XFET®, and Xicor's FGA™ types). The zener diode and its temperature-compensated version are reviewed in the next chapter, while the monolithic references are looked at in more depth in Chapter 14. Even though monolithic voltage references have different topologies, architectures, and characteristics and are produced using different processes, they all share similar characteristics and specifications, which we will look at here.

As mentioned, many A/D and D/A converter products contain their own on-board references, which have similar specifications to the bandgap. Dedicated bandgap reference ICs are available as either two-terminal ("shunt") or three-terminal ("series") devices, while the other types are generally series types (see Figure 12.6). Shunt (parallel) types are configured in a circuit similarly to a zener diode. They are in parallel with the load and usually require a bias resistor, which carries both the reference's varying quiescent current and the maximum load current. In some applications this constant power drain can be a significant factor. However, because they are biased by a resistor, they can be used over a wider range of input voltages, although this leads to a varying quiescent current. They are best suited to working with a constant load, rather than one where the load current varies much.

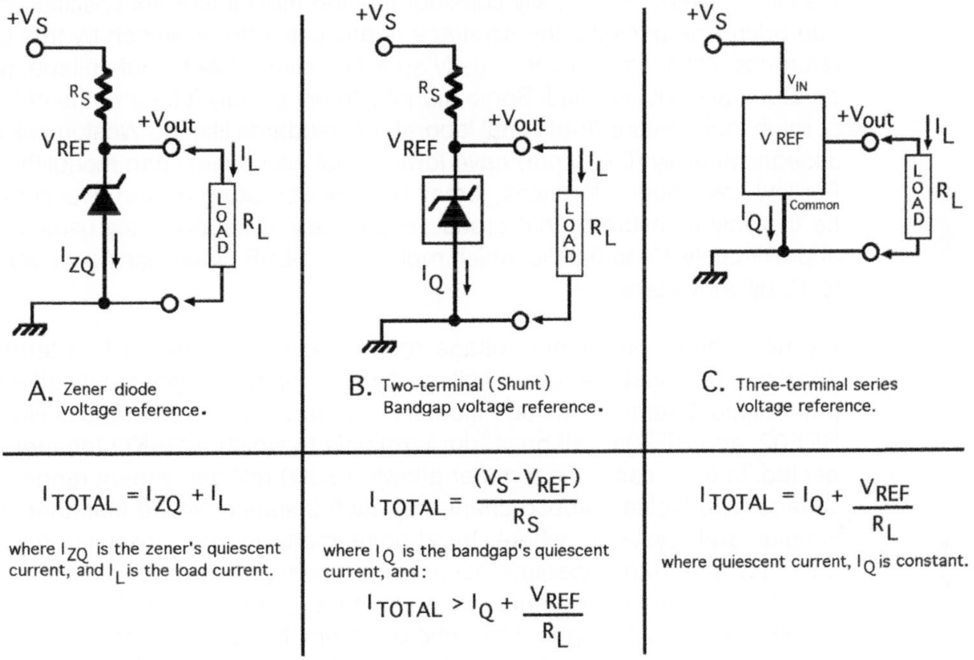

Figure 12.6. Configuring different types of voltage references.

The basic series voltage reference has an input terminal, an output terminal, and a ground return terminal. Series references do not usually need a bias resistor and draw only a constant quiescent current and the load current necessary, irrespective of changes in supply voltage. With either type of reference, the voltage at the output will ideally be held constant, irrespective of changes in input voltage, ambient temperature, or load current. In reality there will be a change in output voltage, but only by a relatively small amount. Typically, series types provide superior performance to shunt types. Some series devices also provide a pair of trim terminals, which allow a 10-KΩ trimmer or resistive network to be connected between them (see Figure 12.5E), which are used for precisely adjusting the output voltage. Others have a compensation terminal, which allows adding an external capacitor for noise reduction (see Figure 12.5D). Others still include a pair of terminals (called sense and force), which allow precise adjustment of the output voltage, where long cable runs are involved (see Figure 12.5F).

The following are some common specifications, which are applicable to all monolithic voltage references.

12.2.1 Initial accuracy (initial error)

This is the output voltage tolerance or error, following an initial application of power to the circuit. Designers usually consider this the *most important* specification for a voltage reference, because the accuracy of any converter is limited by that of its voltage reference. Initial accuracy is usually specified with a fixed input voltage, at room temperature, and with no load. Some manufacturers specify it for a range of load currents, at room temperature. Traditional laboratory standards like the Weston cell (1 ppm) and Josephson array (0.02 ppm) have lower initial accuracies than monolithic references. For low-cost shunt references, it may be specified using a fixed bias current and may be the only measure of that product's accuracy. Initial accuracy usually ranges from approximately 1% to 0.01%, which represents 1-LSB (least significant bit) error in 6-bit to 12-bit converters.

As mentioned, some older voltage references have a pair of trim terminals, which allow one to adjust the output voltage by ±3%, without significantly affecting the output tempco. Examples include industry-standard voltage references like the REF01, REF02, and REF05—all have trim terminals to which a 10-KΩ trimmer can be connected. In each case, the trimmer allows a ±300-mV adjustment range of the output voltage. The typical tempco change in each instance will be less than 1 ppm/°C. In circuits and systems where hand-adjustment is not possible, initial accuracy becomes important. Absolute accuracy is usually of lower priority than initial accuracy, because most designers rely on a final system trim, which tends to balance out any inaccuracies throughout the whole system. It is a whole lot quicker and easier to make just one system adjustment, rather than having to adjust every single precision device in the system.

Table 12.1 provides quick ppm-to-percentage conversion for the most popular voltage references. For example, if one needed to find what 10 ppm would equate to when using a 2.5-volt reference, first locate 10 ppm in the PPM column, and read across to the 2.5-volt column, and you will find that the answer is 25 μV. If you needed to determine what 0.001% of 4.096 volts is, first locate 0.001% in the percentage column. Now read across to the 4.096-volt column, and find that 41 μV is the answer. If you simply needed to know what percentage 10 ppm equaled, simply locate 10 ppm in the PPM column, and read across in the percentage column to find that 0.001% equals 10 ppm.

12.2.2 Temperature drift (tempco, TCV_O)

This is the temperature coefficient of the output reference voltage. It refers to an average of the small changes in the reference's output voltage, caused by changes in temperature ($\Delta V_{OUT}/\Delta T$), and is usually expressed in ppm/°C or in %/°C. Instrumentation designers usually consider this the *second most important* specification for a voltage reference, after initial accuracy. Ideally, the output voltage will not shift by any measurable amount, over a temperature range of –55°C to +175°C (–67°F to +347°F), and a

Reference Volts								ppm	%
1.22V	2.048V	2.5V	3.0V	3.3V	4.096V	5.0V	10.0V	1M	100
0.12V	0.204V	0.25V	0.30V	0.33V	0.41V	0.5V	1.0V	100K	10
12mV	20mV	25mV	30mV	33mV	41mV	50mV	0.10V	10K	1
1.2mV	2.0mV	2.5mV	3mV	3.3mV	4.1mV	5mV	10mV	1K	0.1
120μV	204μV	250μV	300μV	330μV	410μV	500μV	1.0mV	100	0.01
12μV	20μV	25μV	30μV	33μV	41μV	50μV	100μV	10	0.001
1.2μV	2.0μV	2.5μV	3μV	3.3μV	4.1μV	5μV	10μV	1	0.0001
122nV	204nV	250nV	300nV	330nV	410nV	500nV	1.0μV	0.1	0.00001
12nV	20nV	25nV	30nV	33nV	41nV	50nV	100nV	0.01	0.000001
1.2nV	2.0nV	2.5nV	3nV	3.3nV	4.1nV	5nV	10nV	0.001	0.0000001

Table 12.1 PPM to % converter for some popular references.

graph of ΔV_{out} versus temperature would be a straight horizontal line. In reality, though, the output voltage will shift, but only by a relatively small amount. It will be shown on a graph as a series of sharp or shallow slopes, lines, or curves. The sharp 20-degree to 40-degree diagonal lines will apply more to discrete zeners and low-cost bandgap products, whereas the shallow 1-degree to 20-degree slopes will apply to high-precision references. The object here for the instrumentation designer is to choose a device with a low temperature drift, having a shallow enough slope over the temperature range that helps meet or exceed the design goals of the circuit.

Bandgap references have a range of output voltage tempcos of between 2 and 200 ppm/°C. The output voltage tempcos of buried-zener references is probably lowest and shallowest, and ranges from between about 0.5 and 8 ppm/°C. Newer types of references such as ADI's XFET® family, along with Xicor's FGA™ family, also have low output voltage tempcos of between 1 and 10 ppm/°C. Generally, a three-terminal reference will have a lower TCV_O drift but cost somewhat more than a two-terminal bandgap type. However, today's technology has difficulty providing 1 ppm/°C, along with a 1-mV initial accuracy. The bottom line is: the lower the drift (i.e., less than 5 ppm/°C), the more costly the product will be.

As mentioned previously, manufacturers include special temperature-compensation circuits in their reference designs. Devices are measured at the wafer level, and some of the circuit's thin-film resistors are precisely laser-trimmed, before being packaged. Then, after many hours of burn-in, devices are again measured and sorted into different *grades* (i.e., higher/lower percentage, ppm, operating temperature, and package type). The burn-in process tends to reduce stresses within the monolithic chip, which helps stabilize the various drifts and improve the long-term stability. Manufacturers typically use any one of the three following methods to define, test, and graph the output voltage drift over the operating temperature ranges for their products. These are used to test and categorize the products at the factory and also in the printed product specifications for the instrumentation designer.

The Slope method

This is probably the oldest and today the least used of the trio. It refers to a pair of sloping diagonal lines overlaid on a graph, representing the highest and lowest changes over the defined temperature range. This method assumed that the drift was a linear function (it's usually nonlinear) and supports worst-case calculations.

The Box method

This is the most commonly used method today and takes the form of a box shape. The boundaries are formed by the manufacturer's guaranteed minimum and maximum limits for the output voltage (vertical, top, and bottom), over the defined temperature range (horizontal, left, and right). The temperature coefficient is shown by the box's diagonal corners. The box shape provides a better indication of the actual voltage drift error than the Slope method. If the test results are within the boundaries of the box, then the product is within specs. It is commonly seen illustrating the temperature drift of bandgap products, as shown in Figure 12.7. Notice the S-shaped curve that is characteristic of the bandgap.

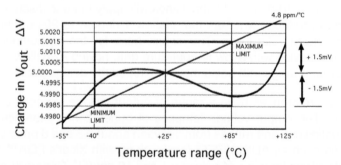

Figure 12.7. Typical output voltage temperature coefficient (TCV_O) for a bandgap reference, graphed with the Box method.

The Butterfly or Bow-tie method

This method takes its name from the shape of these lines created on the graph (it looks like a butterfly or bow-tie). It is a more detailed set of data than the Slope or Box methods, which uses a central reference point of +25°C (room temperature, 77°F, 298°K). It also shows the minimum and maximum slope lines passing through the central reference point, along with various data points along each line (see Figure 12.8). It is frequently used with very-high-precision voltage references.

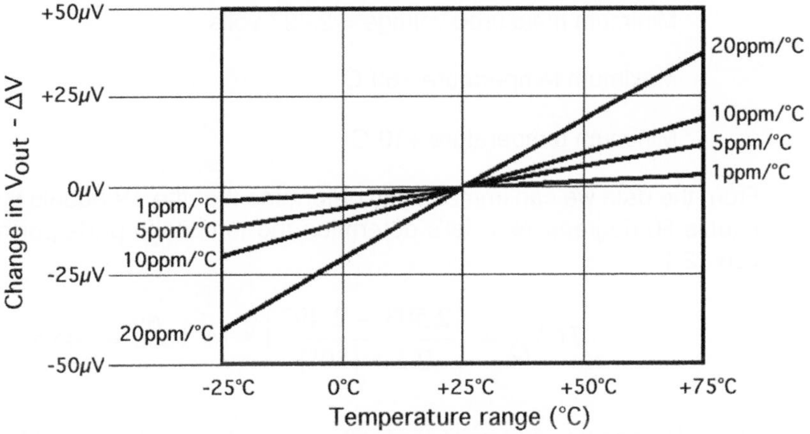

Figure 12.8. Typical output voltage temperature coefficient (TCV_O) graph for a high-performance monolithic voltage reference, using the Butterfly method to illustrate the low tempco of the device. The data is all referenced to +25°C.

Some helpful equations relating to voltage reference temperature coefficient are shown here. Equation 12.1 allows one to determine the reference's tempco in ppm/°C:

$$TCV_O = \left(\frac{V_{max} - V_{min}}{V_{nominal} \bullet \left(T_{max} - T_{min} \right)} \right) \bullet 1 \times 10^6 \qquad \text{(Eq.12.1)}$$

Equation 12.2 allows one to determine the reference's tempco in %/°C:

$$TCV_O = \left(\frac{V_{max} - V_{min}}{V_{nominal}} \right) \bullet \left(\frac{100}{T_{max} - T_{min}} \right) \qquad \text{(Eq.12.2)}$$

Equation 12.3 allows one to determine the voltage drift (i.e., μV/°C):

$$TCV_O = \left(\frac{V_{max} - V_{min}}{\Delta \text{Temp}} \right) \qquad \text{(Eq.12.3)}$$

To show how these equations can be used, let's look at some examples. In the first example, we will assume we are using a precision reference, with the following measured data:

Nominal voltage = 2.500 volts

Maximum measured voltage = 2.503 volts

Minimum measured voltage = 2.497 volts

Maximum temperature +60°C

Minimum temperature +10°C

From the data we can immediately determine that the ΔV equals 6 mV and the ΔTemp equals 50 degrees. Now let's determine the tempco in parts per million, using equation 12.1:

$$TCV_O = \left(\frac{2.503 - 2.497}{2.5 \bullet (50°)}\right) \bullet 1 \times 10^6 = 48 ppm / °C \quad \text{(Eq.12.4)}$$

The reference has a tempco under these conditions of 48 ppm/°C. Now let's calculate what the tempco is in %/°C, using the same data and equation 12.1:

$$TCV_O = \left(\frac{0.006V}{2.500}\right) \bullet \left(\frac{100}{60° - 10°}\right) = 0.0048\% / °C \quad \text{(Eq.12.5)}$$

Finally, let's determine the voltage drift using equation 12.3:

$$TCV_O = \left(\frac{2.503V - 2.497V}{50°}\right) = 120 \mu V / °C \quad \text{(Eq.12.6)}$$

This means that over a 50-degree temperature span, the drift is 120 μV/°C. Now, if the temperature span was restricted to 20 degrees instead, then this would actually yield a higher drift:

$$TCV_O = \left(\frac{2.503V - 2.497V}{20°}\right) = 300 \mu V / °C \quad \text{(Eq.12.7)}$$

Alternately, if the temperature span was expanded to 80 degrees instead, then with the same data this would yield a lower drift, as shown here:

$$TCV_O = \left(\frac{2.503V - 2.497V}{80°}\right) = 75 \mu V / °C \quad \text{(Eq.12.8)}$$

Table 12.2 shows the allowable tempco error requirements with 1/2 LSB resolution, for a range of 4-bit to 24-bit systems. For 1-LSB resolution, simply multiply the ppm/°C or millivolts shown in the table by two.

Bits	Required drift (ppm/°C)	Full-scale ranges 1/2 LSB resolution (mV)			
		1.22V Ref	2.5V Ref	5V Ref	10V Ref
4	312.32	39.04 mV	78.08 mV	156.16 mV	312.32 mV
5	156.16	19.52	39.04	78.08	156.16
6	78.08	9.76	19.52	39.04	78.08
7	39.04	4.88	9.76	19.52	39.04
8	19.52	2.44	4.88	9.76	19.52
9	9.76	1.22	2.44	4.88	9.76
10	4.88	0.61	1.22	2.44	4.88
11	2.44	0.305	0.61	1.22	2.44
12	1.22	0.152	0.305	0.61	1.22
13	0.61	0.076	0.152	0.305	0.61
14	0.305	0.038	0.076	0.152	0.305
15	0.152	0.019	0.038	0.076	0.152
16	0.076	0.009	0.019	0.038	0.076
17	0.038	0.0047	0.009	0.019	0.038
18	0.019	0.00238	0.0047	0.009	0.019
19	0.009	0.0012	0.0023	0.0047	0.009
20	0.0047	0.0006	0.0012	0.0023	0.0047
21	0.0023	0.0003	0.0006	0.0012	0.0023
22	0.0012	0.00015	0.0003	0.0006	0.0012
23	0.0006	0.000075	0.00015	0.0003	0.0006
24	0.0003	0.00004	0.000074	0.00015	0.0003

NOTES: 1. Computed for 100°C temperature span.
2. A tempco of 0.3ppm/°C is needed to maintain 1/2 LSB at 14-bits, over an operating temperature range of 100°. Shorter temperature spans can support a higher tempco.

Table 12.2 Allowable tempco drift requirements for n-bit system

As an example, in order to keep any conversion error to within 1/2-LSB in a true 16-bit converter system, over a 25°C through 125°C temperature range (i.e., a 100-degree span), the (10-volt) reference error will need to be less than 76 μV over this temperature span. From Table 12.2, the 76 μV equates to an allowable tempco of 0.076 ppm/°C. By contrast, in an 8-bit system, over the same temperature range, but with a 1-LSB conversion error instead, the voltage reference will need to have a tempco of less than 39 ppm/°C—a significant difference! This is shown in Figure 12.9, which graphs a part of Table 12.2. For determining the 1-LSB conversion error, simply multiply the figure given in Table 12.2's "Allowable drift (in ppm/°C)" column by two. For determining a 1-LSB conversion error using Figure 12.9, first locate the intersection of the n-bit and T° temperature, and read off the approximate ppm tempco required. (Because this figure is calculated for 1/2-LSB, read off the next bit-line intersection above, to get 1-LSB.)

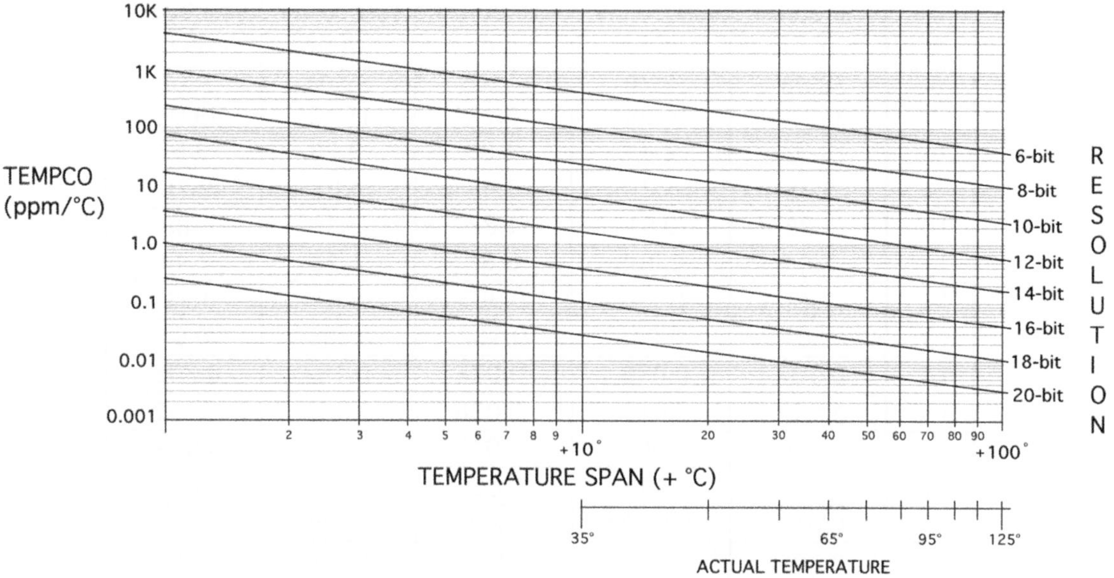

Figure 12.9. *Voltage reference required tempco versus operating temperature range (keeping conversion error to 1/2 LSB).*

12.2.3 Long-term drift

This is a measure of the changes in a voltage reference's output that occur over weeks or months of continuous operation, at room temperature. It is the long-term temperature coefficient of the reference voltage and usually expressed in ppm/1000 hours (1000 hours equals 41 days). Instrumentation designers often consider this the *third most important* specification for a voltage reference. Because any changes are slow, long-term drift is usually considered a form of noise, which is random and unpredictable. Manufacturers typically make the sample measurements for long-term drift over 1,000 hours at room temperature, and then using the statistical data to project longer-term numbers, which are a good indication of how well the reference will perform in the future. Typically, as the reference device ages, its output changes by a smaller and smaller amount, but usually in the same positive or negative direction. This is why many experienced instrumentation designers prefer to additionally burn in their voltage references for over a 168-hour span (168 hours equals 1 week) at 125°C (constantly), before installing them in printed circuit boards. It is believed that this burn-in reduces stresses within the chip, which in turn helps stabilize the various drifts and significantly improves the long-term stability.

Of the several types of monolithic voltage references available, the best for long-term drift are buried-zener references, which have long-term drifts typically between 5 and 20 ppm/1000 hours, while Xicor's FGA™ family have long-term drifts typically around 10 ppm/1000 hours. Analog Devices' XFET® family have long-term drifts typically 40 ppm. Bandgap references generally have long-term drifts between 20 to 120 ppm/1000 hours.

12.2.4 Noise

This refers to any type of voltage noise at the output of the reference, and therefore includes narrowband 1/f noise, as well as wideband/thermal noise. Many manufacturers specify peak-to-peak noise for their products over a 0.1-Hz to 10-Hz bandwidth, whereas others specify it as wideband RMS noise over a 10-Hz to 1-KHz bandwidth. As a result, it may be specified on the product data sheet in either μV peak-to-peak (the most common), in nV/√Hz (as with op amps and discretes), or even μV_{rms}.

Unlike other voltage reference characteristics, output noise evokes in the minds of designers two opposite ideas. Some designers consider noise one of their top three priorities, whereas others put it near the bottom of their list. For those designers who consider it a major priority, their thinking is that no matter how precisely accurate and low drift the reference may be, if it's noisy then they find a better part. For designers who consider it a low priority or who overlook it altogether, the belief is that a precision reference emits such little noise that there is no need to reduce whatever is there to an even lower level—sorry, not true! Low-frequency noise can make it difficult to measure output voltage, while at any frequency a high noise level can obscure low-level signals. Likewise, it can be important in both high-resolution and high-speed converters, because it impacts accuracy. Low-noise references are essential.

Product data sheets often specify noise values in nV/√Hz, which lets you calculate the value according to the bandwidth. This is shown here in equation 12.9:

$$E_{noise} = \frac{V_{REF}}{12 \bullet 2^N \bullet \sqrt{BW}} \qquad \text{(Eq.12.9)}$$

You can convert the nV/√Hz value to the RMS noise voltage, by multiplying it by the square root of the system bandwidth. You can then convert the RMS value to the approximate peak-to-peak value. Typically the peak-to-peak noise is approximately six times the RMS value.

Table 12.3 shows the requirements for some popular n-bit systems using peak-to-peak (pk-pk) noise voltage values, while Table 12.4 shows similar requirements using nV/√Hz noise values instead. As a rule of thumb, the peak-to-peak noise voltage should always be restricted to less than 10% of 1 LSB (as shown in Table 12.3).

Two factors to consider, and which are clear from the tables, are that (1) noise level is linearly proportional to voltage level, and (2) noise requirements become much tighter

Bits	Full-scale ranges @ 1- LSB resolution Noise voltage peak-to-peak (pk-pk)			
	1.22V Ref	2.5V Ref	5V Ref	10V Ref
8	480 μV	970 μV	1.94 mV	3.9 mV
10	120 μV	240 μV	480 μV	970 μV
12	30 μV	60 μV	120 μV	240 μV
14	7.5 μV	15 μV	30 μV	60 μV
16	1.8 μV	3.7 μV	7.5 μV	15 μV
18	469 nV	937 nV	1.8 μV	3.7 μV
20	117 nV	234 nV	469 nV	937 nV
22	29 nV	58 nV	117 nV	234 nV
24	7 nV	14 nV	29 nV	58 nV
NOTES: 1. Computed for 1-LSB resolution.				

Table 12.3 Noise requirements for n-bit systems.

Bits	Full-scale ranges @ 1/2 LSB resolution, and 100KHz bandwidth Spectral noise density in nV/$\sqrt{Hz}$								
	1.024V Ref	1.24V Ref	2.048V Ref	2.500V Ref	3.000V Ref	3.300V Ref	4.096V Ref	5.000V Ref	10V Ref
8	1040nV	1280nV	2048 nV	2576 nV	3088 nV	3392	4208	5152 nV	10304 nV
10	260	320	512	644	772	848	1052	1288	2576
12	65	80	128	160	193	212	263	322	643
14	16	20	32	40	48	53	65	80	161
16	4	5	8	10	12	13.25	16.4	20	40
18	1	1.25	2	2.5	3	3.3	4.1	5	10
20	0.25	0.312	0.5	0.625	0.75	0.828	1.02	1.25	2.5
22	0.062	0.078	0.125	0.156	0.187	0.207	0.256	0.314	0.628
24	0.016	0.0195	0.031	0.039	0.046	0.051	0.064	0.078	0.157
NOTES: 1. Computed for 1/2-LSB resolution.									

Table 12.4 Allowable noise levels for n-bit systems and for 1.024V to 10V ranges

as bit resolution *increases* (i.e., 10-bit, 12-bit, 14-bit, and up) *and* the full-scale voltage *decreases* (i.e., 10V, 5.0V, 4.096V, 3.6V, 2.50V, etc.). This usually leads to additional filtering being required. For example, a 5-volt reference in a 12-bit system (with 1/2 LSB resolution) will have an upper noise limit of approximately 322 nV/√Hz (60 μV pk-pk). In contrast, a 1.2-volt reference in a 16-bit system will impose an upper noise limit of 5 nV/√Hz.

Figure 12.10 shows two good examples of very low noise levels. The following show some examples of noise levels for some popular references. Maxim's MAX6350, an ultra-precision bandgap reference, has a noise voltage of typically 1.5 μV pk-pk over a

0.1-Hz to 10-Hz bandwidth. Another precision Maxim series reference, MAX6325, has similar noise figures. An ultra-precision buried-zener such as an Analog Devices' AD586 has a typical 4-μV pk-pk noise spec over a 0.1-Hz to 10-Hz narrow bandwidth and a typical 100-nV/√Hz spectral density. An excellent bandgap such as Analog Devices' AD780 has similar noise specs. Another high-performance reference like Texas Instruments' REF102, also a compensated buried-zener type, has a typical 5-μV pk-pk noise spec over a 0.1-Hz to 10-Hz bandwidth. Linear Technology's shunt bandgap LT1634 has a 10-μV pk-pk noise spec over the same bandwidth. A popular

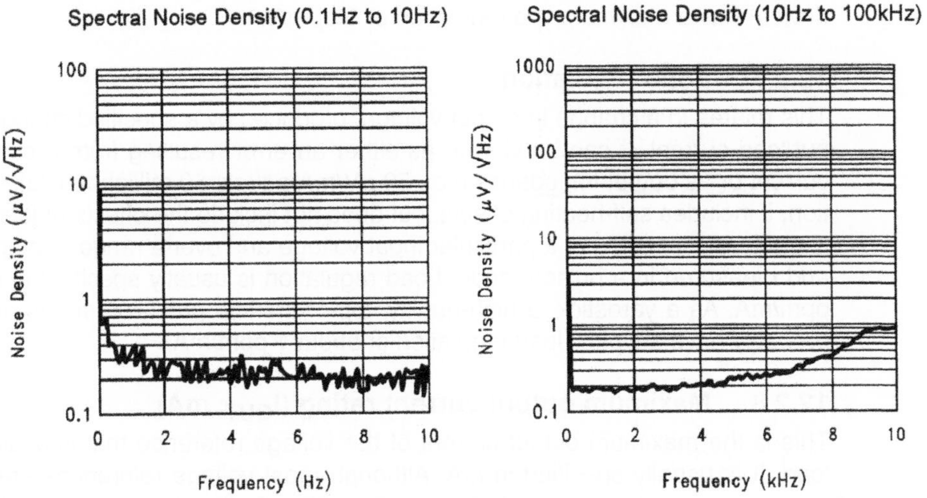

Figure 12.10. Examples of noise density characteristics.

low-cost bandgap reference like National Semiconductor's LM4041AI-2.5 has a typical (wideband) noise voltage of 20 μV pk-pk over a 10-Hz to 10-KHz bandwidth. An Intersil/Xicor FGA™ reference has a noise voltage of typically 30 μV pk-pk over a 0.1-Hz to 10-Hz bandwidth.

12.2.5 Thermal hysteresis

This refers to a change in output voltage in response to a change in temperature, and typically applies to a 25-degree change in temperature (ΔT°) or more. The point is that when the reference returns to the original temperature, it may not have exactly the same initial output voltage as it did originally. This is usually expressed in ppm, a typical value being 25 ppm (0.0025%). It is only specified by some manufacturers, but has become a more important specification in recent years.

12.2.6 Line regulation

This relates to a change in output voltage, produced by a specified change in input voltage. It includes self-heating effects, but does not include the effects of line transients or ripple voltage. It is usually specified in μV/V of input change, but can also be specified in ppm/V or %/V. It is usually measured in a pulsed mode. Manufacturers usually specify it over a particular operating voltage range and operating temperature range. Line regulation degrades with increasing frequency and explains why a voltage reference's input should be well decoupled. Line regulation is also related to dropout voltage. A precision bandgap reference such as Analog Devices' REF195E (a 5-volt device) has a maximum line regulation of 5 ppm/V.

12.2.7 Load regulation

This relates to a change in output voltage, produced by a specified change in the output/load current. It can be viewed as either an error resulting from a change in load current or as output impedance (i.e., 50 μV/mA equals 50 milliohms). Like line regulation, it includes self-heating effects, but does not include the effects of load transients. It is usually specified at a particular input voltage and over a range of output currents, and measured in a pulsed mode. Load regulation is usually specified in μV/mA or in ppm/mA. As a yardstick, a high-quality, top-of-the-line precision series reference will probably have a typical load regulation of about 1 ppm/mA.

12.2.8 Maximum output current rating (I_{OUT}; mA)

This is the maximum output current of the voltage reference that is available for the load. It is usually specified in mA. Although most voltage references are rated for a maximum output current of between 10 to 20 mA, the designer is cautioned not to sink/source too high a current (i.e., keep it at less than 10 mA max or less than 50% of max, whichever is lower). Running the reference at maximum output current may cause self-heating effects and create thermal gradients across the chip, which will degrade the accuracy and stability of the reference. Many references have built-in short-circuit protection that prevents against accidental shorts to the supply rails. Only a few references have higher output current capability, one of which is Analog Devices' REF191 family, which are precision bandgaps, rated at up to 30 mA. In the event that a higher current is required than the reference can provide, consider using an external pass transistor (BJT or DMOS FET), as shown in Figure 12.11. Here the load current is carried by the external transistor, which keeps any heat buildup out of the reference, and maintains its accuracy and drift specifications. One could create a circuit with more than 100-mA capability, but one should use a medium-power device so that the current through it is well below its maximum rating, and it should be located away from the reference IC.

12.2.9 Supply voltage range

This is usually stated on the product's data sheet as part of the absolute maximum ratings. For many older voltage references, this ranged from about 5 volts up to about 36 volts. For many newer low-voltage and low-power products, it is often less than 8 volts

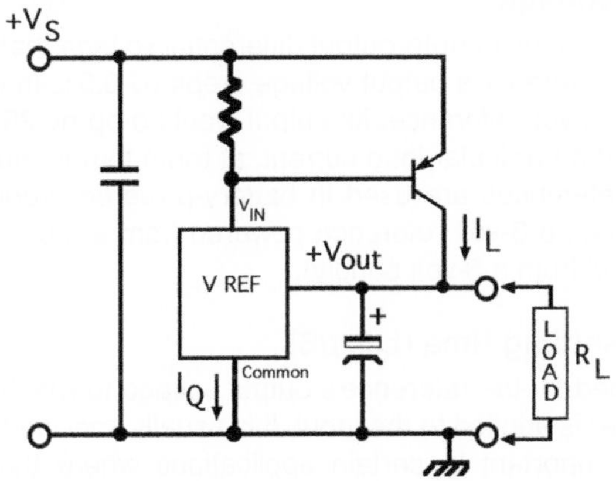

Figure 12.11. *Adding a BJT to the output of a voltage reference keeps the load current out of the reference and maintains its drift specs.*

total. Never exceed this maximum voltage rating, because the device may either be destroyed or some of its main characteristics irreversibly changed. In most cases the reference will perform significantly better if it is supplied from a separate linear voltage regulator, which can take care of external line transients, noise, and so on.

12.2.10 Supply current (I_S) or quiescent current (I_Q)

This relates to the reference's own internal current drain, specified on the data sheet at a particular supply voltage, with no output load. Series references draw only a constant quiescent current, irrespective of changes in supply voltage or load current. Alternately, shunt types have a quiescent current that varies with the input voltage and load current. In battery-powered applications, this constant power drain can be a detrimental factor. Many of today's references would be considered micropower by standards of a decade ago. For example, Xicor's X60008-B50, a precision 5-volt FGA™ reference, has a maximum supply current of 0.8 μA over the –40°C to +85°C temperature range. Many of today's references, including ADI's XFET® family, and most next-generation bandgaps have supply currents of less than 50 μA. By comparison, typical older bandgap types draw over a milliamp (i.e., REF02 has a maximum 1.4 mA). Early buried-zener types that included an integral on-chip heater could draw many milliamps. (National Semiconductor pioneered this particular type of voltage reference, which was invented by Robert Dobkin.) In fact, National's legendary LM199 device, with a total 30-volt supply and at a temperature of –55°C, can draw nearly 30 mA just for the heater (although the reference cell operates at beween 0.5 and 10 mA). It does have an incredible 1 ppm/°C TCV_O drift, over the –55°C to +85°C temperature range, though.

12.2.11 Dropout voltage

This applies to the minimum input-to-output differential voltage that will sustain normal operation, where the reference's output voltage drops by 0.5% from its nominal value (i.e., using a nominal 5-volt reference, its output would drop by 25 mV). It is specified in volts or millivolts, at a particular load current, at room temperature. It is very important where voltage references are used in battery-powered products, particularly in low-voltage circuits (i.e., a 3-volt reference powered from a 3.6-volt supply, or 4.096-volt reference powered from a 5-volt supply).

12.2.12 Turn-on settling time (t_{on}; μS)

This is the time needed for the reference's output voltage to reach within ±0.1% of its final value, after power is applied to the input. It is usually specified in μS. This particular characteristic is important in certain applications where the reference is only powered up briefly.

12.2.13 Turn-on drift ($\Delta V/T$)

This refers to a change in output voltage over a short, specified time after power is applied to the input (typically several seconds). This applies to all monolithic references, except for some buried-zener and bandgap types, which have an integral heater. Such types can take longer to stabilize. It is more important for systems that are only powered up briefly to conserve battery life.

12.2.14 Transient response

This refers to the response of the reference's output to an input voltage transient or to an output current transient. Input voltage transients can be reduced to insignificant levels by powering the reference from the output of a simple linear regulator, which typically has better line regulation, line transient response, and ripple rejection than do most precision references. Adding capacitance at the reference's input and output terminals usually helps performance.

12.2.15 Sleep/Enable

Some voltage references include a sleep or enable terminal, which effectively puts the device into a low-power mode until it is needed. This function is handy for battery-powered applications, and is usually designed to work with common logic levels. It is then easy to have a gate or other logic function control when the reference is active.

12.2.16 Power dissipation

This refers to the power dissipated in the voltage reference, which depends on several factors: input voltage, quiescent current, load current, junction temperature, and ambient temperature. For many reference applications, power dissipation is not even an issue, but where high input voltage, a high output current, and wide operating temperature range are involved, it may become important. Reference manufacturers do not typically specify a power dissipation value, although some do specify a maximum junction temperature temperature and the device's thermal resistance. From these specs it is possible to determine the power requirements involved. The total power dissipation can be found by equation 12.10 as follows:

$$P_{Total} = \left(V_{in\,max} - V_{out\,min}\right) \bullet I_{Load\,max} + V_{in\,max} \bullet I_{Q\,max} \quad \text{(Eq.12.10)}$$

where $V_{in\,max}$ is the maximum input voltage; $V_{out\,min}$ is the minimum output voltage; $I_{load\,max}$ is the maximum load current; and $I_{Q\,max}$ is the reference's maximum quiescent current.

One has to next determine the maximum temperature rise, which can be calculated by equation 12.11:

$$T_{R\,max} = T_{J\,max} - T_{A\,max} \quad \text{(Eq.12.11)}$$

where $T_{R\,max}$ is the maximum temperature rise; T_{Jmax} is the maximum junction temperature; and $T_{A\,max}$ is the maximum ambient temperature.

We can calculate the required value for junction-to-ambient thermal resistance, using P_{Total} and $T_{R\,max}$ as follows:

$$\theta_{(J-A)} = \frac{T_{R\,max}}{P_{total}} \quad \text{(Eq.12.12)}$$

The maximum allowable power dissipation at any ambient temperature can be determined by:

$$P_{Total} = \frac{T_{J\,max} - T_A}{\theta_{J-A}} \quad \text{(Eq.12.13)}$$

This concludes a review of most of the common specifications for voltage references. Next we will look at a few of the practical enhancements that can be made, including bypassing, filtering, trimming, and some mechanical aspects.

12.3 Enhancing the voltage reference design

Although the monolithic voltage reference is a fixed precision component, various measures can be taken by the instrumentation designer to tweak some of its characteristics and enhance its overall performance. These include the following: input and output bypassing, noise reduction, and trimming.

12.3.1 Input and output bypassing

Although not often shown or discussed in many product data sheets, most voltage references are improved by bypassing their inputs. Adding capacitance definitely improves their line transient performance, because most references (2.048-volt and higher) use an internal op amp to amplify the core 1.22 bandgap voltage to the value of the reference output. Typically, a 10-μF electrolytic (with a voltage rating to suit) in parallel with a 0.1-μF disk ceramic will work well. This is shown in Figure 12.12. As mentioned previously, the line regulation, line transient rejection, and ripple rejection will all be improved with the addition of a simple, low-cost, linear voltage regulator at the input of the reference. If the reference is used without the additional linear voltage regulator, then according to the particular reference one may need to add a low-leakage diode, to protect against reverse-bias conditions.

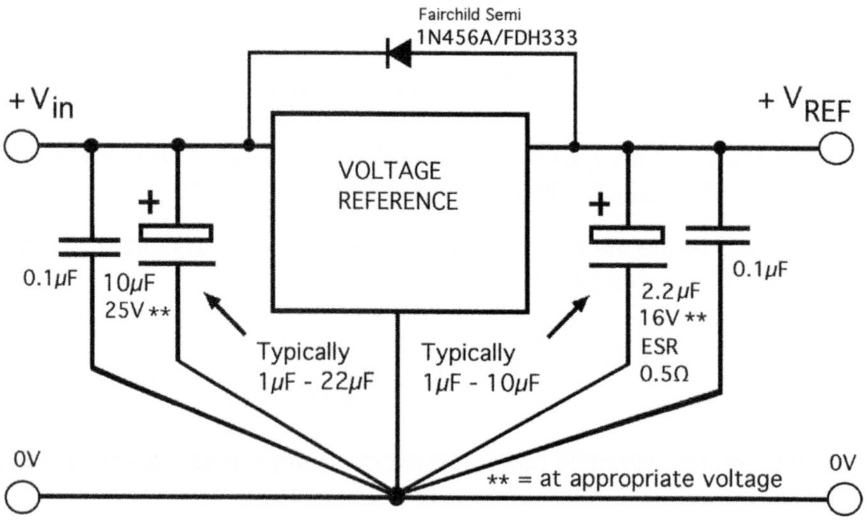

Figure 12.12. A simple voltage reference with input and output bypassing, and reverse-bias protection.

Bypassing the reference's output is a much different situation, because while some references require little or no capacitance, others require up to a specified amount for stability (such as in decoupling the reference from the input of a DAC), and with a range of electrolytic values from 1 μF to 10 μF, but often with a low ESR (Ω) value. However, adding too much capacitance can make the output oscillate! Generally speaking, adding an output bypass capacitor helps reduce noise and load transients and improves stability. One should carefully review the product's data sheet first, to see what the manufacturer recommends regarding output bypassing. What may work well for one type or make of reference may not for another. Even several different makes of bandgaps will have different requirements. In many instances, a value between 1 and 10 μF in parallel with a 0.1-μF disk ceramic will work well (also shown in Figure 12.12), so that high-frequency noise above 1 KHz is significantly reduced.

One should also visually check the quality of the output voltage with an oscilloscope and with different values of added capacitance, looking particularly at what happens to the noise level. In some situations the noise may peak within a certain frequency band. In which case, rather than include regular bypass capacitors, it might be necessary to include a simple RC compensation network or damper between the reference output and ground. (Such a network could also be used with a reference needing an unusually fast start-up feature, with typical values of between 1 to 10 Ω resistance in series with 0.1 to 4.7 μF capacitance to ground.) Many new references are housed in a tiny surface-mount package and are specifically designed to be stable without requiring a bulky output capacitor, which may take up valuable board space. Older references are usually housed in DIP or metal cans and often require more than 1 μF of capacitance for added stability. Because stabilizing the reference output is so important in one's design, it is essential to check the data sheet first or contact the manufacturer's Technical Support/Applications Department for advice regarding this subject.

12.3.2 Noise reduction

The subject of noise as a characteristic of all voltage references was reviewed earlier in this chapter. Here we will look at some practical means for reducing noise levels in 12-bit systems, and higher. As stated previously, some (more expensive) series reference ICs include dedicated noise-reduction terminals. Representative products include Analog Devices' AD586/587/588, and 688, as well as Texas Instruments' REF102 (these are all buried-zener types). Depending on the design and pin-out of the device, one may add a single capacitor between two noise-reduction terminals or add a capacitor from the noise-reduction terminal to ground (see Figure 12.13). Such references include an internal thin-film resistor (low-pass) network, of which the external capacitor is a part. Built-in noise filtering typically provides between 20% and 50% reduction in the noise level. In the event that you decide to use this feature, you should follow the recommendations provided by the reference's data sheet, regarding capacitor type, value, tolerance, drift, and so on. This will likely provide a graph showing output noise voltage versus filter capacitance, or some 'scope pictures of noise reduction.

Because each reference that includes noise filtering capability is unique, there is no single solution for adding noise filtering. Other references that do not include this feature would need some additional external circuitry to accomplish noise reduction. However, in many cases, today's references are only available in small, surface-mount packages, which may not have one or two additional terminals available for noise reduction. They depend instead on advancements in design and processing, together with special internal compensation and noise-reduction circuitry to achieve low noise levels. Several new bandgaps, buried-zeners, and the XFET® can all achieve basic, out-of-the-box noise levels less than 10 μV pk-pk over a 0.1-Hz to 10-Hz bandwidth. These levels can usually be reduced even further with external filtering.

Most products do not include a noise filtering pin, so any filtering must be done directly on the reference output voltage. Typically, the required filter is a simple low-pass RC network. However, before doing anything, one should first be aware of the noise characteristics involved, including the noise requirements of the load (e.g., converter, comparator), and then review the reference's data sheet for any explicit remedial information on this subject. You should also review Tables 12.3 and 12.4. As with bypassing, be aware that what may work well for one type or make of reference may not for another.

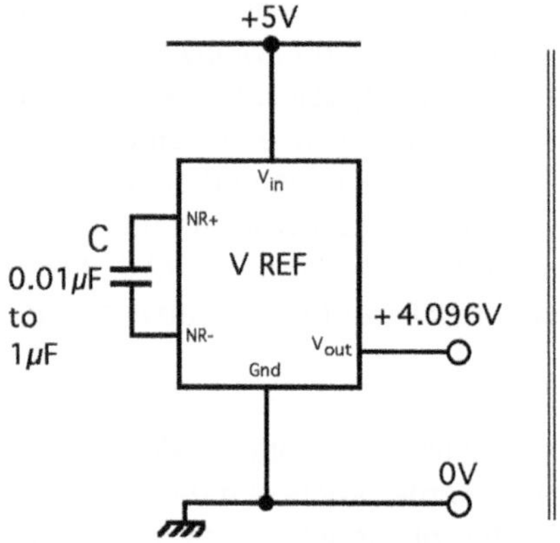

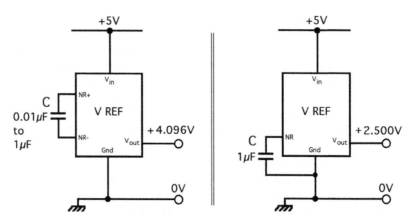

Figure 12.13. *Examples show two different series references, each having noise reduction terminal(s), for adding a single capacitor to that reduces high-frequency noise at the output.*

To reduce the noise level, some manufacturers suggest connecting a simple low-pass op amp filter to the output of the reference. Others discourage that practice, suggesting that putting the RC network directly at the reference's output, together with the op amp's own noise voltage, gain errors, and V_{OS} drift, may even make the circuit's noise level worse, as well as degrading its overall temperature coefficient. They recommend instead using either of these two configurations:

1. Connecting a precision buffer to the reference's output, with a low-pass filter at the buffer's output
2. Connecting the op amp filter inside the reference's feedback loop, assuming that the device has such capability, which only a few older voltage references with force and sense terminals have. Otherwise, by using dual precision op amps, you can build your own.

The first type is shown in Figure 12.14, where a single-pole low-pass filter (R_2C_2) is located at the amplifier's output. (R_1C_1 should be set at twice R_2C_2.) Locating the filter at the amplifier's output, rather than its input, filters the noise from both the voltage reference and the amplifier. The amplifier should be a very low noise, very low offset, low-drift precision type. You can easily add a second precision op amp at the output, if need be, to buffer the load (hence a dual). Capacitors should be high-quality polypropolene, and resistors precision low TC types. This circuit is also enhanced by a simple, low-cost, linear voltage regulator, which helps clean up the supply rail that feeds both the reference and the precision op amp(s). This circuit is likely to be more attractive, because most of today's references have neither force/sense nor noise-reduction terminals.

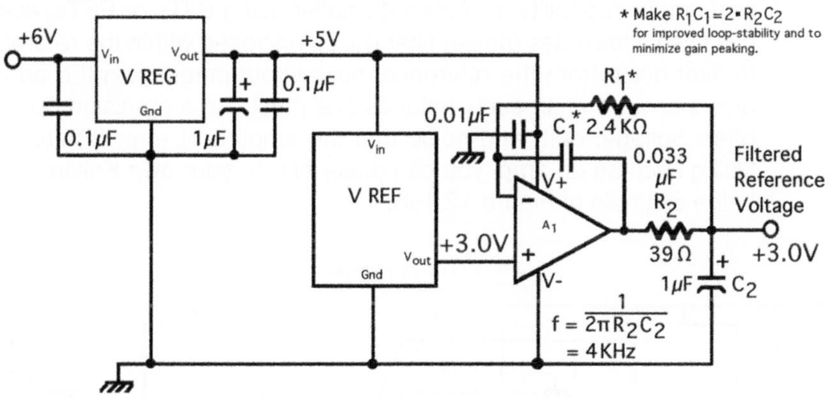

This circuit uses a linear voltage regulator to pre-regulate the positive supply to the reference, as well as an active low-pass filter on the output. The op amp should be a very low noise precision type.

Figure 12.14. Noise filtering the voltage reference.

The second configuration is shown in Figure 12.15, which only applies to a very few high-accuracy (more expensive) references and voltage regulators. This type of reference includes dedicated force and sense terminals, which are primarily intended for compensating for small voltage drops in wiring resistances, connectors, sockets, and circuit board traces between the reference and the load. These kinds of terminals are known as *Kelvin connections*, and manufacturers of this type of voltage reference include Analog Devices, Maxim, Linear Technology, and others. In most designs the load current can produce an I-R error (i.e., voltage drop equals load current times wire/socket/connector resistance). Small amounts of resistance can quickly add up to more than 1 mV of voltage error, which can exceed the allowable error budget in a 12-bit or higher converter. (For example, we can see from Table 12.2 that with a 2.5-volt reference in a 12-bit converter system, the total error budget for 1/2-LSB resolution is only 305 μV. Table 12.3 shows for the same conditions that the allowable noise level should be less than 30 μV pk-pk, while Table 12.4 shows the allowable noise level to be less than 161 nV/√Hz).

Using a voltage reference having these Kelvin connections gets around the problem, because it includes all of these small wiring and circuit trace resistances within the forcing feedback loop of the op amp. The loop forces the output to compensate for all of these errors, and thus provides the exact same voltage to the load as exists on-chip. While that is its primary purpose, it also works in a similar manner when including any filtering and/or buffering (such as in providing a reference voltage with a much higher current output, e.g., 50 mA, 150 mA, 600 mA). As an example, Figure 12.15A shows the general feedback arrangement using a simple buffer, which is shown within the dotted box. This may be replaced by a power op amp buffer with even higher current output capability or a discrete buffer using BJTs or FETs, even an active filter. In this case a low-pass (noise) filter will be enclosed within the reference's feedback loop, to filter noise from the reference, but without introducing the additional op amp gain errors or V_{OS} drift. In each instance one must use a precision op amp with low enough offset voltage, drift, and noise that the amount of error it introduces is insignificant. Using such an op amp, you can easily create your own Kelvin connections, as shown in the example of Figure 12.15B.

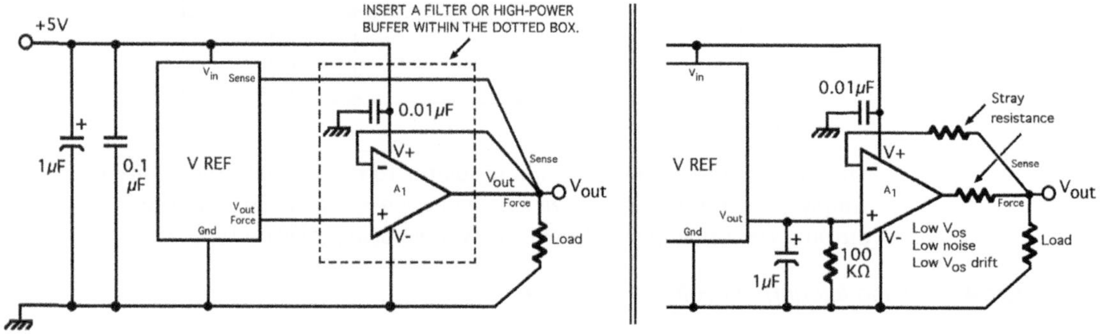

A. Using a high accuracy voltage reference with Kelvin output terminals allows this reference to compensate for voltage drops between its output and the load.

B. Or build-in your own Kelvin connections, using a single precision op amp.

Figure 12.15. Using Kelvin connections can improve performance.

Although active filter design is a little beyond the scope of this book, it may be helpful to quickly review the typical frequency response for a low-pass filter, shown here in Figure 12.16. Most low-pass filters are made to roll off at either 20 or 40 dB per decade. By cascading two such filters together provides even sharper roll-off.

Figures 12.17, 12.18, and 12.19 show some examples of low-pass filters that could be used in the Kelvin-compensated reference circuit. They can be used within the dotted lines of Figure 12.15. In all of these circuits, the filters are best realized using low-noise, very-low-V_{OS}, precision op amps (see Table 11.2 in Chapter 11, which lists many types). In addition, metal-film or thin-film resistors with 1% tolerance or better should be used, along with 1% high-stability capacitors (such as Teflon™, polycarbonate, or silver-mica). The circuit shown in Figure 12.17 uses one op amp to create the

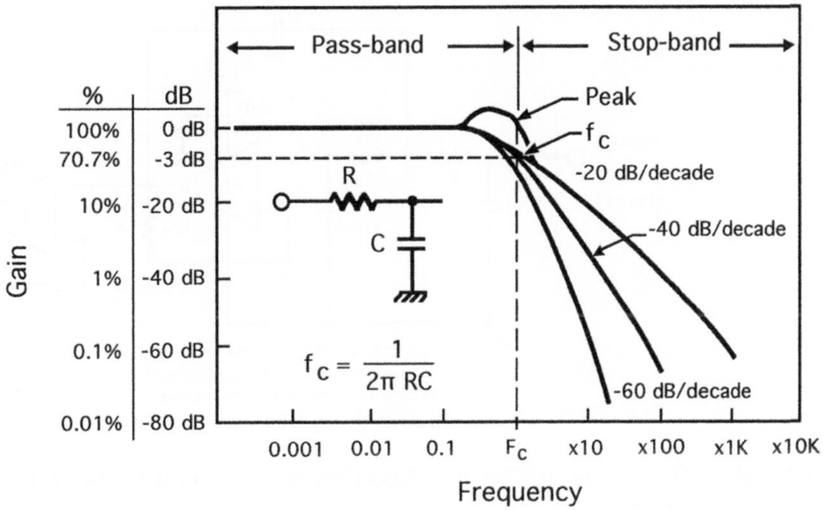

Figure 12.16. Low-pass filter characteristics.

filter and a second op amp as a buffer that directly drives the load. This filter is a modified Sallen-Key, two-pole type with a f_c (3-dB) cutoff at 1.1 KHz.

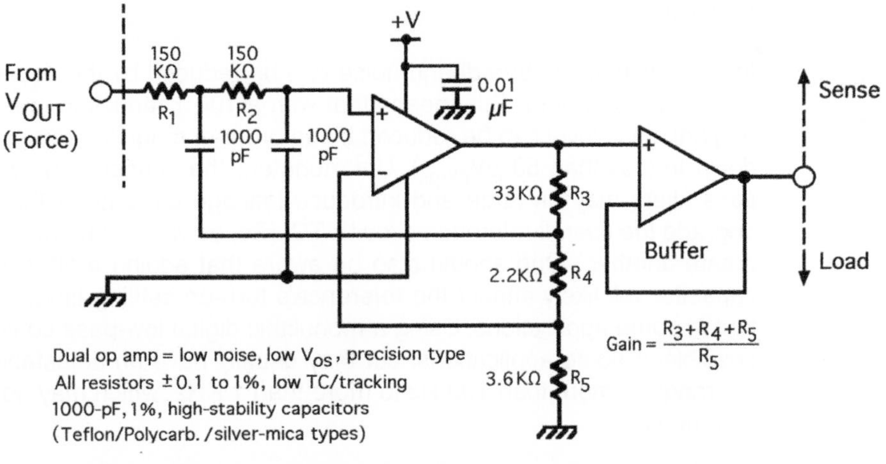

Figure 12.17. Modified Sallen-Key, two-pole, low-pass filter with a 3-dB cutoff at 1.1 KHz.

The next circuit, shown in Figure 12.18, is a three-pole Chebyshev filter, with a 3-dB cutoff at 10 KHz. It has low ripple and a –40 dB/decade roll-off. Like the previous filter, it also uses a dual op amp: one to create the filter and the other as a buffer to directly drive the load.

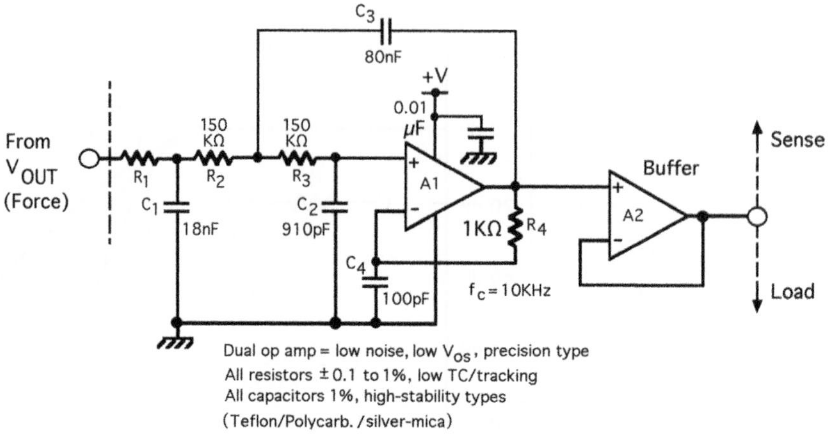

Figure 12.18. Three-pole, 10KHz, low-pass Chebyshev filter with low ripple and a -40dB/decade roll-off.

The last example (Figure 12.19) shows another two-pole low-pass filter, but this time using three op amps. This filter has a constant 180-degree phase-shift beyond the cut-off point.

In many situations, broadband noise can be reduced by the square root of the noise reduction bandwidth. This means that with a 100:1 reduction in noise bandwidth, the original noise level can be reduced by an order of magnitude (e.g., from 500 μV pk-pk down to less than 50 $\mu V_{pk\text{-}pk}$). Unfortunately, the capacitor(s) involved in some low-pass filters may be large and introduce leakage currents of their own, which could degrade the overall reference circuit's TC. (So, as you fix one problem, you unwittingly create another.) One should also be aware that adding a filter or a noise reduction capacitor will likely impact the reference's turn-on settling time, which can be important in some applications. Using a monolithic digital low-pass continuous filter may be possible in some applications, but they usually have an adjustable corner frequency (f_c) range of more than 100 Hz to more than 1 KHz, which may not be low enough for your needs.

12.3.3 Trimming

The voltage reference IC is usually factory-trimmed for a precise output voltage. In some cases though, it may be necessary to adjust its voltage to some precise value or another value altogether. One may even need to adjust it to compensate for a high temperature coefficient or for system calibration. As with noise reduction, some voltage reference products include dedicated trim terminals, but others do not. Those that do not would need additional external circuitry to accomplish any adjustment. Many of

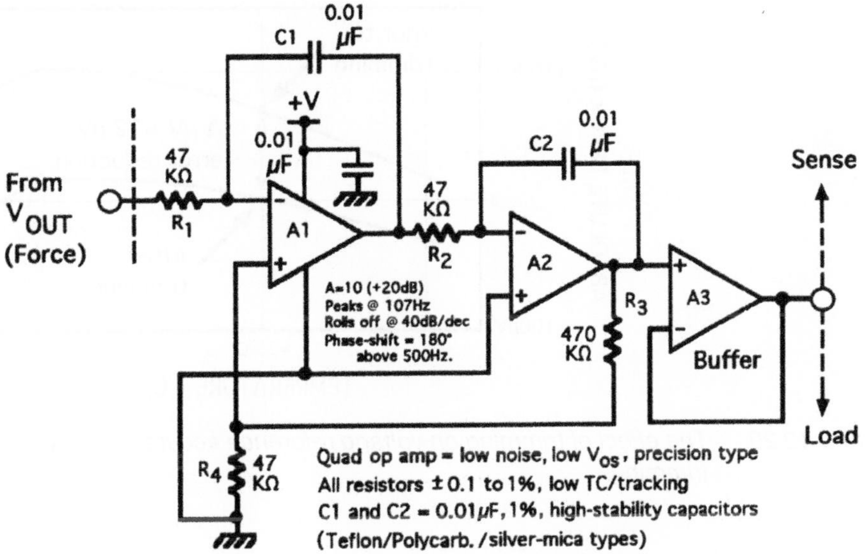

Figure 12.19. Two-pole, 100-Hz low-pass filter with -40dB roll-off and constant 180° phase-shift.

today's references are only available in small, surface-mount packages, which may not have two additional terminals available for trimming. They depend instead on advancements in design and processing, together with special internal compensation circuitry to achieve a precise output voltage over the intended temperature range.

It should be noted that the precise control of room temperature will always impose a practical limit on trimming any circuit's accuracy. However, assuming that you can control the ambient temperature precisely, in many cases trimming can reduce these output errors to a point where only thermal drifts are the remaining sources of error. Figure 12.20 shows an example where a reference's output voltage had an initial error of 72 µV at 25°C, then was trimmed down to zero. Ambient temperature changes, thermal gradients across active components, noise, circuit board layout, or thermal gradients across the circuit board, wiring and a circuit board's track lengths, thermocouple effects, as well as parasitic leakage can all become potential problems, the more precise a reference voltage (i.e., 2.50000V) has to be. In high-performance ADC systems more than 14 bits, especially where a wide operating temperature range is involved, as in the military temperature range, the design (even with trimming) can become extremely challenging

In this example, the applicable error budget for 1/2-LSB is only 76 µV. Small air currents and long-term ambient temperature control are both important factors in any precision measurement/calibration environment. Under those circumstances one will

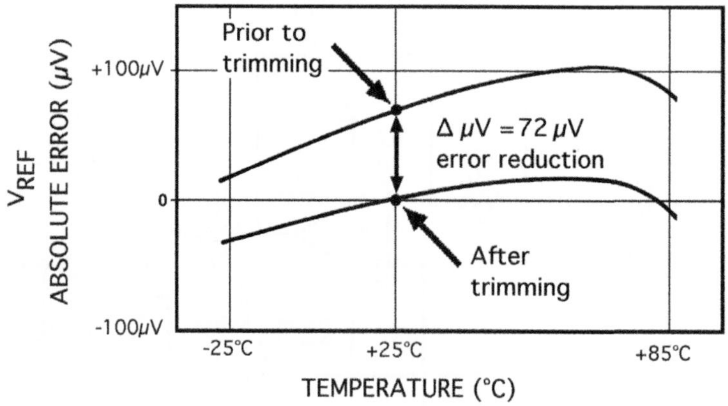

Figure 12.20. The effect of trimming on voltage reference accuracy, shown before and after error trimming.

probably have to resort to various temperature compensation techniques, which we will look at shortly.

For devices that *do* include a trim terminal, they are purposely designed for trimming the output voltage, in most cases by about ±300 mV, but without significantly altering the circuit's tempco. In fact, in such devices the tempco difference is usually less than 1 ppm/°C for a 100-mV adjustment, which is insignificant. Figure 12.21 shows a typical setup for this type of reference adjustment. Many trimmable references are designed to work with either a 10-KΩ or 100-KΩ trimmer, but if you are thinking about using a normal trimmer, remember that the typical cermet trimmer usually has a tolerance of ±10% and a tempco of more than ±100 ppm. A better, but more expensive solution is found by using either (1) a Bourns precision wirewound trimmer with a low ±5% tolerance and a ±50-ppm tempco, or (2) a Vishay metal-foil trimmer that also has a ±5% tolerance but with only a ±10-ppm tempco. You should contact these precision trimmer makers for their product information.

Such devices have very low drift and offer the best solution; in fact, there are no more precise trimmers available. (When one compares a typical 5-volt, 8-bit digital pot to the mechanical precision type, the digital pot has an initial accuracy of around ±30%, along with a tempco of more than 500 ppm/°C and an LSB of almost 20 mV. Although these figures can be reduced by various techniques, it is not easy and further complicates the task of providing a simple, effective, precise means of trimming the voltage reference.)

In using this technique, one should adjust the reference at 25°C, along with the two operating temperature extremes, and a few points in between. To further stabilize the trimmed reference voltage, it will help to provide 168 hours of burn-in (with no power applied), at a constant 125°C/150°C temperature. (You should confirm that the materi-

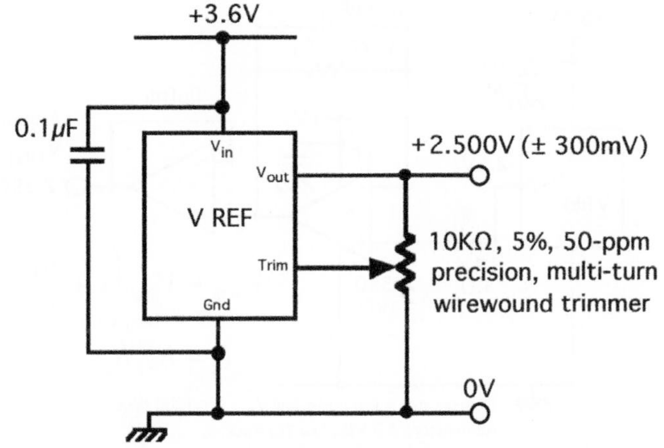

Figure 12.21. Typical series reference with dedicated trim terminal. The 10-KΩ multiturn trimmer allows adjustment of the voltage by ±300mV.

als involved, such as the housing of the trimmer, will survive the 150°C temperature.) The burn-in process tends to stabilize the various drifts and improve the long-term stability. Once this is done you should measure again, and if necessary readjust the reference at 25°C and at the two operating temperature extremes. After that, no further adjustment should be necessary.

For devices that have *no* built-in trim capability, one must use an op amp circuit to make any output adjustment or use some form of compensation circuit. One can easily scale and amplify the reference voltage to another value, with best results achieved by using a thin-film resistor network with a very low tracking TC. Two such circuits are shown in Figure 12.22. Again, the op amp(s) used here must be low-noise, low V_{OS} precision type(s). For example, a 1-ppm/°C change in this 2.5-volt reference circuit is equivalent to an op amp V_{OS} drift of 2.5 μV/°C. You should aim for that level of drift or better, although the resistors will probably be the main error contributors, depending on whether you are able to use a thin-film network (best) or discrete metal-film, foil, or wirewound types.

Figure 12.22A takes the nominal 2.5-volt reference voltage and increases this to a slightly higher voltage (2.6 volts), as required by the particular application. The formula for calculating the output voltage is shown in the diagram. This circuit uses a low single supply and can be enhanced by using a low-dropout reference together with a rail-to-rail input/output, precision op amp. Figure 12.22B shows a means of adjusting the output voltage over a span of ±3 volts in a split-supply circuit. In both circuits, an op amp is used to buffer the load. Remember that with an op amp buffer at the output, the circuit has a current limit as well as both sink or source capability—something that

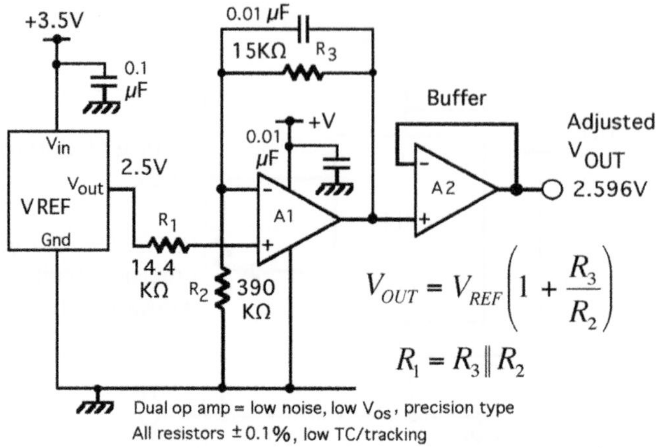

A. Adjusting the reference's output voltage.

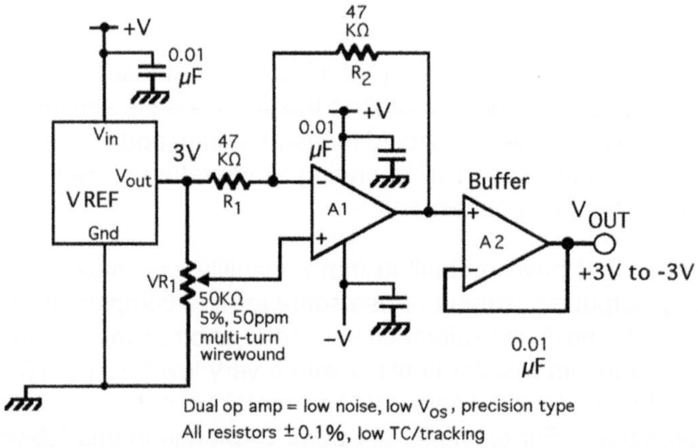

B. Adjusting the reference for a bi-polar output voltage.

Figure 12.22. Low-noise precision op amps provide adjustment of the reference output voltage.

many monolithic references do not have (they can all source, but only about half can actually sink current).

It is sometimes necessary to compensate the tempco of the reference voltage (TCV_0) over a particular temperature range. This can be done in varying ways, from the simple to the highly complex. In any case, it requires that you first graph the existing reference's output voltage over the intended operating temperature range. Using equations 12.1 through 12.3, you should then be able to determine the circuit's tempco in ppm/°C and its drift in μV/°C. Once this data has been calculated, you should review the existing reference's data sheet for any helpful information regarding temperature compensation. This may help you decide on the type of subcircuit needed for any such compensation. It may simply mean that if you were to buy a more accurate, lower tempco reference, it could help avoid this type of compensation and the time and expense involved. (Helpful hint: Do that wherever possible, unless you are a glutton for punishment.)

In each case, whether simple or complex, the circuit will require a monolithic temperature sensor that will generate a current/voltage proportional to absolute temperature (I_{PTAT} or V_{PTAT}). The object here is to create a temperature-sensing circuit that will compensate the reference output voltage with a shallow enough temperature drift, to meet the design goals of the circuit. Needless to say, the temperature sensor should be mounted close to the reference device, so they share the same ambient temperature. Such a circuit is shown in Figure 12.23A, where a popular temperature sensor (an Analog Devices AD590) is used together with some precision low V_{os} op amps to provide an adjustable bipolar voltage range, which can be used to compensate for temperature. The circuit shown has an output capable of between +30 mV/°C and −30 mV/°C, which is controlled by the precision trimmer, VR_1. Other circuits are available that can provide a lower output than this, and probably using fewer parts, but this circuit gives the general idea. The output of the reference could be connected to the inverting input of another op amp, configured as a unity gain inverting amplifier. The output of the compensation circuit could then be fed to the op amp's noninverting input, to provide a subtraction function. Such a circuit is shown in Figure 12.23B. Thus, as the reference voltage gradually increases beyond 25°C, say by 20 μV/°C, the temperature-compensating circuit decreases by exactly the same amount, and so keeps the reference output precisely balanced, with a zero temperature coefficient. This process is shown in Figure 12.23C.

In high-performance ADC systems where a wide operating temperature range is involved, even with the type of trimming just described, the design can still be extremely challenging. Table 12.5 shows just how little headroom there is, as the bit resolution increases. So we will now look at other means of compensating the reference voltage's tempco. One could consider using PTC/NTC thermistors to provide some necessary compensation, but the temperature range will be fairly narrow, and characterizing the reference, thermistors, and precision resistors could be time consuming. The object in any of these circuits is to create a temperature-sensing circuit with an output voltage that accurately compensates for the reference's temperature drift and keeps the output locked at the required voltage. The voltages involved can often be less than 1 μV, as shown in the table.

The following design is more sophisticated and employs *digital correction* to compensate for errant temperature drift, along with more analog circuitry to implement the design, shown in Figure 12.24. Here a monolithic temperature sensor is used with some low V_{os} precision op amps and the voltage reference circuit on one PC board/substrate. On a separate board/substrate, a linear regulator, a small microcontroller, an EEPROM, DACs, and buffers are combined to provide adjustable voltages/currents

Bits	2.5V full-scale range @ 1/2 LSB resolution		
	Resolution (μV)	Allowable drift (ppm/°C)	Allowable noise (μVpk-pk)
12	305	1.22	30
14	76	0.305	7.5
16	19	0.076	1.85
18	4.7	0.019	0.468
20	1.2	0.0047	0.117
NOTES: 1. Computed for 1/2-LSB resolution.			

Table 12.5 Increased Bit-accuracy squeezes allowable drift and noise levels

that are used to compensate for voltage-temperature changes. Using two separate boards keeps any heat from the load drivers or from the regulator away from the reference and its temperature sensor. The circuit depends on the temperature sensor to generate a voltage proportional to ambient temperature—where the output voltage can be in V/°C (lower resolution) or V/°K (higher resolution), as required. The temperature sensor should be mounted next to the reference device, so they share the same ambient temperature. Some older voltage references also include a temperature output pin (Temp) in the package, which may be helpful in some designs.

The temperature voltage from the sensor is converted into a digital word, which is then compared to the data previously stored in the EEPROM. This data forms a lookup table, which contains the correction data used to compensate the voltage reference. The EEPROM data is then applied to the DACs and buffers, which interface with the reference circuit and the load (via Kelvin sense/force connections). The DACs can control amplifier gain, reference trim, reference I_Q, and so on.

The lookup table is created after first plotting and storing the reference output voltages over the entire operating temperature range, and then determining the necessary compensation voltage and storing it in the ROM. Temperature tests should be done in a precisely controlled lab oven. You will first need to determine the following:

- Operating temperature range (e.g., –20°C to +70°C = 90° span)
- Temperature resolution (e.g., every 2°C, 1°C, 0.5°C, 0.25°C)
- Number of bits per correction needed (e.g., 8-bit, 12-bit, 16-bit)
- Various trims/adjustments needed and the type of interface needed (e.g., op amp buffer or discrete BJTs in Kelvin force/sense loops)

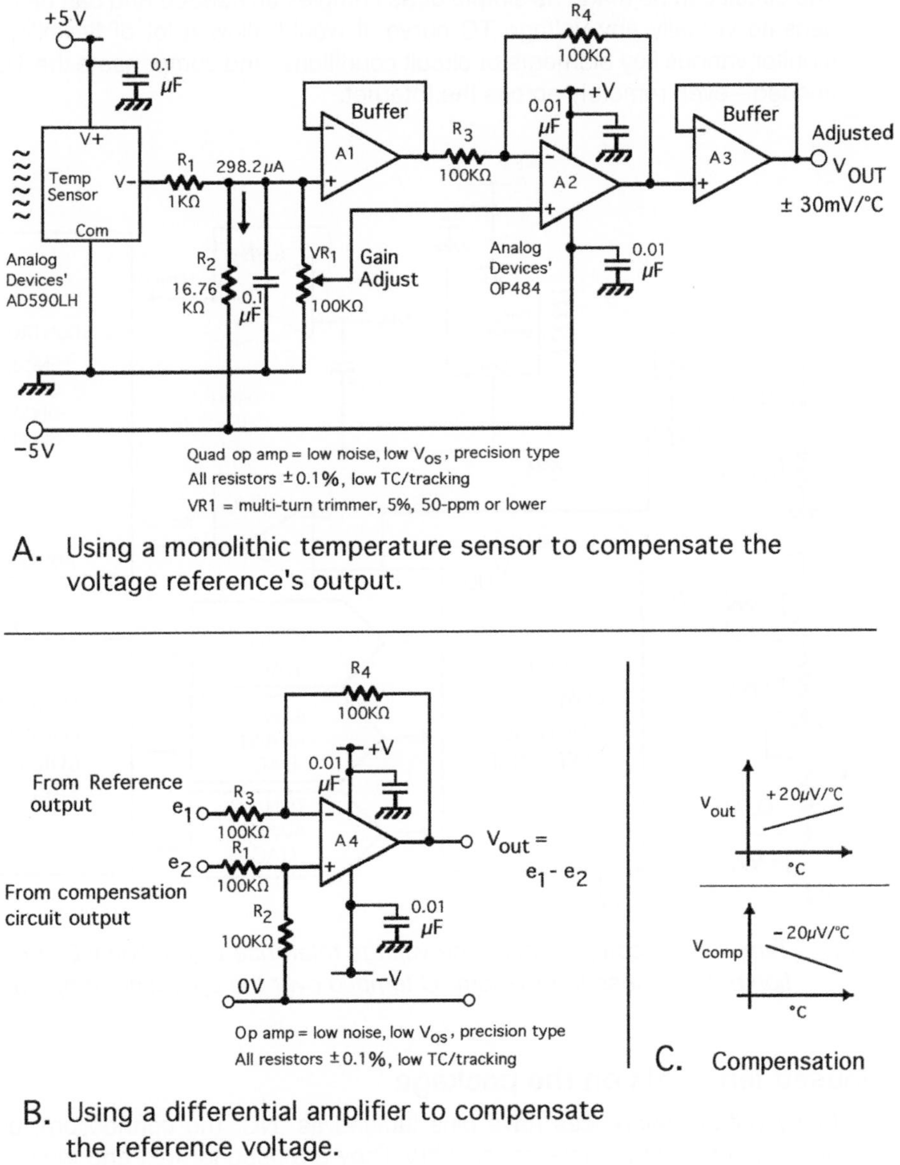

Figure 12.23. *Basic methods of compensation.*

- Reference's untrimmed TC
- Temperature endpoints and any ensuing over-range/out-of-range actions needed

The circuit can be made as simple or as complex as needed and can be made to compensate virtually any voltage TC curve. It would allow a lot of flexibility and let you monitor various key elements or circuit conditions, and compensate the TC as often as needed—even remotely across the Internet.

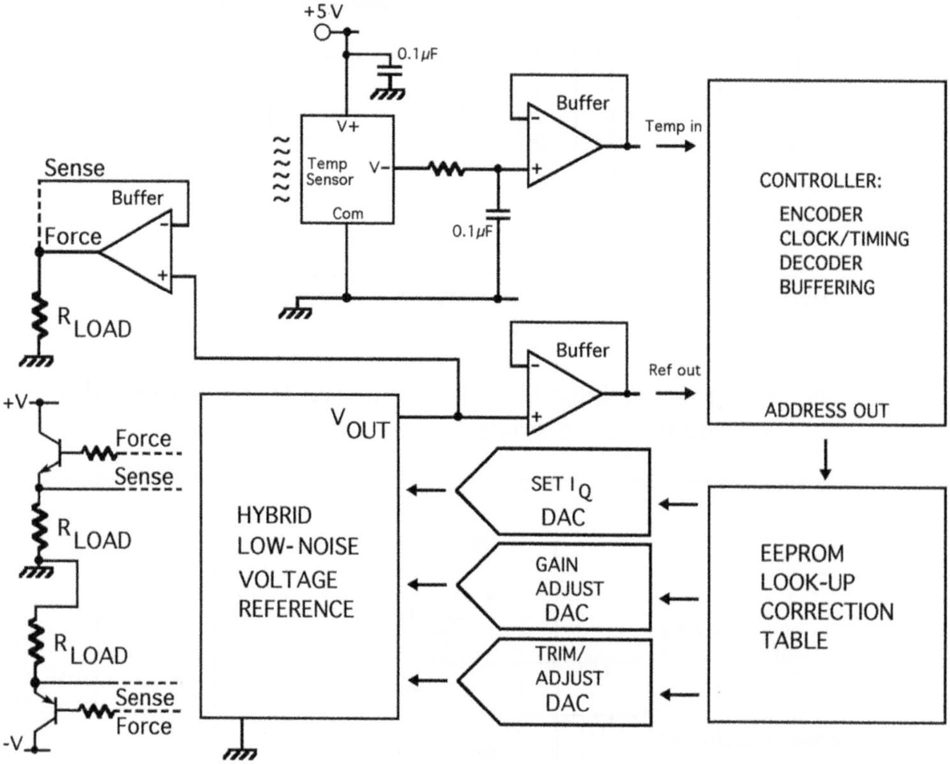

Figure 12.24. *Temperature compensating the voltage reference using digital EEPROM correction provides less than 1-ppm/°C tempco over the operating temperature range.*

12.4 Unused terminals on the package

Many voltage references have pins labeled as "NC" (no connection), but some are actually connected to internal circuitry. They are used for test and trimming purposes at the factory. With some products they should be left unconnected (i.e., they float), whereas others may need to be grounded. In either case you should always consult the data sheet first, or if there is any doubt, contact a factory applications engineer to establish how to deal with this issue before applying power or creating a PC board layout. In some cases it may be necessary to shield these pins from noise pickup or from tiny leakage currents from nearby traces on the circuit board, by using guard rings around them.

12.5 Package types

Today's voltage references range from the simple, low-accuracy zener diode to the latest sophisticated, expensive monolithic IC. There are thousands of different zener diode parts and certainly hundreds of different voltage reference ICs to choose from. Each has its place, and what is good for one design may not be for another. Voltage references are available in many different packages, many of them tiny surface-mount types (see Figure 12.10). The package type not only defines how small the overall circuit can be made, but also governs the power dissipation of the device, as well as the ease of field adjustment and use. Over the past two decades, manufacturers have made voltage references in many different package options, mostly in metal cans, DIL, LCC, and flat packs. Today, however, the market trend is for smaller, surface-mount packages, which has resulted in fewer overall options being available from the manufacturer. An example of a superb voltage reference is Analog Devices' AD588. This is a precision reference suitable for 12-bit tasks, with a programmable output that covers +10-volt, +5-volt, ±5-volt tracking, –5-volt, and –10-volt. It includes uncommitted Kelvin force and sense terminals, a ground sense, along with noise reduction, gain, and balance adjustments, and a resistive divider. The device is available in a 16-pin cerdip, 16-pin side-brazed ceramic DIL, and a 20-pin LCC (surface-mount) package. To show you how much of the monolithic circuit is brought out for the user to access, the 16-pin device has no NC pins, meaning all 16 pins are used. Compare that with one of today's tiny three-pin SMD packages. One type of package is cheaper and easier for the manufacturer to make (the SMD), but harder for the customer to use (microscopic), with far fewer options/functions brought out. The other type of package (DIL) is more costly for the manufacturer to make (it's still expensive for the customer to buy), but much easier and more flexible for the customer to use, particularly in the development phase. *Manufacturers, please bring back the feature-rich DIL package!*

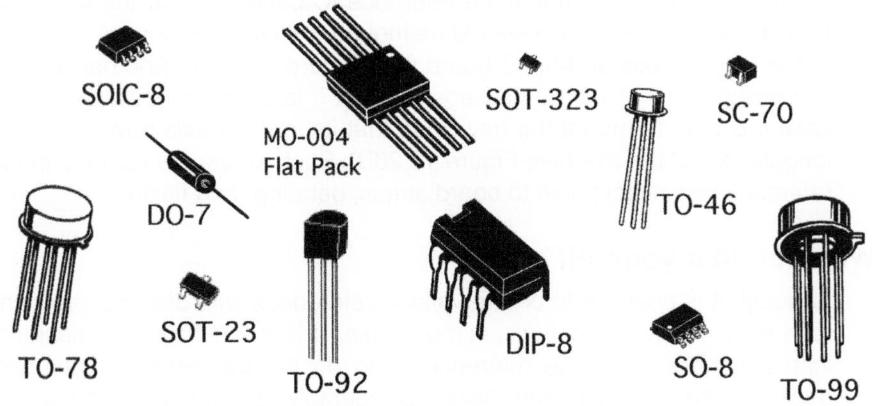

Figure 12.25. Examples of voltage reference packages.

12.6 PCB layout

Poorly printed circuit layout can adversely affect a precision voltage reference by shifting its output, increasing its noise level, or even altering its thermal characteristics. As with any high-precision circuits, one must use very-high-performance devices and components, with well-controlled specs that include an ultra-low-voltage drift (ppm/°C). Thermal drift effects in any type of low-level circuit are always a concern, so the circuit may need to be enclosed in a shielded (but ventilated) box. Using a thermostatically controlled heater/oven in a small enclosed environment may help stabilize and control the ambient temperature around the reference circuit more precisely. A similar technique on a smaller scale has been used for several decades in various older buried-zener voltage reference ICs. In those designs, the device includes an integrated heater, which helps equalize the chip temperature. In the case of National's LM199, this temperature is held at about 90°C.

Having a noisy supply rail too close to the voltage reference output is always asking for trouble, because it could form a high-impedance leakage path, which would seriously degrade the reference voltage. Similar leakage problems can occur unless the PC board is clean and flux-free before applying power. Additionally, noise from that adjacent supply rail can couple over to the reference output and degrade the whole system. A poor ground line is another potential problem. To minimize ground problems, everything connected to the reference should be connected directly to the reference IC's ground pin. If a high current load is involved, it should be returned directly to the reference ground by a separate cable. Another area for potential problems is PC board stress, which can sometimes cause the reference to drift or its noise level to increase. This applies mostly to plastic DIP and surface-mount devices (metal-can types inherently have a more rigid construction), but can be remedied by several means. One is to use a thicker PC board material, which is harder to bend or flex. Another solution is to mount the reference IC package near the board's edges, away from its center. The device should be mounted so that its longest axis is perpendicular to the longest axis of the PC board (see Figure 12.26A). Another remedy is to mill a slot around part of the IC package, so that it is essentially sitting on a tab, in which case it is best to mount the device so that its longest axis now runs parallel with the longest axis of the slot (see Figure 12.26B). Such measures seem to make the voltage reference less susceptible to board stress, bending, and flexing.

12.7 Why not do it yourself?

Although it is possible to create voltage references with discrete components oneself for 4-bit to 10-bit applications, in most cases it is much faster and easier to buy a dedicated monolithic voltage reference IC to do the job. Such devices are specifically designed for the task and have considerably greater accuracy and repeatability because of the use of thin-film resistors, built-in temperature-compensation circuitry, and from being burned-in and precisely laser-trimmed at the factory. Manufacturers carefully measure and characterize their products, so that the product data sheet reflects accurate information. If one tried to build a voltage reference oneself, it would

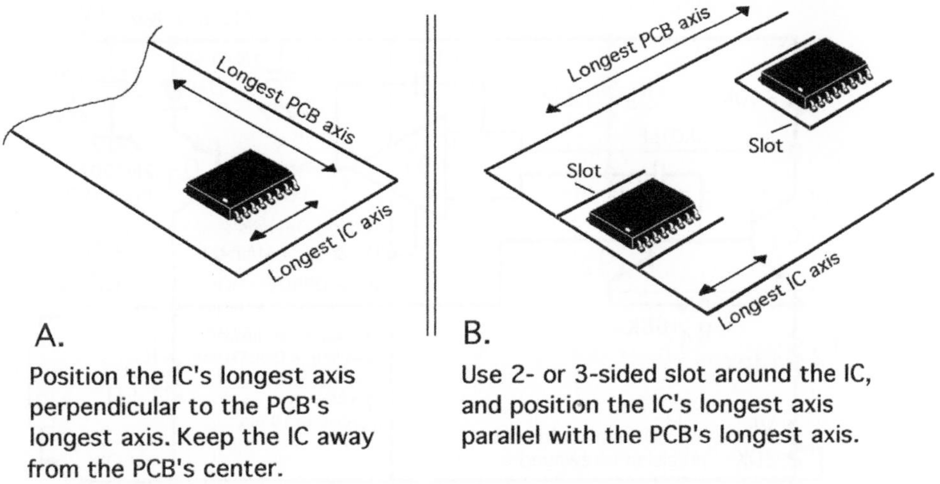
Longest PCB axis
Longest IC axis
A.
Position the IC's longest axis perpendicular to the PCB's longest axis. Keep the IC away from the PCB's center.
Longest PCB axis
Slot
Slot
Longest IC axis
B.
Use 2- or 3-sided slot around the IC, and position the IC's longest axis parallel with the PCB's longest axis.

Figure 12.26. PCB layout can be critical in some high precision reference circuits.

take plenty of time, effort, and cost (factoring in all of the specialized equipment needed to characterize it, including a lab oven) to even come close to matching the same level of accuracy and the level of characterization provided by the manufacturer.

In the event that your voltage reference had to be designed to meet special needs (such as high-voltage, ultra-low-power consumption or some special reference voltage not otherwise available), it would probably be based on the *bandgap* design. (This and other monolithic types are reviewed in greater detail in Chapter 14.) In which case it would consist of a matched-pair of either BJTs, JFETs, or MOS FETs, together with a prime-grade op amp, a few precision resistors, and possibly some additional BJTs or FETs. Such a design for creating a precise 1.23-volt reference voltage is shown in Figure 12.27. This circuit was originally designed by engineers at Precision Monolithics (now a part of Analog Devices Inc.). Unlike most cookbook discrete bandgap circuits, this circuit works well and provides excellent stability and regulation, at a low quiescent current. The 1.23 volts output (the core bandgap voltage) is stable, with a tempco of typically 20 ppm/°C over the temperature range of 0°C to +70°C. Load and line regulation are also good, at approximately 0.001%/mA and 0.01%/V, respectively.

This design is based on using two venerable Analog Devices' parts: a MAT01, an ultra-precision BJT matched-pair, along with an OP90E, a low-voltage, micropower, precision op amp. (See Chapter 5 for more on precision matched-pairs, and Chapter 11 on precision op amps.) The OP90E is a good device, although an OP97E is an even higher-precision product. The circuit runs from a regulated power supply range of between 2.75 to 36 volts, and typically takes less than 30 μA at 30 volts. This supply range is governed by the op amp's maximum voltage rating of 36 volts, not by the matched-pair, which can withstand a higher voltage (BV_{CEO} equals 45 volts max), or

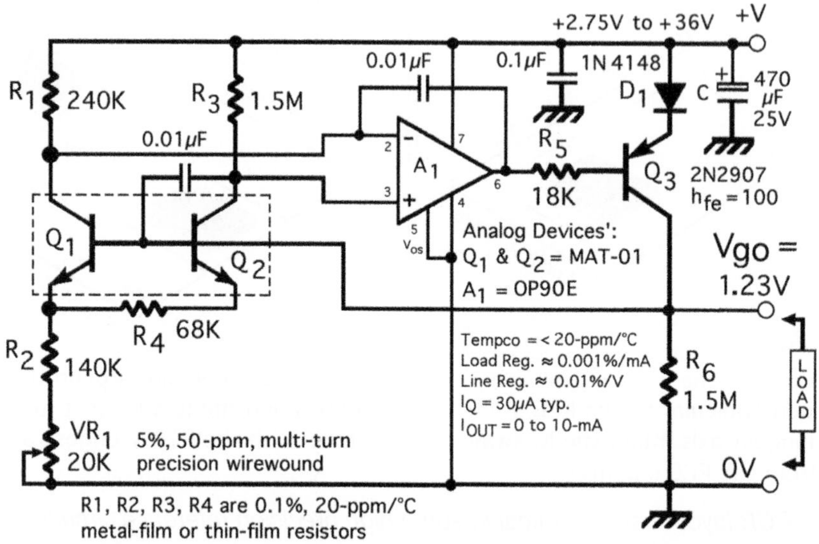

Figure 12.27. A quality bandgap voltage reference built with discrete, precision parts. It provides very good temperature stability, a low quiescent current, and a wide supply range. It is no match for a monolithic reference, like Analog Devices' XFET™ though.

the output transistor Q_3 (BV_{CEO} equals 60 volts min). The circuit permits reliable start-up by grounding one of the op amp's offset-null terminals (pin 5 on the OP90EZ). The other offset-null terminal should be left floating.

The 1.23-volt bandgap-based output voltage appears at the bases of Q_1 and Q_2, and across the output resistor R_6. This 1.23-volt should be precisely adjusted by multiturn trimmer VR_1, at 25°C room temperature to calibrate the circuit. Transistor Q_3 is a general-purpose PNP transistor 2N/PN/MMBT/PZT type, not matched. It provides a nominal 5-mA current to the load, and is in turn regulated by the op amp, via base resistor R_5. This may be easily increased up to 20 mA, if desired, because the 2N2907 has a typical h_{fe} of 100. In the event that the output current is more than 5 mA, this transistor should be located away from the matched-pair, to reduce potential thermal effects, which can degrade the circuit's low-drift performance. The OP90's quiescent current is very low (at less than 30 μA maximum at 30 volts), which supports battery-powered applications. Resistors R_1 to R_4 should all be precision thin-film or metal-film types. The trimming potentiometer effectively allows exact adjustment of the 1.23-volt output voltage, and is shown as a 20-KΩ multiturn trimmer. Remember that a typical cermet trimmer will usually have a tolerance of ±10% and a tempco of more than ±100 ppm, which will virtually negate the benefits of using the precision resistors R_2 and R_4 (i.e., 0.1%, with a 25 ppm/°C tempco, or lower). The combination of these with the trimmer may well equate to a tolerance of ±10%, with an overall tempco of ±150 ppm.

A better but more expensive solution can be had by using either a precision wire-wound trimmer with a low ±5% tolerance (Bourns) and a tempco of ±50-ppm, or by using a metal-foil trimmer (Vishay), also with a ±5% tolerance, but with an even lower tempco of ±10 ppm. The combined tolerances and tempcos of the fixed resistors and the trimmer would have to be measured over the operating temperature range, in order to determine the actual overall stability and drift characteristics of the circuit. Approximate specifications are shown in the circuit (see Figure 12.27), which may be acceptable for your design. Substituting a fixed-value precision resistor instead of the trimmer will definitely improve the circuit's overall tempco and long-term drift. (Substituting for a higher-precision op amp, such as an Analog Devices' OP97, would likely help even further.) This substitute fixed-value resistor would have similar specs (with a 0.1% tolerance, 25 ppm/°C tempco) to the 140-KΩ resistor, R_2. One would probably have to juggle the values of these two resistors to set the 1.23-volt reference precisely.

Although this is a reliable circuit, it may still not be as good as a completely monolithic part. When one compares it to one of today's top-of-the-line Analog Devices new XFET® precision references, you will see the difference immediately (see Table 12.6). The discrete design takes up far more PC board area than either of the small monolithic devices available as XFETs (MSOP-8, SOIC-8). Power consumption is about the same for either circuit, but the discrete design has a wider supply voltage range (up to 36 volts). In every other aspect the XFET® is far better, including low noise characteristics (60 nV/√Hz at 1 KHz and 3.5 μV pk-pk). It also has a wider operating temperature range (automotive: –40°C to +125°C vs. commercial: 0° to 70°C), which is very important. Not only that, but the XFET® is totally repeatable and usable right away, whereas the discrete design has to be built, and then each circuit individually adjusted and tested.

In summary, unless the application is high-voltage or has some unusual requirements, buying a dedicated monolithic voltage reference IC (such as the XFET®) is the only practical solution.

12.8 Comparing precision

As mentioned previously, today's voltage references provide greater precision, better repeatability, improved temperature stability, increased long-term stability, and lower long-term drift. In terms of precision, a typical 500-mW zener diode in a DO-35 glass package will have an output voltage tolerance of ±5% and a temperature coefficient of –0.02%/°C, while drawing about 7 mA from the power supply. The zener is often sensitive to changes in temperature and current, though, as well as being a relatively noisy device. In addition, its specifications can significantly drift over time. A close alternative is to use a special, temperature-compensated zener reference device, which will cost 10 times as much. It will likely provide a nominal 6.2-volt rating (±5%) and have a maximum voltage deviation (ΔV_Z) from its actual value of approximately 10 mV. The device will probably have a tempco of 0.001%/°C on its output voltage, while also drawing approximately 7 mA from the power supply.

Comparing the size of the 2-inch square through-hole discrete design vs. the small smd packages of the monolithic XFET® - there's no contest !

SOIC-8

MSOP-8

Feature:		Dis-crete Band-gap circ.	ADI X-FET®
Reference voltage (min)	Volts	1.23	>2.048
initial error %	%	0.5	0.15
Output voltage tempco	ppm/°C	>20	3
Long term stability (Note 1)	ppm/1K	>200	40 typ
Suitable for n-bit systems	bits	<10-bit	16-bit
Supply voltage range	Volts	3-36	4.1-18
Operating temp. range	°C	0°-70°	-40°/+125

NOTES: 1. Measured over 1000-hours @ 25°C.
2. All semiconductors from Analog Devices Inc.

Table 12.6 Performance comparisons of a discrete bandgap design vs. a monolithic XFET® reference.

By contrast, a typical low-cost, bandgap voltage reference IC can produce a stable output, of say 2.5 volts DC at 5 mA, with an initial accuracy of ±5 mV (of ±0.2%) maximum, and a quiescent current of less than 1 mA. It will probably have a low-frequency noise voltage of less than 25 μV peak-to-peak, along with a temperature coefficient of around 20 ppm/°C, over a –40°C to +85°C temperature span. A higher-performance device also with a 2.5-volt at 5-mA output may typically have a lower initial accuracy of ±2 mV (of ±0.08%) maximum, while drawing less than 100 μA for itself. It will also have a low-frequency noise voltage of less than 25 μV peak-to-peak, but with a lower temperature coefficient of around 5 ppm/°C, over the industrial temperature range. It may well have a low-dropout voltage of about 1 volt or lower, and possibly a low-power sleep/enable pin, a feature that is useful in battery-powered applications.

A very-high-performance 2.5-volt bandgap, built with a bipolar-CMOS process, might offer an even lower initial accuracy of ±1 mV (equals ±0.04%) maximum. It will probably have a low-frequency noise voltage of less than 20 μV peak-to-peak, along with a temperature coefficient of around 5 ppm/°C, over a –40°C to +85°C temperature span. It will likely have a long-term stability of less than 60 ppm/1000 hours, and be available in a tiny surface-mount package. Its specifications are generally suitable for 12- to 14-bit converter systems. Other higher-performance voltage references, as well as devices from different topologies, will probably have better specs than these.

In implementing a voltage reference in a design, one should always consider what level of *accuracy* and *stability* is going to be required, over what *temperature span*, and for what value of *load*. This will determine the *type* and *grade* of voltage reference used, relative to the level of accuracy needed (referenced in digital bits), over the particular temperature range. For example, in order to keep any conversion error to 1/2-LSB in a 16-bit A/D system, over a temperature span of say 100°C (–20°C to +80°C), a voltage reference will need to have a maximum allowable tempco of 0.076 ppm/°C. By contrast, in a 10-bit system, over the same temperature range, but with a 1-LSB conversion error, the voltage reference will need to have a tempco of approximately 8 ppm/°C. At the top end of the voltage reference IC spectrum, devices are laser-trimmed with great precision, burned-in, and tested over the full military temperature range. They have ultra-low drifts and cost a lot more. At the other end, devices receive less attention and have more relaxed specifications, over a much reduced temperature range. They tend to drift more and cost less. Your choice of exactly which product to use may translate into requiring a technician to precisely trim the voltage reference's output, rather than buying a more expensive device for a few dollars more, which eliminates that step.

Most ADCs/DACs containing their own on-board voltage reference (usually a bandgap type) can typically only handle up to 12-bit resolution, even though the converter may have higher resolution. This means that one has to use a high-performance external voltage reference in order to achieve 12-bit and higher resolution. In fact, for 1/2 LSB error in a 12-bit resolution system, the maximum allowable drift is 1.22 ppm/°C, over a 100°C temperature span. Plenty of products are available that will be more than suitable in meeting that specification. In terms of precision, repeatability, temperature stability, long-term stability, and drift, the voltage reference IC offers it all.

Over the past decade, the subject of voltage references has continued to be a much-publicized topic. Mostly what has fueled the continued R&D, growth, and improvement of the precision monolithic voltage reference has been the constant need for supporting higher-resolution A/D and D/A converters. Unlike other parts of the analog semiconductor market (e.g., linear regulators, op amps, comparators), the manufacturers of voltage references can mostly all claim significant contributions to the advances made in this particular field. In the United States, the following semiconductor manufacturers make silicon monolithic voltage references:

- Advanced Linear Devices
- Analog Devices
- Fairchild Semiconductor
- Intersil/Xicor
- Linear Technology
- Maxim
- Microchip Technology
- Microsemi
- National Semiconductor

- ON Semiconductor
- Philips Semiconductors
- Texas Instruments
- Thaler Corp.

It's a tribute to those researchers and innovative engineers from the 1970s and '80s that many of the voltage reference products they created and modeled are still available today—more than 20 years later—and are still finding new applications in today's digital world.

Chapter 13

The zener Diode and the TC zener Reference

13.1 Introduction

The zener diode was the first discrete semiconductor device to be used as a basic voltage reference. It was created in the late 1950s by its inventor Clarence Zener, a Westinghouse researcher. Ever since it became commercially available in the early 1960s, the zener diode has been a popular workhorse in countless industrial and commercial designs worldwide, because of its simplicity, reasonable accuracy, small size, and low cost. Besides providing a simple means of voltage reference, the zener diode is also used for biasing, stabilization, switching, and clamping. Before the arrival of the integrated circuit reference voltage, the zener diode was the only affordable means for designers to create an on-board voltage reference. During the 1960s, engineers began creating small linear regulators and references by combining zeners with BJTs. They also discovered that by characterizing and selecting zeners, and adding one or more series rectifier diodes, they could create temperature-compensated zener references. By the late 1960s, the small glass package zener had been scaled up to provide a power zener, often available in a metal stud package. (In some industrial applications where higher voltages and currents are involved, power zeners are still the only option.) When the silicon bipolar IC became available in the early 1970s, designers began to use the op amp to buffer and amplify the temperature-compensated zener diode, to provide a more stable reference. Figure 13.1 shows the complete zener diode family tree.

After the op amp was first introduced in the mid-60s, the zener was quickly integrated into its design for start-up and biasing purposes and is still used in many of today's analog ICs (see Figure 13.2A and B). As mentioned in the previous chapter, when the first generation of monolithic bandgap references was introduced, they were marketed and referred to as though they were zener diodes (the latter were the established standard at the time). They were even depicted in a circuit diagram with the zener diode's symbol. In fact, it is still common to diagram the two-terminal bandgap voltage reference IC using the zener diode's symbol.

The silicon zener diode followed the introduction of the silicon transistor in the early 1960s. Some of the earliest manufacturers included Texas Instruments, Fairchild Semiconductor, Transitron, International Rectifier, Diodes Inc., National Semiconductor, and Philips Semiconductors. The zener diode also became the basis of the design in the first commercial buried-zener voltage reference (see Figure 13.2F). It was introduced in the early 1970s by National Semiconductor, which at the time also

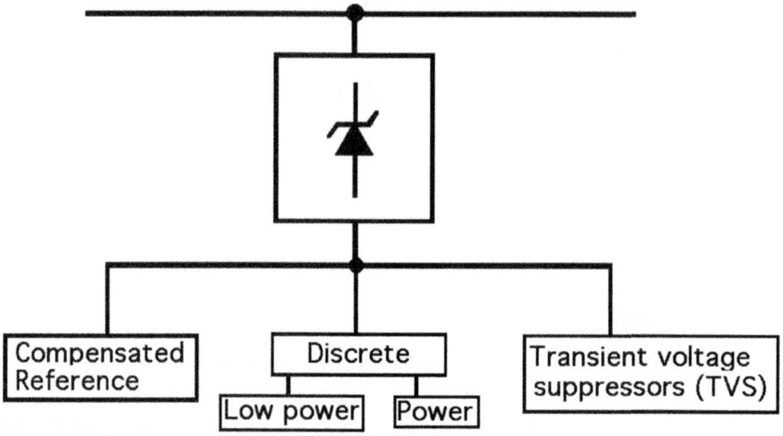

Figure 13.1. *The zener diode family tree.*

made discrete zeners. Because of its well-understood characteristics, the zener was utilized by legendary IC designer George Erdi at Precision Monolithics Inc. at about the same time.

Erdi pioneered a technique for adjusting a precision op amp's input offset voltage (V_{OS}) during manufacture to a very low level. A string of parallel-connected resistors and zeners are trimmed by a computer-controlled laser, which effectively shorts the zener to obtain a precise voltage level (typically less than 50 μV). The technique became known as *zener zapping* (see Figure 13.2D) and has been used industry-wide for many years, to adjust the offsets of op amps and other analog products. The zener has subsequently been integrated into the architectures of some digital logic as well as power MOSFETs, for ESD and gate protection (see Figure 13.2C). In the 1980s, the power zener brought about the silicon transient voltage suppressor (TVS), which is used in today's AC surge protectors and to protect computer modems and communications data lines (see Figure 13.2E).

(*Note*: Although beyond the scope of this book, it should be noted that silicon TVS devices are based on the zener diode structure and operation. The TVS has characteristics that make it ideal for clamping high-energy surges. It is enhanced and carefully crafted to provide both a large surge current capability and peak power dissipation by using larger chips and junctions than a normal zener diode. The regular zener is seldom specified with any surge current value. Although many manufacturers of zener diodes also make TVS, some companies no longer make them in preference for making only TVS devices. One such company is Semtech Inc., and another is Teccor Inc., both of which have developed a wide range of TVS products.)

All in all, the simple zener diode has an impressive résumé. No doubt the zener and its derivatives will continue to find new applications in future products too.

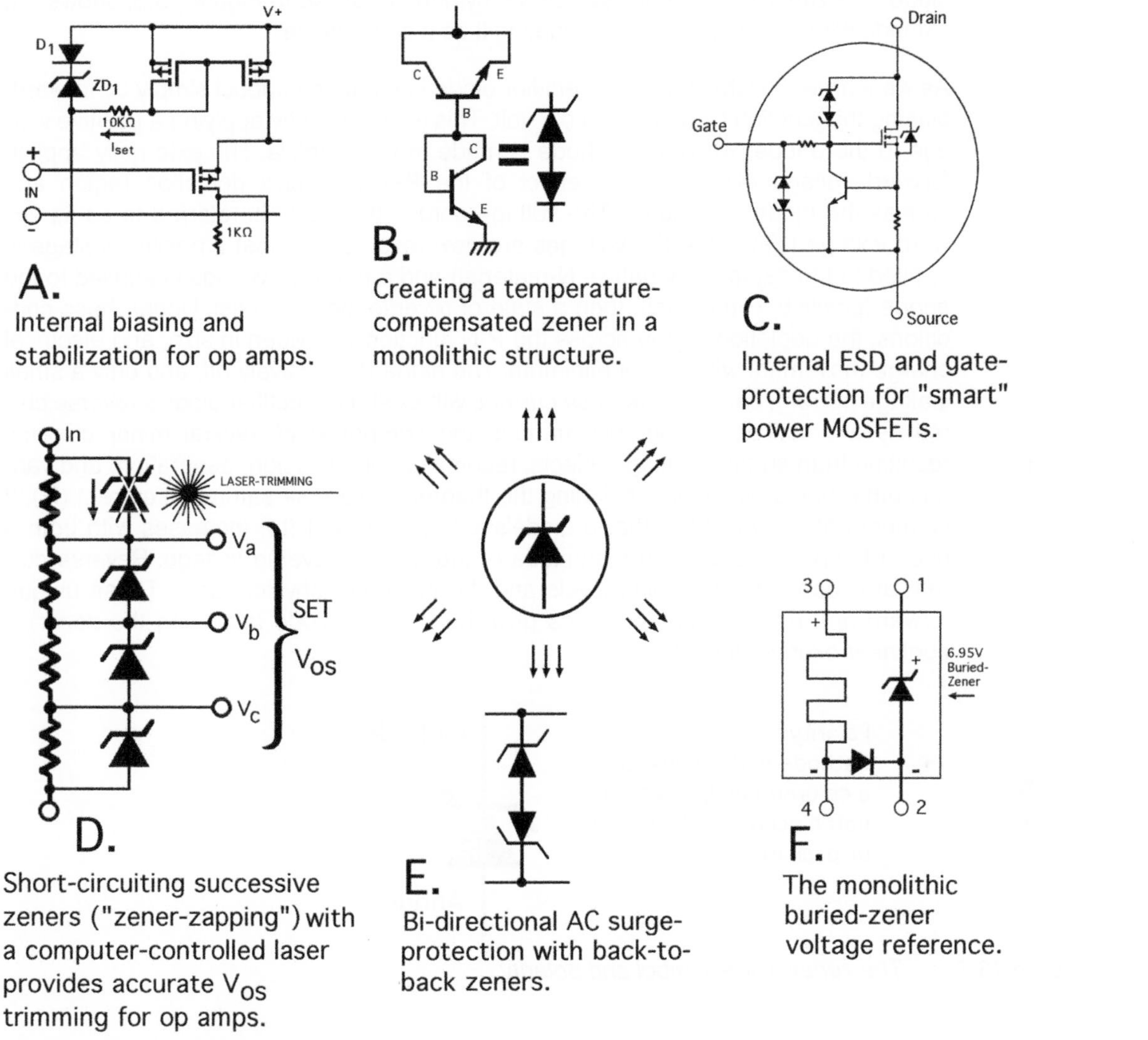
V+
D_1
ZD_1
10KΩ
I_{set}
IN
1KΩ
A.
Internal biasing and stabilization for op amps.
C
E
B
C
B
E
B.
Creating a temperature-compensated zener in a monolithic structure.
Drain
Gate
Source
C.
Internal ESD and gate-protection for "smart" power MOSFETs.
In
LASER-TRIMMING
V_a
V_b
SET
V_{OS}
V_c
D.
Short-circuiting successive zeners ("zener-zapping") with a computer-controlled laser provides accurate V_{OS} trimming for op amps.
E.
Bi-directional AC surge-protection with back-to-back zeners.
3
1
+
+
6.95V Buried-Zener
-
-
4
2
F.
The monolithic buried-zener voltage reference.

Figure 13.2. Common zener applications and spin-offs.

13.2 Characteristics of the zener diode

The zener diode is a two-terminal, single-junction semiconductor device that passes meaningful electrical current in only one direction and simultaneously provides a stable reference voltage between its terminals. It is usually operated so that it is in parallel (shunt) with the load, as with a two-terminal bandgap reference. It is unique among other discrete semiconductors, because its electrical properties are based on operating a diode's P-N junction in the reverse breakdown region. Compared to a rectifier diode, the zener operates in reverse. Its symbol, depicted in Figure 13.3, shows the cathode with a Z-shaped line designating it as a zener diode.

As we learned in Chapter 3, the rectifier diode is made to conduct simply by forward-biasing the junction by more than 0.7 volt. This is achieved by applying a positive voltage to the anode, while the cathode is made more negative. The externally applied forward voltage eliminates the effect of the P-N junction's depletion region and causes the diode to conduct. The voltage across the diode is referred to as its forward voltage (V_F). If the DC voltages are now reversed, so that a positive voltage is applied to the cathode (negative, N-material) and a negative voltage is applied to the anode (positive, P-material), then a state of reverse bias will exist. Under these conditions, the depletion region across the P-N junction will widen in size, and electrical charge movement will be at a minimum. The diode is effectively off, and only a small leakage current, called the *reverse current*, will exist. The rectifier diode's reverse current is a practical, measurable amount and composed of several minor currents resulting from surface leakage effects, recombination, diffusion, generation, and various other nonideal factors, including the theoretical reverse saturation current (I_S). It is important to remember that a diode's leakage current (I_R) increases with both a rise in temperature *and* the magnitude of the applied reverse voltage. Reverse current also depends on doping levels and the junction's physical area. These unique forward and reverse characteristics give the junction diode (including the zener) a nonlinear exponential V-I curve.

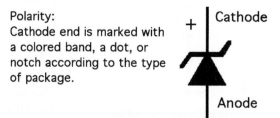

Polarity:
Cathode end is marked with a colored band, a dot, or notch according to the type of package.

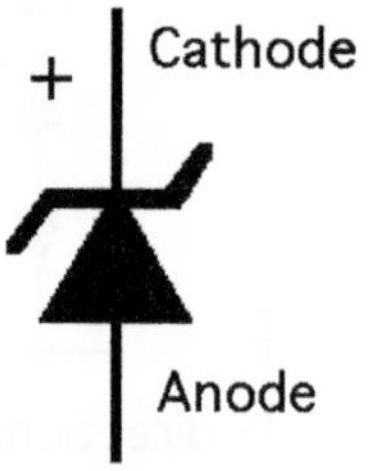

Figure 13.3. The zener diode symbol and polarity.

If the reverse voltage continues to increase, it will reach a point called the *reverse breakdown voltage point* (V_{BR}), shown in Figure 13.4, where the current across the junction increases rapidly. This point, where the reverse current changes from a very low value to heavy conductance, is known as the *knee* and is well characterized for both rectifier and zener P-N junctions. At this point, minority carriers in the semiconductor material gain enough energy to collide with valence electrons and dislodge them from the crystal lattice, causing an avalanche of carriers, and thus an *avalanche current.* A current-limiting resistor is always necessary to prevent the zener diode from being destroyed, because of this excess current flow (which would cause massive self-heating of the chip, before catastrophic failure).

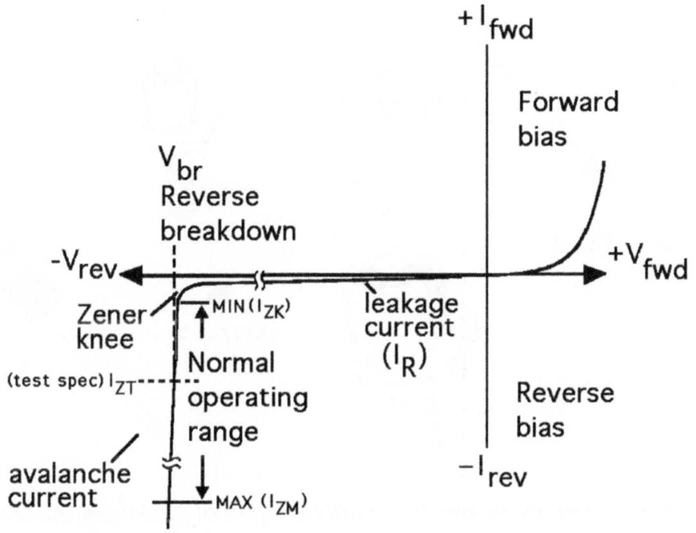

Figure 13.4. A zener diode's typical V-I curve.

When reverse voltages greater than V_{BR} are applied, the voltage drop across the junction remains constant at the value of the breakdown voltage, over a wide range of currents. This characteristic enables the zener diode to be a good voltage regulator and reference. With a zener diode, this breakdown voltage is made to occur at different voltages than with a normal rectifier diode, and in fact two different breakdown mechanisms are involved. At very low voltages (less than 5 volts), the means by which this works is known as the *zener effect*, while beyond that (more than 8 volts), the *avalanche effect* is responsible. Between about 5 and 8 volts, both mechanisms are involved.

In all cases, a *critical electric field* in the junction region is required. This is the mechanism by which zener, avalanche diodes, and silicon transient voltage suppressors all function. Such diodes are specially doped with different resistive materials than rectifier diodes and in different ratios from one another (to obtain different breakdown voltages). They are also processed somewhat differently too. The complete range of voltages available for discrete zener diodes runs from about 1.8 volts to more than 300 volts, with normal tolerances of ±2%, ±5%, and ±10%. Lower tolerances are available by special selection for volume applications. Many axial-leaded zeners (DO-35 and DO-41) have wattage ratings of between 250 mW and 5 Watts, and some are available as small surface-mount types such as SOT-23 and SOT-323. Power devices are available in metal stud-mounted packages capable of more than 50 Watts. Power zeners (and transient voltage suppressors too) can be considered to be bigger versions of the more usual type, but with larger chips, and therefore larger junctions, but in different packages. Some different wattage and package types are shown in Figure 13.5.

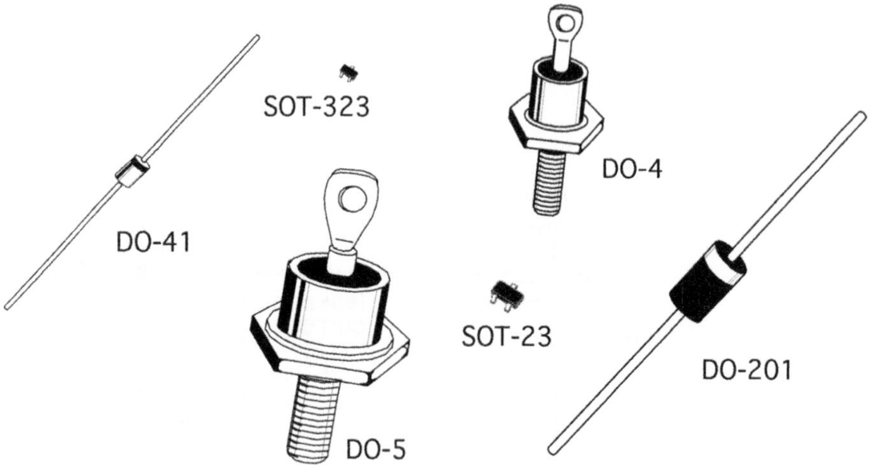

Figure 13.5. *zener diodes are available in various types of packages, as shown here, and include glass, metal, and plastic types.*

As mentioned, discrete zener diodes can be made by various processing techniques. These include mesa etching, or more commonly with the *planar* technique. The necessary dopant is added to the wafer's surface using any one of several different techniques, including ion-implantation and chemical vapor deposition. An important factor in creating a stable zener diode is the *passivation* of its surfaces, using either silicon dioxide or a glass derivative (both insulators). This forces surface contaminants away from the surface and helps create a more stable, reliable, and quality product, with a lower reverse leakage current. Each zener diode's particular specifications are mostly created by differences in processing. Its mechanical construction has the monolithic zener chip welded securely between both internal ends of the leads, sandwich fashion. The axial device can then be encapsulated using either a glass envelope or molded plastic housing. High-power zeners are built in metal packages, using similar techniques as power transistors. Discrete zeners are available in different categories that include general-purpose, low-noise, high-performance, and low-impedance types. Many of these are also industry standard types and are available from multiple vendors.

Although the zener diode appears at first glance to be a relatively simple device, it is not, particularly when you factor in such characteristics as zener voltage, impedance, capacitance, temperature coefficient, and thermal effects. When one looks at any zener diode's specifications, one often encounters some unfamiliar terminology and abbreviations, which we will clarify here. Figure 13.4 shows a few of these abbreviated terms for your reference.

The *nominal zener voltage* is usually designated as V_Z at a particular reference current. The nominal voltage is typically measured with the diode's junction in thermal equilibrium and at a closely controlled +25°C ambient temperature. Measurement is usually made along the device's leads at 3/8 inch (9.52 mm) from the diode's body. For example, a 1N963B is a nominal 12-volt device in a small DO-35 glass package and registered as an industry standard JEDEC "1N" part number. The part has a nominal 12-volt rating at a reference test current of 10.5 mA. The device has a tolerance on this nominal voltage, expressed as a ±%. For example, this nominal 12-volt zener has a ±5% tolerance, designated by the "B" at the end of the part number. In the case of the 1N963B, it means that the nominal 12-volt device can be ±0.6 volt and still be within specifications. A 1N963C has a ±2% tolerance, while a 1N963D has an even better tolerance of ±1%. Be aware that zener voltage is partly temperature dependent.

The *forward voltage* (V_{fwd}) of a typical zener diode is typically less than 1 volt and behaves almost identically to that of a forward-biased rectifier diode. Once the forward voltage exceeds about 0.60 volt, a forward current flows. Increasing the forward bias to more than 0.85 volt boosts the forward current by 10 or 20 times and should be externally current-limited. The zener diode is not usually operated in this manner, because it is cheaper to use a simple rectifier.

Any zener diode's *current range* runs from a minimum at I_{ZK} (the current at the bottom of the knee of the V-I curve) to the maximum allowable DC current for that type of zener at I_{ZM} (or I_{Zmax}). The current at the top of the knee tends to be small and somewhat unstable and has an appreciable noise level associated with it. As the reverse current increases, the noise level tends to decrease significantly. This is why manufacturers specify I_{ZK} at the bottom of the knee in the avalanche area, instead of at the top, because it is more stable. Each zener has an I_{ZM} rating, operating beyond which the device may be damaged or destroyed. The term "V_Z @ I_{ZT}" simply means the zener's nominal voltage at a particular test current, designated as I_{ZT} (the "T" specifies test). This test current is usually between 25% and 35% of the maximum current. The maximum *reverse leakage current* (I_{Rmax}) is usually measured in μA and specified at a particular reverse voltage (V_R). This reverse voltage is often specified at around 80% of the zener voltage, to ensure a low leakage current despite its proximity to the knee region. Remember that the maximum value of leakage current is specified, while the actual value may be an order of magnitude or more lower. Typically, low-voltage zeners have a higher leakage current than high-voltage types. For example, an industry standard 1N5914B (a 3.6-volt part) has a maximum reverse leakage current of 75 μA at a reverse voltage of 1 volt. By contrast, a 1N5923B (an industry standard 8.2-volt part) has maximum reverse leakage current of 5 μA at a reverse voltage of 6.5 volts. A 47-volt part such as 1N5941B has maximum reverse leakage current of 1 μA at a reverse voltage of 35.8 volts. All three are from the same series and packaged in the common DO-41 axial package.

Most data sheets refer to the zener's resistance or *impedance* (Z_Z), which is measured in ohms and indicates the slope of the zener's V-I curve. Using measured current and voltage values allows the circuit board designer to calculate the approximate zener resistance/impedance anywhere along the reverse V-I curve, between I_{ZK} and I_{ZM}. Typically, the lower the zener's impedance, the better the regulation will be. It is usually measured at the factory during a computer-controlled test procedure, using either a 60-Hz or 1-KHz AC signal, and the result must be within predefined levels. This signal is usually 10% in AC_{rms} of the DC current value, and equates to:

$$\Delta I_Z \text{ pk - pk} = 0.282\, I_Z \qquad \text{(Eq.13.1)}$$

One can determine small voltage changes (ΔV_Z), from the nominal zener voltage (V_Z), with small changes in the zener's current (I_Z), using the impedance value given at I_{ZT}. This is shown by:

$$\Delta V_Z = \Delta I_Z \bullet Z_{ZT} \qquad \text{(Eq.13.2)}$$

For larger current excursions, a good approximation is:

$$Z_Z = Z_{ZT} \bullet \frac{I_{ZT}}{I_Z} \qquad \text{(Eq.13.3)}$$

Typically, manufacturers specify a zener diode's impedance at both I_{ZK} (at the bottom of the knee) and at the I_{ZT} test point given for the nominal voltage. Some manufacturers specify further impedance values at other test points. Usually this results in a straight line between these two reference points when plotted on a log scale. Actually, the zener's impedance drops quickly as current increases beyond the knee region (as shown in Figure 13.6). In the example of a 1N963B device, with a test current of 10.5 mA, the nominal voltage is 12 volts. At this level, the maximum zener impedance is specified as 11.5 Ω, while the maximum value at the knee (Z_{ZK} @ I_{ZK}) is specified as being 700 Ω at 0.25 mA. Typically, zeners with voltages less than 5.1 volts have higher impedances than devices with voltages a little higher (measured at the same I_{ZT}). However, because each zener diode's specifications are processed slightly differently from one another (to obtain different breakdown voltages), they exhibit different impedances at the same voltage. If you took six different zeners (from six different families), but each with the same breakdown voltage, little correlation would exist among them in terms of impedance. Larger zeners typically have lower impedances than smaller devices, because of larger junctions. This is one of the major characteristics of the silicon TVS, a close cousin of the zener diode, whose very low impedance in the avalanche region helps clamp and dissipate potentially destructive high-energy surges.

Determining the zener impedance versus zener current is considerably easier than versus zener voltage because of the two different breakdown mechanisms involved and their correspondingly different behaviors. Measuring various different voltage devices (3 to 60 volts), but with their zener currents held constant at either 1 mA, 10

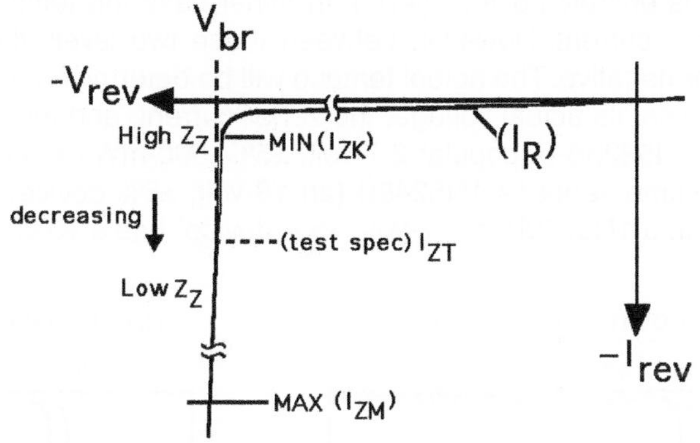

Figure 13.6. The zener diode's impedance decreases with increasing zener current.

mA, or 20 mA, reveals that the impedance level is lowest between 6 and 8 volts. Generally, the zener current must be reduced as voltage is increased, so that power dissipation requirements are met. The zener impedance will increase slightly as voltage is increased, although this is usually much less than the value at the knee.

Another important characteristic mentioned frequently throughout this book, as it applies to current sources and voltage references, is the *temperature coefficient (tempco)*. With the zener diode, this is the tempco of the output reference voltage. It refers to an average of the small changes in the zener's output voltage, caused by changes in temperature ($\Delta V_Z/\Delta T$), and is usually expressed in %/°C. (It can also be expressed in mV/°C or occasionally in ppm/°C.) Ideally, the output voltage will not shift by any measurable amount over the temperature range. In reality, though, the zener's output voltage will shift by a relatively small amount. It would be shown on a graph as a series of sharp or shallow slopes, lines, or curves. Ideally, one should choose a device with a low temperature drift that helps meet or exceed the design goals of the circuit. Discrete zener diodes have a range of output voltage tempcos typically between 0.01%/°C (100 ppm/°C) and 0.1%/°C (1000 ppm/°C). The output voltage tempcos of both temperature-compensated zeners and monolithic buried-zener reference ICs is much lower and shallower. Buried-zener references range from between about 0.5 and 10 ppm/°C (please note that because of their greater precision, they are specified in ppm/°C, rather than %/°C). In contrast, TC zeners have tempcos that range from about 5 to 100 ppm/°C.

Because there are actually *two* different breakdown mechanisms (quantum effects) involved with the discrete zener diode, this can also complicate the subject of the tempco somewhat (see Figure 13.7). Why? Well, as stated previously, below 5 volts the zener effect dominates, whereas above about 8 volts the *avalanche* effect is responsible. Between these two levels, both mechanisms are involved. Actually, below about 3 volts the zener diode's tempco is entirely negative (-TC), and above about 8 volts it is entirely positive (+TC). In either case the tempco is unaffected by changes in zener current. However, between these two levels the tempcos can be either positive or negative. The actual tempco will be determined according to the particular zener device, its actual voltage, the zener current, and the temperature range. For example, a 1N5223B (a popular 2.7-volt, ±5%, 500-mW device) has a –0.08%/°C tempco. In the same series, a 1N5248B (an 18-volt, ±5% device) has a +0.085%/°C tempco. However, a 1N5230B (a 4.7-volt, ±5% device) has a ±0.03%/°C tempco.

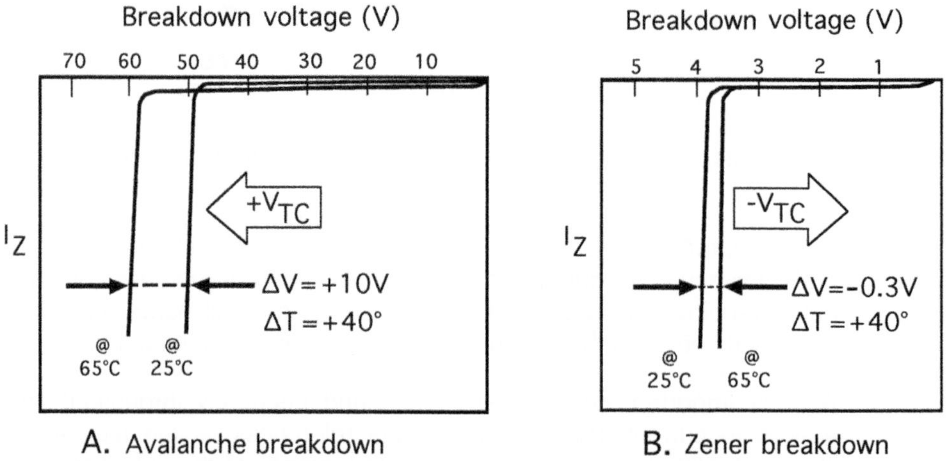

Figure 13.7. *Illustrating the dependance of breakdown voltage on temperature. Notice that low-voltage devices depend on the zener effect, and have a negative tempco, while higher-voltage (>8V) devices depend on the avalanche effect, and have a postive tempco.*

Table 13.1 provides a helpful conversion table for percentage-to-ppm values down to 1 ppm.

For most older discrete zeners, their tempco (TC_{VO} or TC_{VZ}, or a_{VZ}) is not specified on the data sheet at all, although for some newer devices their tempcos are sometimes specified as a typical value or as minimum and maximum values. In order to create some meaningful data, one is often forced to measure and plot a graph of the small changes in the zener's output voltage, resulting from changes in temperature (using a precisely controlled lab oven). From the resulting data, one can determine the tempco of the device. Equation 13.4 allows one to determine the zener's tempco in %/°C:

$$TCV_O = \left(\frac{V_{max} - V_{min}}{V_{nominal}}\right) \bullet \left(\frac{100}{T_{max} - T_{min}}\right) \qquad \text{(Eq.13.4)}$$

Zener Reference Volts				%	ppm
2.7V	3.3V	5.1V	10.0V	100	1M
0.27V	0.33V	0.51V	1.0V	10	100K
27mV	33mV	51mV	0.10V	1	10K
2.7mV	3.3mV	5.1mV	10mV	0.1	1K
270μV	330μV	510μV	1.0mV	0.01	100
27μV	33μV	51μV	100μV	0.001	10
2.7μV	3.3μV	5.1μV	10μV	0.0001	1

Table 13.1 Percent to PPM converter for some popular zener references.

Equation 13.5 allows one to determine the voltage drift in mV/°C:

$$TCV_O = \left(\frac{V_{max} - V_{min}}{\Delta Temp}\right) \qquad \text{(Eq.13.5)}$$

To illustrate how these equations can be used, let's look at some practical examples. In the first example, we will assume we are using a 1N5240, a 10-volt, ±5% tolerance, 500-mW zener, with the following measured data: nominal voltage equals 10 volts; maximum measured voltage equals 10.25 volts; minimum measured voltage equals 9.97 volts; temperature span is +60°C (+5°C to +65°C).

From the data we can immediately determine that the ΔV equals 280 mV and the ΔTemp equals 60 degrees. Now let us determine the tempco in %/°C, using equation 13.4:

$$TCV_O = \left(\frac{10.25 - 9.97}{10}\right) \bullet \left(\frac{100}{60}\right) = +0.0467\% \;/\; {}^\circ C \qquad \text{(Eq.13.6)}$$

Finally, let's determine the voltage drift using equation 13.7:

$$TCV_O = \left(\frac{10.25V - 9.97V}{60^\circ}\right) = +4.67mV \;/\; {}^\circ C \qquad \text{(Eq.13.7)}$$

This means that over a 60-degree temperature span, the drift is less than 5mV/°C. Now, if the temperature span was restricted to a 30-degree span instead, then this would actually yield a higher drift, as shown in equation 13.8. Please note that reducing the temperature span by half, while keeping the same voltage values, effectively doubles the value of drift as shown here:

$$TCV_O = \left(\frac{10.25V - 9.97V}{30°}\right) = 9.33mV\ /\ °C \qquad \text{(Eq.13.8)}$$

Alternately, if the temperature span was expanded to 85 degrees instead, then with the same voltage data, this would yield a somewhat lower drift:

$$TCV_O = \left(\frac{10.25V - 9.97V}{85°}\right) = 3.3mV\ /\ °C \qquad \text{(Eq.13.9)}$$

These examples assume that the zener is being operated at a constant test current (I_{ZT}), which results in these numbers. However, as mentioned previously, below about 3 volts the zener diode's tempco is entirely negative (-TC), and above about 8 volts it is entirely positive (+TC). In either case, the tempco is unaffected by changes in zener current, but between these two levels (because both quantum mechanisms are involved), different zener currents will result in different tempcos. In fact, there is a significant difference in the tempco depending on just how much the device is operated either above or below the I_{ZT} test current. For example, a device exhibiting a 0.01%/°C tempco over the temperature range, with a test current of say 10 mA, may exhibit a 0.0005% /°C tempco at say 14 mA.

Once you know the diode's tempco, you can determine the small change in voltage (ΔV), as it relates to a small change in the junction temperature. It can be calculated by using equation 13.10:

$$\Delta V_Z = TC_{VO} \bullet V_Z \bullet \frac{\Delta T_J}{100} \qquad \text{(Eq.13.10)}$$

The term ΔT_J relates to the zener's junction temperature, but if the zener has already been specified at thermal equilibrium, then the term ΔT_A can be substituted instead for calculating changes in ambient temperature.

Similar to the JFET that we looked at in Chapter 6 with its unique zero tempco point (zero TC), the zener diode has a similar unique point, at which its current varies least with changes in temperature. This is the *zero TC point* for the zener diode, as shown in Figure 13.8.

The zero TC point is not the same as I_{ZK}, or necessarily I_{ZT}, and varies from diode to diode (although devices coming from the same wafer should be virtually identical). Otherwise, one would have to characterize each diode in a precisely controlled temperature chamber, plotting three curves, as shown in the diagram (each temperature curve is plotted against V_Z and I_Z). This will reveal the zener's unique zero TC point,

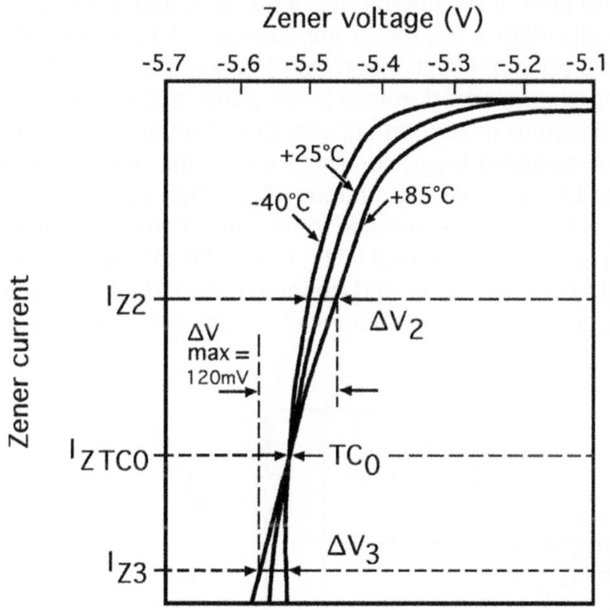

Figure 13.8. Illustrating the zero TC_0 current point that exists for all zener diodes, where temperature changes have minimal effect on current variations, and stability is enhanced.

which occurs at a particular current, I_{ZTCO}, and intersects at a certain zener voltage. Biasing the zener precisely at this point will provide the minimum voltage drift for the required operating temperature range.

Once you know the value of the I_{ZTCO} current, you should consider designing a 10-mA adjustable, precision current source, using either discrete BJTs or DMOS parts, to provide the necessary current. Although this may be helpful for creating a simple low-drift voltage reference with a very light fixed load, it may be totally impractical for references that drive variable loads, because there will be a conflict between I_{ZTCO} and the load current. The use of a buffer, using a BJT or an op amp, can easily provide the necessary drive to the load, without sacrificing this very low tempco. It is easiest to buffer the load from the zener with a precision op amp. Its high-input impedance will not load down the zener; its V_{os} and ΔV_{os} will not add to the zener's accuracy or tempco; its output is current limited; and it can supply several milliamps to the load. Sometimes you may need to increase the voltage, as shown in Figure 13.9. Here the precision op amp provides amplification of the stable voltage reference to a higher level.

In this design, the op amp compares a fraction of the output voltage at D, with the reference voltage at B. The op amp's output drives the BJT at C, which in turn supplies current to the load. It also feeds the JFET at A and zener for improved current and voltage stability. With a low zener impedance (≈10 Ω), current changes from the op amp's output do not affect the circuit's regulation. In this example, the zener is selected and characterized from 6.2-volt parts, and its I_{ZTCO} current should be precisely determined by measurement over three temperature ranges (–55°C, +25°C, and +100°C). The resulting I_{ZTCO} current is set by the JFET current source and can be precisely set by the multiturn trimmer (R_{set}). This zener's I_{ZT} is normally 5 mA, but here its I_{ZTCO} current is determined to be 7 mA. The circuit's output provides a voltage that is not a standard value (8.6 volts) and at 50 mA. By substituting the op amp and BJT for higher-voltage parts and changing the values of R_1, R_2, and R_5, one can upgrade the circuit for higher-reference voltages and at increased current levels.

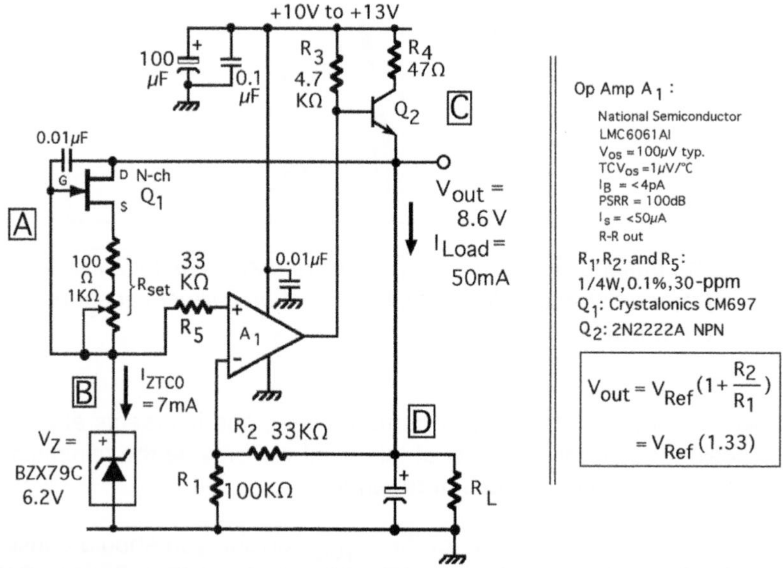

Figure 13.9. A simple zener voltage reference with a near-zero tempco. The zener is precisely set at its I_{ZTCO} by the JFET (at A). The op amp provides gain, and drives the BJT to provide a reference voltage to the load (at D), but with much increased current. The zener is further stabilized by being driven from the circuit's output, rather than the supply rail.

An important consideration for using any zener diode is its maximum power dissipation (P_D). Even though most low-power zener designs are operated well within their safe operating area (SOA), the device still needs to dissipate power (in the form of heat) away from the junction to the outside world. For a zener it is limited by its breakdown voltage, current rating, impedance, power dissipation, package type, and maximum junction temperature. Exceeding any one of these, particularly its current rating, could be fatal to the device. The danger point comes when you consider a simple zener circuit in parallel with its load (see Figure 13.10). Normally, the zener and the load carry only their own respective currents. In the event that the load is disconnected or unexpectedly opens, the zener has to dissipate its normal zener current, plus the entire load current. As a result its power dissipation will soar, and the zener's package will get very hot to the touch. Once the zener's junction temperature reaches beyond 175°C (often the upper operating limit, depending on the device), it is in imminent danger of being destroyed.

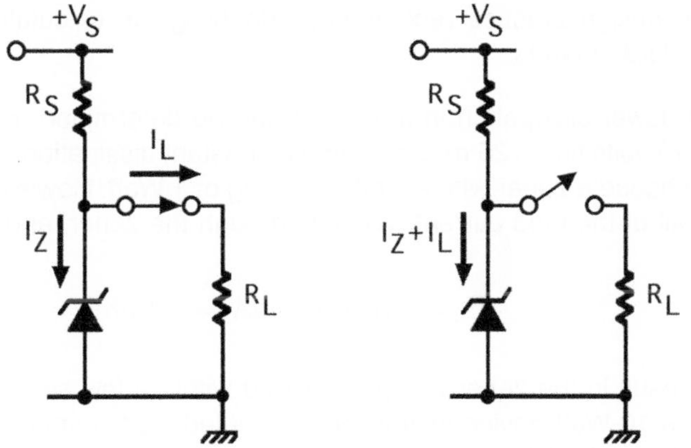

A. Normal operation.

B. Load open circuit, causes the zener to carry the load current as well, causing excess power dissipation.

Figure 13.10. Always design for worst-case power dissipation levels.

To illustrate the point, let's assume that we have a nominal 24-volt power supply, together with a load that requires 18 volts at 200 mA. Let's further assume that we choose an 18-volt zener, with a 26-mA zener current. Under normal conditions, the resistor R_S will drop the 6-volt difference between the supply and the zener voltage. The resistor is always required to limit the current through the zener. (One should always ensure that R_S is small enough to supply the minimum I_Z, even when the supply voltage is minimum and load current is at its maximum.) The total current passing through the resistor is 226 mA (I_L + I_Z). The value of the resistor is given by Ohm's law and will be:

$$R_S = \frac{6V}{226mA} = 26.54\Omega \qquad \text{(Eq.13.11)}$$

The resistor's wattage rating will be:

$$P = V \times I \qquad \text{(Eq.13.12)}$$

which equates to:

$$P = 6V \times 226mA = 1.356W \qquad \text{(Eq.13.13)}$$

The nearest standard value for the resistor will be 27 Ω, with a wattage rating of 1.356 Watts. Good design practice recommends doubling the calculated value up to the nearest standard, 3 Watts.

The normal power dissipation in the zener can be determined from equation 13.12. This will be 18 volts times 26 mA, meaning a constant dissipation of 0.468 Watt, but to be safe we choose a zener with a wattage rating of 1 Watt. However, if the load is disconnected, all of the load current will flow through the zener, and the dissipation will now be:

$$P = 18V \times 226mA = 4.068W \qquad \text{(Eq.13.14)}$$

This would result in the zener being destroyed within a few seconds. However, if we had chosen a 10-Watt device instead and mounted it on a small heatsink, the zener and its circuit would still be working perfectly. Usually a zener will fail as a short-circuit, which in this scenario would then mean that the resistor would be seeing 18 volts across its 27 Ω. This would result in the 3-Watt resistor trying to dissipate around 12 Watts, and it too would soon fail. Hopefully, the regulated power supply would include current-limiting, a crowbar circuit, or a simple fuse to protect the system.

Another scenario is that the zener's load unexpectedly draws 50% more current (i.e., 18 volts at 300 mA). With the values already in place (and R_S previously calculated for a 200-mA load), the 27-Ω resistor is now dropping 8 volts (instead of 6 volts), and so only 16 volts is available to the load. As a result, the 18-volt zener would drop out of regulation.

In summary, always make sure that the series resistor, and the zener are of the correct wattage rating to withstand the worst-case conditions (i.e., open load, maximum voltage supply, maximum current, and maximum temperature) and take into account the variations in power supply voltage and load current.

For power zeners, this maximum power dissipation is just as important as it is for power transistors (a heatsink is likely to be needed). With power zeners, one is often dealing with significant voltages (tens or even hundreds of volts) and currents (Amps). Typically, the zener manufacturer's Power vs. Temperature derating curve (similar to a power BJT or MOSFET's SOA curve) provides information that trades off voltage and current levels. One should always keep below the limits set by the SOA's boundaries for reliable operation. An example of such curves is shown in Figure 13.11. The upper curve is for a 1-Watt zener, while the lower one is for a 400-mW device (similar curves would apply for 10- and 50-Watt devices). In the case of the 1-Watt device, you can see that the device can have a lead temperature of 100°C and still safely dissipate 500 mW. The 400-mW device (shown in the lower curve) can have a lead temperature of 60°C and still safely dissipate 250 mW.

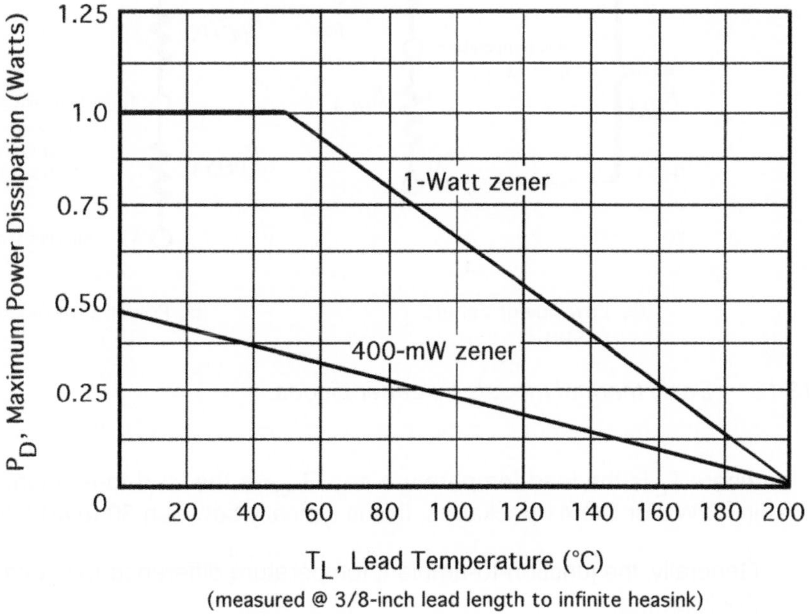

Figure 13.11. Power versus temperature derating curves.

A manufacturer will typically rate the device with a maximum allowable junction temperature, T_{Jmax} (usually specified in the data sheet at 175°C or 200°C), as well as specifying the junction-to-lead thermal resistance ($R_{\theta JL}$), according to the package type. Here it is assumed that the ambient temperature is held at a constant 25°C, and by soldering the device to large square circuit pads, which act as a heatsink (shortening the leads to 3/8 inch or less also helps increase the power dissipation rating, because the thermal conduction path is shorter). General thermal models for zener diodes are shown in Figure 13.12. The left-hand diagram is for low-power zeners, typically axial and surface-mount packages. The right-hand diagram is for a power zener, typically in a metal (heatsinkable) package. The latter has the longer overall thermal path, because of the heatsink.

The low-power zener's lead temperature can be determined from:

$$T_L = R_{\theta LA} \bullet P_Z + T_A \qquad \text{(Eq.13.15)}$$

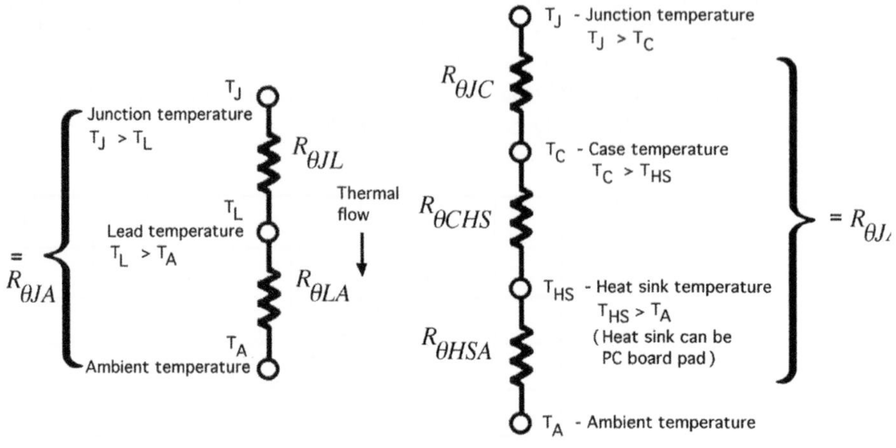

A. Low-power zeners (<5W)

B. Power zeners (10-100W)

Figure 13.12. Basic thermal models for zener diodes.

where T_L is the lead temperature and $R_{\theta LA}$ is the lead-to-ambient thermal resistance in °C/W. For DO-41 packages, this is typically between 30 to 40°C/W.

Generally, the junction-to-ambient temperature difference (ΔT) can be calculated from the following:

$$T_{J\max} - T_A = \Delta T = R_{\theta JL} \bullet P_Z \qquad \text{(Eq.13.16)}$$

where T_{Jmax} is the maximum junction temperature; T_A is the ambient temperature; $R_{\theta JL}$ is the junction-to-lead thermal resistance; and P_Z is the zener power dissipation.

The increase in junction temperature above the ambient temperature (ΔT_{JL}) can be found by:

$$\Delta T_{JL} = R_{\theta JL} \bullet P_Z \qquad \text{(Eq.13.17)}$$

where $R_{\theta JL}$ is the junction-to-lead thermal resistance.

One can determine the junction temperature by:

$$T_J = \Delta T_{JL} \bullet T_L \qquad \text{(Eq.13.18)}$$

Changes in the zener voltage can be found by:

$$\Delta V = V_{ZTC} \bullet \Delta T_J \qquad \text{(Eq.13.19)}$$

where V_{ZTC} is the zener's tempco.

You can determine the maximum allowable power dissipation of a zener diode with equation 13.20, if you know the junction-to-ambient thermal resistance. This will be the reciprocal of the device derating value, usually shown in the data sheet as "derate above 50°C," and then gives a value in mW/°C.

$$P_D = \frac{T_{J\,max} - T_A}{R_{\theta JA}} \qquad \text{(Eq.13.20)}$$

For example, let's assume we need to know the allowable power dissipation for a zener diode with a maximum junction temperature of 175°C, an ambient temperature of 50°C, with a derating value of 6.67 mW/°C above 50°C. We know that the junction-to-ambient thermal resistance is the reciprocal of the device derating value, hence:

$$R_{\theta JA} = \frac{1}{0.00667mW} = 149.92°C \qquad (13.21)$$

Now applying equation 13.20, we have:

$$P_D = \frac{175° - 50°}{149.92} = 0.834W \qquad (13.22)$$

The device is actually rated as a 1-Watt device.

If we took another example, in this case a surface-mount device with an ambient temperature to 25°C and a derating factor of 1.8 mW/°C above 25°C, we could recalculate as follows:

$$R_{\theta JA} = \frac{1}{0.0018W} = 555.55°C \qquad (13.23)$$

Then applying equation 13.20 again, we have:

$$P_D = \frac{150° - 25°}{555.55} = 0.225W \qquad \text{(Eq.13.24)}$$

In this example, the SOT-23 device is actually rated as a 350-mW device. Table 13.2 shows some examples of different packages, wattage ratings, and derating factors.

Package:	Junction temp. max	Total pwr. dissipation	Derate @ mW/°C	above x °C
SOT-23	150°C	350-mW	1.8	25°C
DO-35	175°C	500-mW	3.33	25°C
DO-35	200°C	500-mW	4.0	25°C
DO-41	200°C	1W / 1.5W	6.67	50°C
LL34 (glass leadless)	200°C	500-mW	3.33	50°C

Table 13.2 Examples of various packages, wattage ratings, and derating factors.

For a power zener the calculations are similar, except that some different thermal resistances are involved (junction-to-case, $R_{\theta JC}$, instead of junction-to-lead; also case-to-heatsink, $R_{\theta CHS}$, and heatsink-to-ambient, $R_{\theta HSA}$). Typically the device will need a good heatsink, and for best results require thermal compound and a mica insulator (i.e., a low value for $R_{\theta CHS}$).

Tables 13.3 and 13.4 list some popular zener diode series.

13.3 Some simple zener applications

The discrete zener diode is mostly used for voltage regulation tasks, because it is not as good or as stable a voltage reference as other types reviewed in this book. When compared with today's excellent shunt bandgap references (often considered the lowest-performing IC references), the bandgap excels in virtually every category (e.g., initial accuracy, tempco, noise, quiescent current) over the zener diode. However, the zener can be useful at higher voltages, where the maximum operating voltage limits of IC references cannot operate. In fact, the zener finds much popularity in industrial applications, where higher voltages and currents often apply. The following circuits show how the zener is often used in such applications. In every case, the zener should be current-limited by an external resistor of the correct wattage rating.

Figure 13.13 shows a simple high-voltage application. Using several lower-voltage zener diodes connected in series, as opposed to using just one high-voltage device, has some benefits that include a lower tempco (the tempco rating of any one diode), as well as a significantly lower impedance value. The overall power dissipation is shared among the zeners. If the load ever goes open-circuit, then the zeners will need to be more than 10-Watt devices for safety.

Figure 13.14 shows a shunt regulator, using a combination of a zener diode and an NPN transistor. The BJT is used to improve the power handling capabilities of the nor-

Zener series:	Voltage range:	Wattage rating:	Package:
1N746A - 759A	3.3 - 12V	500-mW	DO-35
1N957B - 992B	6.8 - 200V	500-mW	DO-35
1N2808B - 2846B	10 - 200V	50W	TO-3
1N2974B - 3015B	10 - 200V	10W	DO-4
1N3305B - 3350B	6.8 - 200V	50W	DO-5
1N3506 - 3534	3.3 - 43V	400mW	DO-35
1N3675B - 3710B	6.8 - 200V	1W	DO-41
1N4099 - 4135	6.8 - 100V	250mW	DO-35
1N4158B - 4193B	6.8 - 200V	1W	DO-41
1N4460 - 4496	6.2 - 200V	1.5W	DO-41
1N4614 - 4628	1.8 - 6.2V	250mW	DO-35
1N4678 - 4717	1.8 - 43V	250mW	DO-35
1N4728A - 4764A	3.3 - 100V	1W	DO-41
1N5221B - 5281B	2.4 - 200V	500mW	DO-35
1N5333B - 5388B	3.3 - 200V	5W	DO-41
1N5518B - 5546B	3.3 - 33V	500mW	DO-35
1N5559B - 5594B	6.8 - 200V	1W	DO-4
1N5728B - 5757B	4.7 - 75V	500mW	DO-35
1N5913B - 5956B	3.3 - 200V	1.5W	DO-41
1N5985B - 6031B	2.4 - 200V	500mW	DO-35

Table 13.3 Examples of some popular 1N Series zeners.

mal zener shunt regulator, because the transistor carries the load current instead of the zener. Regulation is improved by this method, because the output impedance (Z_{out}) is reduced by the reciprocal of the transistor's gain ($1/h_{fe}$).

NOTES:

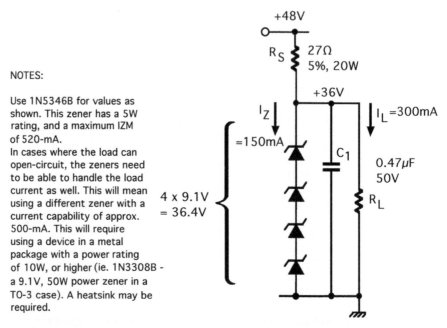

Use 1N5346B for values as shown. This zener has a 5W rating, and a maximum IZM of 520-mA.
In cases where the load can open-circuit, the zeners need to be able to handle the load current as well. This will mean using a different zener with a current capability of approx. 500-mA. This will require using a device in a metal package with a power rating of 10W, or higher (ie. 1N3308B - a 9.1V, 50W power zener in a TO-3 case). A heatsink may be required.

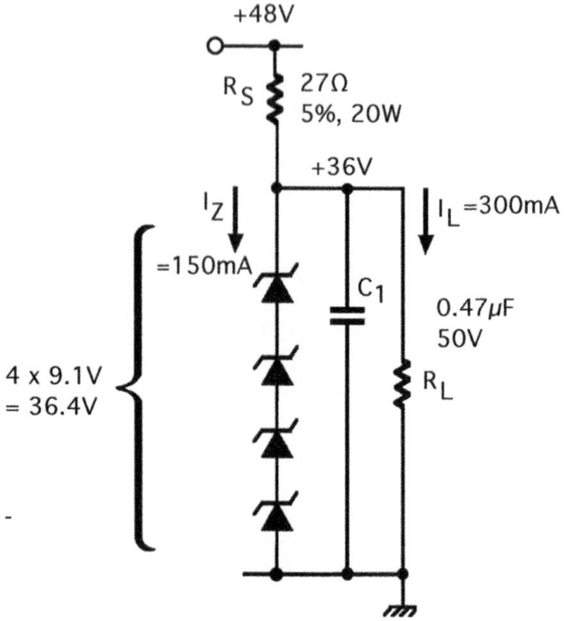

Figure 13.13. *A simple zener shunt regulator, using four low-voltage devices connected in series. Power dissipation is shared between the devices.*

The zener sets the voltage reference and only carries its own (I_Z) current and the transistor's base current. The output voltage is the combination of the zener voltage and the transistor's V_{BE}. The base current should be significantly higher than the current through resistor R_B. These are shown in the following equations:

$$R_S = \frac{V_{IN(\text{min})} - V_{OUT(\text{max})}}{I_{Z(\text{min})}(1 + h_{fe(\text{min})}) + I_{L(\text{max})}} \quad \text{(Eq.13.25)}$$

$$R_B = \frac{V_{IN(\text{min})} - V_{Z(\text{max})}}{I_{Z(\text{min})}} \quad \text{(Eq.13.26)}$$

$$I_Z = I_B + I_{RB} \quad \text{(Eq.13.27)}$$

Zener series:	Voltage range:	Wattage rating:	Package:
BSV55	2.4 - 56V	500-mW	DO-34 LEADLESS
BZX55C	2.4 - 91V	500-mW	DO-35
BZX79C	2.4 - 200V	500-mW	DO-35
BZX83C	2.7 - 33V	500-mW	DO-35
BZX84C	2.4 - 75V	350-mW	SOT-23
BZX85C	3.3 - 100V	1.3W	DO-41
C1Z1	100 - 330V	1W	DO-41
C2Z1	100 - 330V	2W	DO-41
CZ5	6.8 - 200V	5W	DO-201
MLL4	1.8 - 43V	500-mW	DO-34 LEADLESS
MLL5	2.4 - 56V	500-mW	DO-34 LEADLESS
MMBZ5	2.4 - 91V	225mW	SOT-23
MMSZ4	3.3 - 33V	500-mW	SOD-123
MMSZ5	3.3 - 33V	500-mW	SOD-123
MZ4614	1.8 - 10V	500-mW	DO-35
MZD	3.9 - 200V	1.3W	DO-41
MZP	3.3 -100V	1W	DO-41
MZPY	3.9 -100V	1.3W	DO-41

Table 13.4 Examples of some other popular zeners.

The BJT's power dissipation can be determined by:

$$P_D = V_{OUT(\max)}\left(\frac{V_{IN(\max)} - V_{OUT(\min)}}{R_S} - I_{L(\min)}\right) \quad \text{(Eq.13.28)}$$

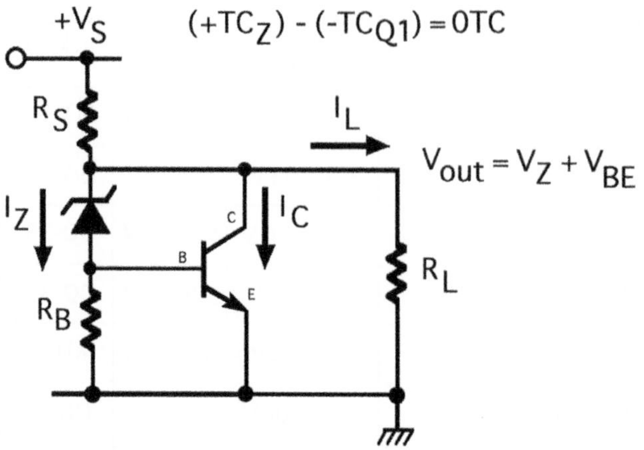

Figure 13.14. Shunt regulator using a BJT to carry the load current. A low tempco is achieved, as the BJT compensates the zener's TC.

The circuit shown in Figure 13.15 shows a simple series regulator, again using a combination of a zener diode and an NPN transistor. Just as with the previous circuit, the BJT is used to improve the power handling capabilities of the zener regulator, because the transistor carries the load current. The zener is temperature-compensated by the rectifier diode, providing a more stable reference at the base of the BJT. The BJT would need a heatsink for the current level shown, or alternately one could use a higher-power transistor, with additional base current.

The simplest current source using an N-channel power MOSFET can be made with a zener diode and a bias resistor on the gate, as shown in Figure 13.16. Simply changing the MOSFET's drain connection enables one to create either a current source or a current sink. In this circuit, the zener's voltage determines the MOSFET's gate voltage, and thereby its drain current, typically ranging from hundreds of milliamps to several amps. Because the FET has a very high input impedance, its gate requires virtually no input current from the bias resistor, hence the resistor needs only to establish the zener's current (I_Z). One can determine the output current and necessary gate-to-source voltage (V_{GS}), using the chosen MOSFET's "*transfer*" and "*output characteristic*" curves from its data sheet.

Although fairly accurate, the simple circuit is not *very* temperature stable, because of the zener's positive tempco. The circuit's overall tempco will be determined according to the quality and tempco of resistor R_1, the particular zener diode, its actual voltage and current, and the operating temperature range. For example, if the voltage chosen gave a positive tempco, this can be compensated for by adding one or more silicon diodes (each of which have a –2.2 mV/°C tempco) and reducing the zener's voltage value accordingly (for each diode's forward voltage drop of approximately 0.7 volt). A

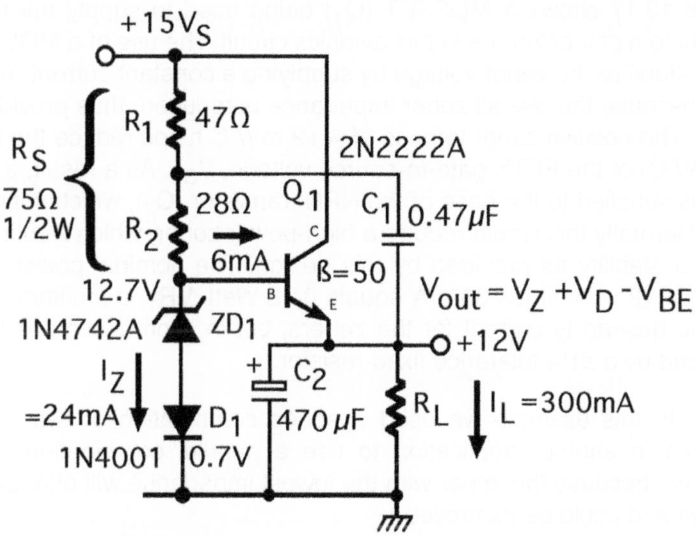

Figure 13.15. A simple series regulator, using an NPN transistor to drive the load and a zener diode as a stable reference.

small heatsink may be required for the MOSFET, depending on the desired current level (I_D), the voltage drop across the FET (V_{DS}), and its package type.

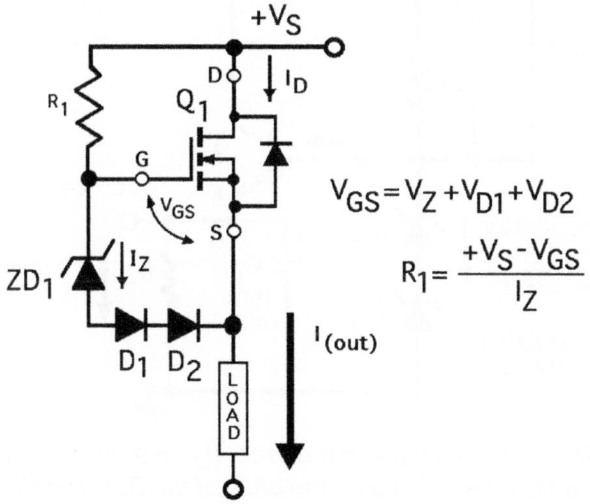

Figure 13.16. In this example, the zener together with the compensation diodes set the gate-to-source threshold voltage for the N-channel power MOSFET, which acts as a high-power current source.

Figure 13.17 shows a MOSFET (Q_1) being used to supply the necessary constant current to a pair of zeners in this avionics circuit. The use of a MOSFET current source helps stabilize the zener voltage by supplying a constant current. Two zeners are used here because the overall zener impedance is reduced, thus providing improved regulation. The positive zener tempco of ≈ +2 mV/°C helps reduce the negative tempco (≈ –5 mV/°C) of the FET's gate-to-source voltage, V_{GS}. As a result, a relatively stable 24 volts is supplied to the base of the NPN transistor (Q_2), which supplies 250 mA to the load. Normally this would require a high-power zener, which would not have the same level of stability as provided by this design. The nominal power dissipation of each zener is 12 volts times 35 mA equals 0.42 Watt. VR_1, a multiturn trimmer, is used to set the desired I_Z current for the zeners, but in a production situation this could be replaced by a ±1% tolerance fixed resistor.

(*Note*: In this example we used a series combination of two zeners. You may be tempted in another application to use a parallel combination. This is *not* recommended, because the zener with the lowest impedance will attempt to carry the entire current and could be destroyed.)

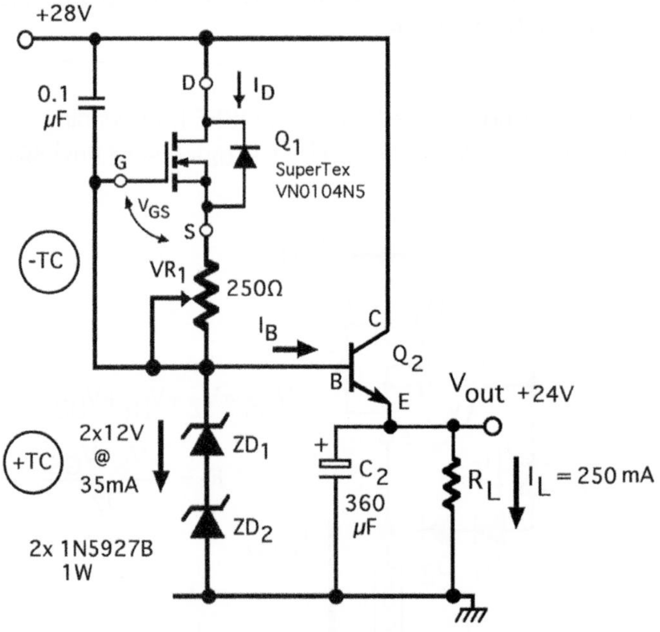

Figure 13.17. A MOSFET supplies the zener current (I_Z) to a pair of zener diodes. The zeners provide a stable reference on the base of the BJT. The BJT buffers the zeners from the load.

The circuit shown in Figure 13.18 illustrates a zener diode/TVS being used to protect an expensive switching power transistor from overvoltage in an ignition circuit. Because of the inductive nature of such circuits, high transient voltages can be generated, which can easily destroy transistors, whether BJTs, MOSFETs, or IGBTs. (Bipolars, TTL, and other devices such as SCRs are usually a little less susceptible than MOS devices.) In this circuit the zener/TVS clamps any transient voltages at about 90 volts, which is 75% of the transistor's rated V_{CE} (120 volts). The zener's clamping voltage is chosen to operate at a safe point within the transistor's SOA. The zener should be mounted as close as possible to the transistor being protected. While the device has a steady state rating of 5 W, its peak power rating is more than 1 KW. The power transistor should be mounted on a good-quality heatsink.

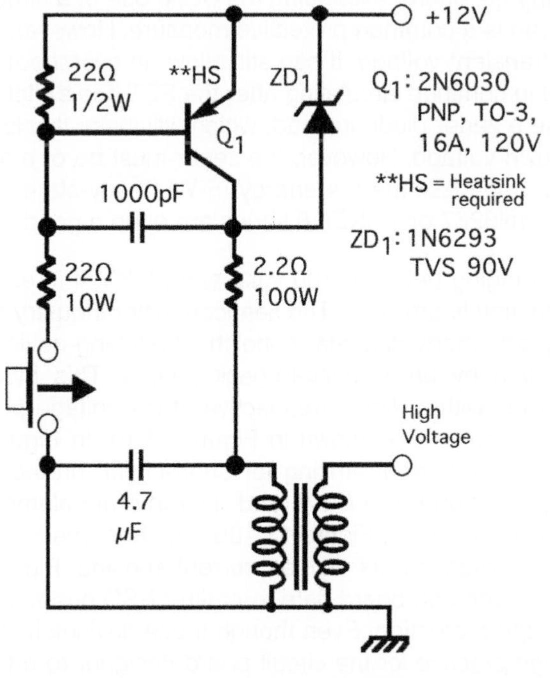

Figure 13.18. The zener diode protects the switching transistor in this ignition circuit.

The zener/TVS excels in transient and ESD protection. This market has grown rapidly over the past 15 years, as a result of the industry-wide research done in this area, particularly for digital and analog CMOS, as well as for discrete MOS transistor protection. In either case, two areas with the MOSFET are particularly vulnerable to voltage transients: the gate and the drain. Because all MOS devices use an insulated gate, they are subject to potential voltage breakdown, which can occur at about 100 volts and higher. Although many devices do include some gate protection in the form of resistor-diode networks, very few guarantee total protection, particularly against ESD, which can destroy a device in just one brief hit, or gradually degrade a device after multiple hits over a period of time. Discrete MOSFET transistors can usually withstand more abuse than MOS/CMOS ICs, because they have larger chips, larger junctions, and therefore have greater input capacitance. This enables them to absorb more energy before being charged to the gate breakdown voltage. However, once this threshold is reached, enough energy is stored in the gate-to-source capacitance to cause punch-through of the gate oxide and destruction of the device.

The second area susceptible to voltage transients, and probably the most common cause of failure, is the MOSFET's drain. This is caused by switching high currents through load and stray inductances, which can contain enough energy to destroy the MOSFET. This is a result of the transient voltage exceeding the maximum drain-to-source voltage, $V_{(BR)DSS}$, and/or exceeding the SOA. Use of a simple diode in parallel with the inductive load is a common protective measure. However, while the diode will clamp most of the transient voltage, it can still allow an overshoot of the V_{DS}, as well as allow the current to continue circulating after the FET has switched off. A much better solution is to use a zener diode instead, which will instantly clamp the transient at the zener's breakdown voltage. However, the zener must be of a sufficient power rating to safely absorb and dissipate the energy (5-W steady-state/1.5-KW peak power devices such as the 1N6267 or 1.5KE6.8 series are often a good choice).

With ESD voltages ranging between 100 volts and 25 KV, it is easy to destroy these devices unless protection is provided. The semiconductor industry takes the subject of ESD very seriously, and many discrete monolithic switching devices now incorporate built-in ESD protection by using back-to-back zeners. This bidirectional structure yields a zener in series with a diode irrespective of the voltage polarity. A few examples of monolithic structures are shown in Figure 13.19. In Figure 13.19A, a smart MOSFET switch includes current limiting for short-circuit protection, gate-to-source clamping for ESD protection up to 2 KV, and gate-to-drain clamping for overvoltage protection with inductive loads. In Figure 13.19B, another smart MOS switch includes drain-source voltage protection, as well as current sensing. Figure 13.19C shows a smart IGBT, which includes on-board gate-to-emitter ESD protection, as well as gate-to-collector overvoltage protection. Even though these devices include built-in protection, it is *good design practice* for the circuit board designer to add additional protection in the form of discrete TVS devices in any designs.

Another area where zeners can provide protection is in some silicon-controlled rectifier (SCR) applications. Here the zener is used to detect the voltage transient and trigger the SCR's gate. Once triggered, the SCR carries the bulk of the shunt current and provides the overcurrent or crowbar function, by tripping a circuit breaker or blowing a fuse. Figure 13.20 shows such a design. Normally both devices are off; the zener is not conducting, and the SCR's gate is biased off by R_2. When a voltage occurs that is equal to or greater than the zener's voltage (V_Z), the zener conducts current, raising the SCR's gate voltage, and triggering it. The SCR now conducts and trips the circuit

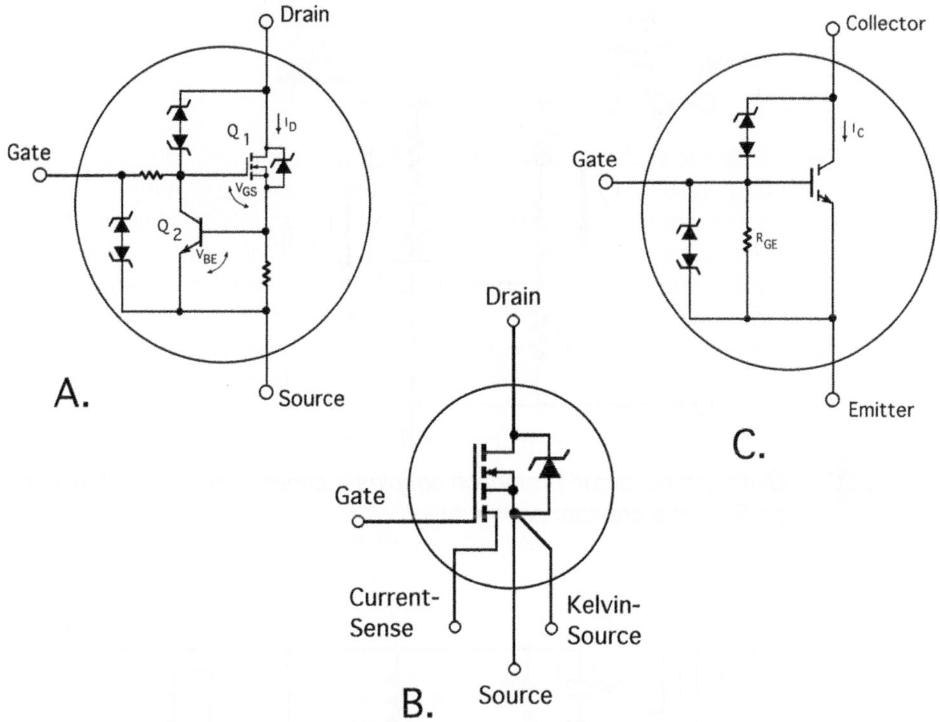

Figure 13.19. The zener diode is used to protect power MOSFETs and IGBTs from ESD and high-voltage transients.

breaker, so protecting the circuit. Depending on the application, the SCR may need to be mounted on a small heatsink.

The simple zener can also be useful in level-sensing applications. Although several excellent voltage detector and watchdog ICs are available from different manufacturers, most devices are intended for relatively low-voltage tasks, such as toggling an I/O pin on a CMOS microcontroller. In industrial applications, where supply voltages are often significantly higher, zeners are an ideal solution. They can be used with BJTs and even with popular ICs such as the ubiquitous 555 timer.

Figure 13.21 shows a simple zener level-detector with a 555 monostable output. The main supply is V_{S2}, which powers most of the circuit. This could practically range from between about 3 to 15 volts (low-voltage CMOS 555 devices are available from makers such as Texas Instruments, ALD, National Semiconductor, and others). The input voltage to the zener detector is a short pulse (less than 1 second), applied via V_{S1}. This pulse voltage could be supplied from momentary relay contacts, a microswitch, or other means. Resistor R_1 determines the input voltage, which could range from about

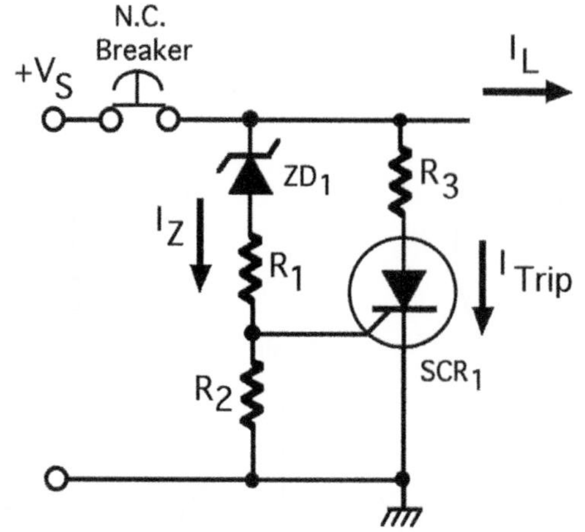

Figure 13.20. *Over voltage circuit protection couples a zener to detect the transient, and trigger an SCR in a crowbar configuration.*

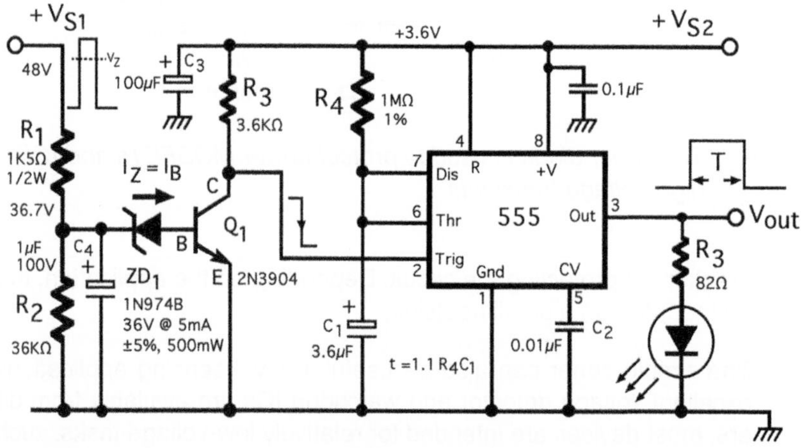

Figure 13.21. *This example shows a zener voltage detector used to trigger a 555-based monostable. In an industrial application, the input voltage to the zener circuit could signal (via a microswitch) that an item is in position, ready for the next stage in a process.*

3 to 300 volts, and current-limits the zener (i.e., 5 to 10 mA). The value chosen for the zener (V_Z) depends on the range of the input pulse voltage (i.e., 48 volts ±10%, 500 milliseconds), along with the value of the transistor's V_{BE} (≈ 0.7 volt). Resistor R_2, which is in parallel with the zener and Q_1's V_{BE}, provides a low current bias point that is much less than I_Z. The NPN transistor Q_1 functions as a general-purpose switch,

with a collector voltage range of approximately +0.5 volt to $+V_{S2}$. This collector voltage is used to trigger the input of the 555 timer. Normally both the zener and the BJT are not conducting, and no current flows through R_3, thus the input to the timer is high ($+V_{S2}$). When a voltage occurs that is greater than the zener's breakdown voltage (V_Z), the zener conducts current to the transistor's base, turning it on. The zener current (I_Z) serves as the transistor's base current (I_B). Current then flows through resistor R_3, pulling the collector voltage down sharply, and triggering the monostable. Capacitor C_1 and resistor R_4 determine the monostable's timing period (i.e., $t \approx 1.1$ RC). The timer's output (which is normally low) goes to approximately $+V_{S2}$ −1 volt, powering the load and the LED indicator during the timing period. In this circuit, the input voltage (pulse) needs to be returned to a voltage lower than is required to make the zener conduct. This will reset the timer, ready for the next pulse. The input pulse width should be less than the monostable's pulse width.

13.4 Temperature-compensated zeners

The temperature-compensated zener (aka the voltage reference diode, or the TC zener) debuted in the late 1960s when engineers began creating voltage references by combining zeners with diodes and BJTs. They discovered that by characterizing and selecting zeners, and adding one or more series rectifier diodes, they could create *temperature-compensated zener references.* The TC zener preceded the first commercial buried-zener voltage reference, which was introduced by National Semiconductor in the early 1970s. (The buried-zener type differs in its architecture and processing from the TC zener, as we will learn in Chapter 14, and offers a significantly higher performance: lower TC, lower noise, lower I_Z, and better load and line regulation.) National Semiconductor, which made discrete zener diodes at the time, played a major role in the development of both types. Over the following decades, because of its well-understood characteristics, the TC zener has subsequently been improved by refinements in processing as one monolithic structure. Its symbol and polarity are shown in Figure 13.22. U.S. manufacturers of TC zeners today include Microsemi and Central Semiconductor, but few others. Both manufacturers offer some excellent products.

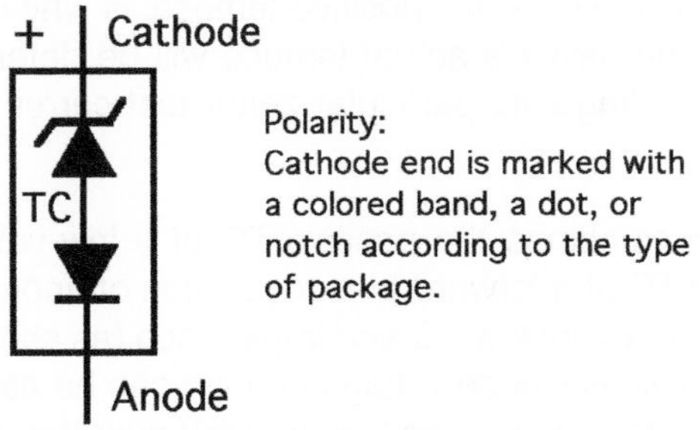

Figure 13.22. The temperature-compensated zener diode; its symbol and polarity.

The monolithic TC zener is processed similarly to the regular zener, specifically using either ion-implantation or chemical vapor deposition. Common types have either a diffused junction or an alloy-diffused vertical junction. As with the regular zener, an important factor in creating a stable device is the passivation of its surfaces, using either silicon dioxide or a glass derivative. This forces surface contaminants away from the surface and helps create a stable, quality product, with a low reverse leakage current. Its mechanical construction has the monolithic chip welded securely between both internal ends of the leads, sandwich fashion. The device is then encapsulated using a glass envelope (see Figure 13.23).

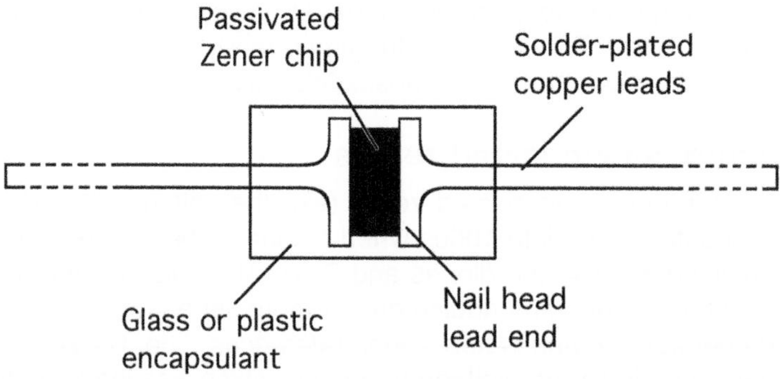

Figure 13.23. The axial-leaded zener diode, the TC zener, and the TVS all share this typical construction.

However, unlike the regular zener, the monolithic TC zener is characterized and tested differently and is available in two voltages: 6.2 volt and 6.4 volt. (Other voltages are available from Microsemi.) As mentioned previously, below 5 volts the zener effect dominates, whereas above about 8 volts the Avalanche effect is responsible. At approximately 5.5 volts, many zeners have only a small positive tempco, while between 5.5 volts and about 4 volts devices have a tempco that can be either positive or negative (below 4 volts the zener has an entirely negative TC). Using the lowest applicable voltage, that has only a small positive tempco, is why the TC zener is centered at about 5.5 volts. The device's actual tempco will be determined according to the processing, its actual voltage, its particular zener test current, and the operating temperature range.

Essentially, the TC zener combines the positive TC of a low-voltage 5.6-volt zener, together with the negative TC of a forward-biased junction of approximately 0.7 volt, to create a 6.2-volt device with a very low TC and impedance (as shown in Figure 13.24). This results in precision devices whose voltage change can be as low as 5 mV over a –55°C to +100°C temperature range, or as low as 2 mV over the shorter 0°C to +75°C temperature range. Devices are usually available with a ±5% tolerance, but 2% and 1% devices are available on special order. Such devices are usually available as either a 400-mW device in a glass DO-204 package, a 500-mW device in a glass DO-213AA package, or 350-mW devices in various surface-mount packages. In addition to a few semiconductor manufacturers, several laboratory voltage standards manufacturers also offer TC zeners.

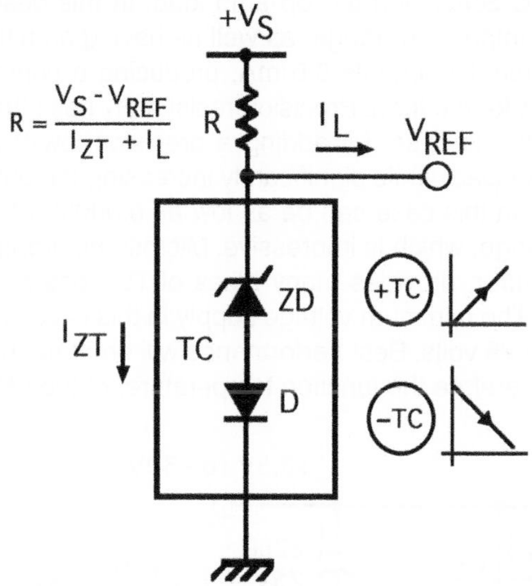

Figure 13.24. A schematic of the temperature-compensated zener diode.

As discussed previously, the zero TC point for the regular zener is not the same as either its I_{ZT} or its I_{ZK}, and varies from diode to diode. The TC zener, by contrast, is specially processed and has its (combined zener and diode) tempco centered very close to the zero TC point. (Otherwise, one would have to do what many standards labs do, which is to characterize each zener and diode combination in a precisely controlled temperature chamber, plotting each temperature curve against voltage and current for –55°C, +25°C, and +125°C.) This will reveal the zener's unique TC_0 point, which occurs at a particular current I_{ZTCO} and intersects at a certain zener voltage. Biasing the zener precisely at this point with the particular I_{ZTCO} current will provide the minimum voltage drift for the required operating temperature range.

Once you know the value of the zener's zero TC current, you can buy or design an adjustable, precision current source using either a monolithic IC, or discrete BJTs, JFETs, CMOS, or DMOS parts, to provide the exact current needed. The current source is important in this design and will provide a precision, low-drift voltage reference suitable for use with a fixed load. The use of a buffer, using an op amp or a BJT, can provide the necessary drive for heavier loads, without sacrificing the ultra-low tempco. TC zeners often receive thousands of hours of burn-in to fully stabilize their characteristics over time, so that a long-term drift of around 100 ppm/month is a practical reality.

Such a design is shown in Figure 13.25, which shows a National Semiconductor/Linear Technology LM134/334 adjustable current source supplying a precise current to a Microsemi 1N4569A TC zener and the op amp load. In this design the TC zener is specified over a wide temperature range, as well as having an ultra-low zener current (I_{ZT}). The TC zener runs at I_Z equals 0.5 mA, producing a constant 6.4-volt output, along with up to 0.5 mA for the load. Precision resistor R_1 (less than 50 ppm) sets the LM134's I_{set} current. In this case, by adding a precision low-V_{os} op amp, one can buffer the zener from the load, while significantly increasing the output current capability. The zener's tempco in this case can be as low as 0.0005%/°C (or 5 ppm/°C) over the –55°C to +100°C range, which is impressive. Microsemi, probably the world's largest TC zener manufacturer, provides many types of TC zeners, in various voltages, grades, and packages. The minimum voltage supply in this circuit is about 8.5 volts but can range to more than 25 volts. Best performance will be at about 9 volts to minimize the voltage drop, and therefore the junction temperature, of the LM134/334.

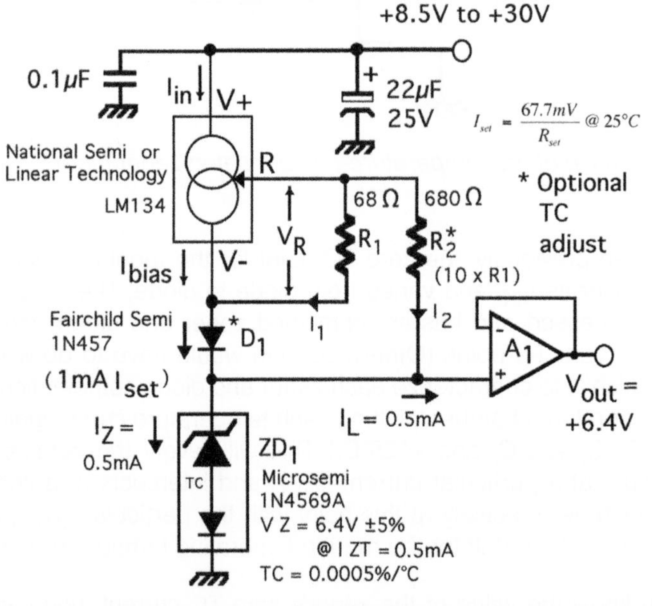

Figure 13.25. The LM134/334 provides a constant current for this temperature-compensated zener reference. R_2 allows optional adjustment of the circuit's tempco. The precision current source is necessary for optimum performance.

Although the LM134/334 can theoretically supply more than 5 mA, in this circuit it should be limited to 1 mA or less, in order to maintain a low overall tempco. As a current source, the best area for operation of the LM134 lies between 5 μA and 1 mA. In fact, beyond 100 μA, internal heating of the chip can have a significant effect on overall current regulation. For example, with I_{set} at the 1-mA level as shown here, each 1-volt increase across the device increases the junction temperature approximately 0.4°C. This, together with the LM134's own tempco of +0.336%/°C, will yield a change in I_{set} of 0.132% (1.3 μA)—a significant amount. To minimize this change, try to minimize the voltage across the LM134, and if possible provide a small heatsink for it. If a zero tempco (TC_0 of less than 25 ppm) is needed, this can be made by adding a second precision resistor (R_2) and a good-quality, low-leakage silicon diode (D_1), as shown in the diagram. Here the positive tempco of the LM134 (+227 μV/°C) is offset by the forward-biased silicon diode's negative tempco (approximately –2.5 mV/°C). The TC zener already has a near-zero tempco, so the circuit will serve as a good stable reference.

Another precision, low-drift voltage reference circuit suitable for use with a fixed load is shown in Figure 13.26. The circuit uses a (different) 6.2-volt TC zener, but with a higher I_{ZT} (7.5 mA), and uses a DMOS FET as a current source to provide this. A DMOS FET current source is used rather than other types (such as monolithic IC current sources, BJTs, and JFETs) because of the higher current required and for design simplicity. The required 7.5-mA zener current is very low for the DMOS FET, so its output conductance will be very low and its quality of regulation high. The op amp used should be a low-V_{os}, low-noise, precision type. It provides the necessary drive for the load (several mA), as well as short-circuit protection, without sacrificing the reference's very low tempco. One could use a dual or quad op amp, in order to amplify the 6.2 volts to another value, or to provide Kelvin connections for superior performance with a remote load. For optimum performance, the 12-volt supply shown here should be well regulated.

The current source is a vital part in designing such a low-drift zener voltage reference. In order to design the precision current source for biasing the TC zener precisely at the required I_{ZTCO} current, you need to know the amount of regulation required. This can be easily calculated, using the information from the data sheet. In this example we will assume we are using a 6.2-volt device such as a 1N827A. This is a wide temperature range device that operates over the –55°C to +100°C range. It has an I_{ZT} current of 7.5 mA and a maximum change in voltage (ΔV_Z) of 9 mV, caused by changes in current. It also has a maximum impedance of 10 Ω at this I_{ZT} and at 25°C. Assuming the impedance value is held close to 10 Ω over the temperature range, we can determine the maximum allowable change in current that the regulator will need to be within, to meet the specifications:

$$\Delta I = \frac{9mV}{10\Omega} = 0.9mA \qquad \text{(Eq.13.29)}$$

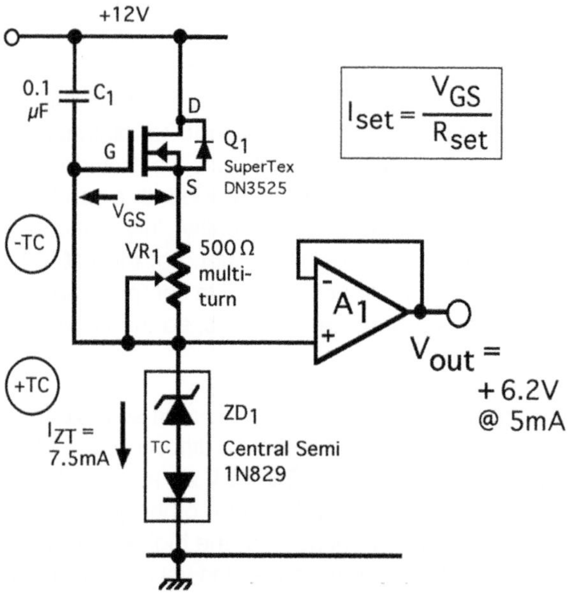

Figure 13.26. A DMOS FET supplies the zener current (I_Z) to a TC zener diode. The zener provides a stable reference to the input of the op amp, which buffers the zener from the load.

The amount of current regulation needed will then be:

$$\% \text{ Regulation} = \frac{0.9mA}{7.5mA} \text{ x } 100\% = 12\% \qquad \text{(Eq.13.30)}$$

This level of regulation is easily met by most precision current sources shown in this book. For the 1N821 series, the current level required is nominally 7.5 mA, which can be provided by discrete DMOS FETs or BJTs, and using low tempco precision resistors. For the 1N4565 series, the current level required is either 0.5 mA or 1.0 mA, which can easily be provided by discrete JFETs, BJTs, or the LM134/334, also using low tempco precision resistors. (The JFET cascode version is also accurate and applies to this current level.)

Table 13.5 provides a listing of the most common monolithic TC zeners, which are essentially two separate series. These are the 1N821/A to 1N829/A series and the 1N4565/A to 1N4574/A. The 1N821 series are 6.2-volt devices at an I_{ZT} of 7.5 mA, but with different tempcos. The 1N4565 series are 6.4-volt devices, at an I_{ZT} of either 0.5 mA or 1.0 mA, also with different tempcos. Both series are hermetically sealed, DO-35 axial glass package types, rated at up to 400 mW. The 1N821 series are specified over the –55°C to +100°C temperature range. The 1N4565 series are available in two temperature ranges: 0°C to +75°C and –55°C to +100°C, with a derating factor of 3.2

Part number:	Voltage: (V_Z ±5%)	@ I_{ZT} (mA)	ΔV_Z max (mV)	@ Temp (°C ±1°C)	Tempco (%/°C)	Z_Z max (Ω)
1N821	6.2V	7.5	96	-55/25/100	0.01	15
1N821A			96		0.005	10
1N823			48		0.002	15
1N823A			48		0.001	10
1N825			19		0.0005	15
1N825A			19		0.01	10
1N827			9		0.005	15
1N827A			9		0.002	10
1N829			5		0.001	15
1N829A	↓	↓	5	↓	0.0005	10
1N4565	6.4V	0.50	48	0/25/75	0.01	200
1N4565A			99	-55/25/100	0.01	
1N4566			24	0/25/75	0.005	
1N4566A			50	-55/25/100	0.005	
1N4567			10	0/25/75	0.002	
1N4567A			20	-55/25/100	0.002	
1N4568			5	0/25/75	0.001	
1N4568A			10	-55/25/100	0.001	
1N4569			2	0/25/75	0.0005	
1N4569A		↓	5	-55/25/100	0.0005	↓
1N4570		1.0	48	0/25/75	0.01	100
1N4570A			99	-55/25/100	0.01	
1N4571			24	0/25/75	0.005	
1N4571A			50	-55/25/100	0.005	
1N4572			10	0/25/75	0.002	
1N4572A			20	-55/25/100	0.002	
1N4573			5	0/25/75	0.001	
1N4573A			10	-55/25/100	0.001	
1N4574			2	0/25/75	0.0005	
1N4574A	↓	↓	5	-55/25/100	0.0005	↓

Table 13.5 Listing of 400-mW, axial DO-35 TC zeners.

mW/°C above 50°C. Both series are also available in the manufacturer's own brand (not JEDEC) of alternative packages, including surface-mount types. Although neither series is specified for either noise voltage or noise density, like most IC references are, it can be assumed that the 1N4565 series probably has a little higher noise level because of its much lower current (0.5 mA vs. 7.5 mA). For more information on this subject, you are recommended to contact the manufacturer's Applications Engineering Department.

These TC zeners are useful in many relatively low-voltage designs, but there are occasions when you might need to create your own TC reference, because of the application's high voltages. This can be done by using either one zener (or several zeners that combine to equal the same voltage), together with one or more standard rectifiers such as the 1N4001-series. In the following example, we need a 65-volt reference voltage, along with a load current of 35 mA. In reviewing various (regular) zener products, we see that in the 1N5333-series, there is a suitable device, a 1N5372B, a 62-volt, 5-Watt device. Looking further in the data sheet, we can see that for this voltage level, the tempco of this device is approximately 10 mV/°C. If we now look at the IN4001-series data sheet, we can see that at a forward current of 20 mA, the forward voltage is approximately 0.72 volts, at 25°C. The device also has an approximate tempco of –2 mV/°C. Five such devices connected in series have a total forward voltage drop of 3.6 volts, but with a combined tempco of –10 mV/°C. Connecting the 62-volt zener in series with the five rectifiers produces a total of 65.6 volts, but with a near-zero TC, as shown in Figure 13.27. In order to get to a practical design, you will need to characterize all of the diodes at the actual zener current, over the temperature range. This should result in a design very close to the one shown. (Be sure to burn in the combination for at least 168 hours, though, and then retest.)

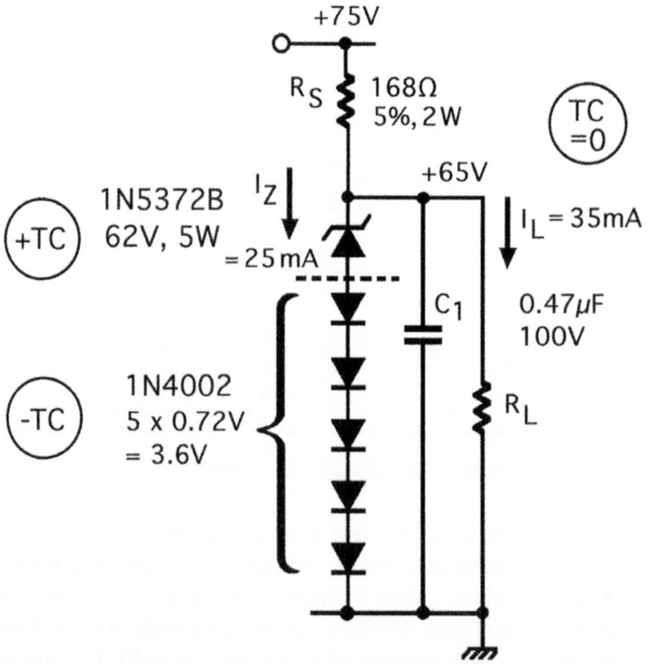

Figure 13.27. Achieving a near-zero temperature coefficient for a high-voltage zener reference. The zener diode's positive tempco is balanced against the negative tempcos of the 1N4002 rectifier diodes.

Because of its limited range of voltage options, together with the fact that it requires a minimum of 7 volts to function, the TC zener has been mostly superseded by monolithic IC references, which offer many more voltage options, lower quiescent currents, and higher overall performance. The buried-zener topology is a good example of this. However, there are still applications that can use the TC zener, in which its inherently small size and fully characterized specifications are important.

Major U.S. manufacturers of discrete zener diodes, TC zeners, and silicon transient voltage suppressors are shown in Table 13.6.

Manufacturer:	Zener diodes:	TC zeners:	TVS:
Central Semiconductor	√	√	√
Diodes Inc.	√		√
EIC Semiconductor	√		√
Fairchild Semiconductor	√		√
Microsemi	√	√	√
ON Semiconductor	√		√
Philips Semiconductor	√		√
Semtech	√		√
Teccor			√
Vishay Semiconductors	√		√

(List compiled July 2004)

Table 13.6 U.S. Manufacturers of Zeners, TC zeners, and TVS.

When you consider what a simple little zener diode can do, it has to be one of the most versatile semiconductor devices ever invented, and certainly an essential building block in all of today's analog electronics. Although at low voltages the zener is outperformed by monolithic IC references, at higher voltages there is little that can match its size and simplicity. It has come a long way since it was invented in the 1950s. No doubt it will be an integral part of our circuit designs for many years to come.

Chapter 14

Characteristics of Monolithic Voltage References

Since the introduction of the discrete zener diode in the early 1960s, monolithic voltage references have evolved over many generations, so that today they are available in various distinct categories. These include the bandgap, the buried-zener, the XFET® from Analog Devices, along with the FGA™ from Xicor/Intersil. IC voltage references are now manufactured using various different processes (e.g., bipolar-FET, complementary-bipolar, bipolar-CMOS, CMOS), which has brought about lower noise, improved initial accuracy, smaller chips and packages, and lower-power devices. As a result, today's voltage references can be subdivided into different types that include shunt, general-purpose, low-dropout, low-power, and ultra-precision types. A diagram showing the different types of voltage references is shown in Figure 14.1. Typically, voltage reference ICs are fixed single devices, but a few dual devices are available. Some devices also have extra features for higher-performance applications. Some even allow one to select the reference voltage, trim it, filter it, and put it to sleep. Chapter 12 covered the reference's more important characteristics, so in this one we take a closer look at the different types and explore the merits of each.

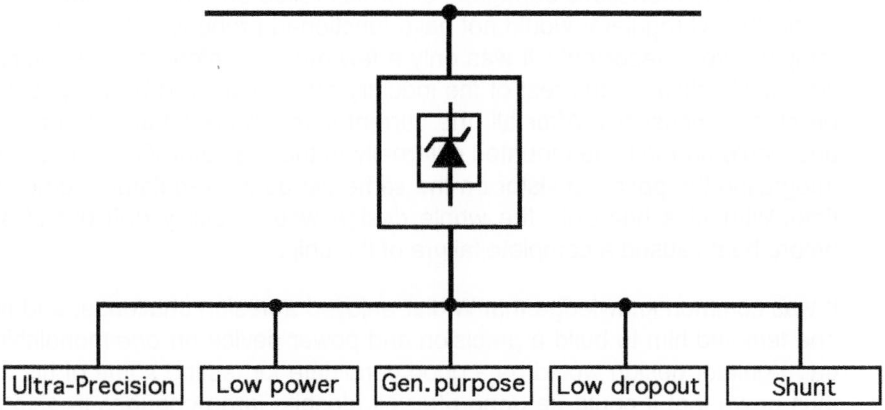

Figure 14.1. Types of monolithic voltage references.

14.1 Bandgap voltage references

While working on further op amp development at National Semiconductor in the late 1960s, the legendary Bob Widlar began looking into the subject of the silicon bandgap (V_{go}) and how currents and voltages were dependent on absolute temperature (today we refer to these as I_{PTAT} and V_{PTAT}). As mentioned in Chapter 4, it is remarkable to note that with the addition of two components, Widlar was able to transform his basic ratioed current source into a simple bandgap voltage reference (see Chapter 4, Figure 4.21).

Widlar not only invented the op amp, the high-power linear regulator, and the very-low-voltage op amp building block, but as a result of his innovative bandgap design, he also laid the foundations for today's bandgap references and monolithic temperature sensors. These are based on the same core design as the bandgap reference, but use either an I_{PTAT} or V_{PTAT} output, which is linear with Kelvin temperature (i.e., 298°K equals 298 μA or 298 mV). The subject of the silicon bandgap was just one uncharted area that Widlar and Dave Talbert, his process engineer, were investigating, along with creating ratioed emitter transistor structures, super-beta transistors, multiple emitters, current-density ratios, current sources and mirrors, temperature-compensation networks, and curve-correction circuits. Widlar was a prolific writer and wrote many papers that he regularly presented at industry conferences or published in various journals. The remarkable fact is that many of these things were first created by Widlar and have become a foundation for much of today's analog technology.

When National introduced the LM109 in 1969, little did anyone realize at the time that it would pave the way for the bandgap voltage reference several years later. Widlar had created the LM109, which was the first monolithic 20-watt, 5-volt linear voltage regulator. Uniquely, the LM109's control circuitry got its temperature compensation from a stable monolithic bandgap cell, which relied on ratioed emitters. Without it, the monolithic IC regulator would not have functioned properly at high currents or at high temperatures. Reportedly, it was only a few months before this discovery that Widlar was as skeptical as the rest of the industry and considered building such a device to be all but impossible. After all, the current-carrying pass transistor(s) would get hot and would normally be mounted externally to the regulator IC and require a heatsink. Integrating the pass transistor on the same die as the regulator seemed crazy at the time. Without a heatsink, the whole design would quickly drift out of specification, before heat caused a complete failure of the chip.

It was common knowledge that Widlar enjoyed a design challenge, and no doubt this one tempted him to build a precision and power device on *one* monolithic chip. A linear regulator design would be a good candidate for testing some of his theories, and thus prove that it could be done after all, despite large changes in chip temperature and current levels. Widlar was the first analog designer to ever do that. In 1971 he co-designed with Robert Dobkin the first dedicated, two-terminal (shunt) monolithic bandgap references, National's LM113, which quickly became a favorite design-in for instrumentation designers.

By the early 1970s, National Semiconductor's analog designers and engineers had characterized and knew enough about their first-generation bandgap references to know that there were some significant limitations. One had to do with the bow-shaped voltage versus temperature curve (which would require special temperature-compensation circuitry), while another area of concern was the noise level, which would only be reduced several years in the future with more advances in processing. Nevertheless, this prompted National to develop a much better product, which Robert Dobkin designed and which National introduced in the mid-70s. This was dubbed the LM199/399, the first buried-zener monolithic voltage reference, and it uniquely contained an on-chip substrate heater to stabilize the chip's temperature. This legendary product offered a 6.95-volt reference, with a 0.3-ppm/°C temperature drift, and a noise spec of less than 10 μV. This was better than anything at the time and is still better than most devices even today. While the buried-zener reference was not one of Widlar's designs, it was a product he may have provided some input for, because he was National's leading analog designer at the time, and clearly considered Bob Dobkin not only a friend, but his protégé.

It is impossible to know why Bob Widlar invented any of the bandgap voltage reference products, but one likely reason is that at the time the only other semiconductor voltage references were zener diodes (or rather zeners created with transistors), which had significant drawbacks for a designer like Widlar. The zener was temperature sensitive, drifted, was noisy, and required more than 5 volts to function. Low-voltage zeners were deemed to be too unpredictable to use in any precision circuitry at the time. Even though Dobkin's revolutionary buried-zener was a low-drift, ultra-low-noise device, it required an even higher voltage to operate. This was not exactly helpful for the low-voltage designs Widlar evidently had in mind, and he often wrote about the issue in his technical papers and application notes.

While his achievements had become legendary, and provided monumental steps forward for the fledgling analog industry, his remarkable low-voltage design (the LM10) was introduced by National Semiconductor in the late 1970s. It is still in production today, more than 25 years later, and combines on one monolithic chip a low-drift (sub) bandgap voltage reference and a very-low-voltage op amp. The LM10 is a low-power audio-frequency op amp with good specifications and is built using standard processing. It is unique because it can run off a total voltage supply of between 1.1 and 40 volts (LM10/10B/10C versions) and with a typical quiescent current of 270 μA. It has extremely low drift characteristics, as well as a very low input bias current and low noise. It was the first very-low-voltage precision op amp device ever created and remarkably included an unconventional 200-mV (buffered) *sub-bandgap* reference as well. Because the normal bandgap voltage is considered to occur at 1.23 volts, this is even slightly higher than the LM10's minimum operating voltage (1.1 volt). Actually, the LM10 will still be working at a supply voltage of 1.0 volt at 1 mA (over a –55 to +85°C temperature range). For this reason, even though the LM10 contains a real bandgap, it is unconventional and could be considered a sub-bandgap version. It's reference drift is typically 20 ppm/°C or 4 μV/°C (between –55°C and +125°C), and line regulation is

0.001%/V. Best operation occurs at supply levels of between 3 to 15 volts for almost 0% change in V_{out}. It is always a potentially useful device in any battery-powered applications.

At about the same time, one of National's many rivals, Analog Devices Inc., introduced its legendary AD580. It was the first precision bandgap reference and based on what is referred to today as the *Brokaw cell* (named after its innovative designer, Paul Brokaw). This product was destined to become one of the most successful voltage reference ICs ever made, and it ushered in a new level of precision for the affordable voltage reference IC. At the time, voltage references were either 10-volt or 6.9-volt buried-zeners, while the remainder were 1.2-volt shunt bandgap devices. Uniquely, the AD580 was a three-terminal (series) 2.5-volt device. It spearheaded much development among rival analog semiconductor makers, and many products today share its basic design, which consists of a bandgap cell, biasing, and an op amp to boost the core 1.21 volts to the required voltage. The AD580 also helped spur the development over the years of many similar but improved products at Analog Devices, which continues to be a world leader in the voltage reference, A/D, and D/A conversion markets.

Ever since the late 1970s, the bandgap voltage reference has been refined and manufactured by various analog product makers. Many of these products were low-cost, two-terminal (shunt) bandgaps, whereas others were more sophisticated series types. Table 14.1 provides a summary of some of the most popular proprietary products over the past 25 years. Some of these products are still available today, but some of these companies no longer exist or have been taken over or merged with others. Over this period of time, the number of reference manufacturers has actually decreased, although there are still many products to choose from, with new products being introduced frequently.

The original bandgap voltage reference that Widlar conceived of is known today by analog IC designers as the ΔV_{BE} reference or the Widlar bandgap reference, and a simplified circuit is shown in Figure 14.2. This type of reference has three identical NPN transistors, with two of them typically operating at either an 8:1 or 10:1 current-density ratio. If Q_1's collector current is made to be 10 times that of transistor Q_2, then the V_{BE} difference (ΔV_{BE}) will appear across R_3. This may be achieved using either 10 times the emitter area or 10 times the current level. At room temperature, and with this 10:1 current density ratio, the ΔV_{BE} will be 59 mV, and with a negative tempco of approximately –0.2mV/°C. This is because two identical junctions with different currents flowing through them will produce different voltage drops. The ratio of the two collector currents determines the value of this offset voltage (ΔV_{BE}), as will be shown in equation 14.2.

So long as the current ratio is maintained at a constant level, then ΔV_{BE} will be linear with absolute temperature (V_{PTAT}), over most of the temperature range. Because of a small amount of nonlinearity at the temperature extremes, today's bandgap references include special circuitry to compensate for this. Q_1 and Q_2's amplified ΔV_{BE} appears

Manufacturers:	Products:
ADVANCED LINEAR DEVICES	ALD-1603
ANALOG DEVICES	AD580, AD581, AD584, AD680
BURR-BROWN	REF-1004
FERRANTI	ZN404, ZN423
GEC-PLESSEY SEMICONDUCTORS	REF12, SR12D
INTERSIL	ICL8069
LINEAR TECHNOLOGY	LT1004, LT1009, LT1019, LT1029
MAXIM	MAX672, MAX673
MICRO POWER SYSTEMS	MP5010
MOTOROLA SEMICONDUCTORS	MC1403, MC1404, MC1503, MC1504
NATIONAL SEMICONDUCTOR	LM113/313, LM136/236/336, LM368, LM185/285/385, LM4040/41
PRECISION MONOLITHICS	REF-01, REF-02, REF-05, REF-43
RCA SOLID-STATE	CA3085
SILICON GENERAL	SG1503/A
TELEDYNE SEMICONDUCTOR	TC04, TC05
TEXAS INSTRUMENTS	TL430, TL431/A
NOTE: This is only a partial listing, and does not reflect current information. Many of these products have been replaced or obsoleted. You should contact the manufacturer for the latest information.	

Table 14.1 Some popular bandgaps from the past.

across resistor R_2, which has a positive tempco. By carefully selecting the ratio of R_2 to R_3, one can sum these opposite sign tempcos to produce a near-zero tempco over the full temperature range. Actually, the resulting tempco will not be exactly zero, but will be typically 20 ppm/°C for the best devices, with lower-grade devices having as much as 120 ppm/°C. Temperature-compensation (resistor) networks help increase the noise level. When transistor Q_3's V_{BE} is summed with the voltage across R_2, it cre-

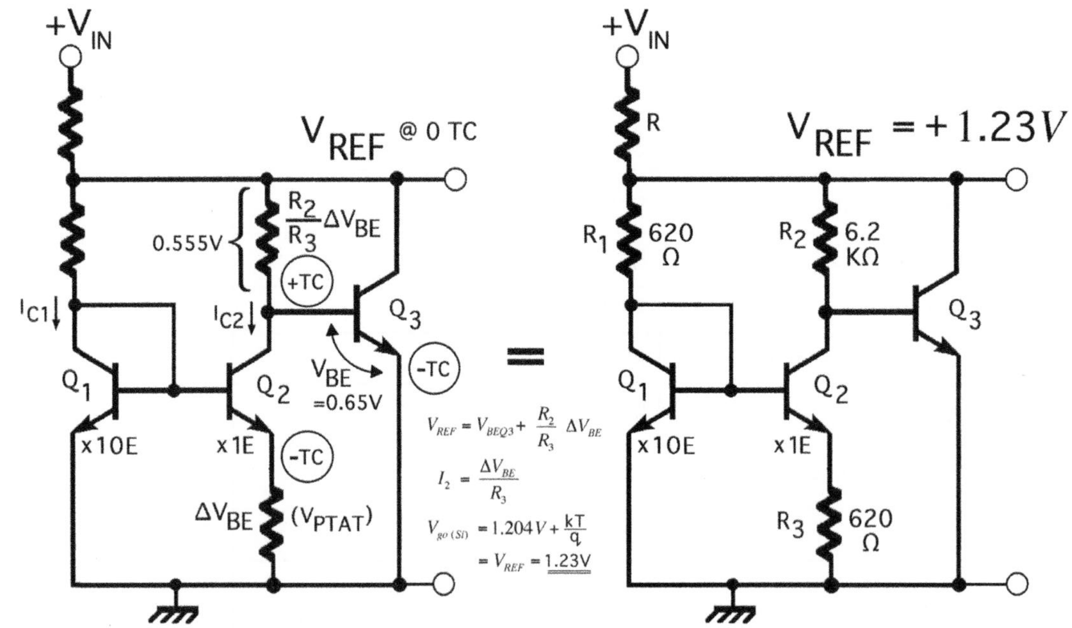

Figure 14.2. Widlar's original bandgap cell, also known as the ΔV_{BE} reference.

ates a constant output voltage that includes 1.205 volts, plus the thermal voltage (kT/q equals 25.7 mV); in other words, V_{REF} equals 1.23 volts, the theoretical bandgap voltage of silicon. (The absolute value varies slightly, according to the processing and ratios used.)

It is relatively easy for the analog IC designer to create such transistors with emitter areas of a particular ratio or a combination of current scaling and emitter scaling. The bandgap design and its operation are unique to analog IC technology and very difficult to re-create using discrete components. It is something that the instrumentation designer can only simulate by using a quality matched transistor pair (see Chapter 5 on matched pairs) and using different currents (as mentioned in Chapter 12, regarding building your own reference). Only the precision matched-pair comes closest to allowing a full re-creation of the basic bandgap cell, because it uses a pair of monolithic transistors with virtually identical specifications.

The bandgap circuit's offset voltage (ΔV_{BE}) can be determined by any of the following equations:

$$\Delta V_{BE} = V_{BE1} - V_{BE2} \qquad \text{(Eq.14.1)}$$

$$\Delta V_{BE} = \frac{kT}{q} \ln \frac{I_{C1}}{I_{C2}} \qquad \text{(Eq.14.2)}$$

where: $\frac{kT}{q} = V_T$ = thermal voltage = $25.7mV$

I_C = collector current

k = Boltzmann' s constant = 1.38×10^{-23} $Joules / K$

T = Absolute temperature in °K (273°K + temp. in°C)

ie. + 25°C = 298°K

q = charge on the electron = 1.6×10^{-19} Coulombs

$$\Delta V_{BE} = \frac{kT}{q} \ln \frac{R_2}{R_1} \qquad \text{(Eq.14.3)}$$

where:

R_2 or R_1 = resistor values (used in a ratio of 8:1, 10:1 etc).

The circuit's temperature coefficient (TC) can be determined by the following:

$$TC\ (\text{in } \mu V / °C) = \frac{\Delta V_{BE}}{\Delta Temp} = \frac{k}{q} \ln\left(\frac{I_{C1}}{I_{C2}}\right) \qquad \text{(Eq.14.4)}$$

where:

$$\frac{k}{q} = 86.25\mu V / °C$$

Figure 14.3 shows how the circuit is compensated to produce a stable output reference voltage of 1.23 volts, with a near-zero tempco. Table 14.2 shows how changing the current-density ratio (Q_1 and Q_2 in Figure 14.2) changes the ΔV_{BE}. The current-density ratio and the emitter area ratio are the key characteristics in the bandgap cell.

In Table 14.2, you can see that using a current-density ratio of 10:1 will produce a ΔV_{BE} of 59.17 mV, which has a corresponding tempco of 199 μV/°C. Applying a gain of 11 to this ΔV_{BE} will provide an amplified ΔV_{BE} of 0.650 volt and a tempco of –2.18

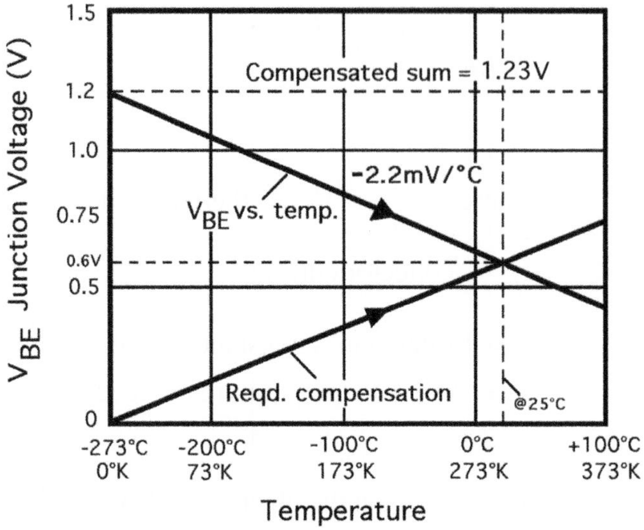

Figure 14.3. *Showing how the bandgap voltage is achieved, by compensating the negative tempco of the transistor's V_{BE} with an equal but opposite tempco. The resulting voltage is approximately 1.23 volts; the theoretical bandgap voltage of silicon.*

mV/°C (using equation 14.3). As a result, it takes an amplified ΔV_{BE} of 650mV with a tempco of +2.18 mV/°C to compensate for Q_3's (with a V_{BE} drop of 0.65 volt) accompanying tempco of around –2.2 mV/°C. Thus, the bandgap circuit has a stable output of 1.23 volts, with a near-zero tempco.

As mentioned previously, the Brokaw cell debuted in the late 1970s in the first precision series bandgap reference: Analog Devices' AD580. The device was available in seven different grades, in a TO-52 metal can. The ground-breaking device was based on a similar 1.2-volt bandgap configuration and consisted of a bandgap cell and its biasing, but with an op amp to boost the core 1.21 volts to the higher 2.5 volts. Adding an external precision op amp with a gain of two times allowed the device to be used in some of the earliest 5-volt DAC and A/D converter applications. Because the AD580 used an amplifier output, it provided better current drive to the load, short-circuit protection, and a significant improvement on the first generation of bandgap references. The circuit was not the same as Widlar's. It used a different circuit configuration, an 8:1 emitter area ratio, along with equal collector load resistors, an op amp with sink/source capability, curve correction circuitry, and laser-trimmed resistors. Figure 14.4 shows the simplified Brokaw cell.

This reference uses two otherwise similar NPN transistors, but typically operating at an 8:1 current-density ratio. Thus with two identical collector currents, if transistor Q_2's emitter area is made to be eight times that of transistor Q_1, then the small V_{BE} difference (ΔV_{BE}) will appear across R_4. (Q_1 operates with an approximate eight times

Ratio:	ΔV_{BE}: (mV)
4:1	35.6
6:1	46.0
8:1	53.4
9:1	56.47
10:1	59.17
11:1	61.62
12:1	63.86
20:1	77.0
measured @ 25°C	

Table 14.2 ΔV_{BE} vs. current-density ratio for bandgap references.

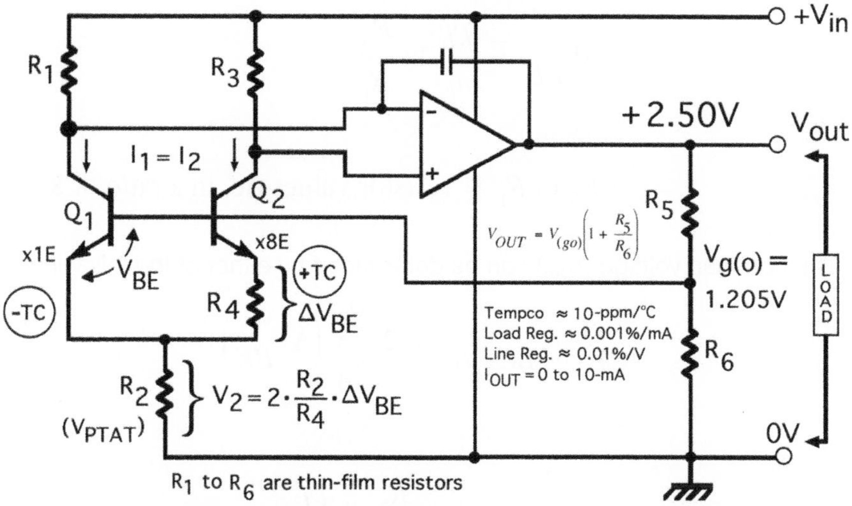

Figure 14.4. The Brokaw cell—used in the first series, precision bandgap voltage reference; Analog Devices' AD 580.

higher current density across its base-emitter junction.) Resistor R_2 produces a voltage drop (V_2) that is proportional to absolute temperature (V_{PTAT}) and will be approximately 0.555 volt at room temperature. Summing the V_{BE} with the V_2 voltage produces the bandgap voltage (V_{go}) at the base of Q_1, and Q_2, and across R6. This effectively becomes part of the op amp's regulating control loop, which continually adjusts this voltage to maintain a state of equilibrium where both collector currents are equal and both amplifier inputs are zero. Summing the V_{BE} voltage, which has a negative tempco, with the ΔV_{BE} voltage, which has a positive tempco, produces a near-zero TC. The op amp's gain is set by the ratio of R_5 to R_6, which boosts the amplifier's output voltage to the desired level. In the case of the AD580, this value is 2.50 volts, but for other products it may be set higher (e.g., 3.0 volts, 4.096 volts, 5.0 volts), according to the supply voltage. The R_5 and R_6 resistors form a precision resistor divider, which in a real reference would be laser-trimmed for a precise output voltage. Again, because of small nonlinearities at the temperature extremes (which give a parabolic type of curve), such a bandgap reference would include special circuitry to compensate for this, as well as other secondary effects.

This bandgap circuit's ΔV_{BE} voltage can be determined by the following equations:

$$\Delta V_{BE} = \frac{kT}{q} \ln \frac{I_{C1}}{I_{C2}} \qquad \text{(Eq.14.5)}$$

$$\Delta V_{BE} = \frac{kT}{q} \ln \frac{R_3}{R_1} \qquad \text{(Eq.14.6)}$$

where:

R_3 or R_1 = resistor values (ie. in a ratio of 8:1, 10:1 etc).

The bandgap voltage (V_{go}) can be determined by either of the following:

$$V_{g0} = V_{BE(Q1)} + 2\frac{R_2}{R_4}\left(\Delta V_{BE}\right) \qquad \text{(Eq.14.7)}$$

$$V_{g0} = V_{BE(Q1)} + 2\frac{R_2}{R_4} \bullet \frac{kT}{q} \bullet \ln(8) \qquad \text{(Eq.14.8)}$$

The reference's output voltage can be determined by the following equation, where R_5 and R_6, together with the bandgap voltage, set the amplifier's gain:

$$V_{OUT} = V_{g0}\left(1 + \frac{R_5}{R_6}\right) \qquad \text{(Eq.14.9)}$$

As an example, we will use Figure 14.5 to show a simple practical circuit based on this Brokaw reference cell and the previous equations. In this circuit, the ratio of the two collector currents (8:1) determines the value of this offset voltage (ΔV_{BE}), as shown in equation 14.6. With an 8:1 current ratio, the ΔV_{BE} will be 53.4 mV (you can check this from Table 14.2), and it will have a positive tempco of approximately 0.2 mV/°C. By carefully selecting the ratio of R_2 to R_4, one can sum these opposite sign tempcos to produce a near-zero tempco over the operating temperature range. The resulting tempco will be typically 10 to 25 ppm/°C for the prime devices. In this circuit, the input voltage supply will be 3.5 volts, while the op amp's control loop holds the output voltage constant at 2.5 volts.

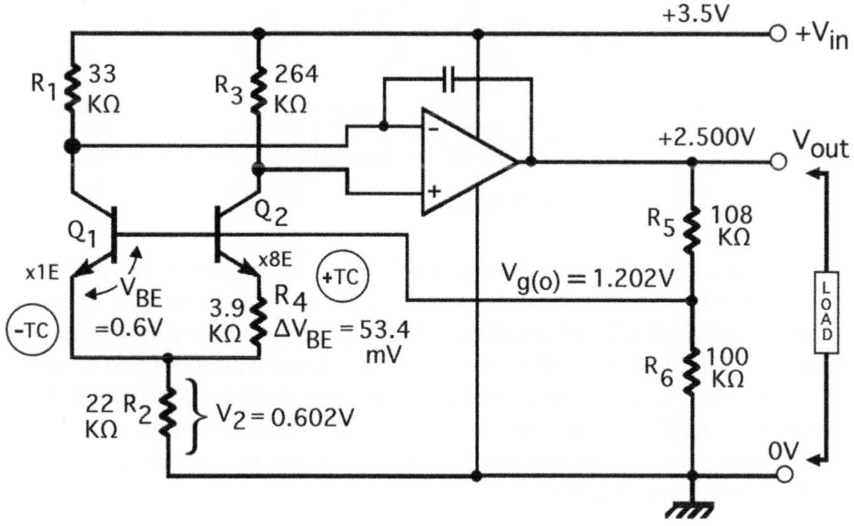

Figure 14.5. *A practical Brokaw cell bandgap voltage reference.*

First we will calculate the circuit's ΔV_{BE} voltage using equation 14.6, as follows:

$$\Delta V_{BE} = \frac{kT}{q} \ln \frac{R_3}{R_1} \quad \text{(Eq.14.10)}$$

where:

$$R_3 : R_1 = 8:1$$

$$\Delta V_{BE} = 25.7mV \ln(8)$$

$$= 0.0257 \bullet 2.079$$

$$= 53.4mV$$

Now we can calculate the bandgap voltage using equation 14.7:

$$V_{g0} = 0.6 + 2\frac{22K}{3.9K}(0.0534V) \qquad \text{(Eq.14.11)}$$
$$= 1.202V$$

We can now determine the reference's output voltage using equation 14.9, where R_5 and R_6, together with the bandgap voltage, set the amplifier's gain:

$$V_{OUT} = V_{g0}\left(1 + \frac{R_5}{R_6}\right) \qquad \text{(Eq.14.12)}$$
$$= 1.202\left(1 + \frac{108K}{100K}\right)$$
$$= 2.500V$$

As a result, this venerable bandgap reference IC can produce a stable output of 2.5 volts at 5 mA, with an initial accuracy of ±10 mV (equals ±0.4%) maximum and a quiescent current of 1.5 mA maximum. It will probably have a low-frequency noise voltage of less than 10 μV peak-to-peak (pk-pk), along with a tempco of less than 25 ppm/°C, over a 0°C to +70°C temperature span. The device will be suitable (with trimming) for up to 8- to 10-bit systems. (National's LM10 should not be overlooked, because it may prove useful in certain low-voltage situations and provide a stable output reference from 0.2 volt upward.)

As you might expect, today's bandgap references provide somewhat higher performance than those devices from 10 or 20 years ago. A similar device with a 2.5-volt at 5-mA output may typically have a lower initial accuracy of ±2 mV (equals ±0.08%) maximum, but now with a lower tempco of around 10 ppm/°C and over the (wider) industrial temperature range, as opposed to the commercial. The device will have a similar low-frequency noise voltage, but now draw a much lower quiescent current (less than 100 μA), as a result of changes in design and processing. Even though various factors have reduced the noise of the core reference to a lower level, the use of added resistor compensation networks for curve and tempco compensation has increased it (resistor noise). Overall then, it is roughly the same for older or newer devices, depending on the grade, and stands typically at between about 5 and 50 μV pk-pk. The device may even have a low-dropout voltage of less than 0.5 volt, and possibly include a low-power sleep/enable pin and/or a noise-reduction pin. This device will probably be suitable with trimming for up to 10- to 12-bit systems. A high-performance 2.5-volt bandgap may offer an even lower initial accuracy of ±1 mV max (equals ±0.04%), along with a tempco of around 5 ppm/°C (over a –40°C to +85°C temperature span). Its long-term stability will be around 60 ppm/1000 hours and be available in a tiny surface-mount package. These specifications are generally suitable for 10- to 12-bit converter systems.

In summary, the bandgap-based reference has improved with advancements over time and with the design efforts of several different manufacturers. This has brought about lower cost, smaller package sizes, better initial accuracy (±%), lower tempco, lower power, and lower dropout. It is available in a two-terminal shunt format for low-cost and small-space applications and for up to 10- to 12-bit applications. The two-terminal type always requires a voltage-dropping resistor at its input terminal, which carries the maximum load current as well as the maximum device quiescent current. The resistor allows the device to operate over a wider operating voltage range. It is also available in a three-pin (or higher) series format, which typically offers a significant increase in performance in most of its characteristics, but at increased cost. More options are available with the series type, as well as requiring no input resistor.

14.2 Buried-zener voltage references

The first commercial buried-zener voltage reference (the subsurface zener) was the LM199. It had been designed by Robert Dobkin and introduced in the early 1970s by National Semiconductor. Subsequent members of the same family included the LM299, LM399, and LM3999, which had less stringent specifications and covered shorter temperature ranges. Uniquely, the reference contained an on-chip substrate heater to stabilize the chip's temperature at 90°C. It also offered a nominal 6.95-volt reference voltage, with a typical 0.3-ppm temperature drift, a long-term drift of less than 20 ppm/month, and a noise spec of about 7 μV rms (10 Hz to 10 KHz). This was much better than anything at the time and is still better than most devices are even today. Although it required more than 7.5 volts to operate the reference and more than 9 volts to operate the heater, it truly was a remarkably low-drift, ultra-low-noise device. It had been designed for the industrial/military markets (10 to 36 volts), and so for applications that needed very high accuracy, low drift, and low noise over a wide temperature span, it was indispensable and virtually unbeatable. Until the LM199 was introduced, the only other references available for instrumentation designers were zener diodes, compensated zeners, a few bandgaps, and expensive hybrids. The LM199 broke new ground, not only with its high-performance specs but also in the design of the chip, the way it was characterized, and even with its special polysulfone thermal shield. National had once again advanced the state of the art with an extraordinary and innovative analog product.

There are some major differences between building discrete zener diodes and ICs, including different crystal doping, processing, testing, and packaging, not to mention different tempcos versus different voltages. However, diodes are one of the essential building blocks of modern electronics, and it is often necessary to create them in the design of integrated circuits, particularly for start-up, current steering, current sources/sinks, and voltage reference circuits. Diodes may be easily created from transistors (BJT, JFET, MOSFETs) in the design of analog ICs, whether bipolar, bipolar-FET, complementary-bipolar, or MOS. There are several advantages for the IC designer in creating diodes from monolithic transistors (e.g., the tight control of the various characteristics, and thereby matching, including thermal matching, and reduced cost). For that reason, and because it's easier, analog IC designers often use a reverse-biased NPN junction transistor to act as a zener diode (see Figure 14.6A). After all, every analog IC circuit is composed of dozens of transistors, no matter the type or topology. Thus, this type of creation became known as the *surface zener.*

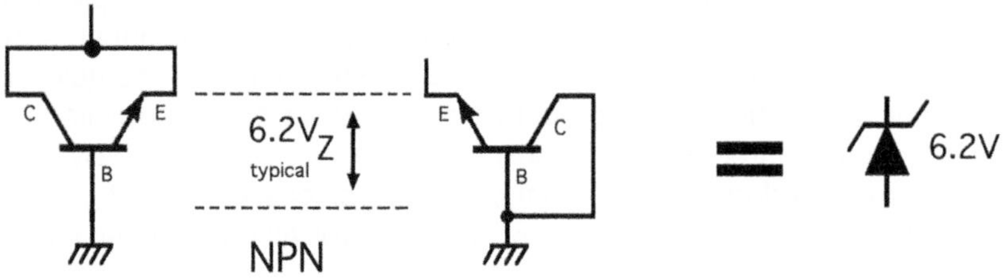

A. Creating a 6.2V, 250mW integrated "surface", zener using standard monolithic transistors.

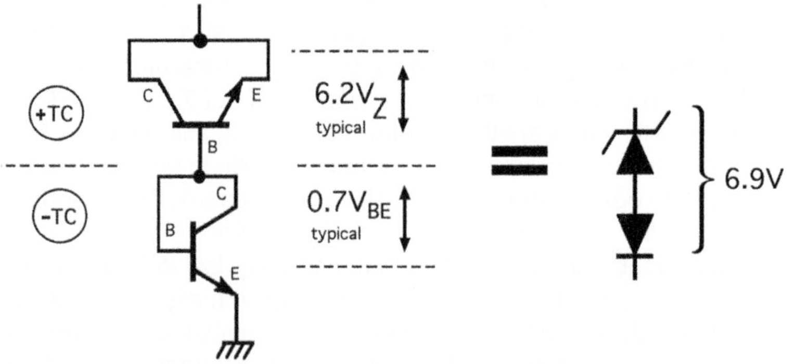

B. Creating a temperature-compensated zener, resulting in a 6.9V device.

Figure 14.6. Creating zeners in monolithic IC form.

With the average process, a reverse-biased BJT breaks down at about 6.2 volts reverse voltage (it would need external current limiting like a zener diode does to be practical though). The transistor's emitter becomes the zener's cathode, and its collector the anode, producing a zener voltage of approximately 6.2 volts ±10%, and with a tempco of around +2.5 mV/°C. To temperature compensate this simulated zener, one could add another NPN transistor connected as a forward-biased (V_{BE}) diode, in series with it (as shown in Figure 14.6B). The forward-biased diode has a tempco of around −2.2 mV/°C. As a result, the two devices' tempcos almost cancel each other out, which is good for stability over temperature. However, while creating a general-purpose zener may not always require low noise and low drift, it does in creating a precision voltage reference. The downside of the surface zener is that because the junction comes to the surface, it is vulnerable to crystal-lattice imperfections, contamination from impurities, and charges on the oxide's surface (see Figure 14.7A). These problems lead to excess noise generation, short-term stability, and turn-on drift, which makes it unsuitable for high-precision/high-bit-resolution tasks where only a very small allowable noise budget may exist.

Low-noise references are essential in high-bit-resolution systems to prevent a loss of accuracy. For example, in a 5-volt, 14-bit system, the allowable noise level for a 100-KHz BW is only about 80 nV/√Hz for 1/2 LSB accuracy. Lower reference voltages and higher bit resolutions require even lower noise levels, so that for a 2.5-volt, 16-bit system, the noise level drops to just 10nV/√Hz!

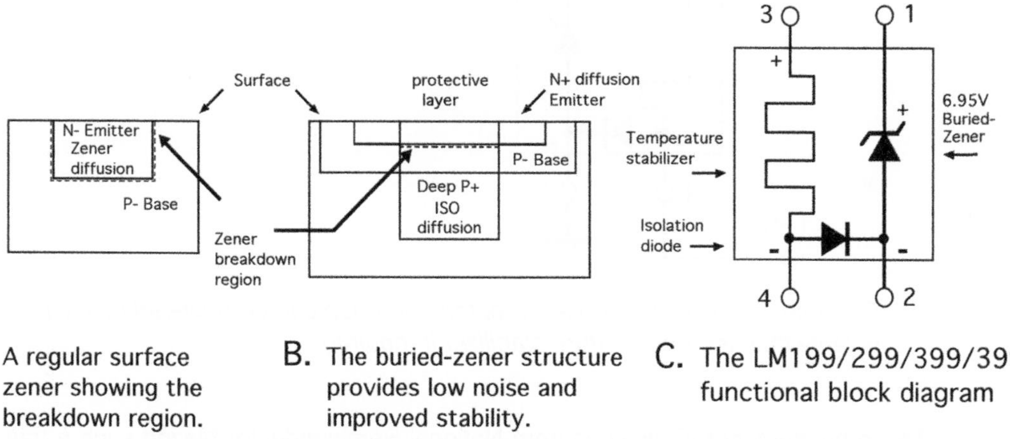

Figure 14.7. Comparing the structure of the common surface zener with the high performance buried zener.

The LM199 circumvented all of those problems by using a unique integrated circuit structure in which the junction was created somewhere below the chip's surface (as shown in Figure 14.7B), and which is covered by an N+ diffusion. The structure that Bob Dobkin had created became known as the *buried-zener reference.* The most important differences in his design were that the noise level, reference drift (tempco), and initial accuracy were all improved over a typical bandgap, by at least an order of magnitude. The LM199 is supplied in a four-pin TO-46 metal can, and its pin-out and schematic are shown in Figure 14.7C. Because the reference and the heater both "float," they can be tied to positive and ground, to negative and ground, or between both supply rails. Figure 14.8 shows such an arrangement where the heater and reference are powered from different supplies. Figure 14.9 shows a design that provides a precision 12-volt reference using an LM199 together with a precision dual op amp. The gain-setting resistors should have excellent 1-ppm TC tracking, so that no additional errors degrade the reference. National soon followed this up with a similar product, the LM129, but this time without the temperature stabilizer. The device was available in either a TO-46 metal can or a plastic TO-92 package.

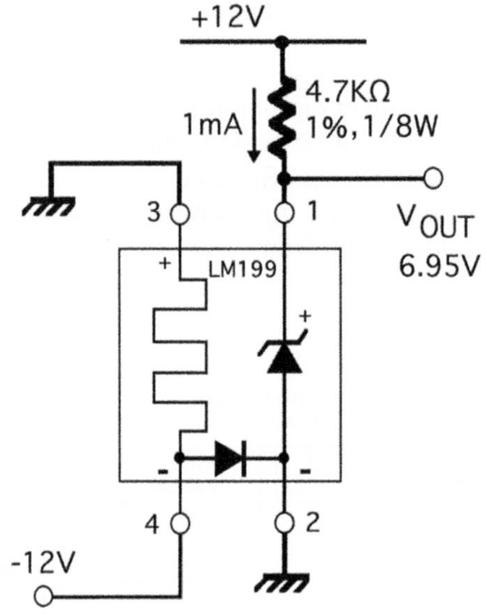

Figure 14.8. *Because the LM199 is so versatile, the reference can be powered by a different polarity than the temperature stabilizer if required.*

These buried-zener IC designs from National Semiconductor blazed such a trail that they encouraged other analog IC makers to develop their own versions. Over time several competitors emerged, including Analog Devices, Burr-Brown, Linear Technology, Maxim, and Precision Monolithics—all had similar or improved buried-zener products available. Like National's LM129, many required no temperature stabilizer. Some examples include Linear Technology's LT1021 family, which was designed by Carl Nelson (who previously designed the LM134 monolithic current source at National Semiconductor in the late '70s, before joining Linear Technology). The LT1021 series were ultra-low-noise (less than 1 ppm pk-pk), ultra-low-drift (2 ppm/°C) references, available in 5-volt, 7-volt, or 10-volt options. Similar devices, the LT1027 and LT1031, had slightly reduced specs and were available as either 5 volts or 10 volts, respectively. Linear Technology also made the ultimate precision reference, with a heater, called

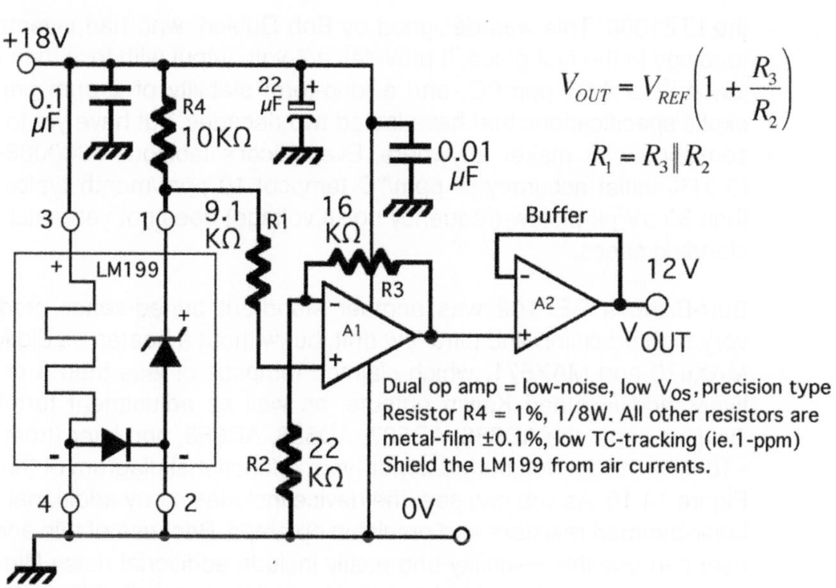

Figure 14.9. *Using the LM199 to create a buffered 12V precision reference.*

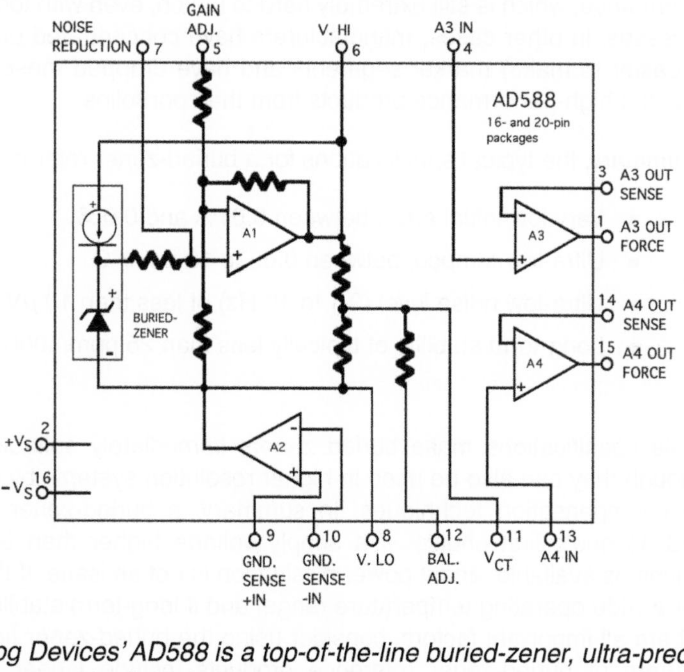

Figure 14.10. *Analog Devices' AD588 is a top-of-the-line buried-zener, ultra-precision voltage reference with all the bells and whistles.*

the LTZ1000. This was designed by Bob Dobkin, who had invented the buried-zener topology in the first place. It provides a 7-volt output with less than 2 μV pk-pk noise, a low drift of 0.05 ppm/°C, and a long-term stability of 2 ppm per month. These are exotic specifications that have lasted two decades, but have yet to be matched by any semiconductor maker anywhere. Even Xicor's fabulous x60008B-50 FGA reference (0.01% initial accuracy; 1 ppm/°C tempco; 10 ppm/month typical stability; and less than 30 μV pk-pk low-frequency noise voltage) does not yet match the LTZ1000's outstanding specs.

Burr-Brown's REF102 was another stand-out buried-zener product. This achieved very high precision and ultra-low drift, but without a heater, as did Maxim's remarkable MAX670 and MAX671, which claimed tempcos of less than 3 or 1 ppm/°C, respectively, and included Kelvin outputs, as well as adjustment terminals. From Analog Devices came the AD586, AD587, AD588, AD688, and later (from PMI) the REF08, a −10-volt/ −10.24-volt buried-zener type. A functional diagram of the AD588 is shown in Figure 14.10. As you can see, the device includes many additional elements, including laser-trimmed resistors and precision op amps. Because of this additional circuitry, the user can use this flexibility and easily include additional noise filtering, precise output voltage adjustment, and Kelvin output sensing—with all of the active components and resistor products needed already on-chip. Note—there are no unused pins.

Many of these different products are still available today; they represent the ultimate in performance, which is still extremely hard to match, even with today's advanced silicon processes. In other cases, manufacturers have concentrated on different (and possibly easier to make) market segments and have dropped these high-priced, hard-to-build, but high-performance products from their portfolios.

In summary, the typical specifications for a buried-zener reference include:

- Very low initial error, between 0.01% and 0.05%
- Ultra-low tempco, between 0.05 to 10 ppm/°C
- Ultra-low noise level (0.1 to 10 Hz) of less than 10 μV pk-pk
- Long-term stability of typically less than 25 ppm/1000 hours

These specifications make buried zeners immediately suitable for 14-bit systems, although they can also be used in higher resolution systems by using special temperature-compensation techniques. In summary, a buried-zener reference should be used as one's first choice, *if* a supply voltage higher than 8 volts (for no heater devices) is available, and if power dissipation is not an issue. If the device has to work over a wide operating temperature range, and if long-term stability, low noise, and low drift are all important factors, consider using the buried-zener first. Today, U.S. manufacturers of buried-zener reference products include Linear Technology, National Semiconductor, Analog Devices, and Texas Instruments.

14.3 The XFET® voltage reference

When Analog Devices unveiled its revolutionary XFET® in 1997, it was a major milestone in the evolution of the monolithic voltage reference. It had been about two decades since a completely new voltage reference topology had last been introduced. In fact, Analog Devices at that time had introduced the series bandgap, in the form of the AD580. The existing topologies available in the mid-90s were shunt and series bandgaps and variations of the buried-zener. Because Analog Devices already made bandgaps, buried-zeners, ADCs, and DACs, it was well aware of the technical limitations of these different products. The XFET® was purposely designed to circumvent many of the limitations of the other types. These limitations had noticeably impacted A/D and D/A converter systems, whose operating voltages were increasingly headed below 5 volts, and whose increased resolutions depended on low-noise precision references.

The main limitations of these references included operating voltage, quiescent current/power dissipation, noise, nonlinear tempcos, all versus cost. Even the best bandgaps and buried-zener products (still) have nonlinear tempcos, particularly at the extremes of their temperature ranges. Even novel attempts to use digital correction techniques, such as a ROM lookup table, proved difficult to implement because of the inconsistencies from one part to another. In contrast, the inexpensive XFET® would provide a low linear tempco and allow for lower voltage operation. It would also feature a lower noise level, a lower quiescent current, lower thermal hysteresis, and a very low long-term drift. In other words, it challenged the well-entrenched and well-understood bandgap and buried-zener types head-on.

After an initial offering of first-generation parts (ADR29x series) and volume manufacturing experience, Analog Devices' designers went back to the drawing board and tweaked the design a little. Then in mid-2001, Analog Devices introduced a family of second-generation XFET® references (ADR42x series). A third generation of XFET® products (ADR43x series) was released in 2003, with a performance upgrade in 2004. These products provided everything their designers intended and more. Most noticeable were an ultra-low noise level of less than 4 μV pk-pk for 2.048-volt and 2.5-volt devices, a very low initial accuracy (less than 0.05%), and a very flat tempco of about 3 ppm/°C. These are all specified over the *automotive* operating temperature range of –40°C to +125°C, with lower-grade devices having lower specs. (Most other competing references are only intended for the industrial temperature range.) These features are in addition to an already low voltage operation and an impressive 40 ppm/1000 hours long-term drift.

As shown in Figure 14.11, the XFET® is based on a unique circuit design. Although it does look at first glance like a relative of the bandgap, this is completely coincidental. Unlike other references, the core of the XFET is based on two P-channel JFETs. One of the two JFETs has an extra ion-implanted gate, which gives rise to its name (eXtra implantation junction Field-Effect Transistor™). In this design the gate-to-source cutoff voltages, $V_{GS\ (off)}$ (aka pinch-off voltage, V_p), are deliberately made different. Both

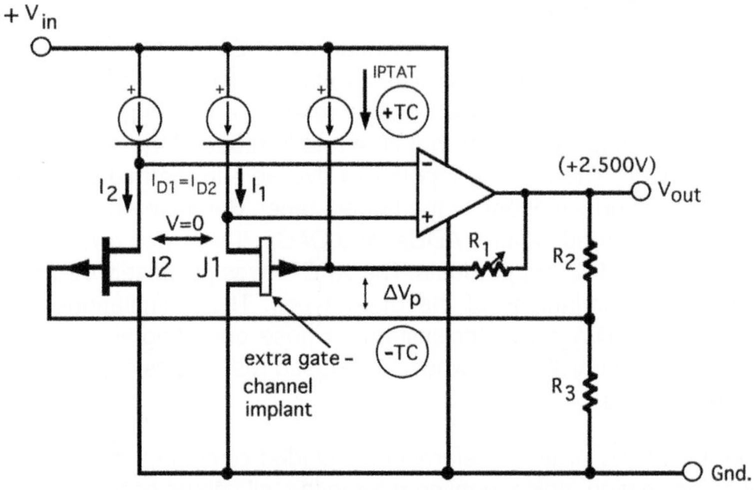

Figure 14.11. Analog Devices' ADR431 uses proprietary XFET® *technology to produce a low-cost but very high performance reference.*

JFETs have otherwise equal source voltages and are driven at the same current level by two matched current sources, but a differential voltage (ΔV_p) occurs between their gates because of the difference in their respective $V_{GS\ (off)}$ values. This can be shown by equation 14.13:

$$\Delta Vp = V_{(gs1)} - V_{(gs2)} \qquad \text{(Eq.14.13)}$$

This basic differential voltage amounts to a ΔV_p of some 500 mV, along with a negative tempco of approximately –120 ppm/°C. The tempco's slope is compensated for by an I_{PTAT} current source and laser-trimmed resistor R_1, as shown in the diagram. Amplifying the constant ΔV_p provides a stable reference output voltage. The op amp's feedback loop keeps both JFET sources at the same potential, while its output drives resistors R_2 and R_3. The resistors are laser-trimmed and provide a range of output voltage options from 2.048 volts to 5.0 volts. The output voltage can be determined by the following equation:

$$V_{OUT} = G \bullet \left(\Delta Vp - R_1 \bullet I_{PTAT}\right) \qquad \text{(Eq.14.14)}$$

where G is the gain of the reciprocal of the divider ratio; ΔV_p is the differential voltage of the two $V_{GS\ (off)}$ values; R_1 is the value of that resistor, and I_{PTAT} is the value of the compensation current, which has a positive tempco. The tempcos involved with ΔV_p and I_{PTAT} tend to cancel out, providing an overall net tempco of less than 3 ppm/°C for B-grade devices (less than 10 ppm/°C for A-grade devices). Because the initial tempco is so much lower compared to most bandgap references, much less correction is required. Because less resistor (compensation) noise is involved, as well as using JFETs, which are inherently low-noise devices, it all helps the device have a significantly lower overall noise level. For example, the typical noise level for an ADR431 (2.5-volt) XFET® is given at less than 4 μV pk-pk (0.1 Hz to 10 Hz), while its noise density at 1 KHz is typically 60nV/√Hz—both superb characteristics. The long-term stability of the device is typically 40 ppm/1000 hours, which is also impressive.

The output of a typical XFET® uses a buffer amplifier, which provides a source capability of up to 30 mA, a sink capability of up to 20 mA, and an indefinite short-circuit to ground current of about 40 mA. The voltage output is specified at ±0.05% (B-grade devices). Some XFET® devices, depending on their output voltage, are specified for operation with supply voltages of between 4.1 and 18 volts. The typical quiescent current will be about 560 μA, while the ripple rejection ratio is typically 75 dB. Line regulation is specified at 20 ppm/V, while load regulation is specified at a maximum of 15 ppm/mA. As with most references, the device should include input decoupling such as a 10-μF electrolytic, in parallel with a 0.1-μF ceramic disk capacitor. The output is perfectly stable without a capacitor, although a 0.1-μF ceramic disk capacitor will help noise suppression. The XFET® also features a very low turn-on settling time of typically 10 μS. (This is a measure of the time required for the output voltage to reach its final value within the specified error band, after a cold start.) Most other references cannot stabilize this quickly.

The XFET® is provided in an eight-pin surface-mount package (SOIC or mini-SOIC) and includes a trim terminal that allows output voltage adjustment over a ±0.5% range. Adjusting the output voltage to another value from nominal allows one to easily trim out system errors. Using this feature has a negligible effect on the device's temperature performance. However, it is prudent to use low tolerance, low TC precision resistors, and a wirewound or metal-foil trimmer. The resistors should be thin-film or metal-film 0.1% types, with a TC of less than 25 ppm/°C. The trimmer should be either a (Bourns) 5% wirewound with a TC of 50 ppm/°C or a 5% metal-foil type (Vishay) with a TC of 10 ppm/°C. Figure 14.12 shows such an arrangement where an ADR431B has its output trimmed to precisely 2.5 volts or to some other value close to that, which could create a deliberate offset. Both input and output are decoupled. Kelvin connections could be added to provide compensation for any stray resistances between the output of the reference and the load. For best performance, use careful layout, quality components, and a single-point ground.

In summary, the XFET® is an outstanding voltage reference with many exceptional features, including a low cost. It also comes from Analog Devices, one of the world's top three analog semiconductor manufacturers (no other company makes as many different types of voltage references). The XFET's unique and innovative design provides very low noise, a very low tempco, a low quiescent current, and a lower thermal hysteresis than most buried-zener references or bandgaps. It also supports a significantly wider operating temperature range than virtually all other types. Its tempco is

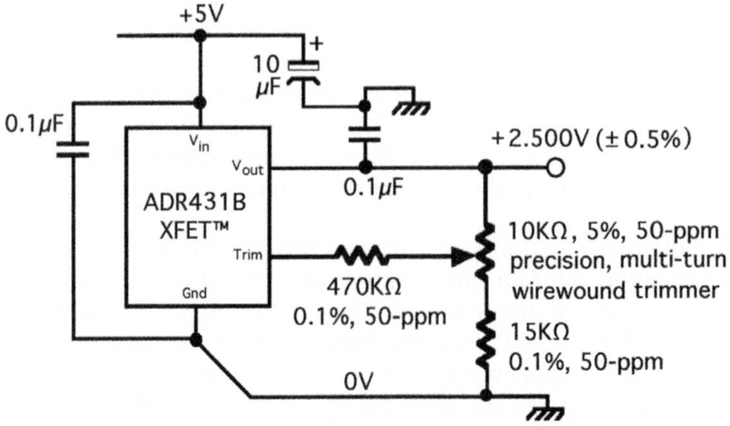

Figure 14.12. Adjusting the output voltage of the XFET® is easily accomplished, and with negligible effect on performance. Use precision resistors though!

flat and linear compared to either the bandgap or the buried-zener, and its long-term stability is better than most other types of reference. Consider using the XFET voltage reference as one's first choice in systems having a supply voltage of between 4.1 and 18 volts. If low noise, high initial accuracy, a low tempco, good long-term stability, a low operating current, a very wide operating temperature range, and a relatively low cost are important factors, then put the XFET at the top of your list. We will review the XFET further in the following chapter.

14.4 The Intersil/Xicor FGA™ voltage reference

Xicor Corporation was co-founded in 1978 by Julius Blank, one of the famous eight young researchers who left Shockley Semiconductor Laboratories to co-found Fairchild Semiconductor, which initially made diodes and BJTs. Although an embarrassed Dr. Shockley referred to them as the Traitorous Eight, these eight talented pioneers are largely responsible for creating today's IC industry (that also includes the most successful semiconductor company of all: Intel, the world leader in microprocessors). Julius Blank was a brilliant mechanical engineer who worked with Sheldon Roberts, Andrew Grove, and Jay Last, and was responsible for designing some of the first semiconductor production equipment (as none was available at the time). Besides optical alignment and assembly equipment, this also included designing furnaces and crystal growers. Although Xicor, the company he co-founded, is not normally considered an analog products company, its years of making programmable mixed-signal, nonvolatile memory and custom products have provided it with some unique capabilities. **As of midsummer 2004, Xicor merged with Intersil Corporation.**

This manifested itself in 2003 when Xicor Corporation announced a completely new type of voltage reference. The patented voltage reference uses Xicor's proprietary

Floating Gate Analog (FGA™) technology in both architecture and CMOS process. It is a cousin of the Electrically Erasable Programmable ROM (EEPROM) memory, which provides nonvolatile storage of digital data. The nonvolatile digital EEPROM was originally pioneered in the 1970s by Intel, Fairchild Semiconductor, General Instrument, Plessey Semiconductors (in the UK), Hitachi (Japan), and others. Today the nonvolatile EEPROM is a commonly used component in computers, microcontrollers, peripherals, and other digitally based equipment. However, the normal EEPROM is a digital device that stores binary data (1's and 0's), whereas Xicor's FGA is a far more complex component, storing *analog* voltage levels—indefinitely.

The core of each FGA™ reference includes a proprietary floating-gate MOS device, analog switches, charge pumps, current sources, a differential charge amplifier, and some special blocking capacitors. These (semiconductor) capacitors are connected in series so that their internal conducting node also forms part of the MOS transistor's floating-gate terminal. Typically, a charge trapped on this floating gate can remain there without loss for more than 10 years. These specially created capacitors form a conduction path with zero leakage current, because they are surrounded by the thick oxide. The device is built as a semiconductor, but it functions as an efficient capacitor, with a leakage current of zero. The process that describes the V-I relationship between the capacitor's current and voltage is known as a form of *Fowler-Nordheim tunneling*. The reference output is actually a buffered version of the ultra-precise floating-gate voltage.

Unlike conventional digital nonvolatile memory technologies, Xicor's proprietary FGA™ technology includes various enhancements that allow smaller threshold voltages to control the status of the floating gate, where the charge remains trapped indefinitely. When used as a reference, this precise voltage translates into improved initial accuracy, a lower tempco, and reduced long-term drift. In order to create a practical manufacturable device, however, there must be built-in provision to increase or reduce the charge on the floating gate during manufacture. Xicor accomplished this by connecting the lower capacitor to a built-in current source, which pulls down to a negative voltage. This limits that capacitor's conduction current to a suitable value and allows charge removal. At the same time, the upper capacitor is connected to a positive voltage source, which enables easy charge addition. This arrangement is shown in Figure 14.13, which shows the main part of the FGA™.

As a result of these two mechanisms, a precise voltage can be set on the floating gate during manufacture, by adjusting either the positive or negative voltage sources. If a positive voltage is applied at V_E, then the upper capacitor (C_1) conducts charge to the floating gate, making it rise. Alternately, if a negative voltage is applied at V_p, then the lower capacitor (C_2) conducts charge from the floating gate, lowering the voltage level. When the precise voltage has been set, both V_E and V_p are then grounded by analog switches. Once set at the factory, the fixed voltage is stored permanently, because there is no loss of charge, and the device is ready to use.

The x60008 FGA™ family is configured as a series reference with a choice of 5.0-volt, 4.096-volt, and 2.5-volt outputs, and is available in various grades. It is intended for operation over the –40°C to +85°C industrial temperature range (a wide 125-degree temperature span). The x60008 series is available in an eight-pin SOIC surface-mount package. (Of these only four pins are used: V_{IN}, V_{OUT}, and two ground pins. The remaining pins are used for factory setting/testing and should *not* be connected but left floating.) Other analog products based on Xicor's proprietary FGA™ technology are future possibilities and already include the ISL60002 and the ISL60007 voltage references, now available from Intersil (see www.intersil.com).

The unique characteristics of the FGA translate into an exceptional voltage reference, which can be used in up to 24-bit systems. Figure 14.14 shows it being used along with a precision op amp, which provides a Kelvin connection to a remote load. Such an arrangement is common in very-high-accuracy reference applications. In many designs the load current can produce an I-R error (i.e., voltage drop equals load current times wire/socket/connector resistance). Small amounts of resistance can quickly add up to more than 1 mV of voltage error, which can easily exceed the allowable error budget in high-performance systems. The force and sense terminals shown in the diagram are compensated by the op amp's feedback loop for any small voltage drops in wiring resistances, connectors, sockets, and circuit board traces between the reference and its load.

The FGA has a very low initial accuracy of less than ±0.5 mV (0.01%), along with an ultra-low tempco typically between 1 and 10 ppm/°C, depending on the grade chosen. The initial accuracy is the best of all voltage references, while its flat tempco is also better than most buried-zener types. The FGA provides an ultra-low long-term drift typically of 10 ppm/1000 hours, and because of its CMOS construction, an ultra-low quiescent current of less than 1 μA. This means that the reference could be left on continuously in some low-power applications, which will provide a more stable reference without the usual warmup drift that affects the initial accuracies and drifts of other references when they are first powered up. The FGA also offers a 10-mA sink/source capability (which many references do not) and a low dropout voltage of less than 300 mV, making it even more attractive for battery-powered applications. Its line regulation is typically less than 100 μV/V, and load regulation is an equally impressive less than 50 μV/mA.

In summary, in systems that have to work over a fairly wide operating temperature range, where initial accuracy, a low tempco, a high bit-resolution (12- to 24-bit systems), an ultra-low long-term drift, a supply voltage of between 5.1 and 9 volts (for 5-volt output), and very low quiescent current are all important factors, then consider the FGA reference *first.* It sets a completely new benchmark for today's high-accuracy monolithic voltage references. We will look at the FGA again in the following chapter.

14.5 Low-voltage considerations

Operation of any voltage reference starts with a stable, well-regulated, low-noise input voltage. The voltage reference has a limiting minimum and maximum input voltage.

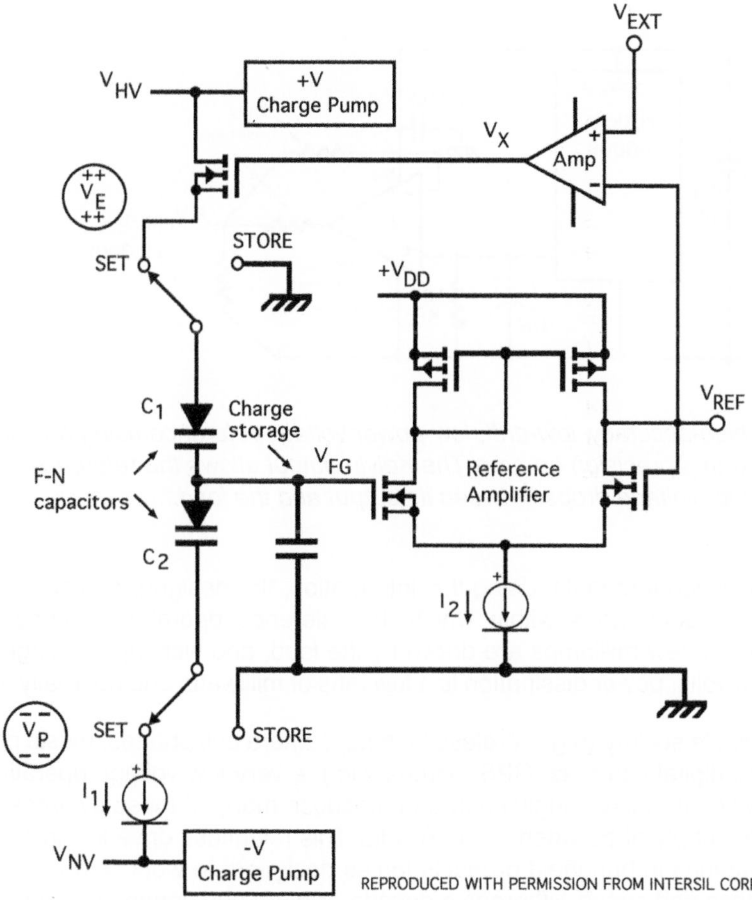

Figure 14.13. Xicor's x60008 Floating Gate Analog (FGA ™) voltage reference achieves its ultra-high precision using a proprietary method of nonvolatile charge storage of analog levels. The internal circuitry enables the voltage level to be precisely set and permanently stored, with zero charge leakage. This results in the x60008 having the best initial accuracy (0.01%), the lowest quiescent current (<1uA), the lowest tempco (1-ppm/°C), and the lowest long-term drift (10-ppm/1000-hours) of any of today's monolithic references.

Too low a voltage, and the reference drops out of regulation and becomes unstable, before it stops working altogether. Too high a voltage, and one runs the risk of over-voltage and permanent damage to the device (not to mention excess power dissipation and thermal instability). Many references, including most bandgaps, specify a dropout voltage, which helps the designer in determining the lowest input voltage that can be applied, often from a nearly discharged battery. The data sheet for any reference invariably specifies the absolute maximum input voltage, with respect to ground

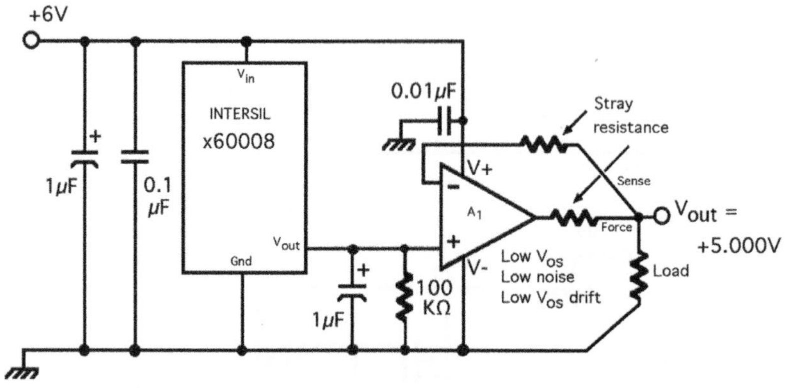

Figure 14.14. Build a high-accuracy, low-drift, low-power voltage reference using an Intersil FGA™ with a precision op amp. The Kelvin output allows the reference to compensate for voltage drops between its output and the load.

and other package terminals. Using this information, the designer can create a set of voltage/current boundaries within which the reference operates properly. In most cases, as only a few milliamps are drawn by the load, and with input voltages generally below 10 volts, power dissipation is a few tens of milliwatts and not really an issue.

In today's mobile society (e.g., wireless laptops, camera cell phones, music pods, CD/DVD players, digital cameras, GPS, games, etc.), a very-low-voltage operation and a long battery life are increasingly important, because many of these systems now use only a single supply of between 1.5 to 6 volts. This relentless drive for using only two AA-sized batteries in handheld products forces designers to work with ever-shrinking supply voltages and space. Whereas a decade or more ago common battery voltages were 9 volt, 6 volt, and 4.5 volt, today's are commonly 1.5 volt and 3 volt. Because batteries are often used in pairs (either in parallel or in series), this means that with a typical 1.5-volt AA alkaline cell having an end-of-life value of approximately 0.9 volt, it results in 3-volt supplies ranging from 3.2 volt (new) to 1.8 volt (discharged), or 1.6 volt (new) to 0.9 volt (discharged).

These supply rails are too low for most voltage references. In fact, only a few shunt and series type bandgaps can operate at these low supply voltages, and only a few of them produce the lowest standard reference output of 1.024 volts. When you consider that the *minimum operating voltage* for most bipolar designs is about 1 volt (a forward V_{BE}, plus a small saturation voltage), this becomes the *limiting factor.* Perhaps future monolithic CMOS designs will follow National's lead with their LM4140. This incorporates a series P-channel, ultra-low threshold, low on-resistance MOSFET at its output instead of a BJT, in order to provide a very low (less than 50 mV) input-output differential (i.e., dropout). This technique is used in some voltage regulators for the same purpose. The difference is that instead of a fairly low 200- to 400-mV dropout, which is typical of many of today's bandgaps, the output transistor's voltage drop would be less than 50 mV at 10 mA. This 150-mV difference could help extend the useful battery life of some products significantly. With many bandgaps, the minimum operating voltage is more than 3 volts, which is unsuitable for lower voltage applications.

A popular Analog Devices product, the AD1580, is a 1.2-volt micropower shunt reference, which can operate with a supply voltage of +1.8 volt and higher. Another is the newer ADR510, a very-low-noise (4 μV pk-pk), precision 1-volt shunt reference that can also operate at below 2 volts.

Thaler Corporation's VRE4100 family includes two low-voltage, high-performance series references in an SOIC-8 package. The VRE4110B (1.024 volts, ±0.05% max, 1 ppm/°C max. tempco) requires a minimum input of 1.26 volts. A second product, the VRE4112B (1.25 volts, ±0.05% max, 1 ppm/°C max. tempco), has a minimum input of 1.49 volts. Both are series types and extremely impressive products.

Maxim also has some excellent low-voltage references, both shunt and series. One is the MAX6018_12, which is a series reference in an SOT-23 package. This device provides a reference voltage of 1.25 volts, for an input as low as 1.8 volts. With a maximum 5-μA quiescent current, the device offers an initial accuracy of 0.2% max and a tempco of less than 60 ppm/°C. Some of the lowest quiescent current references also come from Maxim. One standout family is the MAX6006 series (6006/6007/6008/6009). These are shunt references in an SOT-23-3 surface-mount package, which feature a quiescent current as low as 1 μA I_Q). They are available in four voltages from between 1.25 volts and 3 volts.

National Semiconductor's LM4041, a low-noise, micropower 1.225-volt shunt reference, offers an initial accuracy of 0.1% max and a tempco of less than 100 ppm/°C over the industrial temperature range. The device is available in SOT-23 or SC-70 surface-mount packages and consumes as little as 60 μA quiescent current. It is also useful in 2- to 5-volt applications, and available from National, as well as Maxim and other vendors. National's LM4051, a newer 1.225-volt shunt reference (also second-sourced by Maxim), offers some improved specs, including a maximum 50 ppm/°C tempco that is guaranteed from –40°C to +125°C.

From Linear Technology comes the LT6650, a micropower, adjustable series reference that operates on a minimum 1.4 volts. The reference output is capable of being adjustable from 400 mV and upward.

Semtech's SC4436 is another very-low-voltage reference and is manufactured in an SC-70 surface-mount package. The device is a precision *sub-bandgap* shunt reference and has a 0.6-volt output. It operates from as low as 1.7 volts and draws as little as 10 μA.

In fact, several manufacturers offer references with outputs of less than 1.8 volts, but most require a supply of at least 2 volts in order to function properly. Only a very few references can operate at less than 1.7 volts. The very-low-voltage monolithic voltage reference is referred to in the industry as the *sub-bandgap*—an area currently being researched by most analog reference manufacturers. As you can appreciate, at this voltage level there are numerous pitfalls in developing such products, both in their architectures and processing, but also in their noise levels, response times, output capability, and so on. No doubt some manufacturers will provide sub-bandgap shunt references, while others will offer series devices. For now we can only speculate, but it is certainly a market segment that will get a lot of attention in the future from instrumentation designers as supply voltages for A/D and D/A converters continue to drop.

So what other possibilities exist for working from a 1.5-volt supply? Very few, but don't count out the venerable yet remarkable low-voltage LM10. It combines on one monolithic chip an unconventional 200-mV, low-drift sub-bandgap voltage reference and buffer and a separate low-voltage precision op amp. The LM10 is made by National Semiconductor (Bob Widlar designed it in the late 1970s) and second-sourced by Linear Technology. It was the first very-low-voltage op amp and built using a mature junction-isolated process. Being an older part, it does not use today's much improved PNPs, which is shown by the fact that it has quite a low GBW of only 100 kHz. However, the device is excellent for DC and audio-frequency applications. A newer process using today's improved vertical PNPs (such as National's low-power,

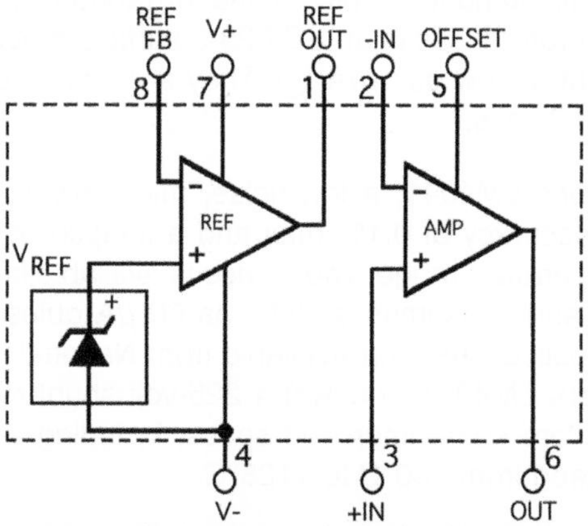

Figure 14.15. The LM10 pinout and functional diagram.

complimentary-bipolar VIP® process) might improve not only the gain bandwidth but also some of the LM10's already good specs.

The LM10 (shown in Figure 14.15) is unique, because it was designed to run off a supply voltage of between 1.1 and 40 volts (LM10/10B/10C versions) and with a typical quiescent current of 270 μA. Some low-voltage versions are LM10BL and

LM10CL, which have a maximum upper voltage limit of 7 volts. Although the best operation occurs at supply levels of between 3 to 15 volts, the LM10 can still be useful in very-low-voltage applications (1- to 3-volt supplies). Actually, the minimum operating voltage for the buffered reference cell over a –55°C to +85°C range is 1 volt at 1 mA (or 0.9 volt at 100 μA at 60°C). At 1.2 volts and higher, the reference amplifier can supply up to 3 mA. The minimum operating voltage for the op amp depends on both load current and the allowable gain error, and is also approximately 1 volt at 1 mA. (However, National only tests and guarantees the specs between 1.2 and 40 volts.) The LM10 has extremely low drift characteristics, as well as a very-low-input bias current. Its reference drift between –55°C and +125°C is typically 20 ppm/°C (or 4 μV/°C), and line regulation is about 0.001%/V. Because of its unique design, it has a much flatter tempco curve than traditional bandgaps and also includes temperature compensation. Figure 14.16 shows some practical low-voltage reference designs using the LM10.

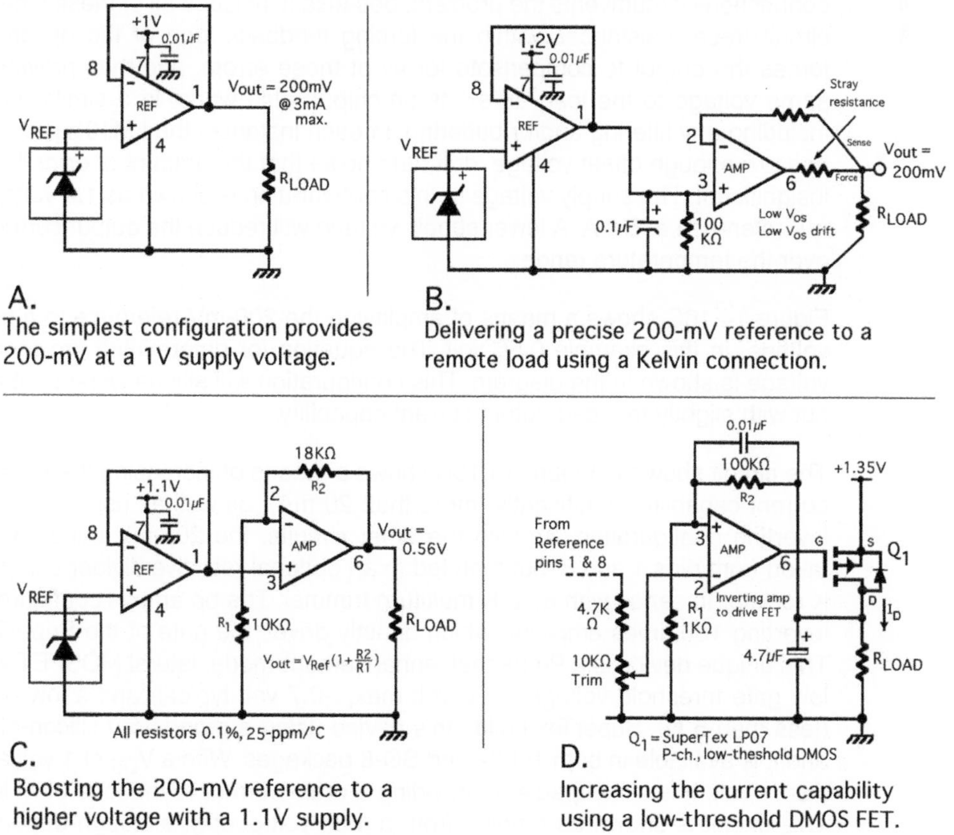

A.
The simplest configuration provides 200-mV at a 1V supply voltage.

B.
Delivering a precise 200-mV reference to a remote load using a Kelvin connection.

C.
Boosting the 200-mV reference to a higher voltage with a 1.1V supply.

D.
Increasing the current capability using a low-threshold DMOS FET.

Figure 14.16. Some LM10 low-voltage reference applications.

Figure 14.16A shows the simplest and lowest voltage configuration, using the buffered reference cell to provide a stable 200-mV reference from a 1-volt supply. While the buffered reference output is capable of 3.5 mA maximum (at room temperature), at a supply voltage of approximately 0.95 volt it can only support a current of 0.1 mA. In the configuration shown, only the buffered reference is used, leaving the op amp available for another task.

Figure 14.16B shows a configuration with the buffered reference supplying the 200-mV reference to the op amp. Here the op amp includes force and sense terminals (Kelvin connections), which compensate for any small voltage drops in wiring resistances, connectors, sockets, and circuit board traces, between the reference and the load. In many designs the load current can produce an I-R error (i.e., voltage drop equals load current times wire/socket/connector resistance). Small amounts of resistance can quickly add up to more than 1 mV of voltage error, which can exceed a converter's allowable error budget. Using a voltage reference with these Kelvin connections circumvents the problem, because it includes all of these small wiring and circuit trace resistances within the forcing feedback loop of the op amp. The loop forces the output to compensate for all of these errors, and thus provides the exact same voltage to the load as exists on-chip. It also works in a similar manner when including any filtering and/or buffering. In each instance, the LM10's precision op amp has low enough offset voltage, drift, and noise that the amount of error it introduces is insignificant. The supply voltage in this configuration is shown as 1.2 volts, with a 200-mV reference at 5 mA. A lower supply voltage will reduce the output current capability, over the temperature range.

Figure 14.16C shows a means of amplifying the 200-mV reference to a higher output voltage, in this example 0.56 volt. The equation for determining the required output voltage is shown in the diagram. This configuration will still be working at about 1 volt, but with slightly reduced output current capability.

The circuit shown in Figure 14.16D shows a means of increasing the reference output current capability significantly (more than 20 mA), as well as using the op amp in an inverting configuration. As with the other circuits, the 200-mV buffered reference is taken from pins 1 and 8, but then fed to an optional resistive divider comprising a 4.7-K resistor in series with a 10-K multiturn trimmer. The op amp is configured as a fixed inverting 100 times amplifier, which directly drives the gate of the SuperTex LP0701. This unique device is a P-channel, enhancement-mode, lateral MOSFET with an ultra-low gate threshold voltage (–1.0 volt max, –0.7 volt typical) and a low on-resistance (less than 5 Ω). SuperTex builds this device using a proprietary silicon-gate process, and it is available in both TO-92 and SO-8 packages. With a V_{DS} of 1 volt and a V_{GS} of also 1 volt, it is still capable of providing a drain current to the load of at least 50 mA. This circuit is shown as running from a 1.35-volt supply, although depending on the MOSFET's exact gate threshold voltage, this could be reduced a little. The op amp's feedback loop includes the load voltage, and the circuit's actual output voltage is determined by the voltage at the trimmer's wiper. All resistors should be high-quality

0.1% metal-film or thin-film types with low TCs or very low tracking TCs if used in a network. The trimmer, if used, should be a 5% wirewound type, with a similarly low TC.

As you can see, the LM10 still proves to be a real workhorse, and it can be helpful in very-low-voltage applications, like those shown. Its relatively low quiescent current and typically low 20-ppm/°C tempco can still be effective in providing a stable reference in very-low-voltage designs, where most other references simply cannot even function. Widlar designed it in 1978, and it's still ahead of the market here in 2005!

14.6 Comparing the different topologies

The lines that have traditionally divided the different topologies have already begun to blur. Some next-generation *super-bandgaps* can match or exceed some of the specifications for the buried-zener, the XFET®, and the FGA™. The voltage reference market is no longer as sharply defined as it was in the past. It is no longer an easy matter to choose between the best of each topology, because many of their key specifications now overlap. It is likely that you will have to carefully prioritize your needs and then compromise on some characteristics such as low noise versus a low quiescent current or low dropout versus operating voltage. The different series reference types are summarized in Table 14.3.

The table compares the specifications for a 5-volt reference only, to provide some equality. Please note that some of these examples are generic and only representative of that reference topology. While this gives some indication of one type's specifications versus another's, you should review the applicable data sheets for exact, up-to-date information (many are available online as Adobe Acrobat .PDF files). Of the different types of series references shown, the types with many of the best specifications are the buried-zener reference and the Xicor FGA™. However, Analog Devices' latest-generation XFET® is very close behind, and the gap is narrowing. The XFET® has some unique advantages—various voltage options, low noise levels, cost, and also runs over a wider temperature range.

Don't rule out the latest generation of bandgaps either, whose specifications (including initial accuracy, tempco, and low noise) are also closing the gap. In every case, the choice of which is best for your application can only be determined by your specific needs and priorities. One often has to compromise in a design, between one's first and second choice, according to those priorities, which usually include cost. Remember that a good reference design keeps it simple, uses worst-case design values throughout, and provides a practical, stable, manufacturable end product. Table 14.4 provides a brief checklist of some of the more important considerations for your design. The first few questions dictate the type of reference required.

Other factors, often hidden, can play a major part in choosing any particular reference for a design. These include the following:

Topology → Specification:	unit	Traditional series bandgap	Super bandgap (Typical)	Buried-Zener (Typical)	ADI XFET® (ADR435B)	Xicor (3) FGA™ (x60008B-50)
Reference voltage	Volts	5.000	5.000	5.000	5.000	5.000
Initial accuracy/tolerance	mV or %	2mV/0.04%	2mV/0.04%	±1mV/0.02%	2mV/0.04%	0.5mV/0.01%
Tempco/drift	ppm/°C	5 max	3 max	2 max	3 max	1 max
Long-term drift	ppm/ 1000-hrs	250 typ	40 typ	20 typ	40 typ	10 typ
Noise voltage (0.1-10Hz)	μV pk-pk	50	10 typ	3 typ	3.4 typ	30
Wideband noise @ 1KHz	nV/√Hz	n/s (2)	26μVrms	2μVrms typ	110 typ	<100 typ
Supply current	μA or mA	50μA max	750μA	2.4mA max	560 typ	0.8μA max
Input voltage range @ 5V$_O$	Volts	5.5 - 15	5.2 - 30	8 - 40	7 - 18	5.1 - 9
Load current max.	mA	25	15	+10/-10	+30/-20	±10 max
Line regulation	ppm/V	4-ppm max	50μV/V max	6-ppm typ	35-ppm max	100μV/V max
Load regulation	ppm/mA	4-ppm max	0.1mV/mA	3-ppm typ	70-ppm max	50μV/mA max
Dropout voltage	Volts	0.5V max	0.2V max	2V typ	2V	0.3V max
Temperature range	°C	-40°/+85°	-40°/+85°	0°/+70°	-40°/+125°	-40°/+85°
Trimmable output (Trim pin)		No	Some	Yes	Yes	No

NOTES: 1. Specifications are for example 5-Volt output only. Specs may be different for other output voltages.
2. n/s = not specified. 3. Xicor is now a part of Intersil Corp.

Table 14.3 Comparing the different series reference topologies.

- *Space and circuit board area.* These can often become a significant factor. Will the location of the reference on the circuit board require special placement away from heat sources and noise generators? Can another nearby circuit board influence or degrade the reference by being too close? Will there be space available that allows the required orientation (for thermal gradient or vibration reasons) or require slots to be specially milled in the PCB? Some of these factors can boost the cost of the product or slow delivery times.
- *Temperature range.* If the reference only has to work over a relatively narrow temperature range, this may translate into being able to use a lower-grade/cheaper reference. An extended industrial or military temperature range may be impossible for many references, including some of the most exotic. Only ADI's XFET®, some buried-zener products, and a few older bandgaps can support the automotive or military temperature ranges. Thus a 3-ppm/°C tempco over a –40°C to +125°C range is probably better than a 1-ppm/°C over a –25°C to +85°C range!

	Requirements :
Supply voltage ?	
Temperature range ?	
Bit-resolution needed ?	
Allowable noise error value ?	
Reference voltage value ?	
Reference initial accuracy ?	
Reference Tempco/drift ?	
Reference noise ?	
Reference Long-term drift ?	
Reference supply current ?	
Reference low dropout ?	
Reference trim terminal ?	
Reference noise terminal ?	
Physical size restrictions ?	

Table 14.4 Checklist for voltage reference design.

- *Circuit complexity.* Will it take additional circuitry such as temperature compensation, Kelvin connections, or additional noise filtering to achieve the accuracy and stability results you need? If so, these can quickly increase the cost of the product, and some may even slightly degrade the overall accuracy or tempco. They may also require a test technician to tweak the design into acceptable limits.
- *Trimming.* In high-resolution systems, it is often necessary to adjust the reference voltage with a built-in trim terminal, to trim out any system errors. This may add to the labor costs in producing the product, as well as adding to the parts cost (e.g., multiturn precision trimmer and precision low-TC resistors, a precision op amp or two). An alternative may be to use a higher-performance device that requires no trimming and saves on labor costs.
- *Filtering.* In high-resolution systems, it is often necessary to filter the reference voltage to meet the noise/allowable error value versus bit resolution of the converter. This may add to the labor costs in producing the product, as well as to complexity of the circuit and the parts cost (e.g., precision op amp(s), high-stability capacitors, and precision low-TC

resistors). An alternative may be to use a lower noise but costlier device that requires no filtering and saves on labor costs. Another option is to use a high-performance device with a built-in noise-reduction terminal. This typically uses only a single capacitor. While convenient, this method may not provide the same level of noise reduction that a quality active (low-pass) filter can.

- *Cost.* Some shunt bandgaps cost less than $1 each at 1,000 pieces, but most other references cost many times more. Precision never comes cheap, so it is wise to budget for more expensive options every time.
- *Availability and lead time.* The downside of designing in a particular high-grade part may result in short supply, a single source, or extended lead time from the vendor. In any event, be sure to specify *exactly* the part required for your design. The difference in a grade-A device versus a grade-B device may be Δ50-ppm/°C, or instead of ±0.01%, it may be ±0.1%. The industry does not standardize on the various grades, so what one vendor calls a grade-A device may be totally different from another. Be sure to clarify for your Purchasing Department what is acceptable for your design and what is not. An unspecified second source or a convenient alternative part in your design might become a major source of problems for your production department, for Q&A/test, and for you.

The following chapter provides a closer look at some of the best voltage references available, as well as how they are used.

Chapter 15

A Review of Some Outstanding Monolithic Voltage References and Their Applications

This chapter takes a brief look at some outstanding monolithic voltage reference products from several leading U.S. manufacturers and how they can be applied. Some helpful design examples are also shown. In many cases the design can be further enhanced by using some of the techniques discussed in earlier chapters, such as filtering, preregulating the power supplies, bypassing, using Kelvin connections, temperature compensation, and so on. As we reviewed in Chapter 14, one first has to review the needs of the reference application, before choosing a particular topology. A quick way to evaluate any reference application is in determining (1) the available supply voltage, (2) the required reference voltage, and (3) the operating temperature range. The supply voltage and its ±% tolerance governs both the applicable output reference voltage and what type of reference you can use. Obviously, you can't expect a 2.9-volt reference from a 3-volt supply, and you shouldn't expect to buy a monolithic 1.23-volt reference with a 0.5-ppm/°C tempco any time soon. A commercial reference IC product that can even run below a +1.0-volt supply is not yet available, although some manufacturers are likely developing such products now for the future.

The lowest input voltage ($V_{in\ min}$) to the reference should be quickly determined with respect to the output voltage. If the difference is less than 1 volt, then consider a low-dropout reference (many of which go down to less than 0.35 volt). The voltage reference has a range of input voltages, which also govern your choice. Some otherwise suitable products may have a minimum supply voltage too high for your needs, while the upper voltage limit may be too low in others. In most cases you should ensure that the incoming supply voltage is well regulated from a nearby linear regulator. The regulator's minimum output voltage to the intended reference device should be a part of your design calculations. In applications with a high bit-resolution, you may need to measure the regulator's output noise-voltage and provide additional bypassing. The regulator will do the heavy lifting and take care of line transients, supply noise, reverse-bias, and overload protection, while supplying a lower-noise, more stable voltage to the reference's input. The linear regulator need only be a small, low-cost, low-noise 50- to 100-mA device. Its bypass capacitors should be 0.1-μF disk ceramic capacitors at both its input and output. Always consult the manufacturer's data sheet for recommendations regarding the regulator's input and output capacitors, because the details vary from product to product.

The operating temperature is a major factor, because all references are affected by temperature, thereby having a tempco and a thermal hysteresis. The operating temperature range will govern your choice of reference, because some otherwise acceptable devices may not be suitable for that particular range (usually at under –40°C and over +85°C extremes). In many cases the device is used to provide a reference scale to an A/D or D/A converter, and the temperature span will directly reflect on the reference's drift. All converters have a bit-resolution and a scale factor tempco (i.e., gain error), which will directly apply to the intended reference, and through a process of elimination, help you decide on potential devices. Once you know the required bit-resolution, you can determine the allowable budget for noise error that applies.

You should also determine the converter's linearity error and its scale factor tempco, so you can calculate the percentage accuracy (or total error) at the maximum temperature. For example, suppose a 16-bit converter has a linearity error of 0.005% and a scale factor tempco of 2 ppm/°C. The reference used has a tempco of 3 ppm/°C, and the maximum operating temperature will be 85°C (or 60° above 25°C room temperature = ΔT°). To determine the accuracy percentage (or error): first multiply the total scale factor tempco by ΔT°, multiply this by 100 to get the percentage, and add the linearity error percentage. (Here the total scale factor tempco includes the scale factor tempco of the converter and the tempco of the reference.) We can now determine the accuracy in this example, using the previous data:

$$Accuracy\ \% = 0.005\% + \left(5ppm \times 60^{\circ}\right) \times 100 = 0.035\% \qquad \text{(Eq.15.1)}$$

One can get a quick approximation of just what the reference can achieve over the temperature range by setting the converter's linearity error to zero, and simply multiplying the reference's tempco by ΔT°, so that:

$$\begin{aligned} Error\ \% &= 3ppm \times 60^{\circ} = 180ppm \qquad \text{(Eq.15.2)} \\ &= 0.018\% = \ > 12\ bit \text{ accuracy} \end{aligned}$$

Other important considerations include the reference's quiescent current (and therefore its power consumption), as well as its long-term drift. Here we will look at some remarkable references according to their topology and how they can be used in various applications.

15.1 Applying the bandgap shunt reference

Today's bandgap-based reference technology has brought about lower cost, smaller package sizes, better initial accuracy (±%), lower tempco, and lower power. The bandgap reference is available in either the three-terminal series format or the two-terminal shunt format. The shunt type, which we will look at first, is for low-cost, low-voltage, small space, and less demanding applications. Unlike the low-voltage zener diode, which it was designed to replace, the bandgap has a lower voltage with a sharper knee, at low current levels. Shunt bandgaps are usually available over the voltage range of between 1.2 volts to 5 volts. Many devices have an initial accuracy of less than ±0.2%, along with a temperature coefficient of less than 100 ppm/°C, over the industrial temperature range (some new devices have much lower tempcos). In terms of digital resolution, the shunt type would typically be used for 8- to 10-bit accuracy, with a few new devices capable of 12-bit accuracy. In many instances, today's shunt bandgap is a low-power device, with a low quiescent current often below 100 μA. Most of the bandgap reference's noise is caused by the use of resistor compensation networks for curve and tempco compensation, resulting in low-frequency noise levels of between about 5 and 100 μV peak-to-peak (pk-pk). Many shunt bandgaps are now available in tiny surface-mount packages like the SOT-23 and the SC70, which makes them attractive for low-power/portable, limited space applications.

The shunt reference (like the zener diode that it is modeled from) *always* requires a voltage dropping resistor (R_S) at the junction of its input terminal (V+, +, V_{in}, cathode). This junction is also the output node to the load. Its anode typically connects to ground or a more negative voltage (and is marked as either "-" or "A" for anode). A typical shunt reference circuit is shown in Figure 15.1.

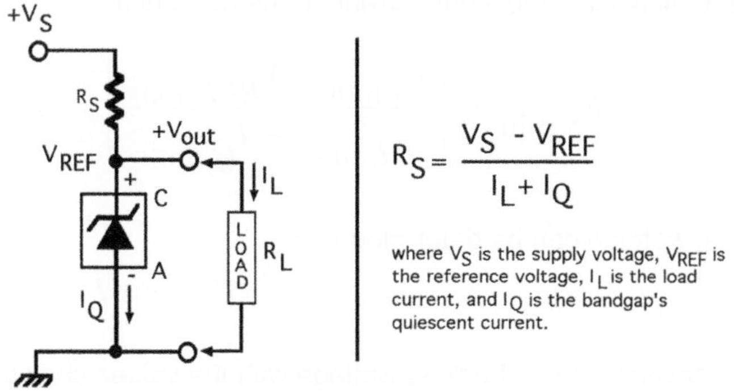
+VS
RS
+Vout
VREF
+
C
A
-
IL
IQ
LOAD
RL
RS = (VS - VREF) / (IL + IQ)
where VS is the supply voltage, VREF is the reference voltage, IL is the load current, and IQ is the bandgap's quiescent current.

Figure 15.1. The two-terminal (shunt) bandgap voltage reference.

The resistor allows the reference to operate over a relatively wide voltage range, which can be important in some applications. The resistor is often given only brief consideration, but it carries the maximum load current as well as the maximum device quiescent current. It also protects the shunt reference from being destroyed. Ideally it should be a good-quality, low-tolerance (less than ±0.5%) precision type, with a low tempco of less than 30 ppm/°C. (See the section in Chapter 2 on precision resistors, networks, and trimmers.) To help reduce the tempco over a wide temperature span, and to avoid self-heating effects, it is advisable to use a higher wattage resistor than needed (i.e., 45 mW required, but a 250-mW device used). The resistor's nominal value should be chosen for the input voltage (V_S), less the reference voltage (V_{REF}), divided by the load current (I_L) and the quiescent current (I_Q) of the shunt reference. It can be described by:

$$R_S = \left(\frac{V_S - V_{REF}}{I_L + I_Q}\right) \quad \text{(Eq.15.3)}$$

However, one must also ensure that the value of R_S is sufficient to limit the quiescent current to whatever the reference's maximum rating is (i.e., less than 10mA, as shown in equation 15.4), when the input supply voltage is at maximum, and both the load current and the reference voltage are at their minimum. This ensures that the resistor's value is always more than shown here:

$$R_S = \left(\frac{V_{S\,max} - V_{REF\,min}}{10mA + I_L\,min}\right) \quad \text{(Eq.15.4)}$$

The resistor's minimum value should be calculated for when the input voltage is at its minimum, the reference voltage and load current (I_L) are at their maximum, and allowing for the minimum quiescent current for the reference.

$$R_{S\,min} = \left(\frac{V_{S\,min} - V_{REF\,max}}{I_{L\,max} + I_{Q\,min}}\right) \quad \text{(Eq.15.5)}$$

Its normal wattage can be determined by:

$$P = V \times I \quad \text{(Eq.15.6)}$$

It is also advisable to recheck its wattage with the values shown in equation 15.4.

Let's take as an example a typical 1.225-volt shunt reference with a 0.5% initial accuracy, a minimum quiescent current of 100 μA, and a maximum capability of 10 mA. In this example the supply voltage is 3 volts with a tolerance of ±5%, and the load current is 1 mA, also with a tolerance of ±5%. This would mean that $V_{S\,max}$ is 3.15 volts, $V_{S\,min}$ is 2.85 volts, $V_{REF\,max}$ is 1.231 volts, and $V_{REF\,min}$ is 1.218 volts. Using equation 15.5, we can calculate the minimum value for R_S as follows:

$$R_{S\,min} = \left(\frac{2.85V - 1.231V}{1.05mA + 0.1mA}\right) = 1.4K\Omega \quad \text{(Eq.15.7)}$$

From equation 15.6, we can see that the actual wattage will be:

$$P = 1.93V \bullet 1.15mA = 2.2mW \quad \text{(Eq.15.8)}$$

Using a good-quality 1.4-KΩ precision resistor with a 0.1% tolerance, (125 mW or higher), and with a known tempco of less than 30 ppm/°C, will provide optimum performance. Rechecking R_S from equation 15.4 and the wattage from equation 15.6 gives a value of 176 Ω for R_S, with a corresponding wattage of 22 mW. Because the actual working value of R_S is much higher (1.4 KΩ and 125 mW), there is no problem.

Shunt bandgaps usually provide only a positive voltage reference (V+) and are depicted in most product data sheets as shown previously in Figure 15.1. Here the resistor and reference provide a stable positive output voltage to the load. There are situations, however, when you need to provide a low *negative* voltage (V-) instead, such as in providing a stable reference to a DAC. Most shunt bandgaps can be used to generate an equal negative voltage reference, simply by grounding the cathode (+ terminal) and taking the output from between the reference's anode (– terminal) and resistor R_S, which should now be connected to the negative supply (-V_{SS}). Most bandgaps are available only down to 1.22 volts. Very few, as mentioned previously (for low operating voltage types), get down below 1.22 volts. However, one device that does is Analog Devices' **ADR510**, a 1-volt precision, very-low-noise shunt reference. The device features an initial accuracy of ±0.35% (±3.5 mV), a maximum tempco of 85 ppm/°C (over the –40°C to +85°C range), and an ultra-low noise specification of 4 μV peak-to-peak (f = 0.1 Hz to 10Hz), or approximately 0.66 μV_{rms}. It has an operating current range of 100 μA to 10 mA, and being stable with any capacitive load, requires no output capacitor. Table 15.1 summarizes the product's main features.

ADR510	Analog Devices Inc.		Precision shunt bandgap reference				
Reference voltage(s):	Best initial accuracy:	Best tempco: (ppm/°C)	LF noise:	Input voltage range:	Supply current (μA)	Temp. range: (°C)	Special features:
1.000V	±0.35% ±3.5mV	70ppm	4μV pk-pk typ.	1.5V min	100 min.	-40°/ +85°C	• low noise • low voltage • low Z_{out} • 3-lead SOT-23 • 10mA capable

Table 15.1 ADR510

A design example is shown in Figure 15.2, where the ADR510 provides a –1-volt (negative) reference for an Analog Devices' AD7533 (a low-cost, 10-bit CMOS four-quadrant multiplying DAC). This circuit runs from regulated ±5-volt supplies. The DAC requires a very low –1-volt negative reference in order to set its positive output range. A 1-volt reference was chosen for reasons of simplicity, over the normal 1.225-volt type. The op amp inverts the DAC's output for a precise 0 to 1-volt output. The ADR510's maximum tempco and low noise voltage makes it compatible in a 10-bit accuracy system. In this example, for a 1-bit LSB error, the temperature span would be about 20°C, and for 1/2-bit LSB error, the span would only be about 10°C. However, this is a small budget-conscious consumer product, which does not require a

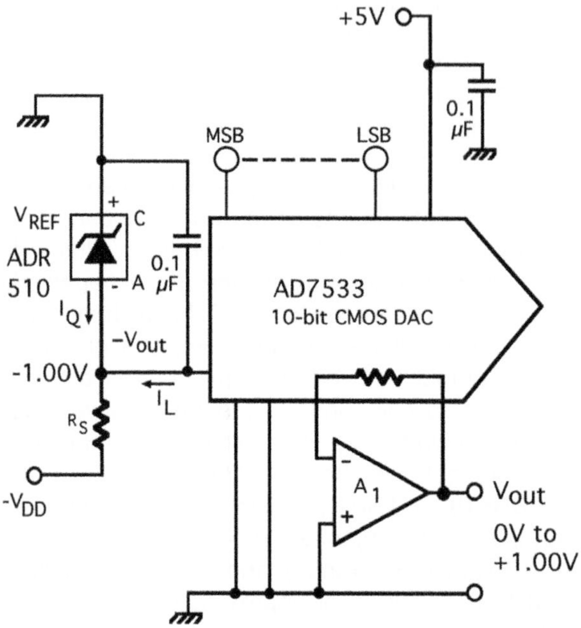

Figure 15.2. The Analog Devices' ADR510 bandgap shunt reference sets the range for this low-cost, 10-bit, CMOS multiplying DAC. The DAC requires a negative input voltage for the correct positive output. This is easily achieved using a shunt bandgap reference.

wide operating range or extreme accuracy. The ADR510 is available in a three-lead SOT-23 surface-mount package.

As of spring 2005, Analog Devices has several good shunt references, but with higher reference voltages than with the ADR510. Other members of the same ADR5xx family include the **ADR520** (2.048-volt), ADR525 (2.5-volt), ADR530 (3.0-volt), ADR540 (4.096-volt), and ADR550 (5.0-volt). These devices are available in either SOT-23 or SC70, three-pin surface-mount packages, and guaranteed over the industrial temperature range (–40°C to +85°C). They feature two performance grades, the best of which has a maximum 40 ppm/°C, an ultra-low low-frequency noise figure of 14 μV pk-pk, and an initial accuracy of ±0.2% max. The devices also have low operating currents that range from 50 μA to 10mA, and run from low-voltage supply rails. Each device also includes an optional *trim pin*, which enables one to adjust the output voltage over a ±0.5% range, without it significantly affecting the tempco. This also allows one to easily trim out any system errors that may be present. Although this device does not require an input or output capacitor, a 0.1-μF capacitor on these pins will help reduce the effects of power supply noise. Table 15.2 summarizes this product line.

ADR520	Analog Devices Inc.		Precision shunt bandgap reference				
Reference voltage(s):	Best initial accuracy:	Best tempco: (ppm/°C)	LF noise:	Input voltage range:	Supply current (μA)	Temp. range: (°C)	Special features:
2.048V 2.500V 3.000V 4.096V 5.000V	±0.2%	40ppm	14μV pk-pk typ.	2.5V min	50 min.	-40°/ +85°C	• low noise • low I_Q • TRIM pin • 3-lead SOT-23 or SC70 • 10mA capable

Table 15.2 ADR520

Figure 15.3 shows an ADR525 shunt device supplying a precise 2.5-volt reference voltage to an Analog Devices' AD5337, a dual 8-bit, voltage-output DAC. The DAC is available in an eight-lead MSOP surface-mount package, and has an operating temperature range of –40°C to +105°C. It operates from low single 2.5-volt to 5.5-volt supply, and consumes 250 μA at a supply of 3 volts. The DAC's output amplifiers feature R-R operation, with a slew rate of 0.7 V/μS. The outputs can range from 0 volts up to the value of the reference voltage. The DAC also has a three-wire serial interface,

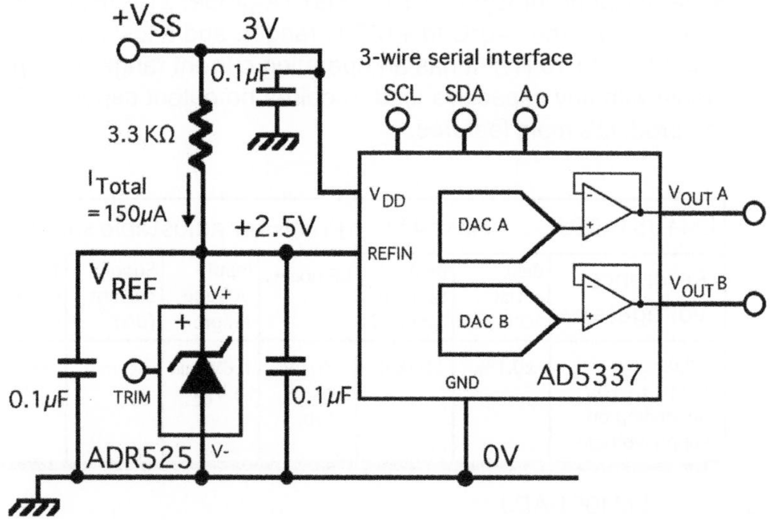

Figure 15.3. In this design, the Analog Devices' ADR525B, a low-power shunt reference, provides a precise 2.5-volt to an AD5337, a dual 8-bit low-power DAC. This low-cost DAC provides rail-to-rail outputs, a simple 3-wire serial interface, and is available in an 8-pin MSOP package. The low noise and low tempco characteristics of the reference make this design attractive for battery-powered tasks.

which can be used at up to 400-KHz clock rates. It also features three power-down modes, which can reduce power consumption down to less than 100 nA at 3 volts, or approximately 1 μW. These devices also feature a power-on reset circuit, which ensures that the DAC outputs go to 0 volt and stay there until a valid WRITE takes place. Additionally, a software CLEAR command will reset all inputs and DAC registers to 0 volt. The ADR525's maximum tempco and low noise voltage makes it compatible in an 8- to 10-bit accuracy system. In this 8-bit example, for a 1/2-bit LSB error, the temperature span could be about 30°C. The design is well suited to battery-operated, portable designs.

Another very popular line of bandgap shunt references comes from National Semiconductor, which first introduced bandgap references and has many good reference products. These include the ever-popular LM4041, a low-noise, micropower 1.225-volt shunt reference that offers an initial accuracy of 0.1% max, and is available in both SOT23 and SC-70 surface-mount packages. It is second-sourced by Maxim and other vendors. National's newer LM4051, a precision micropower shunt reference (also second-sourced by Maxim), offers improved specs that include a tempco of less than 50 ppm/°C over the industrial temperature range. It is available as either a fixed 1.225-volt or as an adjustable version (**LM4051-ADJ**), either of which is in a three-lead SOT-23 surface-mount package. The device features an output voltage tolerance (initial accuracy) of ±0.1% max (A-grade; ±1.2 mV), a maximum tempco of 50 ppm/°C (over the –40°C to +125°C range), and a low noise specification of 20 μV_{rms} (f = 0.1 Hz to 10 Hz). It has an operating current range of 60 μA to 12 mA, and being stable with any capacitive load, requires no output capacitor. Table 15.3 summarizes the product's main features.

LM4051-ADJ	National S C		Precision adjustable shunt bandgap reference				
Reference voltage(s):	Best initial accuracy:	Best tempco: (ppm/°C)	LF noise:	Input voltage range:	Supply current (μA)	Temp. range: (°C)	Special features:
Adjustable 1.24V to 10V depending on supply voltage.	±0.1%	50 ppm	20 μV pk-pk typ.	1.6V min to V_{CC}	60 min	-40°/ +85°C	• Adjustable • 12mA max load • Low noise • No C_{out} needed • SOT-23

Table 15.3 LM4051-ADJ

The design example in Figure 15.4 shows a simple *variable* voltage reference using the LM4051-ADJ, which otherwise has the same specs as the fixed device. For a variety of reasons, it is often useful to have a variable voltage reference. This design can provide an output voltage variable between 1.24 and 10 volts (depending on the supply voltage), and at up to 12 mA. This would put the minimum operating voltage at around 1.8 volts and the absolute maximum voltage at 15 volts.

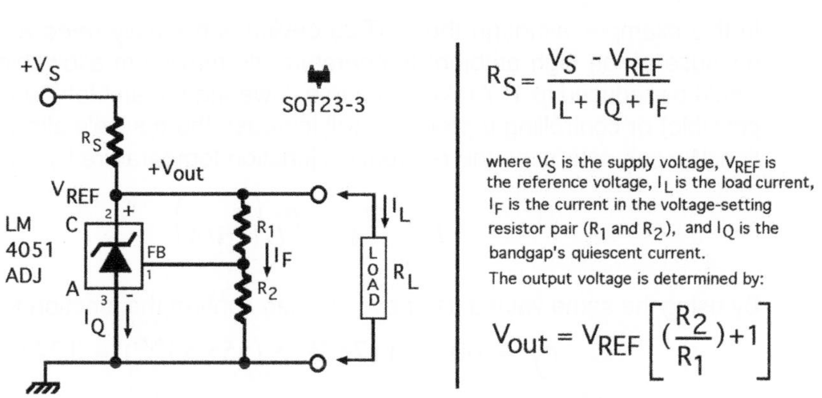
+VS
SOT23-3
RS
+Vout
VREF
LM 4051 ADJ
C
A
FB
R1
IF
R2
IL
LOAD
RL
IQ
RS = (VS - VREF) / (IL + IQ + IF)
where VS is the supply voltage, VREF is the reference voltage, IL is the load current, IF is the current in the voltage-setting resistor pair (R1 and R2), and IQ is the bandgap's quiescent current.
The output voltage is determined by:
Vout = VREF [(R2/R1)+1]

Figure 15.4. *Using National Semiconductor's LM4051-ADJ adjustable, precision micropower shunt reference. The device is available in a 3-lead SOT-23 surface-mount package.*

Remember to check the *power dissipation* specs, particularly when using surface-mount products, when significant supply voltages and load currents are involved. In the case of the SOT-23 package, it is usually rated at around 300 mW, but other smaller packages are often lower. As discussed in a previous chapter, you can determine the maximum allowable power dissipation of any reference by the equation 15.9, if you know the junction-to-ambient thermal resistance. This will be the reciprocal of the device derating value, usually shown in the product data sheet as "derate above xx°C," and then gives a value in mW/°C.

$$P_D = \frac{T_{J\,max} - T_A}{R_{\theta JA}} \qquad \text{(Eq.15.9)}$$

For example, let's assume we need to know the allowable power dissipation for a surface-mount shunt reference rated for a maximum junction temperature of 125°C, an ambient temperature of 60°C, with a data sheet derating value of 1.8 mW/°C above 50°C. We already know that the junction-to-ambient thermal resistance ($R_{\theta JA}$) is the reciprocal of the device's derating value, hence:

$$R_{\theta JA} = \frac{1}{0.0018W} = 555.55°C \qquad \text{(Eq.15.10)}$$

Now applying equation 15.9, we have:

$$P_D = \frac{125° - 60°}{555.55} = 117mW \qquad \text{(Eq.15.11)}$$

In this example, although the SOT-23 device is normally rated as a 300-mW device, because of the high ambient temperature, its maximum allowable power dissipation would be reduced to 117 mW. Of course, lowering the ambient temperature (were that possible) or controlling it precisely will increase the device's allowable power dissipation. We can determine the reference's junction temperature by equation 15.12:

$$T_J = T_A + P_D\left(R_{\theta JA}\right) \qquad \text{(Eq.15.12)}$$

By using the same values as above, we can confirm the junction temperature:

$$T_J = 60° + 117mW \bullet (555.55°C) = 124.99°C \qquad \text{(Eq.15.13)}$$

In any case, it is always worth checking the power dissipation in every reference design; don't just take it for granted.

Another popular source of shunt references comes from Maxim Integrated Products, which has many excellent voltage references. Maxim has been making voltage references for more than 20 years and has some of the greatest experience in this field in the analog semiconductor industry. Apart from being a major A/D and DAC manufacturer, Maxim's experience includes both top-of-the-line monolithic and hybrid buried-zener products (to MIL specs) and both types of bandgaps. Maxim's voltage references today reflect these two well-understood topologies, which are now mostly built as surface-mount products. As far as shunt references are concerned, one of Maxim's best is the **MAX6138** series. These are micropower references, produced with a proprietary BiCMOS process, and available in the tiny SC-70 package. The series includes the following range of voltages: 1.225, 2.048, 2.5, 3.0, 3.3, 4.096, and 5 volts. Taking the 1.225-volt device as an example, it runs from an input as low as 1.8 volts. With a 65-μA quiescent current and a maximum current capability of 15 mA, the device offers an initial accuracy of 0.1% max (i.e., 1.22 mV max) and a maximum tempco of ±25 ppm/°C over the –40° to +85°C range. The MAX6138 series has slightly differing noise levels, according to the reference output voltage. Devices with outputs of 2.5 volts and lower have low-frequency noise levels of approximately 20

MAX6138	Maxim IP		Precision shunt bandgap reference				
Reference voltage(s):	Best initial accuracy:	Best tempco: (ppm/°C)	LF noise:	Input voltage range:	Supply current (μA)	Temp. range: (°C)	Special features:
1.225V, 2.048V 2.500V, 3.000V, 3.300V, 4.096V, 5.000V	±0.1%	25 ppm	28 μV rms typ.	1.8V min to Vcc	65 typ.	-40°/ +85°C	• 7 voltage choices • Tiny SC70-3 • 15 mA max load • Low I_Q=60μA • No C_{out} needed • Low tempco

Table 15.4 MAX6138

μV_{rms} at f = 0.1 Hz to 10 Hz, which for micropower shunt references is impressive. Table 15.4 provides a quick summary of the main features.

The next design, shown in Figure 15.5, illustrates a simple level-sensing application, using a voltage reference to set a precise level for a comparator. Although it seems simple at first glance, it would be worth reviewing this carefully. The key requirements in this application include:

- Wide temperature range (–20°C to + 45°C, i.e., a 65-degree range)
- Nominal 3.3-volt single supply, with a 10% tolerance (i.e., +3 volts to +3.6 volts)
- Very low current consumption (less than 1 mA)
- Fixed reference voltage of +1.22 volts nominal (1.20 volts min to 1.24 volts max)
- Tempco of less than 100 ppm/°C
- Fairly low noise level of less than 50 μV_{rms}
- Very limited circuit board area (1/2-inch square)
- Using only surface-mount parts (SOT-23 or smaller)
- High volume/low cost (i.e., less than $4 total at 2,500 pieces, parts only)

Two suitable devices include a Maxim MAX6138_EXR12_T, a low-cost, micropower shunt reference, and a Maxim MAX9030, a low-cost, low-power comparator. Both devices are intended for a greater temperature range than required by the application (shunt ref. has –40°C to +85°C; comparator has –40°C to +125°C). They both offer a very low quiescent current (shunt ref. has 65 μA; comparator has 55 μA), which is also much lower than required. Both devices are in tiny SC-70 surface-mount packages and will work from the required 3- to 3.6-volt single supply. All of the parts in the design, including the precision resistor and capacitors, total less than $3.35 at that volume, which is less than the $4 budgeted. All of the parts fit within a 1/4-inch-square circuit board area, which is significantly less than required.

Additionally, the 1.2205-volt shunt reference has a 0.1% initial accuracy, which means that it will range from 1.219 volts to 1.2217 volts. This is better than the required ±1.6% or ±20-mV tolerance. It also has a typical noise level of 20 μV_{rms} and a tempco of ±25 ppm/°C, which both meet the criteria. The MAX9030 is a single, micropower comparator featuring a supply current of 55 μA max. It can run on supplies of between +2.5 volts and 5.5 volts and has a shutdown mode that reduces its supply current to less than 1 μA. In normal operation, it has a fast 230-nSec propagation delay, a rail-to-rail output, and a built-in 4-mV hysteresis on the switching levels. The incoming voltage supply should best be regulated, but should at least be bypassed by a small value electrolytic in parallel with a 0.1-μF disk ceramic capacitor. The dropping resistor

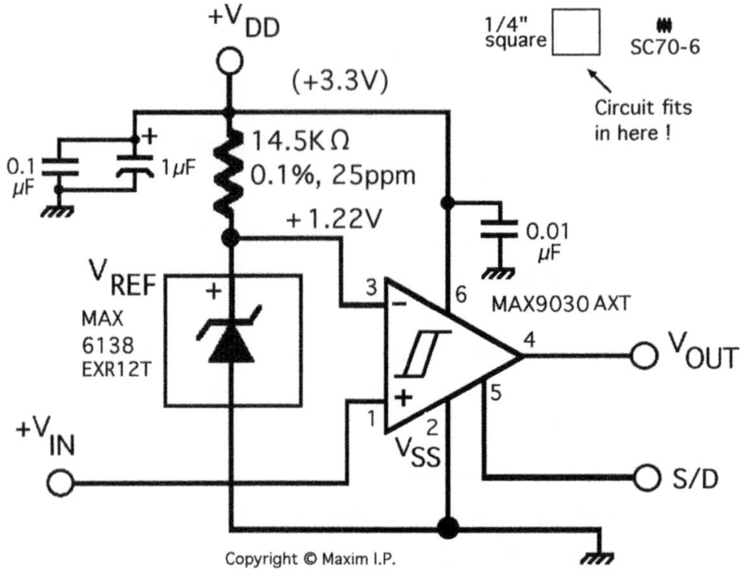

Figure 15.5. *Combining a 1.22-volt MAX6138 precision shunt reference (0.1%, 25ppm/°C), with a low-power comparator, MAX9030, provides a good, low-cost solution using SC70 parts.*

needs to be a good-quality, low-tolerance (less than ±0.5%) precision type (e.g., metal-film, wirewound, thin-film), with a low tempco of less than 30 ppm/°C and a higher than needed wattage rating (i.e., 100 mW).

If you are looking for a shunt reference with a very low quiescent current, Maxim has some of the lowest power references available in the industry. One such family is the **MAX6006** series of shunt references, which feature a minimum I_Q of less than 1 μA and an operating current range of 1 μA to 2 mA. This low I_Q, together with being available in three-pin SOT-23 packages, makes them perfect for low-power, portable products. The MAX6006 series includes the MAX6006 (1.25 volts), the MAX6007 (2.048 volts), the MAX6008 (2.5 volts), and the MAX6009 (3 volts). The premium-grade products have a maximum initial accuracy of 0.2%, along with a maximum tempco of 30 ppm/°C. The main features of this family are summarized in Table 15.5. One would expect for such an ultra-low quiescent current that the noise level would be high. That is usually the case with other low-power products, where low quiescent current is traded off against a high noise level. In the case of the MAX6000 series, however, that is not the case, because the devices all have a respectable less than 60 μV pk-pk (approximately 10 μV_{rms}) rating. In many cases, the devices require an output capacitor of 0.01 μF or higher for added stability.

MAX6006 - 6009 Maxim IP			Ultra low-power shunt bandgap reference				
Reference voltage(s):	Best initial accuracy:	Best tempco: (ppm/°C)	LF noise:	Input voltage range:	Supply current (μA)	Temp. range: (°C)	Special features: • Ultra low power • Low tempco • 3-pin SOT-23 • 4 voltages available
1.225V 2.048V 2.500V 3.000V	±0.2%	30 ppm	30 μV pk-pk typ.	1.8V min to Vcc	<1 max	-40°/ +85°C	

Table 15.5 MAX6006-6009

Another leading analog products manufacturer is Linear Technology Corp., which has many excellent leading-edge products, including voltage references. LTC has been making voltage references for a very long time and has enormous experience in this field, making top-of-the-line buried-zener products, as well as shunt and series bandgaps. In fact, one of LTC's co-founders is the legendary designer Robert Dobkin, who not only co-designed the first commercial bandgap reference, the LM113 (while at National Semiconductor), but also designed many other popular references, including the industry-standard LM185/385. As far as shunt references are concerned, LTC has some exceptional products available.

One outstanding product family is the **LT1634** series. These are micropower, precision types available in various voltages from 1.25 volts to 5.0 volts. All devices are available in SO-8 and TO-92 packages, with the 1.25- and 2.5-volt versions also available in an MSOP package. Some outstanding features of this series are their low operating current (I_Q =10 μA), a very low drift of 10 ppm/°C max, an initial accuracy of 0.05%, and devices characterized for both the industrial and commercial temperature ranges. To get this level of initial accuracy, the reference uses trimmed, precision thin-film resistors, while LTC's advanced curvature correction techniques are responsible for the very low tempco. It has an operating current range of 10 μA to more than 10 mA, and being stable with any capacitive load, requires no output capacitor. The typical low-frequency noise specs for the LT1634 series are impressive and range from about 10 μV pk-pk to about 35 μV pk-pk, the amount varying a little according to the reference voltage chosen. Table 15.6 provides a summary of the main features of this family. These specs make the LT1634 series some of the lowest noise shunt references currently available.

Figure 15.6 shows the LT1634-5, a 5-volt shunt device, supplying not only the precise 5-volt reference voltage, but also the supply voltage to the LTC1448, a 12-bit, dual CMOS DAC.

Yet another standout product series is Linear Technology's **LT1389**. These are nanopower precision shunt bandgaps, available in the SO-8 surface-mount package. The series covers 1.25 volts, 2.5 volts, 4.096 volts, and 5 volts. These have an initial accu-

LT1634	Linear Technology		Precision shunt bandgap reference				
Reference voltage(s):	Best initial accuracy:	Best tempco: (ppm/°C)	LF noise:	Input voltage range:	Supply current (μA)	Temp. range: (°C)	Special features:
1.250V 2.500V 4.096V 5.000V	±0.05%	10 ppm	10 μV pk-pk typ.	1.75V min	10 min.	-40°/ +85°C	• very low noise • ultra-low power • low tempco • SO-8, MSOP-8, or TO-92 • 20mA capable

Table 15.6 LT1634

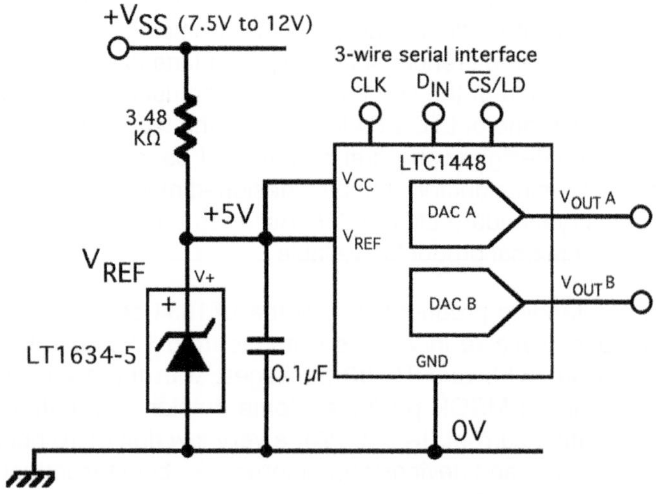

REPRODUCED WITH PERMISSION FROM LINEAR TECHNOLOGY CORPORATION.

Figure 15.6. In this design, the Linear Technology micropower shunt LT1634-5 provides both a precise 5-volt reference (0.05% initial accuracy, 10-ppm/°C) and the +V_{CC} supply to an LTC1448, a dual 12-bit micropower CMOS DAC. This low-cost DAC provides rail-to-rail outputs, a simple 3-wire serial interface, and is available in either an 8-pin SO or PDIP package.

racy of 0.05% max, an impressive 10 ppm/°C tempco over the 0°C to +70°C range, and a low-frequency noise performance of between 25 and 100 μV pk-pk (depending on the reference voltage). Incredibly, the 1.25-volt device has a maximum 0.60 μA quiescent current (the other voltages have slightly higher I_Qs, but still below 1 μA). The main features are summarized in Table 15.7.

Although the device does not require an output capacitor, a 1-μF capacitor on the output pin will help reduce the effects of power supply noise to less than 0.01%. With

LT1389	Linear Technology		Ultra low-power shunt bandgap reference				
Reference voltage(s):	Best initial accuracy:	Best tempco: (ppm/°C)	LF noise:	Input voltage range:	Supply current (μA)	Temp. range: (°C)	Special features:
1.250V 2.500V 4.096V 5.000V	±0.05%	10 ppm	10 μV pk-pk typ.	1.75V min	<1 μA	0°/ +70°C	• 0.05% accuracy • ultra-low power • low tempco • low noise • SO-8 • 2 mA capable

Table 15.7 LT1389

this unusual combination of impressive specs (i.e., low initial accuracy, low tempco, low noise, and ultra-low quiescent current), there is no doubt that the LT1389 family will find a lot of popularity with designers of battery-powered instruments, as well as in consumer products such as cell phones, music pods, PDAs, digital cameras, and laptops.

The circuit of Figure 15.7 shows an LT1389-2.5 reference providing a precise 2.5-volt supply to a compound current source, using two matched MOS pairs from Advanced Linear Devices. The N-channel FETs are a dual ETRIM™ pair (ALD110902A, covered in Chapter 9), with an electrically trimmed gate threshold of 0.20 volt. The P-channel pair (ALD1102A) is slaved to the N-channel pair to create a mirrored 3-μA current source for a load. The circuit is enhanced by the 10-ppm/°C tempco of the LT1389, together with its 0.05% initial accuracy and its minuscule 800-nA operating current. The active devices shown have an operating temperature range of 0° to +70°C, and the circuit draws about 20 μW. It may be run at lower voltages by using the 1.25-volt version of the LT1389 instead, but the gate threshold (V_{TH}) of the ALD1102A, P-channel pair is a maximum 1 volt (0.7 volt typical), which would likely be a limiting factor.

In summary, the shunt reference is a popular workhorse in analog electronics, partly because of its small size and ease of use. It is best suited to working with a constant load, rather than one where the load current varies much. In some applications, the constant power drain can be a significant factor, but because the shunt reference is biased by a resistor, it can be used over a wider range of input voltages, although this leads to a varying quiescent current. U.S. manufacturers include Analog Devices, Intersil, Linear Technology, Maxim, National Semiconductor, and Semtech. As a result of these companies' continued development and technical improvements, the shunt reference now competes with its sibling, the series bandgap reference, which we will look at next.

15.2 Applying fixed-series bandgap references

Although many improvements have occurred over the past five years with the bandgap shunt reference, even more development has focused on the series type. This being the result of the individual design efforts of several leading U.S. analog semi-

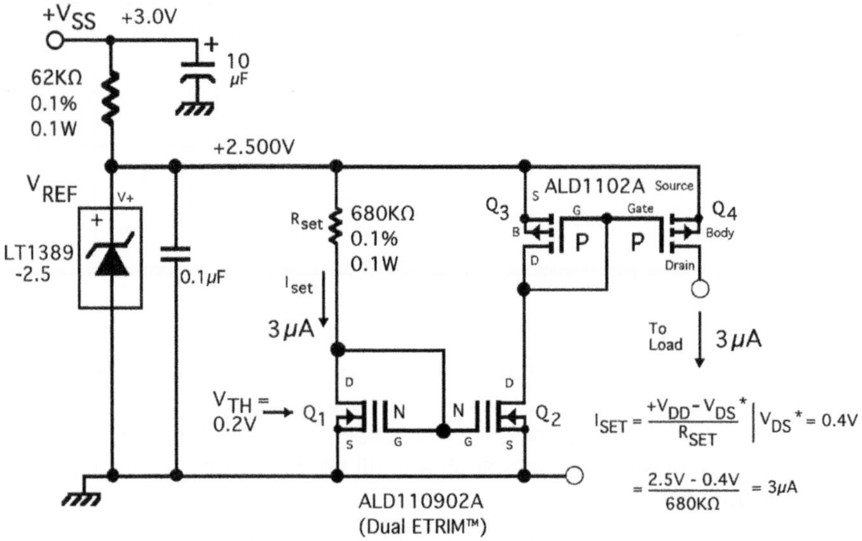

Figure 15.7. *Combining Linear Technology's LT1389 nano-power voltage reference with two dual CMOS matched transistors from Advanced Linear Devices creates a precision 3μA current source. The circuit runs from a single 3-volt supply and draws less than 8 μA. Note that the ETRIM™ pair have a gate threshold of only 200 mV.*

conductor manufacturers, including Analog Devices, LTC, Maxim, Microchip, National Semiconductor, Intersil, Texas Instruments, Semtech, Thaler, and Microsemi. Typically, series types provide superior performance to shunt types, in most of their characteristics, and in fact some of the newest series bandgap products now have features that challenge the high-performance topologies (i.e., the buried-zener, Analog Devices' XFET® and Xicor's FGA™). These improvements with the series bandgap have brought about better initial accuracy (±%), lower temperature coefficients/drift, lower noise levels, lower dropout voltages, smaller package sizes, lower supply voltages, and lower power consumption. Improved products bring about higher accuracy and therefore better designs. The series bandgap reference is available in a three-pin (or more) format and in various package types that include DIL, throughhole, and surface mount. It also has more options available and is suitable for up to 14-bit applications. Unlike the shunt type, no input resistor is required, and low dropout devices are available for low-voltage/battery-powered applications.

The basic series voltage reference has an input terminal, an output terminal, and a ground return terminal, and is shown in Figure 15.8. Because series references do not need a bias resistor, they draw only a constant quiescent current, and the load current necessary, irrespective of changes in supply voltage, ambient temperature, or load current. With the series bandgap reference, the voltage at its output will be held constant and will change only by a relatively small amount. Some series devices also provide a TRIM terminal that allows a trimmer (or a precision resistor network) to be connected to it and used for precisely adjusting the output voltage and zeroing out any offset. Some series bandgaps include a noise-reduction terminal, which allows one to add an external capacitor. The capacitor works with an internal resistor network, to form a low-pass filter. Most references require an input bypass capacitor, and some require a particular value of output capacitor for stability and improved performance.

A few products even include sense and force terminals. For applications that have long cable runs/circuit board traces or that involve very high accuracy and high bit resolutions, such terminals are vitally important. They compensate for any small voltage drops and put the exact reference voltage at the load. Some series bandgaps include an enable/shutdown terminal, which when used will put the device into a low-power mode. For example, a device drawing a quiescent current of, say, 250 μA may draw less than 10 μA when in shutdown mode. Some references have significant turn-on settling times (e.g., 400 μS), but others are enhanced for a fast response (less than 40 μS).

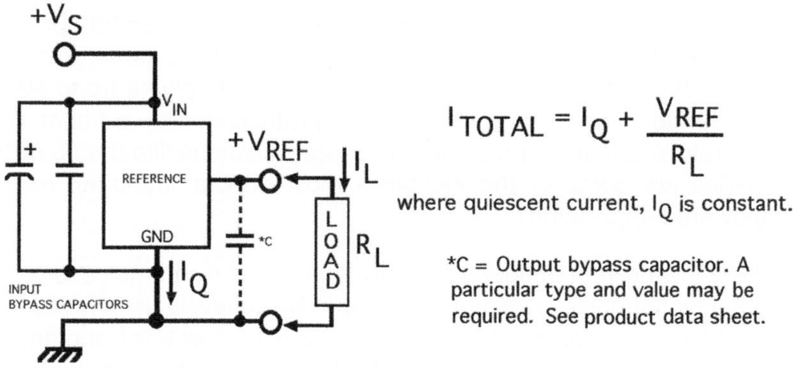
+VS
VIN
+VREF
IL
REFERENCE
GND
*C
IQ
LOAD
RL
INPUT
BYPASS CAPACITORS
ITOTAL = IQ + VREF/RL
where quiescent current, IQ is constant.
*C = Output bypass capacitor. A particular type and value may be required. See product data sheet.

Figure 15.8. *The basic 3-terminal series voltage reference.*

The series bandgap market is quite competitive and has more vendors than any other part of the IC reference market. As a consequence, there are many good products. We will look at a few of the most unusual and exceptional ones here. The first one is a family of outstanding precision references from Linear Technology in the form of its **LT1461** series. These high-performance micropower bandgap references are built with a proprietary bipolar process and are available in the SO-8 surface-mount package. They combine high initial accuracy, a low tempco, a low dropout voltage, along with low power consumption. The LT1461 series provide reference voltages of 2.5 volts, 3 volts, 3.3 volts, 4.096 volts, and 5 volts in several performance grades, the best of which has a low initial accuracy of ±0.04%. Premium-grade products also feature a maximum tempco of 3 ppm/°C, over the industrial temperature range (–40°C to

LT1461	Linear Technology		Precision series bandgap reference				
Reference voltage(s):	Best initial accuracy:	Best tempco: (ppm/°C)	LF noise:	Input voltage range:	Supply current (μA)	Temp. range: (°C)	Special features: • High accuracy • low tempco • low power • low dropout • High I_{out} >50mA • $\overline{SHDN}$ pin
2.500V, 3.000V, 3.300V, 4.096V, 5.000V	±0.04%	3ppm	8μV pk-pk typ.	V_{out} + 0.5V to 20V	50 typ.	-40°/ +85°C	

Table 15.8 LT1461

+85°C), although this value is typically about 1 ppm/°C. This very low tempco is actually *guaranteed* up to +125°C, which is impressive. The main features of this series are summarized in Table 15.8.

Equally impressive is the typical long-term stability figure of 60 ppm/1000 hours. The LT1461 series also include a low maximum dropout voltage of less than 400 mV at 10 mA output, and a very low quiescent current of less than 50 μA (with no load). Additionally, they feature a load regulation of less than 30 ppm/mA and a very good line regulation figure of less than 8 ppm/V. Unlike most competitors' series bandgaps, the LT1461 family are also capable of supplying up to 50 mA at their outputs (featuring internal thermal overload protection, like a linear regulator), and they include short-circuit protection. With specifications like these, no wonder the LT1461 series are some of the industry's most popular bandgap references for battery-powered applications.

Very-low-power bandgaps typically have a significant noise level, because they trade off a higher noise level for lower power. Noise levels usually reduce as quiescent current increases, but the LT1461 series do not follow this trend and have an impressive low-frequency noise voltage (0.1 Hz to 10 Hz), of typically 8 μV pk-pk. These specifications make the LT1461 series generally suitable for 12- to 14-bit converter systems. Like other LTC reference products, they include proprietary tempco curvature correction and precise laser trimming of their thin-film resistors. The LT1461 series also include a shutdown terminal to put the device into a low-power mode when taken to a low level. In shutdown mode, the reference draws less than 35 μA. Under normal conditions the pin can be left floating or can be pulled up through a resistor to a voltage of 2.4 volts or higher. These devices require a 2.2-μF capacitor on the output to help reduce the effects of power supply noise and improve load transient response. Similarly, a parallel combination of 0.1-μF and 1-μF (or higher) capacitors located at the input terminal of the reference will help improve stability under a wide range of load conditions.

Figure 15.9 shows a circuit where an LT1461AISB-2.5 is used to provide a 2.5-volt reference to a remote A/D converter using a low-power, RRIO precision op amp, an

LT1218. The A/D converter is an LTC1285, a 3-volt micropower 12-bit SAR type. In this design the reference's maximum tempco of 3 ppm/°C and low noise level enables it to provide 12-bit accuracy (with a conversion error of less than 1 LSB), over the application's temperature range of –20°C to +50°C (this temperature range is significantly less than the devices are specified for). The three devices chosen all have SO-8 packages, work with a 3-volt regulated supply, and together draw less than 800 μA.

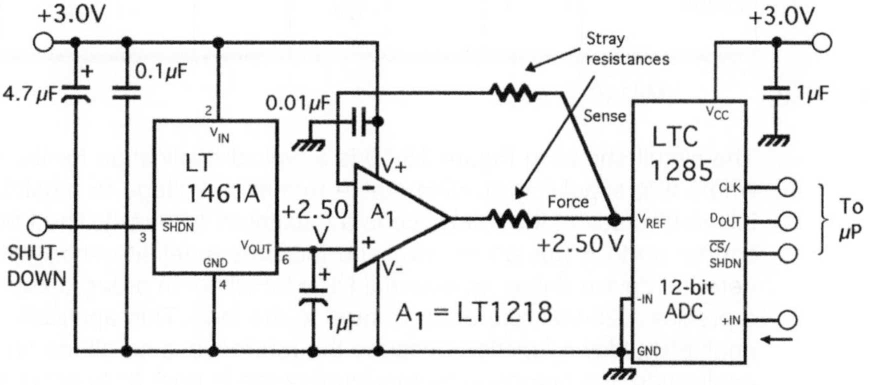

Figure 15.9. This circuit uses a Linear Technology LT1461A reference, together with an LT1218, a low-power op amp to provide a Kelvin connection to a remote A/D converter (an LT1285). The reference provides a very high accuracy of ±0.04%, a 3 ppm/°C tempco, and an 8 $\mu V_{pk\text{-}pk}$ noise level, together with a low dropout, and a low quiescent current over the industrial temperature range.

Next we will look at National Semiconductor's **LM4140**, a family of high-performance bandgaps, built with a proprietary CMOS process, and available in the eight-pin SO surface-mount package. These products combine high accuracy, ultra low noise, very low tempco, and low power. The LM4140 series provides a range of reference voltages that include 1.024 volts, 1.25 volts, 2.048 volts, 2.5 volts, and 4.096 volts in three performance grades, the best of which has a low initial accuracy of ±0.1%. The 1.25-volt reference, for example, has a low-frequency noise voltage of 2.2 μV peak-peak, along with an impressive tempco of around 3 ppm/°C over the commercial temperature range (0°C to +70°C). Table 15.9 summarizes the main features of this series.

These devices also include a very low dropout voltage and an enable pin, which when grounded reduces the quiescent current to about 1 μA. These impressive specifications are generally suitable for up to 14-bit systems. Unlike other voltage references, this series uses a combination of CMOS DACs and an EEPROM for tempco curvature correction and precise laser trimming. Very few of today's high-performance bandgap references have a combination of specifications better than these.

LM4140 National S C			Precision series bandgap reference				
Reference voltage(s):	Best initial accuracy:	Best tempco: (ppm/°C)	LF noise:	Input voltage range:	Supply current (μA)	Temp. range: (°C)	Special features:
1.024V, 1.250V, 2.048V, 2.500V, 4.096V	±0.1%	3 ppm	2.2 μV pk-pk typ.	1.8V min to 5.5V	230 typ.	0°/ 70°C	• Ultra-low noise • low power • low dropout • uses advanced EEPROM + DAC for tempco compensation

Table 15.9 LM4140

The circuit shown in Figure 15.10 is a typical application for the LM4140 reference, where it is supplying an ultra-stable reference voltage to multiple 12-bit DACs (the initial accuracy of the reference is a maximum 1.25 mV). The LM4140 can normally source up to 8 mA on its own and includes indefinite short-circuit protection, but here the device drives an external NPN transistor in order to supply a total 36 mA at a precise 1.25-volt reference voltage to the load. This application does not use the enable/shutdown function because the reference is on all the time. In this particular application, the operating temperature range is from 0° to 50°C, which allows the (3

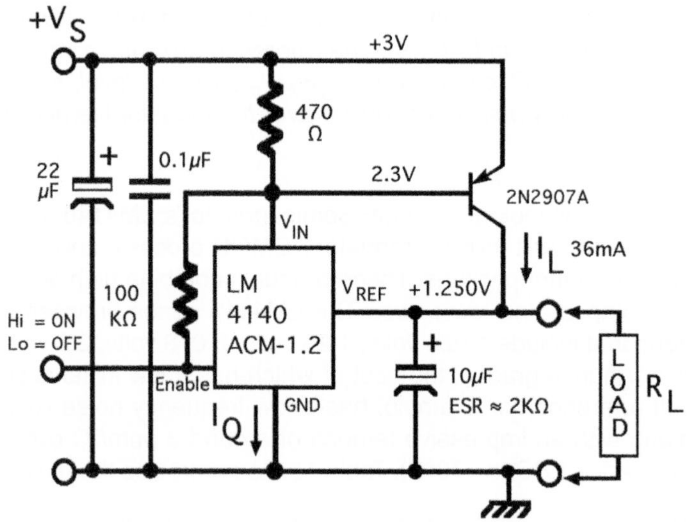

Figure 15.10. Adding a BJT to the output of this National Semiconductor LM4140-1.25V precision voltage reference boosts the load current while maintaining the excellent 3ppm/°C tempco and ultra-low noise characteristics.

ppm/°C max) reference to keep the 12-bit DACs' conversion error to within 1/2 LSB. Now that's impressive!

Another family of outstanding precision references from National Semiconductor is the **LM4130** series. These high-performance bandgap references are also built with a proprietary CMOS process and are available in the five-pin SOT-23 surface-mount package. They combine high initial accuracy with a very low tempco, a low dropout voltage, and low power. The LM4130 series provides a reference voltage of either 2.5 volts or 4.096 volts in five performance grades, the best of which has a low initial accuracy of ±0.05%. The references have an impressive tempco of 10 ppm/°C max, over the industrial temperature range (–40°C to +85°C). Noise voltage is typically 150 μV pk-pk. They also include a low dropout voltage of less than 275 mV at a 10-mA output, and a low quiescent current of less than 75 μA. They are also capable of supplying up to 20 mA at their outputs and include short-circuit protection. The main features of this series are summarized in Table 15.10.

LM4130	National S C		Precision series bandgap reference				
Reference voltage(s):	Best initial accuracy:	Best tempco: (ppm/°C)	LF noise:	Input voltage range:	Supply current (μA)	Temp. range: (°C)	Special features: • High accuracy • low power • low dropout • uses advanced EEPROM + DAC for tempco compensation
2.500V 4.096V	±0.05%	10 ppm	150 μV pk-pk typ.	2.8V min to 5.5V	<75 typ.	-40°/ +85°C	

Table 15.10 LM4130

These specifications make the LM4130 series generally suitable for up to 14-bit systems. Like the LM4140 series we looked at previously, the LM4130 series also use a combination of CMOS DACs and an EEPROM for tempco curvature correction and precise laser trimming. Although these devices do not require an output capacitor, a 0.1-μF capacitor on the output pin will help reduce the effects of power supply noise and load transient response. Similarly, a small 0.1-μF capacitor at the input of the reference will help improve stability under a wide range of load conditions.

The circuit in Figure 15.11 shows an application where the LM4130A-2.5 is used to provide precise ±2.5-volt references from ±3-volt supplies. The low dropout voltage of the LM4130 is a big asset in this circuit. The +2.5-volt reference is filtered, inverted, and buffered by the National Semiconductor LM6144A, a low-voltage rail-to-rail quad op amp. The reference voltage is filtered by a low-pass active filter and available from the buffer, A_3. The filtered reference voltage is inverted by A_2 and output as –2.5 volts, via the A_4 buffer. Both the op amps and the reference are rated for operation over the –40°C to +85°C industrial temperature range and available in surface-mount packages (the quad op amp is available in 14-pin DIL and SO packages). This results in a high-accuracy 0.05% bipolar reference circuit, with a maximum tempco of 10 ppm/°C over the temperature range. Using precision matched resistors (±0.1%) with very low tempcos (less than 30 ppm/°C) or low TC-tracking is essential here.

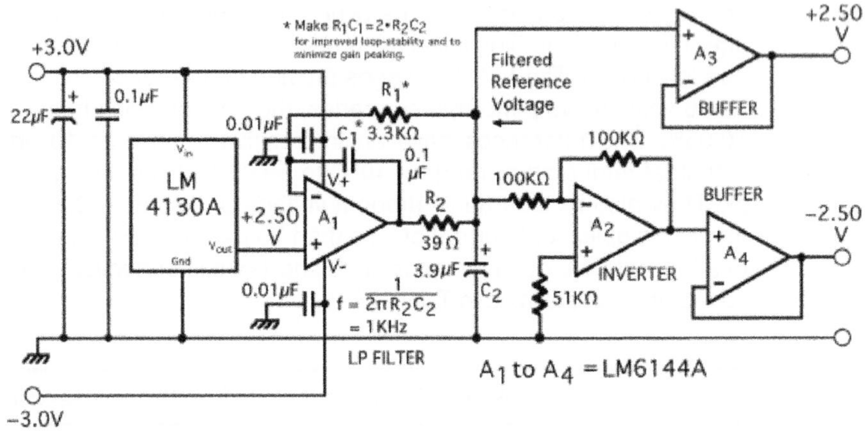

Figure 15.11. *This circuit provides a very high accuracy 0.05%, 10 ppm/°C filtered, bipolar reference over the industrial temperature range, by using a combination of two National Semiconductor products: the LM4130 CMOS voltage reference and the LM6144A, the low-voltage quad op amp.*

If you are looking for a high-performance series reference with a combination of low initial accuracy, low tempco, low quiescent current, low noise voltage, low dropout voltage, a small surface-mount package, with a shutdown feature, and that works over the *automotive* temperature range (–40°C to +125°C), then look closely at Analog Devices' **ADR390 series.** They provide an output current up to 5 mA with reference voltages of 2.048 volts, 2.5 volts, 4.096 volts, and 5 volts in two performance grades, the best of which (B-grade) has a low initial accuracy of ±0.16%. They also have a very low tempco of 9 ppm/°C over the wide temperature range, along with an ultra-low low-frequency output noise voltage of 5 µV pk-pk (0.1 to 10 Hz). These products also feature a maximum supply current (I_Q) of 120 µA, which drops to a maximum of 3µA when in shutdown mode, and they also have a low dropout voltage of 300mV. The ADR390 family's main features are shown here in Table 15.11.

For a 2.048-volt reference, the input voltage can range from 2.5 volts to less than 18 volts. Because of this combination of good specifications, these devices are ideal for most battery-operated applications, particularly because they are available in the tiny five-pin TSOT surface-mount package. The ADR391B, a precision 2.5-volt reference built using a proprietary Analog Devices' complimentary-bipolar process, is shown in Figure 15.12.

ADR390	Analog Devices		Precision series bandgap reference				
Reference voltage(s):	Best initial accuracy:	Best tempco: (ppm/°C)	LF noise:	Input voltage range:	Supply current (μA)	Temp. range: (°C)	Special features:
2.048V 2.500V 4.096V 5.000V	±0.19%	9 ppm	5 μV pk-pk typ.	V_{out} + 0.3V to 18V	<120	-40°/ +125°C	• Ultra-low noise • low tempco • low power • low dropout • Wide op. temp • $\overline{SHDN}$ pin

Table 15.11 ADR390 series

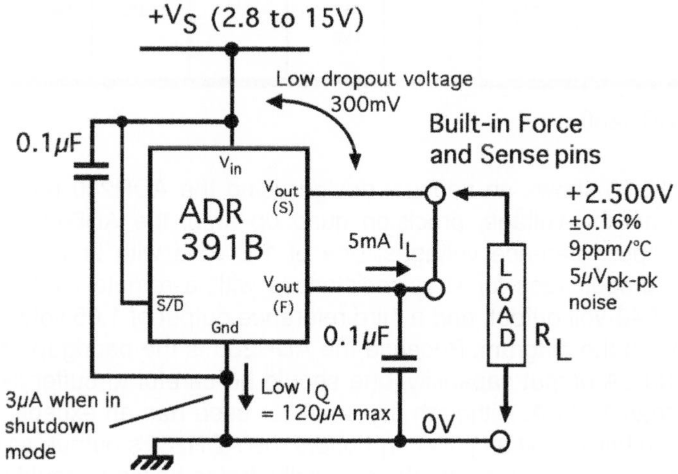

Figure 15.12. Analog Devices' ADR390 series provides an outstanding set of specifications in a tiny 5-pin TSOT package.

Another standout series bandgap reference from Analog Devices is its **ADR280**. The ADR280 is a 1.2-volt reference that features ultra-low power consumption and high power supply ripple rejection. It is supplied in either a three-lead SOT-23 or an SC70 surface-mount package. It is intended for operation from 2.4-volt to 5.5-volt supplies and over a temperature range of –40°C to +85°C. Unlike other series references, which include a gain stage that sets the output voltage, the ADR280 includes only the bandgap *core.* This allows designers to customize the reference's voltage, current, or transient response by using their choice of op amp to suit the design. The reference can drive a load of up to 100 μA on its own. Beyond that it requires an additional external buffer to drive a load. The device has an initial accuracy of ±0.4% (5 mV) max, a typical noise level of 12.5 nV/√Hz, a very low supply current of less than 16 μA, and a maximum tempco of 40 ppm/°C over the temperature range. It also features excellent line regulation typically of around 2 ppm/V and a power supply ripple rejection ratio of

–80 dB at 220 Hz. These are enhanced characteristics, which are especially helpful in applications where heavy dynamic supply variations are involved, such as data converter references in GSM, GPRS, and 3G mobile stations. Like most other references, it should be decoupled with 0.1-μF ceramic disk capacitors on both the input and the output. The main features of the ADR280 are summarized in Table 15.12.

ADR280	Analog Devices		Precision series bandgap reference				
Reference voltage(s):	Best initial accuracy:	Best tempco: (ppm/°C)	LF noise:	Input voltage range:	Supply current (μA)	Temp. range: (°C)	Special features:
1.200V	±0.4% ±5mV	40 ppm	12.5μV rms typ.	2.4V to 5.5V	16 max.	-40°/ +85°C	• Ultra-low power • Low noise • low tempco • good line reg. • High PSRR • SC70 or SOT-23

Table 15.12 ADR280

Figure 15.13 shows an unusual design using the ADR280 reference, along with an exceptional low-voltage, precision quad op amp, the AD8609. The circuit provides three output reference voltages: one of 1.2 volts with buffered Kelvin connections (which compensates for stray resistances with a remote load), a second reference with a 2.048-volt output, and a third reference output of 1.65 volts. The scaling formula is shown in the diagram. Because the ADR280 is the bandgap core and has a maximum 100 μA output capability, one should be careful to buffer its output to any load higher than 100 μA. Although the amplifier used has an exceptionally low input bias current, in this circuit amplifier A_1 buffers the ADR280's output, as it provides a convenient test point. In any event, the circuit illustrates how one could provide multiple outputs using the ADR280 as the central reference.

The circuit shown is a low-voltage, micropower design, which requires one to use a low-voltage, micropower op amp. It is also a low-noise *precision* design, thus a low-noise *precision* op amp is required. Op amps that can meet these diverse specs are uncommon (because low noise levels are usually traded off against low power), but the one chosen can easily meet these requirements. Analog Devices' AD8609 is a quad micropower, low-noise precision CMOS type, with R-R inputs and outputs. It is every low-voltage circuit designer's dream, because it combines many otherwise conflicting characteristics into one device. (Actually, it is part of a family of similar spec op amps: the AD860x series.) It is a single supply, low-voltage, low-power type: it can operate from single supplies of between 1.8 volts and 5.5 volts, drawing less than 60 μA per amplifier. It is very low noise and has a low-frequency noise specification of less than 3.5 μV pk-pk, and a typical voltage noise density at 1 KHz of about 25 nV/√Hz. It also has a low offset voltage of less than 50 μV, with a V_{OS} drift of less than 4.5 μV/°C. It has an ultra-low input bias current; its input bias current over the industrial temperature

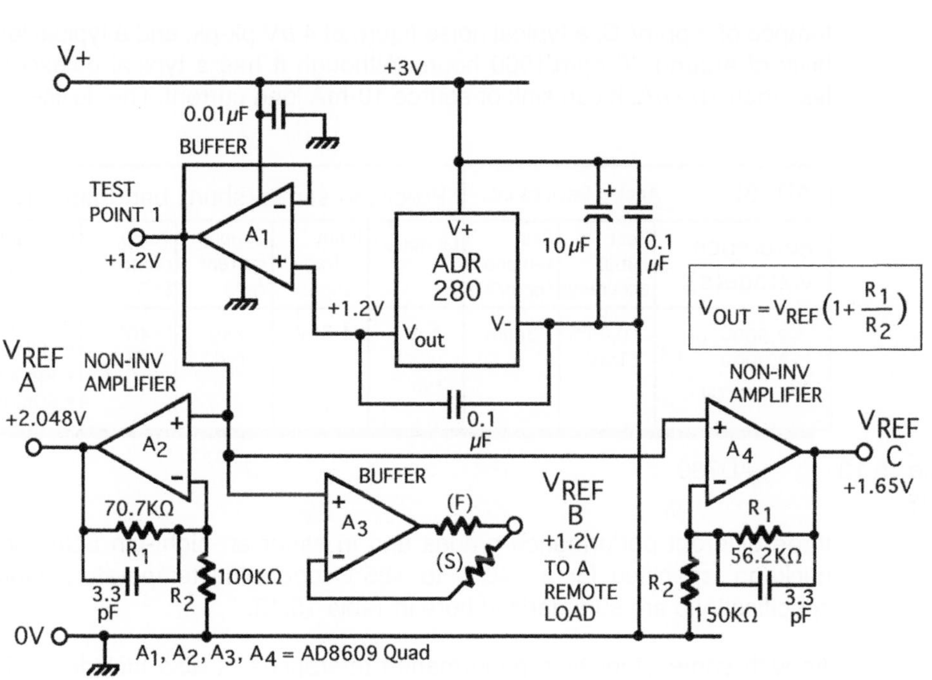

Figure 15.13. A micropower, low-noise triple voltage reference, using an Analog Devices' ADR280 with an AD8609, a low-voltage CMOS precision quad op amp. Buffer A_1 provides a test point. Amplifier A_2 provides a precise 2.048V reference, Amplifier A_4 provides a 1.65V reference, while buffer A_3 provides a Kelvin connection for a precise 1.2V output.

range (–40°C to +85°C) is less than 50 pA. At room temperature, this is typically less than 1 pA. It can sink or source over 10 mA with R-R performance.

The quad device is also available in either a 14-pin SOIC or TSSOP surface-mount package, and it is a low-cost device too. As a consequence, this circuit features low noise (approximately 25 nV/√Hz), low supply voltage (2.4 volts to 5.5 volts), low ground current (less than 20 μA with a 10-μA load), high accuracy (±0.4%), a low tempco (less than 40 ppm/°C over the industrial temperature range), and good line regulation (2 ppm/°C typical) and good power supply ripple rejection (–80 dB). With no load, the complete circuit draws less than 300 μA, and because of its surface-mount parts, it takes up minimal circuit board area. Combining two outstanding products like these can produce an exceptional design.

Another truly outstanding and unusual series bandgap is Analog Devices' **AD780**. This is a 2.5-volt/3-volt pin-programmable high-precision reference with specifications that include a maximum initial accuracy of ±0.04% (±1 mV), a maximum tempco of 3 ppm/°C, a typical noise figure of 4 μV pk-pk, and a typical long-term stability of around 20 ppm/1000 hours. Although it has a typical quiescent current of less than 700 μA, it can sink or source 10-mA load current. The device is available in

AD780	Analog Devices Inc.		Precision series/shunt bandgap reference				
Reference voltage(s):	Best initial accuracy:	Best tempco: (ppm/°C)	LF noise:	Input voltage range:	Supply current (μA)	Temp. range: (°C)	Special features:
+2.500V or +3.000V Pin-selectable	±0.04% ±1mV	3ppm	4μV pk-pk typ.	4-36V	750 typ.	-40°/ +85°C	• drives large C_{out} • series or shunt mode • TRIM pin • V_{PTAT} pin • SOIC-8/PDIP-8

Table 15.13 AD780

three different performance grades and in either an eight-pin SOIC or plastic DIP package, specified for a –40°C to +85°C operating temperature range. The key specifications are summarized here in Table 15.13.

As with some other high-performance references, it also includes a TRIM pin for adjusting the reference output voltage (by ±0.1 volt) to a precise value. The AD780 also includes a TEMP pin that outputs a linear voltage (approximately 500 mV to 620 mV), which is proportional to absolute temperature (V_{PTAT}). At room temperature this voltage is typically 560 mV, with a positive tempco of about +1.9 mV/°C. This voltage may be used to monitor the system temperature, adjust the TRIM voltage with external circuitry, or compensate another circuit. Because this voltage comes directly from the bandgap's core, it is imperative to use a low-input bias current op amp to buffer the TEMP output and avoid overloading it. Such a circuit is shown in Figure 15.14A.

Some of the AD780's additional features include (1) operation from a 4-volt to 36-volt supply voltage, which is much wider than most references; (2) its ability to be used in either series or shunt mode (when used in shunt mode, it can also provide a negative reference if desired, which is shown in Figure 15.14B, with similar specifications to its normal series mode); and (3) its ability to drive virtually any value of capacitive load. Many references require a particular value of bypass capacitor at their output for optimum stability, transient rejection, and so on, particularly when driving some types of A/D converters like sigma-deltas (Σ-Δ). This is because a small value capacitor on the reference output does not have enough storage capacity to maintain a stable reference voltage for the ADC during its conversion period, and errors can result. Transients will occur when the Σ-Δ's internal input circuitry switches, unless the capacitor is large enough to suppress them. The AD780 is stable under any capacitive load with a 1-μF bypass capacitor on its input and a small optional capacitor (C_2) on its TRIM pin (the value of which is proportional to the magnitude of the capacitive load). The prod-

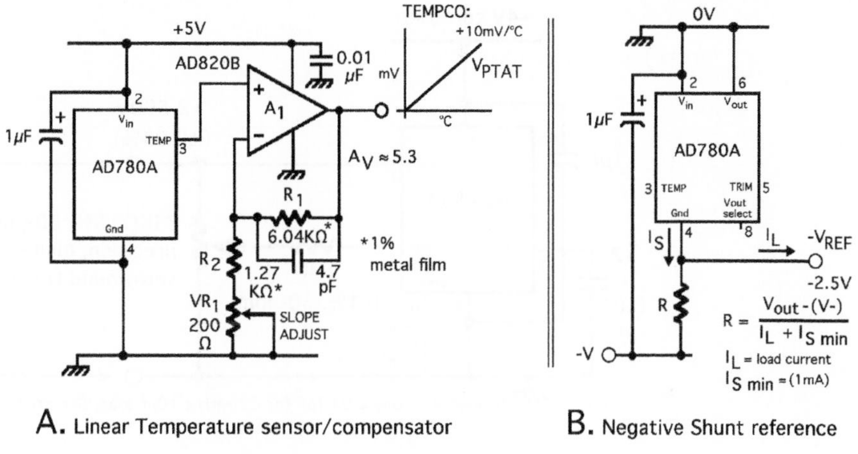

Figure 15.14. *The versatility of the AD780 is illustrated here. In Figure A the device is being used as a temperature sensor. The V_{PTAT} voltage at the TEMP pin is amplified and adjusted for a +10mV/°C output that can be used to determine the chip temperature, or for compensation purposes. In Figure B, the device is connected so as to emulate a shunt reference, and with a negative -2.5V output.*

uct data sheet includes a handy graph for determining the optimum value for C_2. For example, a 0.1-μF capacitor on the TRIM pin will allow the reference to drive a 100-μF capacitive load. Figure 15.15 shows the enhanced circuit when dealing with a large 10-μF capacitive load, and it also provides fine adjustment of the output voltage. Once enhanced as shown here, it can be used in converter systems of up to about 20 bits.

One of the industry's lowest-power series bandgaps is Maxim's **MAX6129 family**. These high-performance bandgap references are built with a proprietary BiCMOS process, and are available in the five-pin SOT-23 surface-mount package. The MAX6129 series provides reference voltages of 2.048 volts, 2.5 volts, 3 volts, 3.3 volts, 4.096 volts, and 5 volts in two performance grades. They combine ultra-low power with an initial accuracy of ±0.4% max and a low tempco of 40 ppm/°C maximum over the industrial temperature range. They also feature a low supply range of 2.5 volts to 12.6 volts and a low dropout voltage of 200 mV max (at 4-mA load current). The reference has an output source capability of 4 mA, a sink capability of 1 mA, and short-circuit current limiting. The main features of the MAX6129 family are summarized in Table 15.14. The major feature of this series is their ultra-low quiescent current of 5.25 μA, making them some of the lowest-power series bandgaps ever produced, and in a SOT-23 package. In order to get very low power consumption, the reference noise level is often compromised, but in this case the devices have low-frequency noise

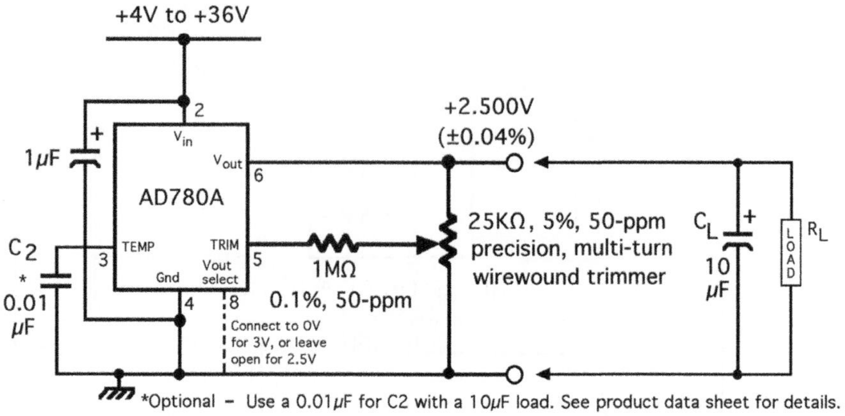

Figure 15.15. Unlike other references, which can have instability problems driving large capacitive loads, Analog Devices' AD780A multifeatured precision bandgap reference is stable under all load conditions. It also has a very low noise characteristic and a low tempco.

specs that typically range from about 30 μV pk-pk for a 2.048-volt device to around 90 μV pk-pk for a 5-volt device.

MAX6129	Maxim IP		Precision series bandgap reference				
Reference voltage(s):	Best initial accuracy:	Best tempco: (ppm/°C)	LF noise:	Input voltage range:	Supply current (μA)	Temp. range: (°C)	Special features:
2.048V, 2.500V, 3.000V, 3.300V, 4.096V, 5.000V	±0.4%	40 ppm	30 μV pk-pk typ.	2.5V to 12.6V	<6 typ.	-40°/ +85°C	• 6 voltage choices • Ultra-low I_Q • low dropout • No C_{out} needed • miniature SOT23

Table 15.14 MAX6129

The circuit shown in Figure 15.16 uses a dual supply of ±3 volts and shows a MAX6129_EUK25_T being used as a reference for a 10-bit CMOS ADC. Here the reference's output voltage is filtered by a single-pole low-pass filter (RC), before being fed to a pair of CMOS op amps. The buffer provides the positive reference voltage to the converter's $+V_{REF}$ input. The second op amp inverts the filtered reference voltage and provides this to the converter's $-V_{REF}$ input. The bipolar input range for the ADC is set for ±2.5 volts. In this circuit, the reference's initial accuracy of 0.4% max, 40 ppm/°C maximum tempco, and industrial temperature range makes it a perfect match for the converter, providing a conversion error of better than 1 LSB. Its low dropout voltage and minuscule quiescent current are major advantages, and using very-low-power CMOS parts makes this circuit attractive for battery-powered applications. Maxim makes a wide variety of excellent op amps and A/D converters from which to choose.

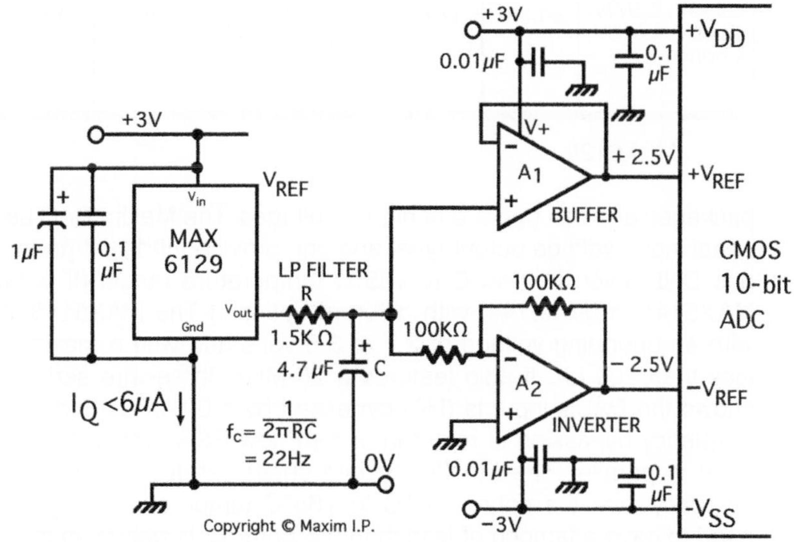

Figure 15.16. *The bipolar range for this A/D converter is provided by Maxim's MAX6129 ultra-low-power reference, a simple passive low-pass filter, and two low-power R-R precision op amps.*

Another exceptional series bandgap reference family from Maxim is its **MAX6126 family**. These BiCMOS products are available as 2.048-volt, 2.5-volt, 3-volt, 4.096-volt, and 5-volt devices, designed to work over the *automotive* temperature range (–40°C to +125°C). They are available in four performance grades, all in a choice of SO-8 or tiny μMAX® surface-mount packages. The best grade features a low initial accuracy of 0.02%, an ultra-low tempco of 3 ppm/°C over the –40°C to +85°C range, and an ultra-low noise voltage of typically 1.3 μV pk-pk (at 0.1 Hz to 10 Hz). They also have a low 200-mV maximum dropout voltage and a low quiescent current of typically less than 400 μA at room temperature. The MAX6126 series have Kelvin connection outputs (including a ground sense connection) and an optional noise-reduction feature as well. The main features of the MAX6126 family are summarized in Table 15.15.

The circuit shown in Figure 15.17 shows the MAX6126A21 providing a reference for a MAX5143, a 14-bit DAC, in a single 3-volt system. It is quite remarkable when you consider that both the reference and the DAC are in tiny 5mm-by-3mm μMAX® packages (a unique Maxim surface-mount package). They are also both available in SO

MAX6126		Maxim IP	Precision series bandgap reference				
Reference voltage(s):	Best initial accuracy:	Best tempco: (ppm/°C)	LF noise:	Input voltage range:	Supply current (μA)	Temp. range: (°C)	Special features:
2.048V, 2.500V, 3.000V, 4.096V, 5.000V	±0.2%	3 ppm	1.3 μV pk-pk typ.	V_{OUT} + 200mV to 12.6V	380μA typ.	-40°/ +125°C	• Ultra-low noise • Ultra-low tempco • high intl. accuracy • low dropout • Force/sense outputs

Table 15.15 MAX6126

packages and can operate at higher voltages. The Maxim DAC used in this circuit is a serial input, voltage output type, and can provide full 14-bit performance (±1LSB, INL and DNL) over the –40°C to +85°C temperature range. (It is part of a DAC family MAX5141 to MAX5144, with different options.) The MAX5143 is a low-power DAC, with an operating voltage of 2.7 to 3.6 volts and with a maximum supply current of less than 200 μA. It also features a 25-MHz, three-wire serial interface. The circuit shows the DAC's input is (HF) bypassed by a 0.1-μF ceramic disk capacitor. If low-frequency bypassing is required, a 1-μF low-ESR tantalum or film capacitor can be used in parallel. The MAX5143 data sheet carefully reminds us that to get 14-bit accuracy over the entire –40°C to +85°C temperature range, the reference would need to have a tempco of less than 0.5 ppm/°C. However, in this circuit's application, the ambient temperature is carefully controlled to be within +3°C to +30°C (37°F to 86°F). Under those conditions the circuit can provide 14-bit accuracy with conversion error held to less than 1 LSB.

15.3 Applying adjustable-series bandgaps

As mentioned previously, it is often necessary to have an adjustable voltage reference. The term *adjustable* can be ambiguous though. Some high-performance references have a dedicated TRIM pin, whereas others have an ADJust pin. What's the difference? The TRIM pin works with an external trimmer and allows one to trim the nominal reference voltage, usually over a small range of several hundred millivolts (i.e., 5 volts ±300 mV). It allows the designer to trim out any *system errors*, by deliberately offsetting the reference voltage to a slightly different value from the nominal. Other series references have a dedicated adjustment pin (ADJ), which when used in conjunction with an external resistor network, allows one to adjust the reference voltage in a range from hundreds of millivolts to many volts. This allows the designer to provide a *custom* reference voltage or to be able to have a *wider* trim range than is otherwise available. Next we will look at some outstanding adjustable series types.

The first one we will look at is a remarkable new product from Linear Technology, the **LT6650**. This is a versatile, adjustable micropower 400-mV (sub-bandgap) reference, which includes a rail-to-rail buffer amplifier. Its schematic is shown in Figure 15.18A. It is provided in a low-profile (1-mm) five-lead SOT-23 surface-mount package. Unlike

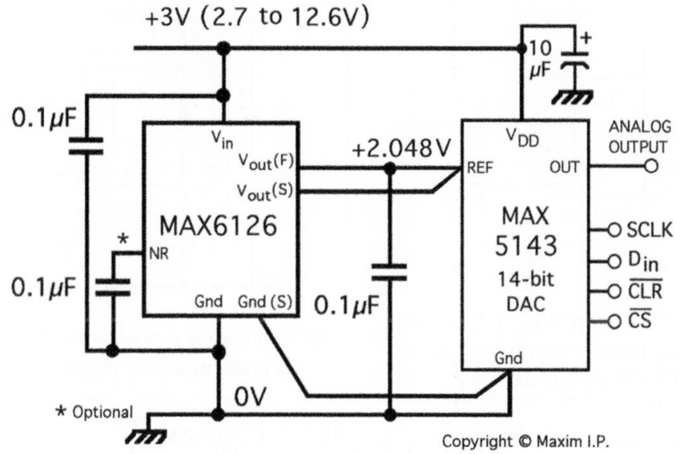

Figure 15.17. Maxim's MAX6126 provides an ultra-low noise reference voltage for this low-power 14-bit DAC. The device has a 1.3 μV pk-pk noise voltage, and a 3 ppm/°C tempco.

most bandgaps, the LT6650 operates down to a supply rail of 1.4 volts and typically draws less than 6 μA. It can be configured to work as either a fixed or adjustable *shunt* reference, in either positive or negative versions. As a *series* reference, it can be con-

LT6650	Linear Technology		Adjustable series /shunt bandgap reference				
Reference voltage(s):	Best initial accuracy:	Best tempco: (ppm/°C)	LF noise:	Input voltage range:	Supply current (μA)	Temp. range: (°C)	Special features: • Micropower • Low voltage • High accuracy • low tempco • low dropout • series or shunt mode
+0.4V adjustable up to higher voltages.	±0.5%	30ppm typ.	20μV pk-pk typ.	1.4V to 18V	<6	-40°/ +125°C	

Table 15.16 LT6650

figured to work either as a buffered 400-mV reference, or using its internal amplifier with external precision resistors, it can provide any voltage from 400 mV up to almost the supply rail (the maximum supply is 18 volts). The output can sink or source up to 200 μA over the *automotive* temperature range (–40°C to +125°C). With a 5-volt supply rail and the buffered 400-mV output, its maximum initial accuracy is 0.5% (2 mV max). It has a typical tempco of 30 ppm/°C and a low-frequency noise voltage of around 20 μV pk-pk. The LT6650's main features are summarized in Table 15.16.

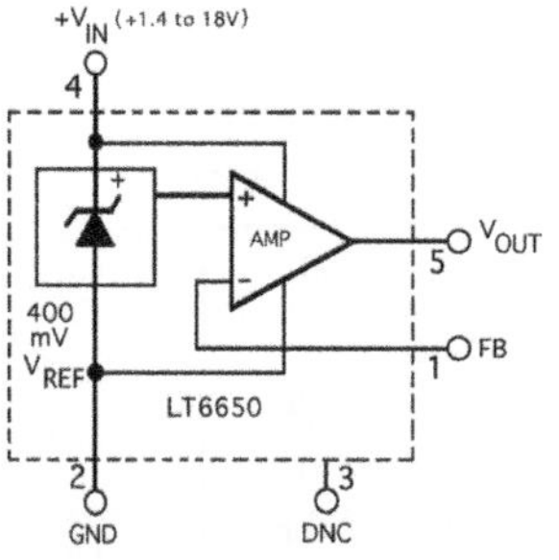

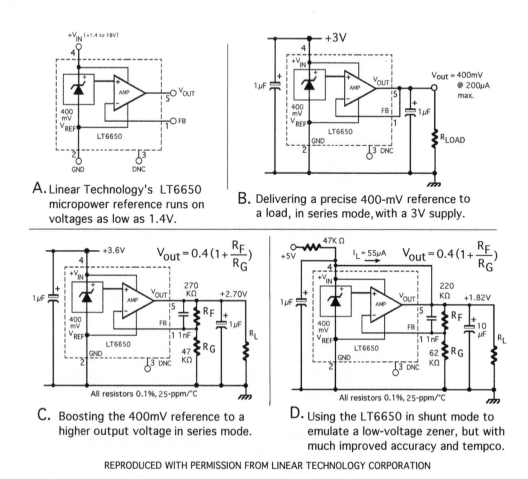

A. Linear Technology's LT6650 micropower reference runs on voltages as low as 1.4V.

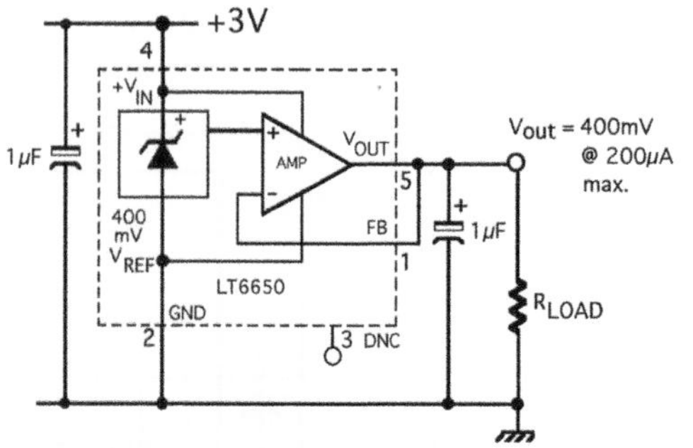

B. Delivering a precise 400-mV reference to a load, in series mode, with a 3V supply.

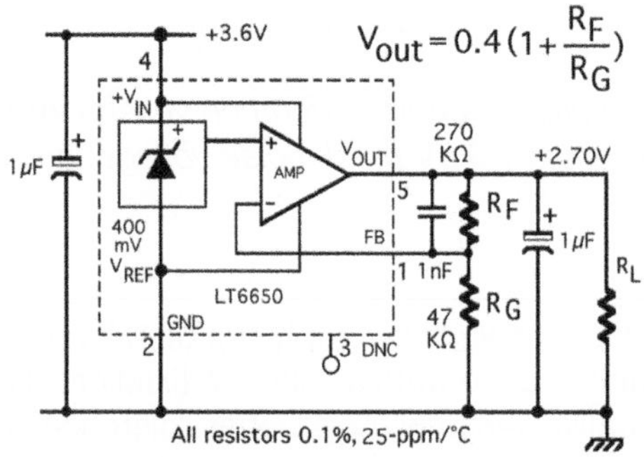

C. Boosting the 400mV reference to a higher output voltage in series mode.

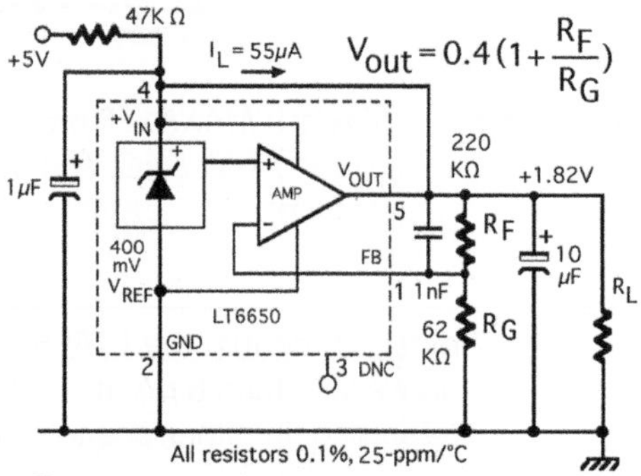

D. Using the LT6650 in shunt mode to emulate a low-voltage zener, but with much improved accuracy and tempco.

Figure 15.18. Some LT6650 low-voltage reference applications.

Figure 15.18 shows some design examples for the LT6650. Figure 15.18B shows a basic configuration where the internal reference is buffered by the op amp to provide a stable 400-mV reference. For optimum stability, both the input and output should be decoupled with 1-μF (or higher) capacitors. Figure 15.18C shows how to provide a higher reference voltage, as could be used for biasing purposes. In the example shown, an output voltage of 2.7 volts is created by using a pair of resistors (RF and RG), which should be either metal-film or thick-film types, with a 0.1% tolerance, and a tempco of less than 25 ppm/°C. The voltage setting equation is shown in the diagram. The 1000-pF capacitor is included for added stability purposes. Figure 15.18D shows just how versatile the device is. In this example, the device is used as a shunt reference with an output voltage of +1.82 volts. Compared to a typical low-voltage zener diode, the LT6650 has a temperature-compensated output. It has a much higher initial accuracy (0.5% vs. the zener's usual 5%), a significantly lower tempco (more than 300 ppm/°C for the zener vs. approximately 30 ppm /°C for the LT6650), a much lower noise voltage (probably 10 to 30 times lower), and a 100 times lower quiescent current (I_Q). It can also provide a similar performance as a *negative* shunt reference. For applications having very low supply voltages or for creating a special low reference voltage, the LT6650 may be your only choice.

In the next example, we will look at the Maxim **MAX6325** family, which includes three voltages (2.5 volts, 4.096 volts, and 5 volts) in different performance grades. This gives an example of the TRIM function. These devices are examples of the highest precision bandgap references available anywhere, with specifications that include a maximum initial accuracy of 0.02%, a maximum tempco of 1 ppm/°C, and an ultra-low noise figure of 1.5 μV pk-pk. Such figures are normally associated only with buried-zener types, but because Maxim has had long experience in creating that topology, it has now been able to re-create similar specs with a leading *super-bandgap* product. In order to achieve those dazzling specifications, it comes at a higher cost, a little higher operating voltage (8 to 36 volts), a higher dropout voltage, and with a higher quiescent current (a maximum of 2.7 mA) than other bandgaps. However, these devices are

MAX6325		Maxim IP	Ultra-precision series bandgap reference				
Reference voltage(s):	Best initial accuracy:	Best tempco: (ppm/°C)	LF noise:	Input voltage range:	Supply current (μA)	Temp. range: (°C)	Special features:
2.500V 4.096V 5.000V	±0.02%	1 ppm	1.5 μV pk-pk typ.	8V to 36V	1.8mA typ.	-40°/ +85°C	• High intl. accuracy • ultra-low tempco • ultra-low noise • Noise redctn. pin • TRIM pin • SO-8 or PDIP-8

Table 15.17 MAX6325

intended for professional top-of-the-line equipment, which usually requires those particular characteristics and where the reference's ambient temperature can be tightly controlled. The main features of the MAX6325 family are shown in Table 15.17.

The MAX6325 family are available in three temperature ranges (commercial, industrial, and military) and in various package options that include eight-pin DIL and the SO-8 surface-mount package. They also include a dedicated TRIM terminal that allows trimming of the reference output over a typical ±50-mV range, and without it affecting the ultra-low tempco. This is because Maxim's proprietary tempco curvature correction circuitry also includes the adjustment feature, and so does not degrade the tempco as one might expect. Adjustment can be made by using either a 10-KΩ precision metal-foil or wirewound trimmer or a low TC-tracking thin-film resistor network. The devices also include a dedicated noise-reduction pin, to which one can connect a high-stability capacitor (up to 1 μF). This reduces any wideband noise emanating from the power supply and coupling through the reference.

Figure 15.19 shows a very-high-performance circuit, which would be used in a 16-bit converter system. Normally, a bandgap circuit is only suitable in up to 14-bit systems at best. Here the reference is buffered from the load by an equally high-performance op amp, the MAX400. This device is no ordinary precision op amp, but one of the best ever made. It has a maximum ultra-low offset voltage of 10 μV, an offset voltage drift of less than 0.3 μV/°C, and an ultra-low noise figure of less than 0.6 μV pk-pk. The MAX400 also has a wide operating voltage of 6 to 36 volts and an astonishing 0.2 μV/month long-term V_{OS} stability, all of which makes it a perfect companion for any of these ultra-precision Maxim references. Because both ICs have an operating temperature range of –40°C to +85°C, then for this temperature span the devices will provide a 12-bit system with accuracy to 1/2 LSB. For a 14-bit system they can provide an accuracy to 1 LSB, over a –40°C to +60°C range. For a 16-bit system they can provide an accuracy to 1 LSB, but over a shorter –40°C to +40°C range; all of which is very impressive. The MAX6325 family raises the bar and sets a new benchmark for high-performance references. One could consider them *super bandgaps*.

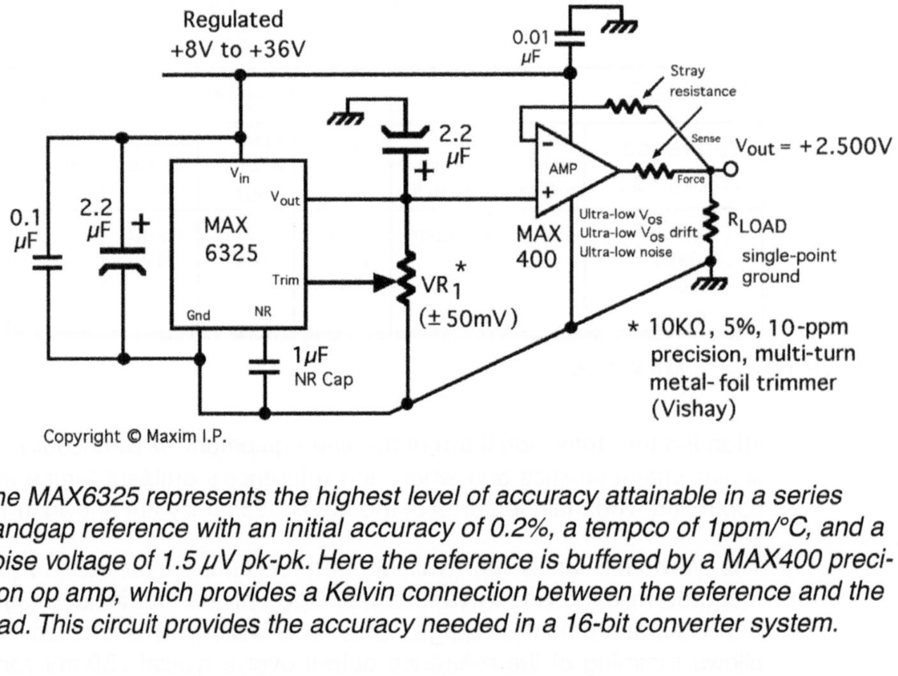

Figure 15.19. The MAX6325 represents the highest level of accuracy attainable in a series bandgap reference with an initial accuracy of 0.2%, a tempco of 1ppm/°C, and a noise voltage of 1.5 μV pk-pk. Here the reference is buffered by a MAX400 precision op amp, which provides a Kelvin connection between the reference and the load. This circuit provides the accuracy needed in a 16-bit converter system.

In the next example we will look at an adjustable reference, in the form of another Maxim product, the **MAX6037-ADJ**. The MAX6037 is a family of high-performance bandgap references also built with a proprietary BiCMOS process and available in the five-pin SOT-23 surface-mount package. They combine high initial accuracy with a low tempco, a low dropout voltage, and low power. The MAX6037 series provides a reference voltage of 1.25 volts, 2.048 volts, 2.5 volts, 3 volts, 3.3 volts, or 4.096 volts in three performance grades, the best of which has a low initial accuracy of ±0.2%. The premium-grade references have a maximum tempco of 25 ppm/°C, over the *automotive* temperature range (–40°C to +125°C). They also include a low dropout voltage of less than 100 mV max, a low quiescent current of less than 275 μA, and a shutdown feature that will reduce the quiescent current to less than 0.5 μA when used. All versions can work from a supply voltage as high as 5.5 volts, and the output can sink or source up to 5 mA. The main features of the MAX6037 family are shown in Table 15.18. The MAX6037-ADJ version can be adjusted over the 1.184-volt to 5-volt range and uses a resistor divider network to set the output voltage.

MAX6037		Maxim IP	Precision series bandgap reference				
Reference voltage(s):	Best initial accuracy:	Best tempco: (ppm/°C)	LF noise:	Input voltage range:	Supply current (μA)	Temp. range: (°C)	Special features:
1.250V, 2.048V, 2.500V, 3.000V, 3.300V, 4.096V	±0.2%	25ppm max	6μV pk-pk typ.	2.5V to 5.5V	<275 max.	-40°/ +125°C	• Low power • low tempco • low noise • wide temp range • shutdown pin • low dropout

Table 15.18 MAX6037

Unlike the previous example, where the reference could be adjusted over just a small ±50-mV range, the MAX6037-ADJ (and other similar adjustable references) allow one to adjust the output voltage over a range of several volts (with this particular product it is approximately 3.8 volts). This allows the designer to create either a particular custom reference voltage, which is not available as a standard fixed voltage value (e.g., 1.25 volts, 2.5 volts, 3 volts), or for making a large system trim. In this example, we require a reference voltage of 1.8 volts running from a nominal supply voltage of 3 volts and sourcing a load at 2 mA, over a temperature span of 0°C to +45°C. Figure 15.20 shows the application circuit using the MAX6037-ADJ, which can provide the accuracy needed for a 10-bit system, to less than 1-LSB resolution, over the required temperature span. Resistors R_1 and R_2 are largely responsible for setting the required output reference voltage, as shown in the diagram, and the small current in this resistor-divider adds to the overall quiescent current. These resistors should be quality low-TC precision types (i.e., ±0.1%; less than 30 ppm/°C). For best transient response, the input should be decoupled with a 0.1-μF to 1-μF ceramic capacitor located as close as possible to the reference. The output should also be decoupled with a 0.022-μF to 1-μF capacitor, which will enhance frequency stability. The shutdown terminal can be tied directly to the supply rail if unused, or alternately should be pulled up through a resistor, as shown in the diagram. A low voltage on this terminal will disable the output and put the device into a very-low-power mode, where typically less than 500 nA is drawn.

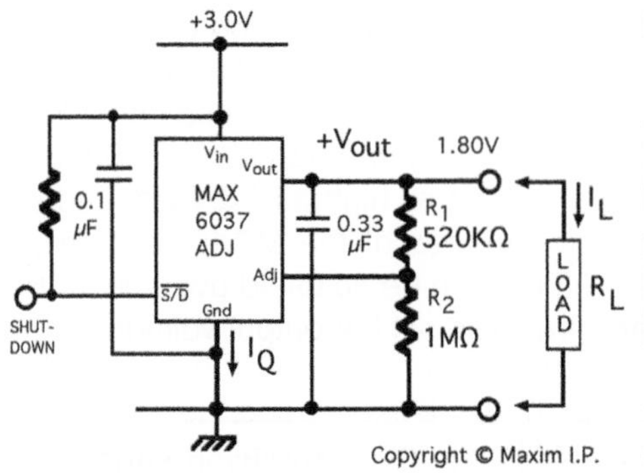

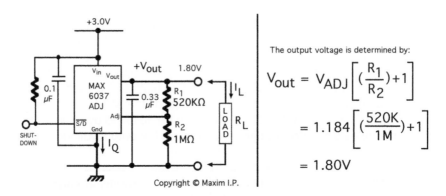

The output voltage is determined by:

$$V_{out} = V_{ADJ}\left[\left(\frac{R_1}{R_2}\right)+1\right]$$

$$= 1.184\left[\left(\frac{520K}{1M}\right)+1\right]$$

$$= 1.80V$$

Figure 15.20. Using Maxim's MAX6037-ADJ adjustable, precision micropower reference, in order to easily create a custom reference voltage.

In the third example of an adjustable-series voltage reference, we will look at a novel Intersil/Xicor product, the **x60250**. This is a micropower, digitally programmable bandgap reference. It is built with a proprietary Xicor CMOS process and available in an eight-pin TSSOP package. Besides an on-board EEPROM, it includes a two-wire serial bus interface, an 8-bit (256-tap) 100-KΩ digitally controlled, nonvolatile pot, and a temperature-compensated bandgap reference. Being a CMOS product, the x60250 typically has a quiescent current less than 25 µA. The main features of the x60250 are shown in Table 15.19.

x60250	Xicor/Intersil		Digitally-programmable bandgap reference				
Reference voltage(s):	Best initial accuracy:	Best tempco: (ppm/°C)	LF noise:	Input voltage range:	Supply current (µA)	Temp. range: (°C)	Special features:
Adjustable: 0V to 1.25V or 0.625V to 1.25V	±1%	20ppm typ.	100µV pk-pk typ.	2.7-5.5V	20 max.	-40°/ +85°C	• 8-pin TSSOP • 2-wire interface • 8-bit, 100KΩ internal digital pot • Non-volatile EEPROM

Table 15.19 x60250

The actual core reference voltage of 1.25 volts is directly available at the V_{REFOUT} terminal, which can be useful in some applications, so long as the 400-µA current output is not exceeded. The digital SCL clock frequency can be as high as 400 KHz, and a worst-case nonvolatile write cycle can be as high as 10 mS. The EEPROM consists of two separate 8-bit registers, one of which holds the reference voltage setting. The second register can be used for other purposes, such as storing a date, time, serial number, or even a temperature setting. The V_{OUT} terminal should be connected to a high

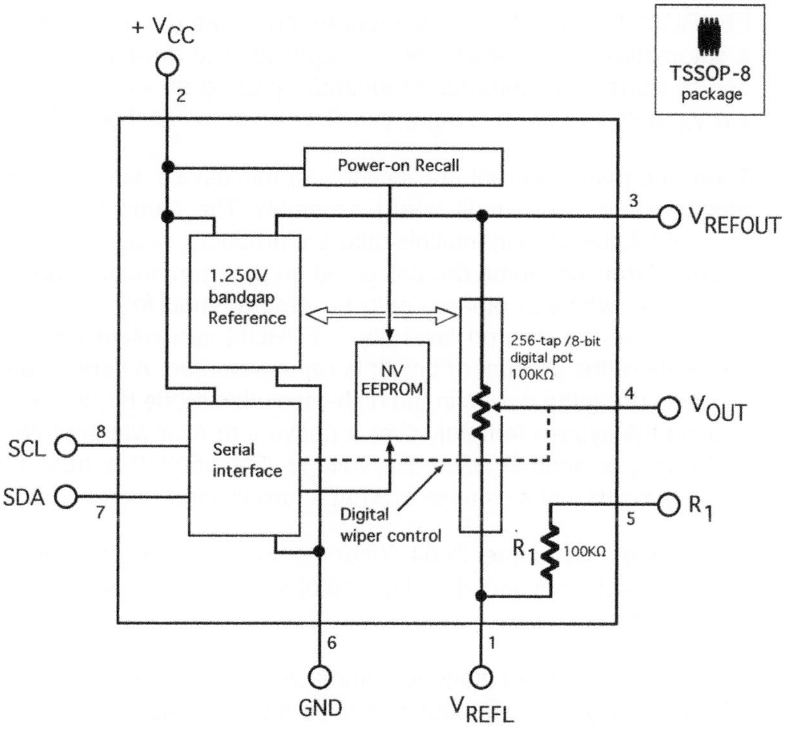

REPRODUCED WITH PERMISSION OF INTERSIL CORPORATION

Figure 15.21. Xicor's x60250, a CMOS micropower, 8-bit digitally programmable voltage reference with nonvolatile EEPROM, allows easy in-circuit programming.

impedance load, such as an op amp (buffer) input. The x60250's functional diagram is shown in Figure 15.21.

The x60250 interfaces to a digital system via its bidirectional two-wire serial interface, and can be reprogrammed as many times as needed. Normally, with the device's V_{REFL} grounded, the reference can be digitally adjusted over a range of 0 volt to 1.25 volts (with 8 bits or 255 steps over a 1.25-volt range, equates to a resolution of 4.9 mV). In this configuration the internal 100-KΩ digitally controlled potentiometer is between V_{REFOUT} and ground, with the wiper setting available at V_{OUT}. When the device's R_1 terminal is grounded instead, this allows the reference to be adjusted over a range of 0.625 volt to 1.25 volts. In this configuration an additional (internal) matched 100-KΩ resistor is now in series with the digitally controlled potentiometer. This doubles the output voltage control resolution, and thereby increases the accuracy of the reference's output voltage (with 8 bits or 255 steps over a 0.625-volt range, now provides a resolution of 2.45 mV). Here the output setting accuracy can be as high as

0.1%. In either case, the setting will be retained indefinitely in the nonvolatile EEPROM. This enables one to program and store a voltage level indefinitely and recall it automatically, whenever power is applied. The device can provide higher voltages and/or higher load currents if required, by using an external precision op amp with a low V_{OS} drift. Intersil have many excellent op amps to choose from.

There are many different applications for this device, such as in programming the reference value *after* circuit board assembly. The same reference device and circuit board could be used in multiple different products, which could save having to buy different references. Some devices could be programmed for operation at, say, 1 volt for Product A, whereas others could be programmed for, say, 0.75 volt for Product B. Once set at the desired level, the EEPROM can retain this setting throughout the entire life of the product or until it is reprogrammed. A very different application could be when using the device in the high-resolution mode (R_1 terminal grounded) and calibrating the system remotely over a network or even via the Internet for, say, a 1-volt, 1.024-volt, or another reference voltage. The x60250 is truly a remarkable product, and it gives us just a glimpse of what future voltage reference products may be like.

(*Note*: As of midsummer 2004, Xicor Inc. was acquired by Intersil Corporation. For the Xicor products mentioned in this and other chapters, please check Intersil's Web site at *www.intersil.com.*)

One last example of a standout trimmable bandgap reference is from Analog Devices' **ADR01 family**. Why? Because most series bandgap products range from 2.048 volts to 5 volts. The ADR01 family is no exception, but it also includes a 10-volt device. It is sometimes necessary to have a *high-voltage* output reference and one that can run from a higher supply voltage (e.g., 15 volts, 18 volts, 24 volts, 28 volts). The ADR01 is a 10-volt reference that can operate from between 12 and 40 volts. It is available in either a five-lead SC70, a five-lead TSOT, or an eight-lead SOIC surface-mount package, and it is intended for operation over the *automotive* temperature range (–40°C to +125°C). It is available in two performance grades, the best of which has a 0.05% initial accuracy and a 3 ppm/°C tempco. It also has a low 20 μV pk-pk noise characteristic (0.1 to 10 Hz) and a low quiescent current of less than 1 mA. The device can also source up to 10 mA at its output. Table 15.20 shows the main features of the ADR01 family. This is one of the few bandgaps available that truly emulates a buried-zener reference in its voltage, accuracy, tempco, and noise level. If you did not know it was a bandgap, you would probably think it was a buried-zener type.

Being a precision reference, the device also features a TRIM pin for precise adjustment of the output voltage. This allows one to trim out any system errors by offsetting the nominal voltage by a small amount. It can also be used to set the output voltage to 10.24 volts if desired. The 10-volt output can be adjusted over a range of approximately ±350 mV. The SOIC-8 version of the ADR01 is a drop-in upgrade for the industry-standard REF01 reference. Consider this: the REF01 had specs of 10 volts, with an initial accuracy of 30 mV (0.3%), versus the ADR01's 0.05%. The REF01 had a

ADR01 family Analog Devices Inc.			Precision series bandgap reference				
Reference voltage(s):	Best initial accuracy:	Best tempco: (ppm/°C)	LF noise:	Input voltage range:	Supply current (μA)	Temp. range: (°C)	Special features: •5-lead SC70,TSOT, or SOIC-8
+2.50V +3.00V +5.00V +10.00V	±0.05%	3ppm	6-20μV pk-pk typ.	+4.5 to +40V	650 typ.	-40°/ +125°C	•TRIM pin •V_{PTAT} TEMP pin •Wide temp. range

Table 15.20 ADR01 family

maximum quiescent current of 1.4 mA, whereas the ADR01's quiescent current is significantly lower. The best-grade REF01A has a maximum tempco of 8.5 ppm/°C and a typical spec of 3 ppm/°C. Compare that with the ADR01B's spec of 3 ppm/°C max and a typical spec of 1 ppm/°C. As with most references, the ADR01 should have its input and output decoupled with a 0.1-μF disk ceramic capacitor. For best transient response, a small value (10-μF) electrolytic capacitor should be added at its input.

The ADR01 also includes a TEMP pin that outputs a linear voltage (approximately 440 mV to 760 mV), that is proportional to absolute temperature (V_{PTAT}). At room temperature, this voltage is typically 550 mV, with a positive tempco of about +1.96 mV/°C. The SOIC-8 package versions of the ADR01, ADR02, and ADR03 are pin-compatible with the SOIC-8 versions of the industry-standard series bandgaps: REF01 (+10 volts), REF02 (+5 volts), and REF03 (+2.5 volts). Their exact specs may be a little different though. This TEMP pin duplicates the same pin-out on the original PMI-designed REF02 and REF03, which have a similar pin, and can be used as a temperature sensor. (The REF02 voltage is a little different though, because at room temperature this voltage is typically 630 mV, with a tempco of +2.1 mV/°C.)

Figure 15.22 shows a design using the ADR01 together with an Analog Devices' AD820B op amp. The circuit provides an adjustable 10-volt output. The op amp is a R-R type with very low input bias current (25 pA max), low offset voltage (0.8 mV max), low drift (1 μV/°C typical), and wide operating supply range (5 to 36 volts). The op amp buffers the reference's TEMP output because this voltage is directly from the bandgap's core. It is imperative to use a low-input bias current op amp such as the AD820 to avoid loading the TEMP pin. The TEMP voltage at 25°C is approximately 550 mV and has a positive slope of approximately +1.96 mV/°C. This can be used to monitor the system temperature and even adjust the reference voltage if necessary, with additional circuitry. Although a good-quality precision trimmer (wirewound or metal-foil types) allows convenient adjustment, for a lower overall tempco, substitute the trimmer for 0.1% tolerance, 25-ppm/°C precision resistors, or better.

Also falling into this category of trimmable/adjustable references, consider this: For a very-high-performance design one can digitally compensate such a reference's output

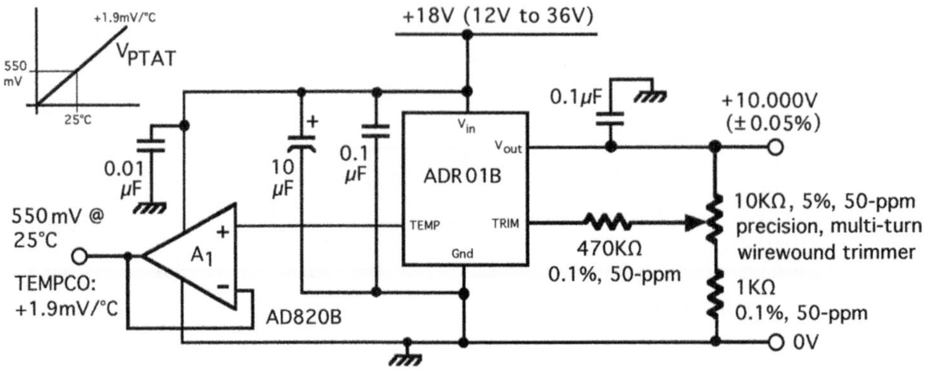

Figure 15.22. *Adjusting the output voltage of the ADR01B series reference is easily accomplished, with negligible effect on tempco or performance. The trimmer allows adjustment by about ±350mV. Use precision low-TC resistors though. The Temp output produces a linear V_{PTAT} voltage, which is buffered by the op amp, and can be used for monitoring the system temperature, or for compensation purposes.*

voltage (V_{OUT}) and drift (ΔV_{OUT}) over the temperature range. This would mean that a design that started out with a 0.1% initial accuracy, a 10 ppm/°C tempco, and a long-term stability of 100 ppm/°C could be significantly improved. The resulting reference, using **digital compensation**, would be close to having a 0.01% initial accuracy, a 1 ppm/°C tempco, and a long-term stability of less than 10 ppm/°C. In other words, an order-of-magnitude improvement could be made, or maybe more. A higher than desired noise level could be reduced with a separate (buffered) analog low-pass filter. To do all this, one would need to start with an exceptionally good reference having a low initial accuracy, a low tempco, a low noise figure, and the additional TEMP and TRIM pins. (This applies not only to bandgaps but to all high-performance references, such as Analog Devices' AD780, ADR01, REF02, ADR43x XFET®; Maxim's MAX6325, MAX6143, MAX6173 to MAX6177; Texas Instruments' REF102; Linear Technology's LT1021, and others.) Without a TEMP pin, one would need to add a monolithic temperature sensor with a V_{PTAT} output voltage, mounted next to (or underneath or on top of) the voltage reference. In every such case here, the reference device would need a TRIM adjustment pin. Best performance is obtained with a reference with *both* pins!

It is advisable to first burn in your prototype circuit and any additional voltage reference ICs, to ensure added long-term stability. An initial burn-in at 125°C will age any circuit and should force any errors, drifts, and mechanical stress factors to occur within the first week (168 hours) during burn-in. Following the circuit's return to normal room temperature, you should power up the circuit (as shown in Figure 15.23), with the mechanical trimmer installed. Now measure the voltages at the V_{OUT} and TRIM terminals. The point of this exercise is to determine how much trim voltage is needed to change the output voltage, as the trimmer's wiper is moved up or down. Record all this data (V_{TRIM} vs. V_{OUT}). Temperature tests in still air should be done next, in a precisely controlled environmental chamber or lab oven.

The temperature range, such as –25°C to +85°C, should be a little more than the application requires (e.g., –10°C to +75°C). The data collected will be programmed into the next prototype's EEPROM. In this second prototype, the reference's output voltage, V_{OUT}, and its TEMP voltage will be converted into separate digital words by the circuit's A/D converters. This data forms a lookup table, which will contain the correction data used to compensate the voltage reference. The data consists of the exact temperature, the corresponding V_{TEMP} voltage, and reference V_{OUT} voltage. From the data recorded from adjusting the trimmer in the first prototype, one can calculate the necessary TRIM voltage that will compensate the temperature drift. This data must also be programmed into the EEPROM. (In normal operation, the circuit is working constantly, with the voltages being compared with the data stored in the EEPROM. Any correction data necessary is then applied via the DAC and buffer, which interface with the reference's TRIM pin.) The design shown in Figure 15.23 functions as though you were constantly adjusting the TRIM terminal. In more sophisticated designs, one could use multiple DACs, which could control the reference trim, the reference I_Q, the amplifier gain, and so on. The circuit can be made as simple or as complex as desired and can be made to compensate virtually any shape of voltage TC curve. It would allow you to compensate the reference's TC as often as needed and in tiny increments.

Once the second prototype circuit's EEPROM has been programmed, power up that circuit and test it over the intended temperature range, monitoring the three voltages (V_{OUT}, V_{TEMP}, and TRIM). Any corrections should now be programmed into the EEPROM and the circuit retested over the temperature range. Once the design has reached this stage and has passed all its tests, it should be ready for use. In a mass-production situation, you may want to build several more prototypes, but after the burn-in stage use this latest EEPROM data. Check those prototypes over the same temperature range, and look for any changes using Prototype 2 as the yardstick (reference). Review the data from all the prototypes, and if necessary create an average of the correction values. Program a new EEPROM (version 3), and test some more prototypes over the same temperature range using its data. If they all pass 100%, your design is almost ready for handing over to your Production department. Copy several EEPROMs with version 3 data. The design will still need thorough documentation to support its manufacture, such as burn-in and assembly instructions, voltage levels, the EEPROM copies and ROM codes, temperatures, test procedures, and so on. Even though this may seem to be a costly design because of the labor/testing involved, the end result will be reliable and an excellent foundation for the end product. Unfortunately, there is no cheap shortcut to creating high-precision products at this level (i.e., more than 14-bit resolution).

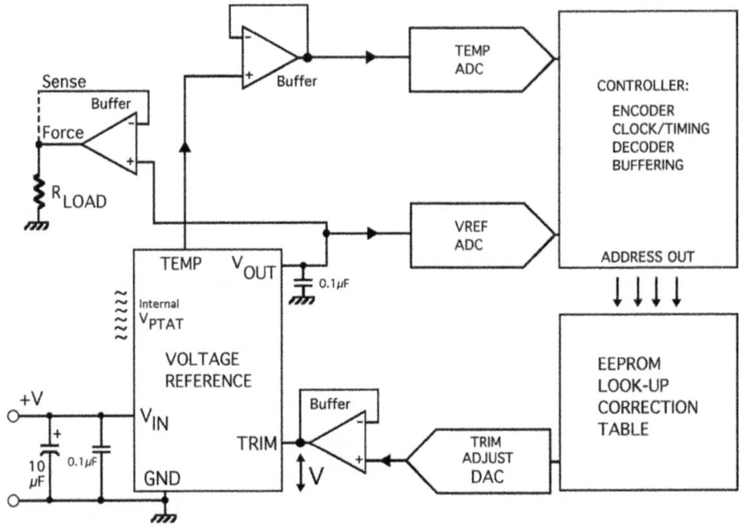

Figure 15.23. *Using EEPROM correction to temperature-compensate the voltage reference. In this circuit the reference's V_{PTAT} voltage and V_{OUT} voltage are used by the look-up table to provide the required TRIM correction voltage. Such techniques can provide 0.01% accuracy and <1-ppm/°C tempco over the operating temperature range.*

15.4 Using the Analog Devices' XFET® reference

As discussed in Chapter 14, the XFET® is an outstanding voltage reference with many excellent features, including low cost. Most noticeable are its ultra-low noise level, a very low initial accuracy (±0.05%, B-grade devices), low quiescent current (less than 800 µA), a very flat 3 ppm/°C tempco, and a wider operating temperature range (*automotive*, –40°C to +125°C), than most bandgaps or buried-zener types. Its tempco is flat and linear compared to either the bandgap or buried-zener, and its long-term drift is better than most other types of reference. Because its tempco is inherently lower than most bandgap references, much less compensation is required. Because less resistor noise (used for compensation) is involved, as well as using inherently low-noise JFETs at its reference core, this all helps the device have a significantly lower overall noise level. For example, the noise level for an **ADR431** (2.5-volt) XFET® is given as less than 4 µV pk-pk (at 0.1 Hz to 10 Hz) typical, whereas the noise density at 1 KHz is typically 60nV/√Hz—both excellent characteristics. The long-term stability of the device is typically 40 ppm/1000 hours, which is also impressive.

The output of a typical XFET® uses a buffer amplifier, which provides a source capability of 30 mA, a sink capability of 20 mA, and an indefinite short-circuit current to ground of about 40 mA. This is significantly higher than most competing devices. Lower reference voltage XFET® devices are specified for operation with supply voltages of between 4.1 and 18 volts. (The 5-volt device is specified for operation with an input of more than 7 to 18 volts.) The typical quiescent current will be less than 800 μA, whereas the ripple rejection ratio is typically 70 dB. Line regulation is specified at a maximum of 20 ppm/V, and load regulation is specified at a maximum of 15 ppm/mA. The main features of the ADR43x family are shown in Table 15.21.

As with most references, the device should include input decoupling such as a 10-μF electrolytic, in parallel with a 0.1-μF ceramic disk capacitor. The output is perfectly stable without a capacitor, although a 0.1-μF ceramic disk capacitor will help noise suppression. Uniquely, the XFET® also features a very fast turn-on settling time, typically about 10 μS. (This is a measure of the time required for the output voltage to reach its final value, within the specified error band, after a cold start) Most other references cannot stabilize this quickly.

ADR43xB	Analog Devices Inc.		Precision series XFET® reference				
Reference voltage(s):	Best initial accuracy:	Best tempco: (ppm/°C)	LF noise:	Input voltage range:	Supply current (μA)	Temp. range: (°C)	Special features:
2.048V, 2.500V, 3.000V, 4.096V, 4.500V, 5.000V	±0.05%	3ppm	3.5 μV pk-pk typ.	+4.1 to +18V	560 typ.	-40°/ +125°C	• Ultra-low noise • ultra-low tempco • wide temp. range • high intl. accuracy • fast turn-on time • TRIM pin • SOIC-8/mini SOIC

Table 15.21 ADR43xB

The XFET® is provided in an eight-pin surface-mount package (SOIC or mini-SOIC), which also includes a TRIM terminal that allows output voltage adjustment over a ±0.5% range. (For example, the 2.5-volt reference provides an adjustable range of ±2.5 mV.) Adjusting the output voltage to another value from nominal allows the designer to easily trim out system errors. Using this feature has negligible effect on the device's temperature performance. However, it is prudent to use a low-TC, multiturn, precision trimmer (Bourns, 5% wirewound, with a TC of less than 50 ppm/°C; or a Vishay metal-foil trimmer, 5% type with a TC of 10 ppm/°C). The precision resistors should be thin-film or metal-film 0.1% types, with a TC of less than 50 ppm/°C (or low tracking TCR if used in an array/network). An improved tempco can be achieved by using a precision, fixed-value, thin-film resistor array, rather than a trimmer, although the latter is a convenient way to precisely adjust the output voltage.

Figure 15.24 shows an application circuit in which the **ADR431A**, a 2.5-volt XFET®, is used as a precision reference for another outstanding Analog Devices' product (AD7701), a very high-performance 16-bit sigma-delta (Σ-Δ) A/D converter. This ultra-low-power CMOS A/D converter contains a low-pass digital filter for both the analog signal input and the external reference, also an on-chip clock generator, and a two-wire serial interface among other features. It offers true 16-bit resolution, along with an

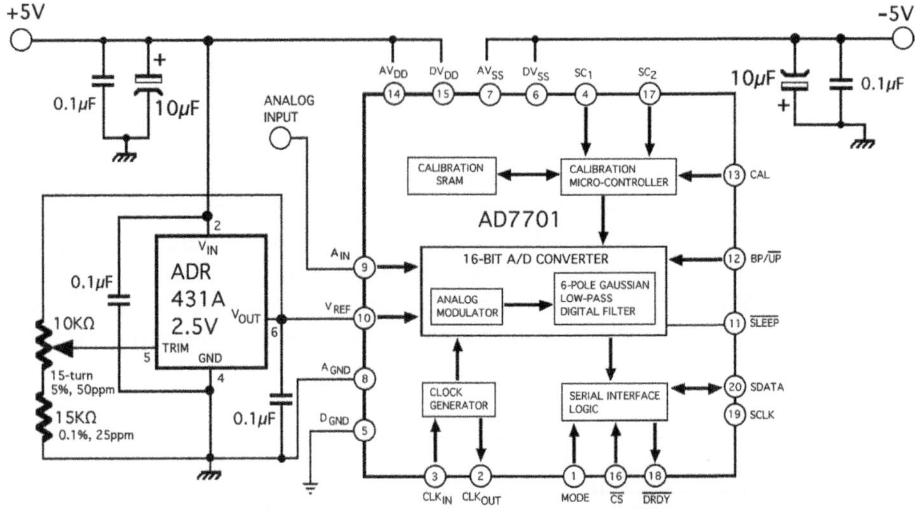

Figure 15.24. In this example, an Analog Devices' ADR431A, a 2.5-volt XFET® provides an ultra-low-noise reference voltage to a high-performance, 16-bit sigma-delta A/D converter.

outstanding 0.0015% accuracy (linearity error). The converter's on-chip self-calibration feature can be used to eliminate the effects of temperature drift, even those of external circuits where gain errors and system offsets can be a problem. It also includes its own SRAM memory to store the necessary calibration data and offsets involved. The AD7701 is intended for measuring a wide dynamic range and low-frequency signals, such as those usually involved with medical, industrial, and scientific applications.

The ADR431A is chosen here as the reference companion for this A/D converter, because of its unique combination of low noise (less than 4 µV pk-pk), very low tempco, high initial accuracy, low quiescent current, fast turn-on settling time, and very wide operating temperature. The overall combination provides a subsystem having low power, very low noise, and high accuracy. The combined operating temperature range of these two devices (–40°C to +125°C) is another advantage too. This circuit shows an output adjustment capability using a precision 10-KΩ multiturn potentiometer, which helps trim out any errors. A 0.1-µF bypass capacitor is shown on the output of the XFET®, which helps reduce any transients occurring on the Σ-Δ ADC's REF input. This is a common condition for some Σ-Δ A/D converters, which use a switched capacitor input. As a result, noise can be generated as the ADC's internal capacitor switches, which can cause conversion errors. (As we saw earlier in this chapter, another Analog Devices' product, the AD780, has the ability to drive virtually any value of capacitive load, particularly when driving Σ-Δ A/D converters.)

Figure 15.25 shows a novel design in which the ADR435B, a 5.0-volt XFET®, is used to create an adjustable, bidirectional Howland current pump. An Analog Devices' OP1177, a single precision op amp (A_1), is used together with a pair of low-TC resistors (or a precision trimmer) to create dual reference voltages of ±2.5 volts. These voltages are applied to each end of the precision potentiometer (VR_1), so that the trimmer's midpoint is effectively 0 volt. Amplifier A_1 drives the ground pin of the XFET®, thereby creating a stable –2.5 volts. Selecting the voltage at VR_1's wiper (V_{REF}), which ranges from +2.5 volts to –2.5 volts, controls the magnitude and polarity of the circuit's output current. Amplifier A_2 is used to buffer the V_{REF} voltage, whereas amplifier A_3 drives the load at up to 10 mA. Amplifiers A_2 and A_3 are an Analog Devices' OP2177, a precision dual op amp. Because the circuit relies on a bipolar reference voltage to set the current level, it is necessary to use a split power supply in this design, which can range from ±8 volts to ±15 volts.

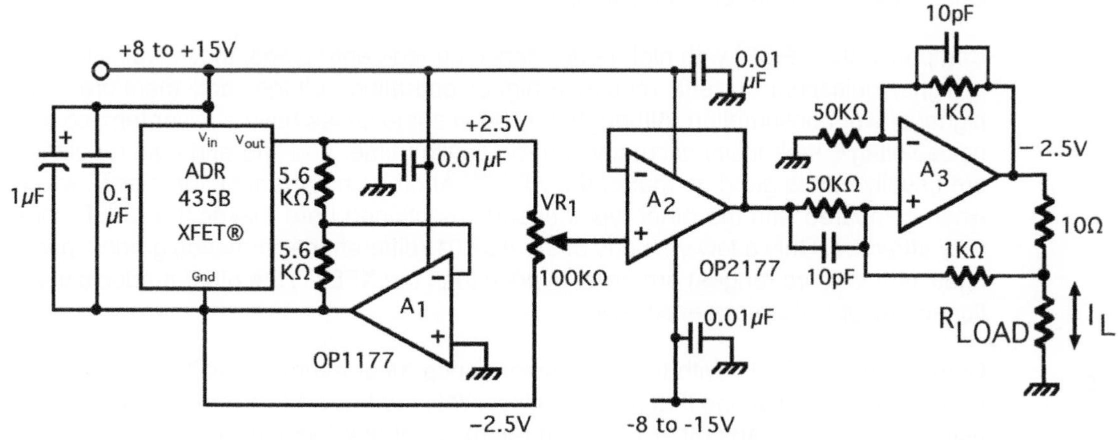

Figure 15.25. Creating a bidirectional Howland current pump using the XFET® to create a precise bipolar voltage reference. A precision trimmer and some high-performance op amps control the polarity and size of the output current.

When comparing the key specifications of the XFET® with some of the best bandgap references, the XFET® is better than any shunt type and almost all series products. Of these, only a handful of *super-bandgaps* can equal or better the XFET®, but only in some of their characteristics. Those products were reviewed earlier in this chapter. However, few of those super-bandgaps are specified over the *automotive* temperature range (–40°C to +125°C), whereas the XFET® is. The Analog Devices' AD780 is a super-bandgap, with a few marginally better specs than the XFET®, but its operating temperature range is significantly shorter at –40°C to +85°C, and it offers just two voltage options. The same applies to Maxim's MAX6325, probably the best bandgap of all, but in this case it has a choice of only three reference voltages, as opposed to the six offered by the XFET®. National Semiconductor's outstanding LM4140 CMOS reference is another competitor, but its operating temperature range is even shorter (0°C to +70°C), its input voltage range is also smaller at 1.86 volts to +5.5 volts, and its load and line regulation are not as good. Another super-bandgap is the ADR01 from Analog Devices, with several key specs that equal the XFET®'s. However, its noise level is significantly higher, at about 20 μV pk-pk, versus the XFET®'s less than 4 μV pk-pk, and it too has an inferior load and line regulation to the XFET®'s. Then there is Maxim's MAX6126, another outstanding bandgap, but in this case its operating temperature range is shorter at –40°C to +85°C, it provides an accuracy of 0.2% versus 0.05% for the XFET®, and it too has load and line regulation that are not as good as that of the XFET®. In fact, the XFET® features excellent load and line regulation that is better than the competition, with the exception of the AD780's marginally better line regulation of less than 10 μV/V, and the MAX6325's slightly better load regulation (less than 7 ppm/mA). However, both of those products reference their load and line specifications over a shorter operating temperature range.

Comparing the XFET® with high-performance buried-zener types, those typically have a higher quiescent current, require a higher operating voltage, and therefore have higher power consumption. Although they do in some cases have a lower tempco and noise voltage, their initial accuracy, turn-on settling time, and line and load regulation are usually not as good as that of the XFET®. Another important factor is cost, which when compared with the other types already mentioned here makes the XFET® look very attractive. Only a few versions of the ADR01 (different performance grades, packages, temperature ranges) are lower in price than the XFET®; the other products mentioned are typically more expensive.

Comparing the XFET® with the high-performance Xicor/Intersil FGA™, both products have many specifications that are very close. Yes, the FGA™ is a remarkable product, better than virtually any other precision reference of any topology, anywhere. In comparison with the XFET®, three important considerations are as follows:

1. The XFET® has a significantly wider operating temperature range (–40°C to +125°C), to which its specs are referenced (the FGA™ has a –40°C to +85°C operating temperature range).
2. The XFET® has a choice of six output voltages.
3. Even though both devices have current-sink and source capabilities, the XFET® has at least twice the output current capability of the FGA™.
4. The XFET® has a significantly lower noise level.

Cost of the XFET® is another consideration, which is less than most versions of the FGA™. When you compare all of these factors with all of the different topologies and types previously mentioned, it makes Analog Devices' XFET® a very attractive option.

15.5 Applying buried-zener references

As we saw in Chapter 14, the buried-zener reference is one of the most mature voltage reference topologies, and it was designed to meet an extremely high level of precision. Only in recent years have any of the other topologies come close to matching some of its outstanding characteristics. The buried-zener product was expensive to build, characterize, test, and burn in (many products received many hours of burn-in at 150°C as part of the preconditioning process), and so it resulted in being an expensive product. The earliest products used an on-board substrate heater, which required a significant amount of power to operate. Subsequent designs did not include a heater, and instead used various compensation techniques and laser trimming to attain similar (but slightly lower) specifications. Although many devices provided 6.9-volt and 10-volt outputs, newer devices brought the reference voltage down to the more popular 5 volts. Over time various products became available from National Semiconductor (which first introduced the topology), as well as Analog Devices, Burr-Brown, Linear Technology, Maxim, Precision Monolithics, and others. At its peak came various expanded buried-zener products in larger DIL packages (rather than the more usual eight-pin TO-99 metal can), which included additional features such as laser-trimmed resistors, precision op amp/buffers, Kelvin sensing, noise filtering, output voltage adjustment, and so on. Both Analog Devices and Maxim introduced such products at the time. (Maxim's MAX671MDD was a +10-volt precision buried-zener reference with ±1 mV initial accuracy, a 1 ppm/°C tempco, and a 12 μV pk-pk low-frequency noise voltage. It was provided in a 14-lead side-brazed hermetic CerDIP package, and was characterized over the –55°C to +125°C *military* temperature range. The device also provided two fine adjustment pins and six Kelvin force/sense terminals.) A few of these types of products are still available; after all, they represent the ultimate in performance, which even today are still extremely hard to match.

Over the past decade or so, many manufacturers have concentrated on different market segments or on emerging markets with different requirements, and have dropped these hard-to-build, high-performance products from their portfolios. Compared with making buried-zener products, other types are usually easier to make. This has manifest itself in the emergence of the *super-bandgap* class of products, which have some of the same characteristics as the buried-zener. We have already seen some examples of these in this and previous chapters, and they include Analog Devices' ADR01, AD780, and ADR390; Linear Technology's LT1461 series; Maxim's MAX6325, MAX6143, MAX6173 to MAX6177; National Semiconductor's LM4130A and LM4140A. Today only a few analog semiconductor manufacturers still offer buried-zener products, most notably National Semiconductor, Analog Devices, Linear Technology, Maxim, and Texas Instruments.

Typical key specifications for any buried-zener reference include a very low initial error of between 0.01% and 0.05%; a tempco of between 0.05 ppm/°C and 5 ppm/°C; an ultra-low noise level of less than 10 μV pk-pk (0.1 to 10 Hz); and a long-term stability of typically less than 20 ppm/1000 hours. These specifications make buried-zeners suitable for 14-bit systems, although they can also be used in more than 14-bit systems by adding special temperature compensation networks, or in some cases depending on their pin-out, by combining *digital-compensation* techniques as discussed previously. One should only consider using a buried-zener reference if a supply voltage higher than 8 volts is available and if power dissipation is not an issue. Additionally, a buried-zener should be used if the device has to work over a wide operating temperature range, and if a very low tempco, long-term stability, and low noise are all vitally important factors (such as with high bit-resolution applications).

Here we will look at some exceptional buried-zener products beginning with Analog Devices. Analog Devices makes several well-characterized buried-zener products that include the AD586 (+5 volt), AD587 (+10 volt), AD588 (a pin-programmable output that includes +10 volt, +5 volt, ±5 volt, –5 volt, –10 volt), and the AD688 (±10 volt). A diagram of the AD588 was shown previously in Chapter 14, while a functional diagram of the AD688 is shown in Figure 15.26. As you can see, the AD688 (like the AD587/8) includes many additional circuit elements, including precision laser-trimmed thin-film resistors and precision op amps. Because of this additional circuitry, one can use this flexibility and easily include additional noise filtering, output voltage adjustment, and output Kelvin sensing—with all the active components and resistor products needed *already on-chip*, and sharing a virtually identical chip temperature.

The **AD688** is intended as a bipolar reference, having ±10-volt outputs, with initial accuracy of ±0.02% (±2 mV) maximum. The device is available in three different performance grades, each in a 16-pin CerDIP package. The AD688BQ's output voltage tempco is ±3 ppm/°C over the –40°C to +85°C industrial temperature range (or ±1.5 ppm/°C over 0°C to +70°C). Tracking between the two outputs is less than ±1.5 mV (±0.015%) maximum. Optional gain and balance adjustments are provided on-chip, which allow precise trimming and centering of the ±10-volt outputs. The device also features an exceptionally low noise voltage of typically 6 μV pk-pk (0.1 Hz to 10 Hz) or approximately 1 μV_{rms}. Its spectral noise density at 100 Hz is typically 140 nV/√Hz. A built-in noise-reduction pin is also provided, which allows addition of a single capacitor to further reduce noise. The long-term stability of the voltage reference at room temperature is specified at a typical 15 ppm/1000 hours. The main features of the AD688 are shown here in Table 15.22.

These out-of-the-box specifications make the AD688 buried-zener immediately suitable for 12-bit to 14-bit systems. For 12-bit/10-volt systems, with 1/2 LSB conversion error over a 100°C temperature span, the required resolution is 1.22 mV; the allowable noise voltage is 60 μV pk-pk (0.1 Hz to 10 Hz); and the allowable drift is 1.22 ppm/°C. The unadjusted device can provide the required accuracy, noise level, and (nearly) the low tempco needed, but over a shorter 20°C span. Although its ±2 mV initial accuracy can be trimmed to zero, and its noise level is quite acceptable, its tempco is slightly higher than required. For 14-bit/10-volt systems, the unadjusted specs of the AD688 can limit the temperature range to about 6°C for 1/2 LSB conversion error. The reference can be used in 16-bit systems and higher, by trimming, or by adding special temperature compensation networks, adding a low-pass filter at the output, using the

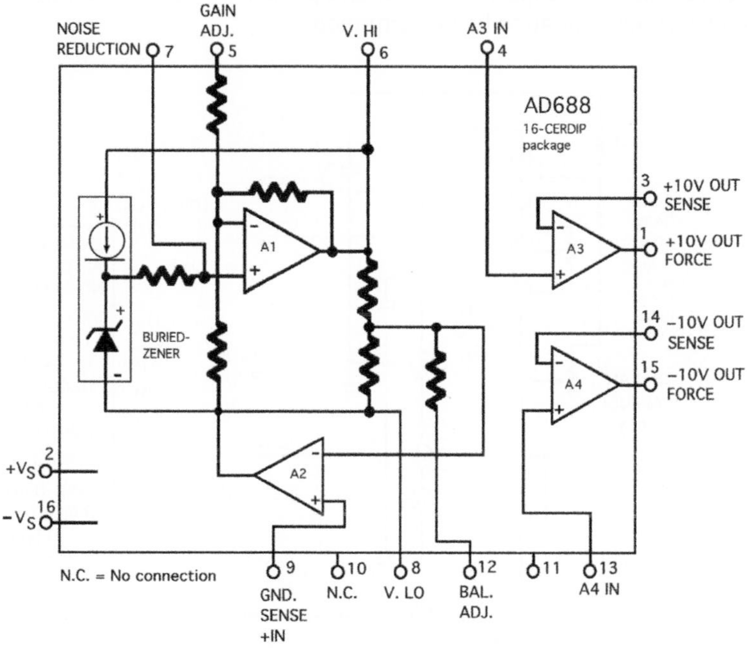

REPRODUCED WITH PERMISSION FROM ANALOG DEVICES, INC.

Figure 15.26. Analog Devices' AD688 is a precision buried-zener voltage reference with ±10V tracking outputs. The device features Kelvin connections, an initial error of 0.02%, a low drift of 1.5-ppm/°C, a very low noise figure of 6 μV pk-pk, and a long-term stability of approximately 15-ppm/1000-hours.

AD688	Analog Devices Inc.		Ultra-precision ±10V buried-zener reference				
Reference voltage(s):	Best initial accuracy:	Best tempco: (ppm/°C)	LF noise:	Input voltage range:	Supply current (μA)	Temp. range: (°C)	Special features:
±10V	±0.02%	3ppm	6 μV pk-pk typ.	±15V	<12mA	-40°/ +85°C and MIL versions	• Ultra-low tempco • ultra-low noise • high intl. accuracy • Force/sense out • Gain/Balance adj. • TRIM and NR pins • CERDIP-16

Table 15.22 AD688

Kelvin connections, or by using digital compensation. In this way the initial error can be virtually canceled out, and the noise level can be reduced to less than 1 μV pk-pk. An enhanced circuit is shown in Figure 15.27. Stabilizing the ambient temperature, preregulating the supply voltages, and protecting the circuit from large or small air currents will help enhance its performance.

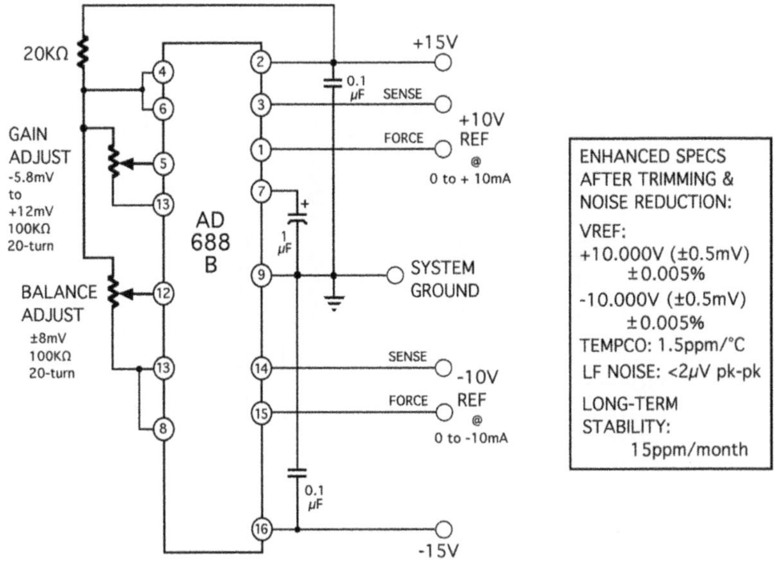

Figure 15.27. Showing how the AD688 buried-zener voltage reference can be enhanced for operation in 16-bit and higher converter systems. Few references can match the performance of the AD688 over the -40°C to +85°C industrial temperature range.

Another standout Analog Devices' buried-zener part is the **AD586**. The AD586 is a +5-volt reference with some of the same features as the AD688, but in a smaller eight-pin package. Like the AD688 and other high-performance buried-zener devices, it has a TRIM adjust pin and a noise-reduction terminal. It has a +5-volt output, with an initial accuracy of ±0.04% (±2 mV) maximum. The device is available in different performance grades, three different temperature ranges, and an eight-pin SOIC, an eight-pin PDIP, or an eight-pin CerDIP package. The AD586BM's output voltage tempco is 2 ppm/°C over the –40°C to +85°C industrial temperature range. An optional TRIM adjustment is provided on-chip, which allows precise trimming of the output by as much as 300 mV with an external multiturn trimmer. The device also features a very low noise voltage of typically less than 4 μV pk-pk (0.1 Hz to 10 Hz), or approximately 1 μV_{rms}. Its spectral noise density at 100 Hz is typically 100 nV/√Hz. A built-in noise-reduction pin is also provided, which allows addition of a single quality capacitor to further reduce noise. The long-term stability of the voltage reference at room temperature is specified at a typical 15 ppm/1000 hours. The device requires a supply voltage of between 7 volts (minimum) and 35 volts. Its quiescent current is typically 2 mA, and it is capable of sinking or sourcing up to a 10-mA load. The main features of the AD586 are shown in Table 15.23, and its functional diagram is shown in Figure 15.28.

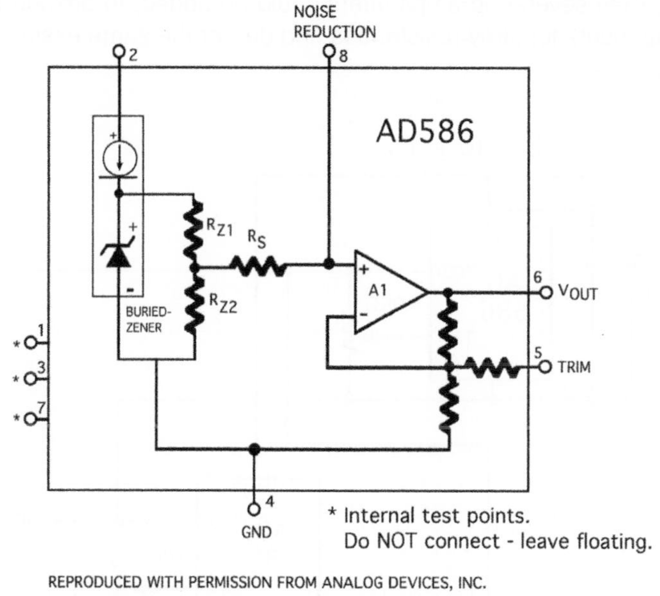

Figure 15.28. Analog Devices' AD586 buried-zener reference uses ion-implantation, laser-trimming, and thin-film resistors to achieve outstanding performance at low cost.

AD586	Analog Devices Inc.		High-precision +5V buried-zener reference				
Reference voltage(s):	Best initial accuracy:	Best tempco: (ppm/°C)	LF noise:	Input voltage range:	Supply current (μA)	Temp. range: (°C)	Special features:
+5.000V	±0.04% (±2mV)	2 ppm (0-70°C) 10ppm (-55°C to +125°C)	4 μV pk-pk typ.	+7V-36V	<3mA	0°C/ +70°C and MIL versions	• Ultra-low tempco • ultra-low noise • high intl. accuracy • TRIM and NR pins • CERDIP-8/DIP-8

Table 15.23 AD586

It is sometimes necessary to create more than one reference voltage in an application. The design shown in Figure 15.29 shows three AD586 devices used to create three different reference voltages in a stack. In this circuit, the reference voltages are +5 volts, +10 volts, and +15 volts. Because the 5-volt reference receives its input from the 15-volt output, its line regulation is improved. In fact, a change of input voltage of +18-volt to +46-volt produces a change in output that is below the devices' noise level. One can add more references, but the limiting factor is that the maximum load current is not exceeded. Each reference uses the TRIM capability to precisely set its output voltage. Depending on the loads, each output may require bypassing. If long traces are involved, then several op amp/buffers could be added, to provide Kelvin connections that compensate for stray resistances and deliver the same exact reference voltage to the load.

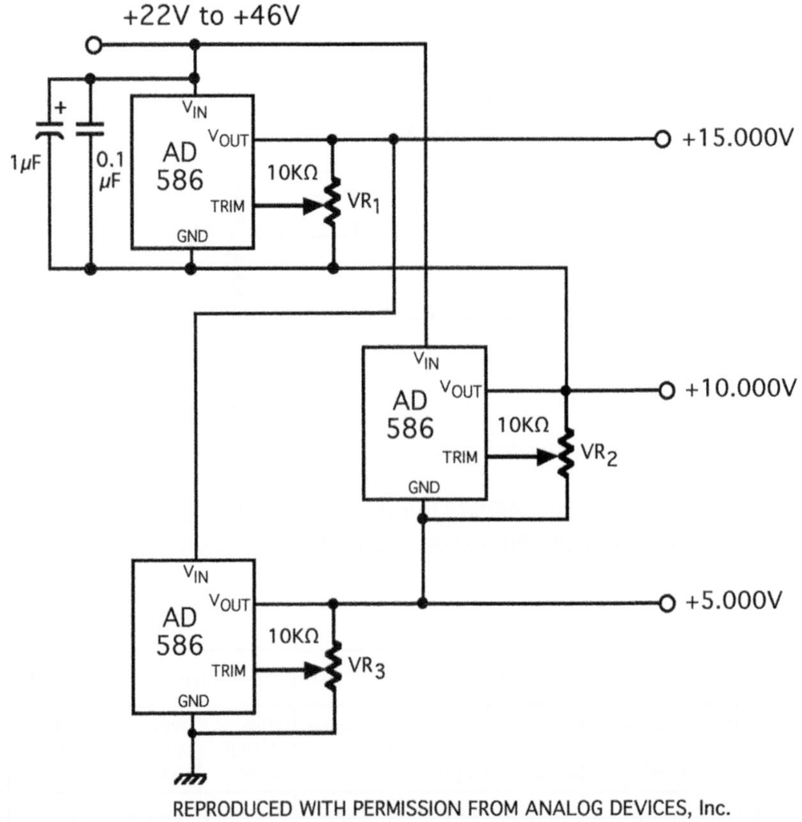

REPRODUCED WITH PERMISSION FROM ANALOG DEVICES, Inc.

Figure 15.29. Stacking three AD586 on top of one another provides three reference voltages of +5.000V, +10.000V, and +15.000V. Each output can be trimmed precisely by the 10KΩ trimmer.

Another exceptional buried-zener part is the **REF102** from Texas Instruments. The REF102 was originally designed by Burr-Brown several years ago and is now part of Texas Instruments' Analog/Power Management portfolio, following TI's acquisition of Burr-Brown in 2000. The REF102 is a classic very-high-precision reference, with some excellent features. It is a precision +10-volt reference, with an initial accuracy of

±0.025% (±2.5 mV) maximum. The device is available in three different performance grades, in three different packages (SOIC-8, DIP-8, and TO-99 metal can). Even though the device has eight pins, only five are used. It is available in two different operating temperature ranges: –25°C to +85°C, and the military range of –55°C to +125°C. The REF102's best output voltage tempco is ±2.5 ppm/°C over the military temperature range. The reference can source 10 mA or sink 5 mA with the load. An optional TRIM adjustment is provided on-chip, which allows precise trimming of the +10-volt output, using an external 20-KΩ trimmer. The device also features an exceptionally low noise voltage of typically 5 μV pk-pk (0.1 Hz to 10 Hz). A built-in noise-reduction pin is provided, which allows addition of a single (1-μF) capacitor to further reduce noise. The long-term stability of the voltage reference at room temperature is specified at a typical 5 ppm/1000 hours. The REF102 can operate from a supply voltage as low as 11.4 volts to a maximum +36 volts and has a maximum quiescent current of 1.4 mA. It also has an exceptionally impressive line regulation of 1 ppm/V maximum and an equally impressive load regulation of 10 ppm/mA maximum. The main features of the REF102 are shown in Table 15.24.

REF102	Texas Instruments, Inc.		High-precision +10V buried-zener reference				
Reference voltage(s):	Best initial accuracy:	Best tempco: (ppm/°C)	LF noise:	Input voltage range:	Supply current (μA)	Temp. range: (°C)	Special features: •Ultra-low tempco • ultra-low noise
+10.000V	±0.025% (±2.5mV)	2.5ppm (-55°C to +125°C)	5 μV pk-pk typ.	+11.6 to 36V	<1.4mA	-25°C/ +85°C and MIL versions	• line reg.1ppm/V • TRIM and NR pins • SOIC-8/DIP-8 and TO-99

Table 15.24 REF102

The circuit in Figure 15.30 shows how the REF102 can be enhanced with an adjustment trimmer, a Kelvin connection, and additional noise filtering. The TRIM pin and an external trimmer provides adjustment of the 10-volt output. Using the trimmer, one can adjust the output to precisely 10.24 volts if required, or to exactly 10 volts, or else one can use it as a system (offset) trim. The minimum amount of the trim range is ±300 mV, when using a 20-KΩ linear, multiturn, precision trimmer (trimming changes the output tempco by approximately 0.008 ppm/°C per millivolt trimmed). To minimize any changes to the output tempco, one should ideally use a precision trimmer (Bourns, wirewound 5%, 25 ppm/°C; or Vishay metal-foil 5%, 10 ppm/°C). Noise filtering is provided by the 1-μF tantalum capacitor connected to the noise-reduction terminal and reduces the typical 5 μV pk-pk low-noise voltage to less than 3 μV pk-pk.

For many applications in which this high-performance reference would be used, the use of Kelvin connections would ensure that any slight voltage drops that occurred between the reference output terminal and the load would be compensated for. Being a market leader in this field, Texas Instruments has a wide array of high-performance op amps to choose from. The op amp required would need to be a low-noise precision type, with a low offset voltage (V_{OS}) and a low offset voltage drift (ΔV_{OS}). Some examples of Texas Instruments' precision single op amps include TLE2021 (15 nV/√Hz; V_{OS} equals 0.12 mV; supply range 4 to 40 volts); TLC2201 (supply range 4.6 to 16 volts; noise 8 nV/√Hz; R-R output); and TLE2027 (2.5 nV/√Hz; V_{OS} less than 0.10 mV; GBW 13 MHz typical). Most of these products are also available in dual and quad versions.

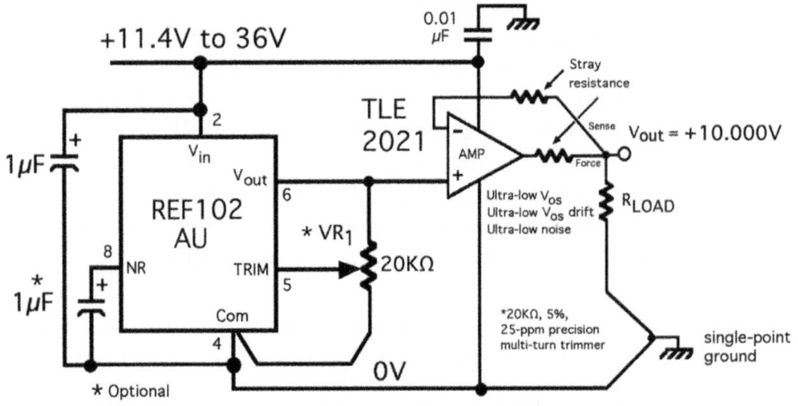

Figure 15.30. *Texas Instruments' REF102 precision buried-zener reference provides an ultra-low 5µV pk-pk noise voltage, and a 2.5 ppm/°C tempco. The device features a maximum 1 ppm/V line regulation, and long-term stability of 5 ppm/1000 hours. The 20 KΩ trimmer facilitates easy adjustment of the reference voltage. The op amp provides a Kelvin connection to the load for optimum accuracy.*

In the next example we will look at Linear Technology's **LT1021 family** of high-performance buried-zener references. These are available in either an eight-pin DIP or an SO-8 package and represent some of the best buried-zener products ever made. They combine an ultra-low tempco with high initial accuracy, very low long-term drift, and ultra-low noise level. The LT1021 series provides a reference voltage of 5 volts, 7 volts, and 10 volts in three performance grades, the best of which has a low initial accuracy of ±0.05%. The premium-grade (MIL-grade) references have a maximum output voltage tempco of 2 ppm/°C, over the military temperature range (–55°C to +125°C). Other devices are specified over the 0°C to +70°C range, with slightly reduced specs. All devices feature an exceptionally low noise voltage (0.1 Hz to 10 Hz), typically 3 µV pk-pk for the 5-volt version, 4 µV pk-pk for the 7-volt version, and 6 µV pk-pk for the 10-volt version. Long-term stability is typically 15 ppm/1000 hours for the 5-volt and 10-volt parts, and typically 7 ppm/1000 hours for the 7-volt device. They also include a quiescent current of less than 1.5 mA (typically between 0.8 to 1.2 mA), and the 5- and 10-volt versions include a TRIM feature that allows precise adjustment of the output voltage. The 7-volt version does not provide a TRIM pin, so trimming must be done externally with a pair of resistors and a precision trimmer.

All versions can sink or source 10 mA with a load. Load regulation is less than 25 ppm/mA maximum in all cases, whereas line regulation is typically less than 4 ppm/V at 25°C. These impressive out-of-the-box specifications are generally suitable for 12- to 14-bit converter systems. Unlike some other buried-zener voltage references, this series does not use a substrate heater, but instead uses advanced curvature correction techniques. The devices also feature the industry-standard REF02 pinout, which makes for easy upgrades, but with significantly improved tempco, noise level, and long-term stability. The main features of the LT1021 family are shown in Table 15.25. Although the devices are normally used in series mode, because of their architecture they can also be used in shunt mode, with virtually identical characteristics.

LT1021	Linear Technology		High-precision buried-zener reference family				
Reference voltage(s):	Best initial accuracy:	Best tempco: (ppm/°C)	LF noise:	Input voltage range:	Supply current (μA)	Temp. range: (°C)	Special features: •Ultra-low tempco •ultra-low noise
+10.000V +7.000V +5.000V	±0.05%	<2 ppm (-55°C to +125°C)	3 μV pk-pk typ. (5Vout)	V_{ref} + 1.5V to 35V	<1.4 mA max.	0°C/ +70°C and MIL versions	•high intl accuracy •Sink/source 10mA •TRIM pin •SO-8/DIP-8

Table 15.25 LT1021

The circuit in Figure 15.31 shows the LT1021 reference supplying a low-noise, ultra-stable reference voltage to a load. This application uses the TRIM function to adjust the reference voltage to exactly 10 volts. The trimmer should be a high-precision, multiturn, wirewound or metal-foil type. The LT1021 can normally source up to 10 mA on its own, but here the device drives an external NPN transistor in order to supply a total 100 mA at a precise 10-volt reference voltage to the load. In this particular application, the operating temperature range is from 0° to 50°C, which allows the (2 ppm/°C max) reference to keep the conversion error to within 1 LSB for a 14-bit DAC. That's very impressive!

The most remarkable reference ever made is the venerable **LTZ1000**, from Linear Technology. This is an ultra-precision 7-volt buried-zener reference, which provides an ultra-low tempco of 0.05 ppm/°C, a 1.2-μV pk-pk noise level, and a long-term stability of 2 μV per month. The LTZ1000 is packaged in a standard eight-pin metal can, whereas the LTZ1000A is the same product but with a proprietary high thermal resistance die attached, which eases thermally insulating the device. The LTZ1000 can provide the ultimate in voltage reference performance and was designed by Robert Dobkin, who invented the topology when he was at National Semiconductor. At National, Dobkin designed the LM199/399, the first ever buried-zener product. He

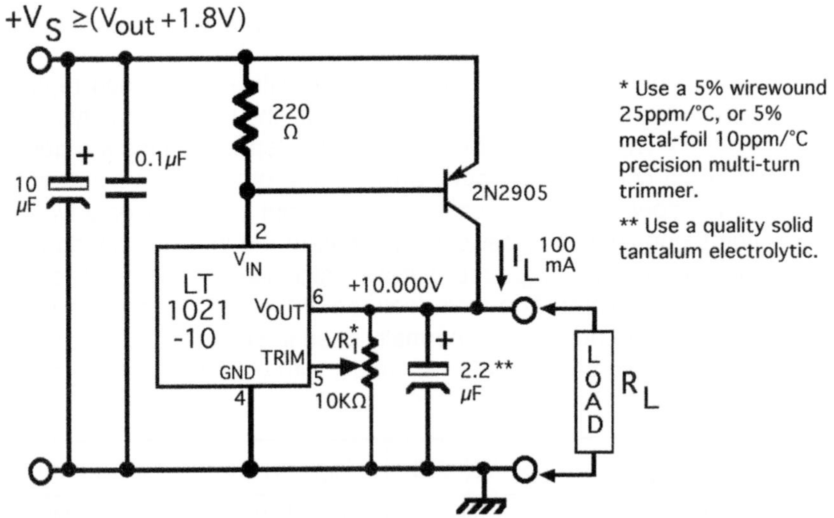

REPRODUCED WITH PERMISSION FROM LINEAR TECHNOLOGY CORP.

Figure 15.31. Adding a BJT to the output of this Linear Technology LT1021A-10 high-precision voltage reference boosts the load current to 100-mA while maintaining the ultra-low 1 ppm/°C tempco, and excellent 6 $\mu V_{pk\text{-}pk}$ noise characteristics, over the military temperature range.

designed the LTZ1000 later at LTC, and it provides a superior performance over his original design. By contrast, the LTZ1000 has a similar reference voltage, but with a 10 times lower tempco, a 10 times lower long-term drift, and about a six times lower noise specification.

Both devices uniquely employ a substrate heater to minimize temperature effects and drifts. (With the LM199/399, this temperature is approximately 90°C, whereas with the LTZ1000 it is approximately 60°C.) Both designs are bipolar and use a junction isolated process. Unlike other buried-zener references, the LTZ1000 provides only the core circuitry (shown in Figure 15.32), which includes the buried-zener and its compensating transistor, the temperature-sensing transistor, and the substrate heater. Please notice the elegant simplicity of the design (something that Bob Dobkin shares with his former colleague, Bob Widlar, both masters of the "art of the transistor").

The main features of the LTZ1000/A are shown in Table 15.26. Please note that the devices are specified over the full military temperature range (–55°C to +125°C). Typically the LTZ1000/1000A can be used in 14- to 20-bit systems, because of its ultra-low tempco and very low noise level, not to mention its almost immeasurable 1 μV/month long-term stability—the three most important characteristics for any precision voltage reference.

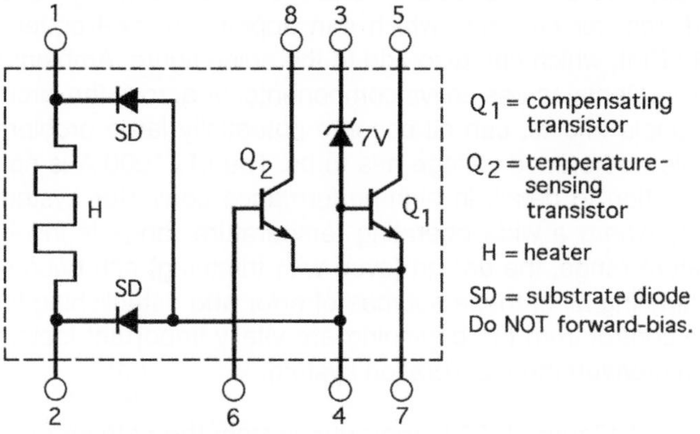

Figure 15.32. A block diagram of Linear Technology's LTZ1000/A buried-zener reference.

LTZ1000	Linear Technology		Ultra-high-precision buried-zener reference				
Reference voltage(s):	Best initial accuracy:	Best tempco: (ppm/°C)	LF noise:	Input voltage range:	Supply current (μA)	Temp. range: (°C)	Special features: •Ultra-low tempco •ultra-low noise •2ppm long-term drift •temperature-stabilized @ 65°C
+7.000V	±0.05%	0.05ppm (-55°C to +125°C)	1.2 μV pk-pk typ.	10V to 35V	<0.2mA max.	MIL -55°C/ +125°C	

Table 15.26 LTZ1000

Additional circuitry is needed to take advantage of the LTZ1000's superior characteristics, but the circuit's *layout* is critical to its performance. The circuit should use a single-point ground connection, with the load connected via a separate return cable directly to ground. The circuit's PC board should use the same size of pads and track lengths, and with the temperature of the LTZ1000's pins held at identical temperatures, to eliminate thermocouple effects. Thermal EMFs, or thermocouples as they are more commonly known, are parasitic effects that degrade the overall performance of any reference or high accuracy circuit. They include any connections/junctions involving different metals at different temperatures, such as with copper and solder alloy used in PC boards, or the tinned metal leads of an IC and the copper circuit track, and steel-to-gold connections used in plugs, sockets, and connectors (the latter are the most difficult to control, because their resistance can also change each time you connect or disconnect). The circuit may also require guard rings to keep parasitic leakages in check.

Additionally, the circuit should be enclosed in a thermally constant environment and shielded from air currents, which can appear as low-frequency noise voltage, and stray RFI/EMI, which can also add to the noise figure. Ambient temperature changes, thermal gradients across active components, or across the circuit board, together with thermocouple effects, can all become potentially large problems, the more accurate and stable a reference voltage has to be. The LTZ1000/A is no exception in requiring great attention to detail. In high-performance converter systems more than 14 bits, especially where a wide operating temperature range is involved, as in the military temperature range, the design (even with trimming) can become extremely challenging. Eliminating all of these sources of error and establishing long-term ambient temperature control from the beginning are vitally important factors in the design of any precision measurement/calibration system.

The circuit of Figure 15.33 is reproduced from the data sheet and illustrates how the LTZ1000 should be used. Because the reference and the heater both float, they can be tied between positive and ground, between negative and ground, or between both supply rails. (Caution must be used to ensure that the device never becomes reverse-biased, otherwise permanent beta-degradation will occur. Neither should the substrate diodes shown in Figure 15.32 ever be forward-biased.) The zener is a 6.2-volt buried-zener structure with a tempco of around +2.2 mV/°C. Transistor Q_1 temperature-compensates the zener with its forward-biased V_{BE}, having a tempco of around –2.2 mV/°C. As a result, the two devices' tempcos virtually cancel out. The temperature-sensing transistor Q_2 constantly monitors the die temperature, and together with amplifier A_1 and the 2N3904 transistor, controls the heater element in a feedback loop. The second op amp, A_2, controls the reference current (as you will notice, the reference and its support circuitry are supplied from A_2's output). You will see from the diagram that Kelvin connections are used to sense the zener voltage. This is to minimize any stray wiring resistance and the ensuing voltage drops. In this example, the circuit runs from a +15-volt single supply, which should be well regulated and bypassed, and outputs +7 volts at the Zener+ Sense terminal. This voltage can be amplified if desired to provide other voltages.

In summary, the LTZ1000/A is a remarkable low-drift, ultra-low-noise device. It stands as a tribute to both the brilliance and imagination of its designer Robert Dobkin and the skills and expertise of its maker Linear Technology. The LTZ1000/A has withstood the test of time (it has been available for about two decades), and yet its ultra-high stability is still unmatched by any other semiconductor reference—even today (in mid-2005).

15.6 Applying the Intersil/Xicor FGA™ X60008

We first reviewed the Intersil/Xicor FGA™ x60008 in Chapter 14, and found it to be an outstanding voltage reference with many excellent features. The most remarkable are its ultra-low quiescent current (0.5 μA typical), ultra-low initial accuracy (±0.01%, A- and B-grade devices), and very low 1 ppm/°C tempco (A-grade), all specified over the industrial temperature range (–40°C to +85°C). The long-term stability of the device is typically 10 ppm/1000 hours, which is also impressive. Its quiescent current is the low-

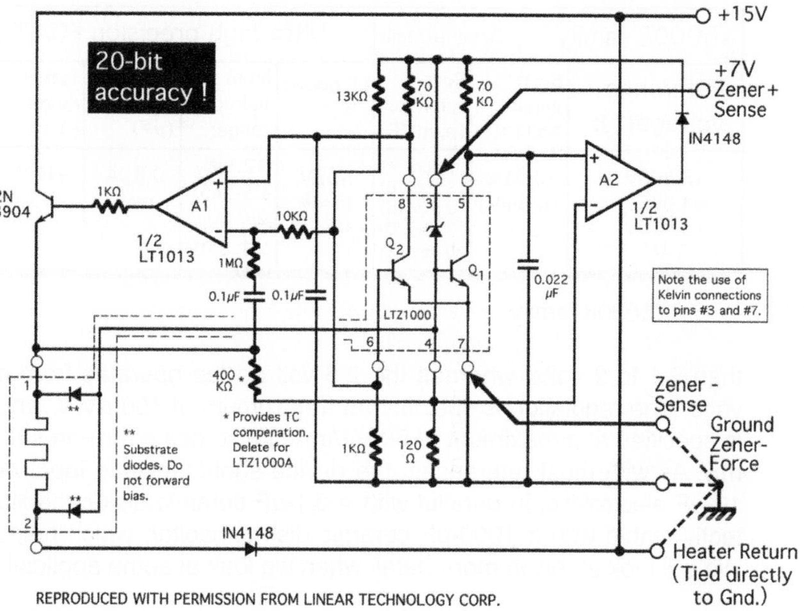

Figure 15.33. *The ultimate buried-zener voltage reference using Linear Technology's LTZ1000/A. With careful layout, thermal symmetry, a single-point ground, and a shielded enclosure, this circuit can provide 7 volts with a 0.03 ppm/°C tempco, a noise level of 1 μV pk-pk, and long-term stability of about 1 μV/month.*

est of any other reference topology, making it ideal for battery-powered applications. In many instances, this ultra-low power drain translates into being able to leave the device on continuously, which enhances its low drift and overall stability characteristics. It also makes it easier to use where conversions are performed periodically, because there is no need to allow for turn-on settling time or power-up drift. The output of a typical x60008 uses an internal buffer amplifier, which provides a source/sink capability of up to 10 mA and an indefinite short-circuit current to ground of about 50 mA. The device also has a low dropout voltage of typically 150 mV (but less than 300 mV) and a low-frequency noise voltage of typically 30 μV pk-pk. The FGA™ is provided in an eight-pin surface-mount package (SOIC-8), which includes input and output terminals and two ground terminals. The remaining terminals are used by the factory during manufacture and must otherwise be left floating by the user. The main features of the x60008 family are shown in Table 15.27.

The **x60008** family is available in three different voltages (2.5 volts, 4.096 volts, and 5 volts), with five different performance grades (A, B, C, D, and E). The devices are specified for operation with different supply voltages according to the reference voltage. For example, the 5-volt reference is specified for operation with an input of more

x60008 family Xicor/Intersil			Ultra-high precision FGA™ series reference				
Reference voltage(s):	Best initial accuracy:	Best tempco: (ppm/°C)	LF noise:	Input voltage range:	Supply current (μA)	Temp. range: (°C)	Special features:
+2.500V +4.096V +5.000V	±0.01% (0.5mV)	1 ppm (A) 3 ppm (B) 5 ppm (C) 10 ppm (D) 20 ppm (E)	30 μV pk-pk typ.	@2.5V$_O$ 4.5-6.5V @5.0V$_O$ 5.1-9.0V	0.8 μA max.	-40°/ +85°C	• Ultra-low tempco • Ultra-low power • Very high initial accuracy • Very good long-stability • SOIC-8

Table 15.27 x60008 family

than 5.1 to 9 volts, whereas the 2.5 volt device operates from more than 4.5 to 6.5 volts. Line regulation is specified at a maximum of 100 μV/V, whereas load regulation is specified at a maximum of 50 μV/mA. It also has a turn-on settling time of about 10 mS. As with most references, the device should include input decoupling such as a 10-μF electrolytic, in parallel with a 0.1-μF ceramic disk capacitor. The output is perfectly stable with a 1000-pF ceramic disk capacitor, which helps noise suppression. We will look at this in more detail when we look at some application circuits.

Chapter 14 showed the x60008 used in series mode with a Kelvin connection, but Figure 15.34A shows an application circuit in which a 2.5-volt FGA™ is used as a precision negative shunt reference, with an output of –2.5 volts at 2.5 mA. The application runs from a –5-volt supply, where the reference voltage and the precision resistor set the output voltage. The main specifications are the same in shunt mode as with series mode (initial accuracy, tempco, noise, long-term drift). The formula for calculating the resistor's value is shown in the diagram.

Figure 15.34B shows a circuit using the reference in series mode, where the output voltage (V_{out}) is set by a Xicor/Intersil programmable digital potentiometer. The output from the pot is buffered by the op amp. The x9119 is a single supply, low-power CMOS, 1024-tap (10-bit) nonvolatile digital pot. The circuit's tempco and bit resolution is limited by the characteristics of the digital pot, but does allow easy digital control via its two-wire serial interface. Buffering the output is an Intersil op amp, such as an EL8186 or a CA3420.

The noise voltage of the x60008 is typically 30 μV pk-pk in the 0.1- to 10-Hz bandwidth and is typically 400 μV pk-pk in the 10-KHz to 1-MHz bandwidth. Using a 1000-pF capacitor on the output of the reference, the noise voltage in the 10-KHz to 1-MHz bandwidth can be reduced to about 50 μV pk-pk. Although noise in the 1-KHz to 100-KHz bandwidth can be reduced further by using a 0.1-μF capacitor on the output, noise in the 1-Hz to 100-Hz bandwidth may increase, as a result of instability of the very-low-power internal amplifier with a 0.1-μF load. The circuit of Figure 15.34C shows the recommended way to provide filtering of any x60008 voltage reference (regardless of grade or output voltage), so that noise is reduced to less than 40 μV pk-pk over the entire 1-Hz to 1-MHz bandwidth.

Figure 15.34D shows a circuit that provides ±5-volt reference voltages. You will notice that the +5-volt reference is used in series mode, whereas the –5-volt reference is used in shunt mode (performance is virtually identical though). The formula for determining the negative voltage is shown in the diagram. One should ensure that the output current is limited to 10 mA or less.

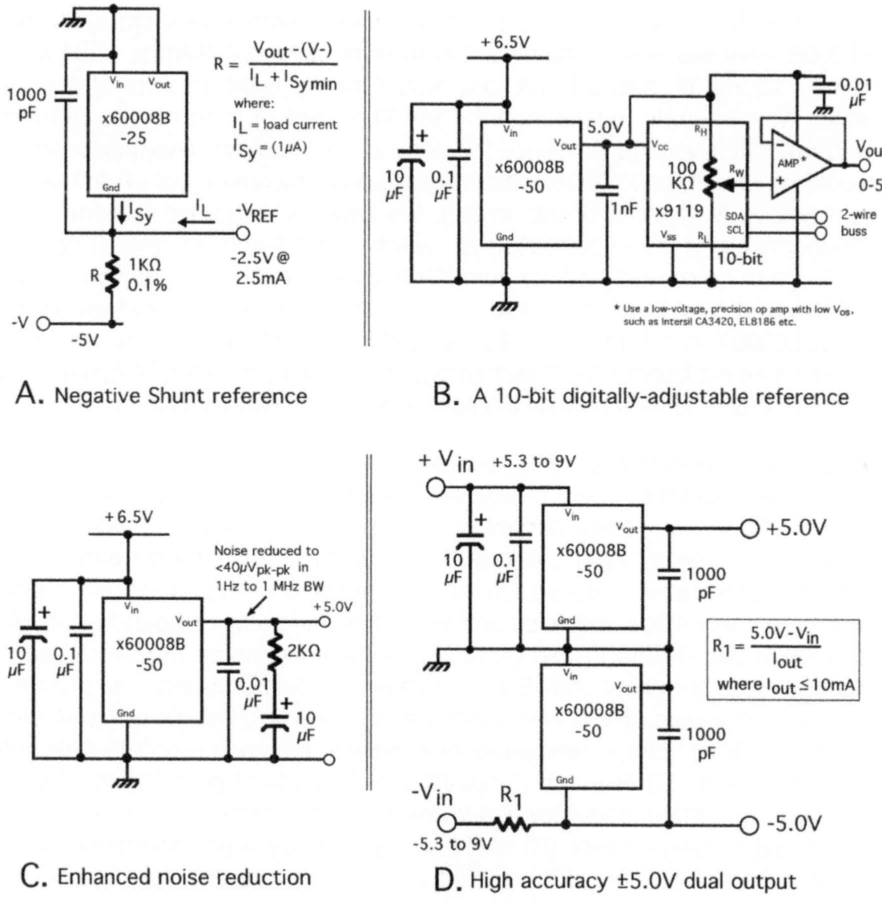

Figure 15.34. The versatile, ultra-high precision x60008.

This book is exclusively about **current sources** and voltage references and their interrelation with one another. In that regard, you can make an accurate (but inexpensive) current source, using lower-grade x60008 devices such as the C, D, or E versions. In any event, the reference I_Q is still only about 500 nA, which is likely to be insignificant (0.05%) in a 1-mA or higher current source. The noise level is also the same at about 30 μV pk-pk. A big advantage to using the x60008 rather than another type of reference is that most bandgaps have too high a quiescent current, which also varies with output current. With a constant current source, this varying quiescent current is significant and creates an error, which adds to the total I_{Load} current.

Other references have a significant quiescent current of several hundred μA or more. For example, a top-of-the-line bandgap device, with specs close to those of the x60008, may well have a quiescent current of several milliamps. Thus for creating a current source of, say, 2.5 mA, one would ideally want an error of less than 0.1%, which would equate to less than 2.5 μA. This level of accuracy would be impossible with any other voltage reference topology at the moment. The only practical solution would be to use the x60008, which would add a maximum 0.8 μA to the desired current output. At the time of this writing, the lowest-voltage and low-cost device in this product family is the x60008DIS8-25, which is a 2.5-volt, ±1 mV, 10 ppm/°C version. An E-grade device (±5 mV, 20 ppm/°C) is available, but currently only applies to the 4.096-volt or 5-volt versions. (You should also consider the brand-new **ISL60002** and the **ISL60007**, which provide 1.25-volt and 2.5-volt voltages. They are also members of the same exclusive FGA™ technology that created the x60008 family. Please check Intersil's Web site for the latest availability and pricing information.)

As you will see in Figure 15.35, two simple current source circuits are shown, which rely on the x60008's superb characteristics for their very high accuracy. The design in Figure 15.35A shows the current source using a single JFET, while Figure 15.35B shows an improved circuit, using a JFET cascode. (See Chapter 6 for more on JFET current sources and cascodes.) In both examples, the circuit's tempco depends on the tempco of the voltage reference and the precision resistor. Depending on which grade of x60008 you decide to use (A to E), this will determine the overall accuracy and tempco of the current source. Even the lowest grade (E) version can provide an excellent current source, the characteristics of which would be difficult to duplicate using other monolithic voltage references or even with matched monolithic discretes. Using A-grade (1 ppm/°C), B-grade (3 ppm/°C), or C-grade (5 ppm/°C) x60008 devices will provide even greater accuracy and lower drift, but at higher cost. In early development the current-setting resistor (R) could be replaced by a potentiometer, just for convenience. Once the value of the output current (and thereby the resistance) is determined, the potentiometer should be replaced by a precision resistor. Choosing a high-quality precision resistor (less than 0.1%) with a very low tempco (i.e., less than 10 ppm/°C) will help further minimize drift. A 1000-pF capacitor on the output of the reference can provide added stability, if desired.

The circuit also provides a reasonably low noise voltage, most of which is created by the noise from the voltage reference and the resistor. JFETs are inherently low-noise devices, so their noise will be significantly lower than the other components used in these two circuits. In fact, some of the JFETs listed in Figure 15.35 have noise levels of less than 5nV/√Hz! The initial accuracy (aka tolerance) of either current source is directly dependent on the voltage reference, which in circuit A is approximately ±0.122% (±5 mV), whereas circuit B is approximately ±0.02% (±0.5 mV). In both circuits the JFET's gate current is of the order of a few nano-amps, which is irrelevant to the loading of the reference output. Although the voltage reference has a maximum source capability of 10 mA, the lower the load current (i.e., 50 μA to 2.5 mA), the better the overall current regulation will be. For best results, the JFET's I_{DSS} should be rated at many times the actual load current. This is why I have specified a medium/high current JFET in the designs, even though the actual load current is significantly lower. This will produce a lower output conductance (g_{os}) value and much better regulation. For example, if the required load current is 1 mA, the JFET's minimum I_{DSS} should be rated at more than 10 mA.

Another important factor is that if there is any change in the JFET's gate-source voltage (ΔV_{GS}), it will be compensated for by the supply rejection of the x60008 voltage reference. The JFET's gate-source voltage (V_{GS} equals 1 to 3 volts) also helps keep the voltage reference biased well above its dropout level. In either circuit, the JFET's drain-source voltage rating (V_{DS}) allows the circuit to operate over a higher supply voltage than the x60008 normally allows (the cascode provides the largest allowable voltage). The long-term stability/drift of either circuit is again dependent on that of the voltage reference device, which is typically 10 ppm/1000 hours. The circuits shown are intended for operation generally in the 50-μA to 8-mA range. The lower limit is determined by the reference IC's quiescent current (0.8 μA max), in proportion to the desired constant current value. A ratio of 60:1 seems good, whereas 200:1 is much better. The upper limit is determined simply by the reference IC's maximum output current capability (10 mA).

The circuit in Figure 15.35A shows a simple current source using a single N-channel JFET. This provides an output current of 410 μA (±0.12%), good current regulation, low noise (less than 50 μV pk-pk), and an output impedance (Z_{out}) typically more than 100 MΩ. This particular circuit runs on a minimum supply voltage of approximately 8 volts and a maximum depending on the JFET's BV_{DSS} (which is typically 20 volts or higher). The reference used here is a 4.096-volt E-grade version, but it could be lower if desired (which would allow a lower supply voltage). Substituting a 2.5-volt B-grade device would enable the current source to have a 0.02% initial accuracy and a 3 ppm/°C tempco (best possible), but at higher cost. The JFET should be a low-noise, low output conductance, surface mount type, with a low $V_{GS(off)}$ of 0.5 volt to 4 volts, and with an I_{DSS} of more than 5 mA. Using an inexpensive JFET with the x60008 provides a current source that is simple, stable, and very accurate.

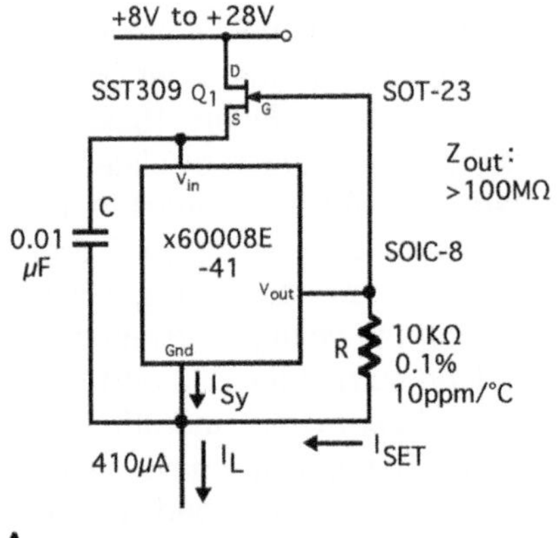

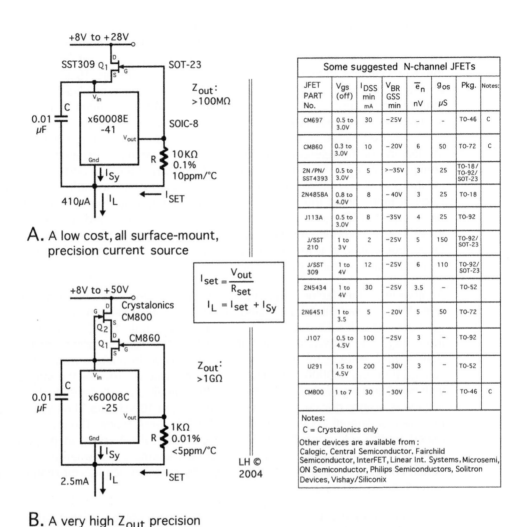

A. A low cost, all surface-mount, precision current source

Some suggested N-channel JFETs							
JFET PART No.	V_{gs} (off)	I_{DSS} min mA	V_{BR} GSS min	$\overline{e}_n$ nV	g_{os} μS	Pkg.	Notes:
CM697	0.5 to 3.0V	30	−25V	–	–	TO-46	C
CM860	0.3 to 3.0V	10	−20V	6	50	TO-72	C
2N/PN/ SST4393	0.5 to 3.0V	5	>−35V	3	25	TO-18/ TO-92/ SOT-23	
2N4858A	0.8 to 4.0V	8	−40V	3	25	TO-18	
J113A	0.5 to 3.0V	8	−35V	4	25	TO-92	
J/SST 210	1 to 3V	2	−25V	5	150	TO-92/ SOT-23	
J/SST 309	1 to 4V	12	−25V	6	110	TO-92/ SOT-23	
2N5434	1 to 4V	30	−25V	3.5	–	TO-52	
2N6451	1 to 3.5	5	−20V	5	50	TO-72	
J107	0.5 to 4.5V	100	−25V	3	–	TO-92	
U291	1.5 to 4.5V	200	−30V	3	–	TO-52	
CM800	1 to 7	30	−30V	–	–	TO-46	C

Notes:
C = Crystalonics only
Other devices are available from: Calogic, Central Semiconductor, Fairchild Semiconductor, InterFET, Linear Int. Systems, Microsemi, ON Semiconductor, Philips Semiconductors, Solitron Devices, Vishay/Siliconix

$$I_{set} = \frac{V_{out}}{R_{set}}$$

$$I_L = I_{set} + I_{Sy}$$

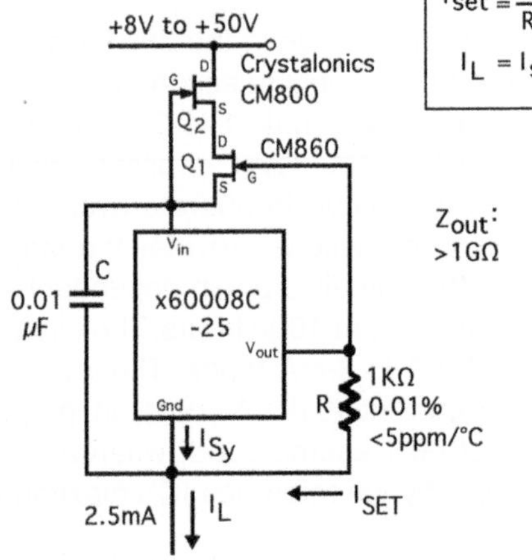

LH © 2004

B. A very high Z_{out} precision current source

Figure 15.35. Creating precision current sources with the x60008 reference.

An improved circuit design, using a JFET cascode (instead of a single JFET), is shown in Figure 15.35B. Cascoding generally improves high-frequency operation, provides even greater output impedance (Z_{out}), increases high-voltage operation and compliance, and reduces the circuit's overall output conductance (g_o). The low g_{os} is achieved by the two-JFET cascode, caused by degenerative feedback, and the JFETs' combined forward transconductance. Here the circuit's output conductance is much less than the g_{os} of the single JFET circuit—about 100 times lower! Both JFETs are N-channel types, whereas the voltage reference device is a 2.5-volt type. Here the circuit provides an output current of 2.5 mA (±0.02%), improved current regulation, low noise (less than 50 μV pk-pk), and a much increased output impedance (Z_{out}), typically more than 1 GΩ! Because of the inclusion of a cascode, the circuit runs on a higher supply voltage than the previous circuit and has a minimum of approximately 8 volts and a maximum depending on the combined JFETs' BV_{DSS} (which will be typically more than 40 volts). The lower JFET (which is the actual current source) should have a relatively low $V_{GS(off)}$ of 0.5 volt to 3 volts and an I_{DSS} of more than 10 mA. The upper JFET should have a higher $V_{GS(off)}$ and a higher I_{DSS} rating than the other JFET. Both JFETs must be operated with adequate drain-to-gate voltage (V_{DG}), so that V_{DG} is ideally more than two times $V_{GS(off)}$, or else the circuit's output conductance will not be optimized. The cascode provides much sharper regulation than that of the single JFET. In comparison, it has a higher operating voltage, a much lower output conductance (10 times lower), and a much higher output impedance (10 times higher). Figure 15.35 also shows a short list of suitable and commonly available JFETs that can be used in these applications.

Note: Two newly released FGA™ products from Intersil are the **ISL60002** and the **ISL60007**. The ISL60002 is available as either a 1.25-volt or a 2.5-volt reference and in either an SOT-23 or an SOIC-8 surface-mount package. Both feature a 20 ppm/°C tempco, and with either ±2.5 mV or ±5 mV initial accuracy. The ISL60007 is a 2.5-volt device in an SOIC-8 package, with a choice of 0.5 mV or 1 mV initial accuracy, and with a tempco of either 3, 5, or 10 ppm/°C. Specifications are otherwise very similar to the x60008 series. These products may also be used to create precision current sources, but allow lower JFET gate voltages.

In summary, the FGA family (x60008, ISL60002, and ISL60007) is still evolving, and now with available reference voltages that range from 1.25 up to 5 volts. It is hard to believe that this topology is still less than three years old. They offer ultra-low power dissipation (the lowest of all reference topologies), good initial accuracies (0.05% to 0.1%), a range of tempcos of between 1 to 20 ppm/°C, excellent 10 ppm long-term stability, and all intended for use over the industrial temperature range. No doubt as they evolve further, their noise levels will be further reduced and their outputs will be able to drive higher capacitive loads. Their present specifications challenge the best buried-zener references and exceed virtually all others. In any event, the FGA family are truly remarkable products. Besides establishing itself as a technical benchmark, the FGA is also a major milestone in the overall evolution of the monolithic voltage reference.

15.7 Multiple voltage references and multiple loads

It is sometimes necessary to provide multiple voltage references in an application. In some cases you may have to generate the *same* voltage to *multiple* loads, whereas in others several *different* voltages may be required. Over the course of the past 25 years, a few monolithic reference devices have been introduced that have provided two different simultaneous voltages. Few of these devices are still available though. For one thing, producing two different reference voltages on the same chip is expensive for manufacturers. In addition, testing and trimming two or more reference voltages to very high accuracy may result in one reference passing, whereas the other fails.

For most designers, creating a custom circuit using two or more individual references has been the only possible solution. One can choose different references from the same family and manufacturer or choose a particular reference from supplier A, another from supplier B, and yet another from supplier C. They can work constantly and independently of one another (as shown earlier in Figure 15.29) or be individually selectable, so they only work some of the time. Using different voltage devices from the same family makes it relatively easy in one's design, because many of the same basic specifications will be the same, such as tempco, dropout voltage, noise level, line/load regulation, I_Q, thermal hysteresis, turn-on settling time, power-up drift, and so on. Using different voltage devices (perhaps even using different topologies) and from different vendors can make one's design somewhat more challenging.

Figure 15.36 shows a design where five different reference voltages are required. The common requirement is that each reference has an Enable pin, so that it can be individually selected or be put into a low-power mode. The basic idea shown here uses a simple CD4028 CMOS decoder to select each reference. However, in your design you would more likely use a microcontroller, where each reference could be selected using a dedicated I/O line. In this design, though, a logic "1" selects the reference, together with two diodes to fulfill the truth table. The CD4028B is a popular CMOS BCD-to-decimal decoder consisting of four inputs, decoding logic gates, and 10 output buffers. A BCD code applied to the four inputs results in a high level at the selected output. Here the code "0000" disables all of the references. In this application, when output 1 is selected, all other references are disabled. The 0.625-volt reference could be provided by using a Linear Technology LT6650 (which was reviewed earlier in this chapter). This device requires less than 10 μA and has a minimum input requirement of only 1.4 volts. It could be powered directly from the decoder's output, and thus would not require an Enable pin. When output 2 is selected, only the 1.25-volt reference is selected. Similarly, when outputs 3 or 4 are selected, only that particular reference (either the 2.5-volt or 5-volt reference) is selected. When output 5 is selected, both 5-volt references are selected, to create a 10-volt output (one is stacked on top of the other).

The minimum supply voltage should be more than V_{out} plus $V_{dropout}$. You probably noticed that the range runs 10 volts, 5 volts, 2.5 volts, 1.25 volts, and 0.625 volt—each an exact half of the voltage above it—and could be used as an instrumentation calibrator. A small linear regulator (V_{REG}) is used to power the 2.5-volt and 1.25-volt reference ICs. The decoder's outputs may need to be inverted, and you would need to choose references of the correct voltage having an Enable/Shutdown pin. Some of the devices reviewed earlier in this chapter would be very suitable (i.e., Linear Technology LT1461A: 2.5 volts and 5 volts; National Semiconductor LM4140A: 1.25 volts and 2.5 volts; Analog Devices ADR390: 2.5 volts and 5 volts; Maxim MAX6037: 1.25 volts and

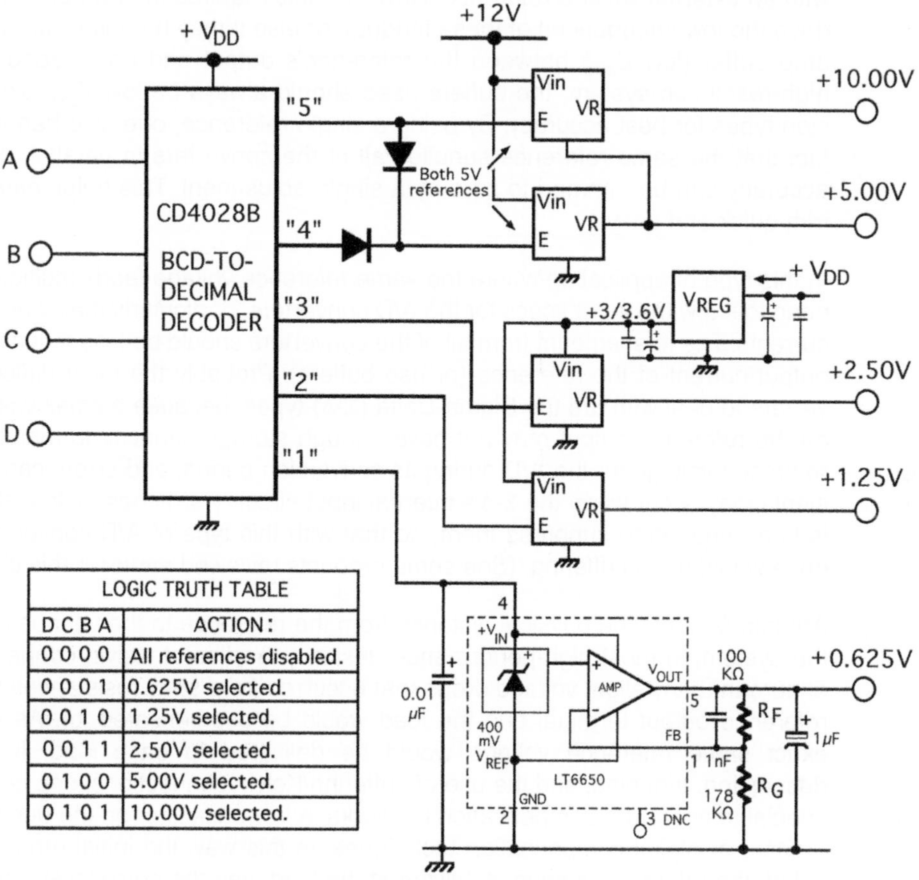

LOGIC TRUTH TABLE	
D C B A	ACTION:
0 0 0 0	All references disabled.
0 0 0 1	0.625V selected.
0 0 1 0	1.25V selected.
0 0 1 1	2.50V selected.
0 1 0 0	5.00V selected.
0 1 0 1	10.00V selected.

Figure 15.36. *Digitally selectable, 5-reference system calibrator.*

2.5 volts). Be sure to check their specs, including their turn-on settling time, to match your needs.

As mentioned previously, it is also necessary sometimes to provide the same voltage but to multiple loads. For example, equipment using large numbers of A/D converters includes data acquisition, optical networking, and imaging systems used in military, satellite, medical, scientific, and industrial instrumentation. In this type of equipment, the number of individual converters can often range from tens to hundreds of devices or channels and is often referred to as an *array.* Many converters have an on-board reference; however, in a multichannel system this can lead to poor matching between channels. For highest accuracy, it is more practical to disable any internal (bandgap) references on-board the converters and use a more precise external reference. Some converters allow you to directly drive their internal reference ladders with an external voltage reference. However, this requires the reference to be able to drive the low impedances of these ladders, or else it may be necessary to use an op amp buffer (low Z_{out}) between the reference's output and each ladder input. In a high-resolution system, the buffers used should always be low V_{OS}, low ΔV_{OS} precision types for best accuracy. By using a single reference, one also benefits from the fact that the same reference supplies all of the converters in parallel, and its initial accuracy can be trimmed to zero by a single adjustment. This helps make a system trim quick and easy.

In this type of application where the same reference voltage feeds multiple loads, you need to know the input specs for the A/D converters, particularly their worst-case input currents. The total amount from all of the converters should be less than the maximum output current of the reference (or use buffers). Probably the most difficult A/D converters to deal with are the Sigma-Delta (Σ-Δ) types, because a small value capacitor on the reference output may not have enough storage capacity to maintain a stable reference voltage for the A/D during its conversion period, and errors can result. Transients may occur when the Σ-Δ's internal input circuitry switches, unless the capacitor is large enough to suppress them, so that with this type of A/D converter, the reference would need buffering. (See some products reviewed earlier in this chapter).

Another factor is the physical distance from the reference to the various converters in the system. In most high-performance designs, the use of Kelvin connections would ensure that any slight voltage drops that occurred from stray resistances between the reference output terminal and the load would be compensated for. As a result, the exact same reference voltage would be delivered to each load. In addition to decoupling, trimming, and the use of buffering/Kelvin connections, one could also add special temperature compensation networks, a low-pass filter on the output of the reference, or digital compensation techniques. In this way, the initial error can be canceled, the reference voltage stabilized at the load, and the noise level reduced. Stabilizing the environment's ambient temperature, preregulating the supply voltages, protecting the circuit from large or small air currents, and careful circuit layout will all help enhance performance. In Figure 15.37A, a single reference drives many A/D converters. Here the reference voltage is fixed and not trimmable. A simple low-pass filter is used between the reference output and the input to all buffers. Use only quality, precision, low TC parts. A small capacitor (0.1 μF) may be required at the reference input of each converter. The buffer must be capable of driving this capacitive load.

Figure 15.37B shows a slightly different circuit, in which the reference is a trimmable type. This allows one to adjust the reference output to an exact voltage, without disturbing the filter response. The output of the reference is then connected to a low-pass active filter, before being fed to the buffers. Unlike circuit A, in this design the single-pole low-pass filter (R_2C_2) is located at the amplifier's output. Locating the filter at the amplifier's output, rather than at its input, filters the noise from the voltage reference as well as the amplifier's own noise. R_1C_1 should be set at twice R_2C_2 for improved loop stability and to reduce gain peaking. The amplifier should be a very-low-noise,

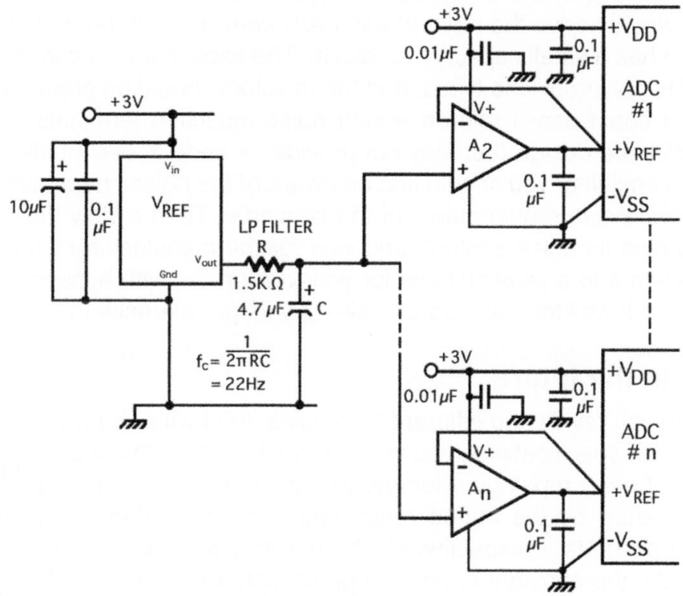

A. The reference is filtered and fed by Kelvin connections to the ADC's input pin for external references.

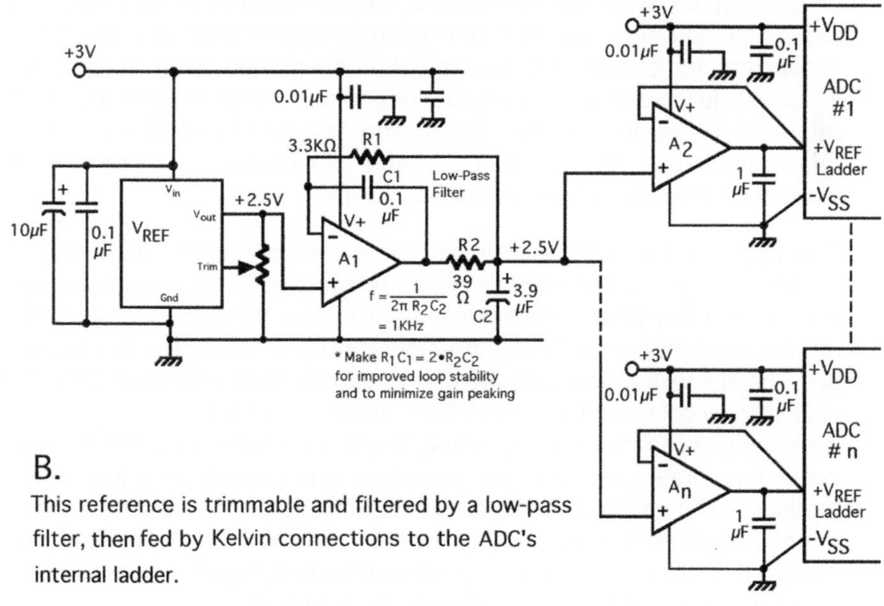

B.
This reference is trimmable and filtered by a low-pass filter, then fed by Kelvin connections to the ADC's internal ladder.

Figure 15.37. One reference—multiple loads.

very-low-offset, low-drift precision type. The buffers connect to the output of the filter and directly drive the input of each A/D converter's internal reference ladder (each of which has a small capacitor across it). The capacitors used in the filter should be high-quality polypropylene types, and the resistors should be precision low TC types. Either circuit could use a reference with noise-reduction terminals if desired, which could simplify the design (but may not provide as good a level of filtering). However, before doing anything, you should first be aware of the noise characteristics involved, starting with the noise requirements of the converter. Then review the reference's data sheet regarding its noise content, and look for the manufacturer's recommended means of reducing it to a level suitable for your design. As with bypassing, be aware that what may work well for one type or make of voltage reference may not for another.

15.8 A look to the future

When you review the different topologies, the available products within each topology, and their specifications, you would probably agree that there is a *convergence*, as the best of each topology attempts to compete with the most exotic specifications first established by the buried-zener. Generally, the other topologies are becoming far more accurate, irrespective of whether they are shunt or series types. Over the past decade, the constant need to support higher-resolution A/D and D/A converters in a myriad of digitally based products has fueled the growth and improvement of the precision monolithic voltage reference. Although many converters incorporate their own on-board (bandgap) voltage reference, this is limited to about 12-bit resolution, even though the converter may be capable of much higher resolution. For the most part it is convenient and enables a design to get off the ground faster. In order to achieve true 12-bit resolution and higher, it forces one to use a high-performance external voltage reference. Remember that for 1/2 LSB error in a 12-bit resolution system over a 100°C temperature span, the maximum allowable tempco is only 1.22 ppm/°C (along with an allowable noise spec of less than 60 μV pk-pk).

The 1-ppm/°C tempco specification over a product's temperature range has become the industry-wide benchmark for series references. Besides some buried-zener products, only a few other products have yet broken through this 1-ppm/°C barrier and reached that milestone. These are precision *super-bandgaps* like Maxim's MAX6325 and Thaler's VRE4100 series, as well as Xicor/Intersil's x60008A FGA™. Other products, including Analog Devices' XFET®, are close behind and are likely to show further improvements as they evolve, so that they too reach the 1-ppm/°C barrier. The buried-zener has already reached that milestone and peaked as a topology, so that new products are unlikely as manufacturers concentrate their efforts on easier-to-make, more marketable topologies. Improvements in the other topologies will come as a combination of advancements in processing and design. Key areas of development for future references will doubtless include the following:

- Lower operating voltages
- Lower reference voltages

- Lower dropout voltages
- Lower quiescent currents
- Lower tempcos
- Lower noise levels

In order to accommodate such improvements, it will require a low-power bipolar, CMOS, or BiCMOS technology, as used with precision op amps. So far, low-voltage bipolar technology can produce low-noise reference products, whereas CMOS technology can produce lower-power devices, but usually with higher noise levels (there is always some trade-off between low power versus noise). Nevertheless, lower-noise devices are required, which can provide less than 7.5 μV pk-pk (this level is the allowable noise level for a 2.5-volt reference for 1/2-LSB error, at 14-bit resolution, over a 100°C temperature span). It's of little use to have a ±0.01% initial accuracy, with a tempco of less than 5 ppm/°C, if the actual noise level is more than 50 μV pk-pk. The noise level in this example is the limiting factor, thus a very low noise figure is essential.

A dedicated noise-reduction terminal on the reference is helpful but may not do quite as good a job as an external active filter. Using the noise-reduction terminal usually takes just a single small capacitor to reduce the noise level by approximately half. Although one can add external filtering to reduce the noise level, it adds to the complexity of the circuit, its physical area, and the cost. Ultra-low-noise devices inherently require no external filtering, take up less board space, and cost less than their noisy counterparts.

When you look at the range of operating (input) voltages for series-mode topologies other than buried-zener types, they typically range from several volts to several tens of volts. Linked to that is the low dropout characteristic of some references, which can achieve less than 150 mV. In the future, references with lower operating voltages will be needed, as lower-voltage converters become available. Converters that presently run on 3 volts and use external references of less than 2.5 volts will soon be running on less than 2 volts and require a reference voltage of less than 1.5 volts. As we have seen in this and previous chapters, such devices are presently scarce. A few new products have been introduced that enable *sub-bandgap* voltages (output voltage of below 1.23 volts), but more are needed. Most of those that are available do not yet have the ultra-low tempcos or low noise levels needed. Most of the super-bandgaps require several volts to function, not to mention more than 1 mA quiescent current. Unfortunately, where very low operating voltages are concerned, other characteristics are affected, such as noise, accuracy, tempco, stability, dropout voltage, and so on. Nevertheless, it won't be long before the market shifts to requiring precision voltages of between 0.25 volt and 1.2 volts, and with similar (excellent) specs to which we have all become accustomed.

Note: Maybe someone should check out how the venerable National Semiconductor LM10, which was designed in the mid-70s by the legendary Bob Widlar (and built by Dave Talbot using a standard bipolar process), manages to run on as little as 1.1 volts (it contains both a good audio-frequency op amp and a 200-mV sub-bandgap reference). How did they do that *then*, and yet we don't seem to be able to improve much on it even *now*? Perhaps the key lies somewhere in the magic of the processing.

Present-day devices do a pretty decent job for output current capability and the ability to either source (usual) or sink the load current (e.g., ±10 mA). Typically, the output is from an internal op amp, and its ability to deal with either a highly capacitive load or with load transients varies from one product family to another. As we have seen in this and previous chapters, many reference products require a minimum output capacitance for stability, whereas others have both a minimum and a maximum. A few others, like Analog Devices' AD780, have virtually no limits in terms of handling a large capacitive load. In the future, reference makers will need to provide products that can drive larger capacitive loads than they do at present. The op amp's output capability is most important, as well as its ability to swing rail-to-rail. This goes back to the operating voltage and the processing and what is entailed there. The output could be built with Kelvin connections, which would be more practical. However, instead of one output terminal, this would require two or more. Maxim has already introduced a family of ultra-high precision bandgap products (MAX6126 series), which incorporate force and sense Kelvin outputs in both their proprietary μMAX-8 and SO-8 packages.

When it comes to low power, this aspect has already been given high priority by reference makers for some time now. Many low-power products are available on the market, where the current drawn by the reference is very low. So far, the lowest-power devices have been Intersil/Xicor's CMOS-based x60008, ISL60002, and ISL60007 series, with their maximum 0.8 μA current drain. These products have such a low power requirement that they can be left on all the time, and therefore do not require a Sleep/Enable pin. There are now many products available that have less than 20 μA current drain, although there is actually little practical reason to get below 1 μA. Perhaps for some products, this ultra-low current drain could be traded off for a lower noise level, even if it comes at the expense of a little higher current drain (e.g., 20 μA). After all, a lower-noise product supports a higher bit-resolution.

Intertwined with both operating voltage and low current drain is the dropout voltage. Most designers working on battery-powered circuits are interested in this particular characteristic. Not only must the product run from a low supply voltage and draw as little current as possible, but an ultra-low dropout is significant because it allows the circuit/product to operate longer. As with other types of products, most notably linear regulators and op amps, a MOSFET pass transistor does a much better job than a BJT when it comes to dropout voltage. The P-channel pass transistor does the best job of all and depends on a very low gate threshold voltage and a low on-resistance (see Figure 15.38). For a voltage reference, its dropout voltage can be a major factor but requires a low-power CMOS technology to be most effective. One such example that we looked at previously was National Semiconductor's LM4140, a CMOS precision reference that has an ultra-low dropout voltage of typically 20 mV at 1 mA load current. Few references come close to that particular specification, but it is something uniquely possible in CMOS, and after all National Semiconductor is one of its original pioneers.

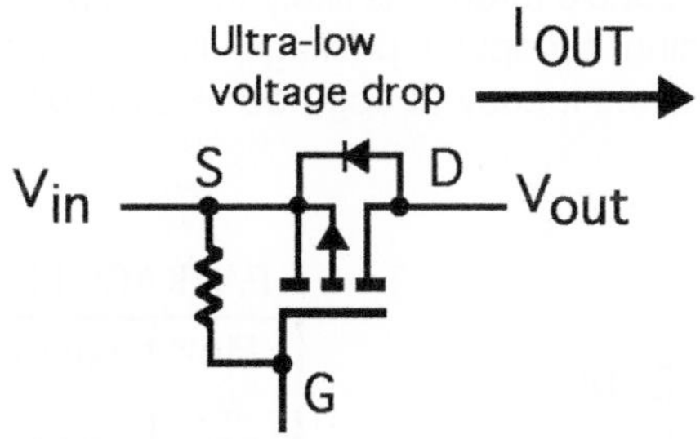

Figure 15.38. The P-channel MOSFET serves as a very efficient low-dropout pass transistor in both linear regulators and voltage references.

As we have learned in this and earlier chapters, an important function available on some reference products is a dedicated TRIM terminal. This ability to trim or adjust the output reference voltage is important because it can provide a system offset/trim. Future references should continue this capability, because it also allows for digital compensation techniques to directly adjust this terminal in a feedback loop, and so alter the reference output voltage as needed.

Digitally programmable references are on the horizon. The first to appear is the Intersil/Xicor x60250, which we reviewed earlier in this chapter. The reference is presently limited to 8-bit resolution by the tempco of its internal digital pot. Industry-wide improvements in digital pots, or in the manner they are used inside a reference, are required. When they can combine this programmability and nonvolatility with high analog accuracy, an ultra-low tempco, very low noise, a low supply voltage, and a low dropout voltage, then we will see future programmable reference products that are attractive. Nevertheless, Intersil/Xicor should be congratulated for this innovative product, as well as for their contribution of the newest topology, the FGA™.

When it comes to packages, most shunt references favor the tiny SC70 or the SOT-23 surface-mount types. The series reference is a different matter though. The low-end series reference can probably best use the three- or five-pin SC70 or SOT-23 packages, but the high-performance reference requires more terminals to be effective. This results in the SO-8/SOIC-8 being the common package of choice used by each of the four topologies. The package is relevant to both the manufacturer and the user. The package should have good thermal characteristics and preferably have eight pins, some of which can only be used by the manufacturer during assembly and test. Fewer than five pins may not be suitable for a high-performance product, and greater than 10 pins is a waste. Whichever of these types is used is no more difficult for the user to implement. Eight pins, therefore, seem to be the best compromise at present. An eight-pin package provides a voltage input, a ground, a reference output, a TRIM terminal, and an enable or noise-reduction terminal, leaving three terminals for use by the manufacturer for postassembly trimming and testing. The SOIC-8 is shown in Figure 15.39, and for the foreseeable future it is likely that this package and pin-out will be retained industry-wide. Using a popular package like this and having an industry-standard REF02 pin-out makes it easy to upgrade existing designs, as well as implement new ones.

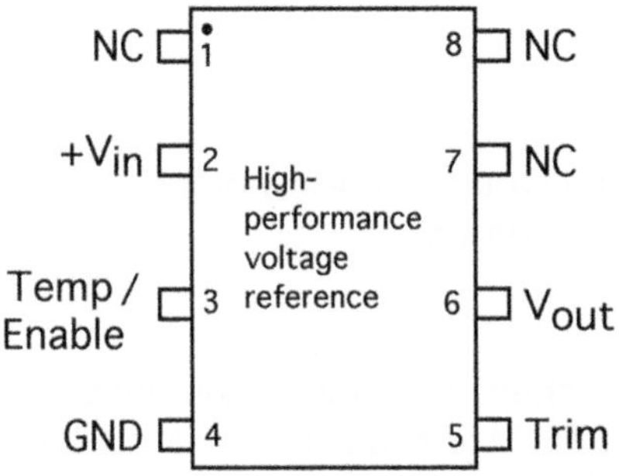

SO-8/SOIC-8/DIP-8 packages

PACKAGE PIN DESIGNATIONS

Pin #:	Function:	Notes:
1	NC	Mostly used as a factory Test Point. Must float.
2	$+V_{in}$	Universally designated as the input voltage pin
3	Temp	Used as a TEMP output or Enable pin, but not always
4	GND	Universally designated as ground
5	Trim	Often used as the Trim pin, else must float.
6	V_{out}	Universally designated as the reference output
7	NC	Mostly used as a factory Test Point
8	NC	Mostly used as a factory Test Point. Must float.

Figure 15.39. The pinout for high-performance voltage references varies from product to product, but is essentially based on the industry-standard REF02 pinout.

Perhaps a starting point in designing the next great voltage reference chip (that has some of the necessary improvements already mentioned) is to first use a well-proven dual precision R-R op amp. One op amp would be used to amplify the core reference voltage, whereas the second op amp provides the necessary output capability for driving large capacitive loads. Reserve space on the chip for some trimmable precision thin-film resistors, then begin designing the voltage reference. Although such a design would be unconventional, sometimes that is what it takes to make the next step forward—innovation. This book acknowledges several such products, which flew in the face of conventional wisdom and changed history.

In summary, the lines have begun to blur as far as the different topologies are concerned. Some next-generation super-bandgaps are already intruding on territory held exclusively by more exotic types. The voltage reference field is no longer as sharply categorized as it was a decade or two ago. It is no easy matter to choose between an XFET®, an FGA™, or a super-bandgap type, because many of their key specifications now overlap. It is likely that you will have to carefully prioritize your needs, and then compromise on some characteristics such as low noise versus low quiescent current or low dropout versus supply voltage. Using this book will help you navigate through the different topologies and their specs, and help with your next current source or voltage reference project. Alternatively, perhaps it will be you who will make the next breakthrough, and create the next generation of voltage references. In the words of Nobel Prize winner and genius, Albert Einstein:

"Imagination is more important than knowledge!"

Appendix A

References and Tables

A.1 Powers of 10 and Equivalents

Power of 10	Base 10	Prefix	Symbol
10^{15}	1,000,000,000,000,000	1P/peta	P
10^{14}	100,000,000,000,000		
10^{13}	10,000,000,000,000		
10^{12}	1,000,000,000,000	1T/tera	T
10^{11}	100,000,000,000		
10^{10}	10,000,000,000		
10^{9}	1,000,000,000	1G/giga	G
10^{8}	100,000,000		
10^{7}	10,000,000		
10^{6}	1,000,000	1M/mega	M
10^{5}	100,000		
10^{4}	10,000		
10^{3}	1,000	1K/kilo	K
10^{2}	100		
10^{1}	10		
10^{0}	1	1	1
10^{-1}	0.1		
10^{-2}	0.01		
10^{-3}	0.001	1 milli	m
10^{-4}	0.0001		
10^{-5}	0.00001		
10^{-6}	0.000001	1 micro	μ
10^{-7}	0.0000001		

Power of 10	Base 10	Prefix	Symbol
10^{-8}	0.00000001		
10^{-9}	0.000000001	1 nano	n
10^{-10}	0.0000000001		
10^{-11}	0.00000000001		
10^{-12}	0.000000000001	1 pico	p
10^{-13}	0.0000000000001		
10^{-14}	0.00000000000001		
10^{-15}	0.000000000000001	1 femto	f

NOTES:

1. The number multiplied by the power of 10 is the **coefficient**.
2. Typically only one number is shown to the left of the decimal point (e.g., 2.356^{-8}).
3. Numbers less than 1 are shown with negative exponents (e.g., 1-milliamp (0.001A) = 1×10^{-3} A).
4. To **add or subtract** two numbers expressed as powers of 10, both numbers should have the same exponents, in which case the exponent stays the same. Add or subtract the coefficients, for example:

   $2.3 \times 10^3 + 3.2 \times 10^3 = \mathbf{5.50 \times 10^3}$

   $6.9 \times 10^4 - 2.5 \times 10^2 = \mathbf{6.88 \times 10^4}$

   $2.4 \times 10^4 + 2.1 \times 10^5 = 0.24 \times 10^5 + 2.1 \times 10^5 = \mathbf{2.34 \times 10^5}$

5. To **multiply** two numbers expressed as powers of 10, multiply the coefficients, and add the powers, for example:

   $1.4 \times 10^{-6} \times 3.9 \times 10^3 = \mathbf{5.46 \times 10^{-3}}$

6. To **divide** two numbers expressed as powers of 10, divide the coefficients, and subtract the powers, for example:

   $1.4 \times 10^{-6} \div 3.9 \times 10^3 = \mathbf{3.59 \times 10^{-10}}$

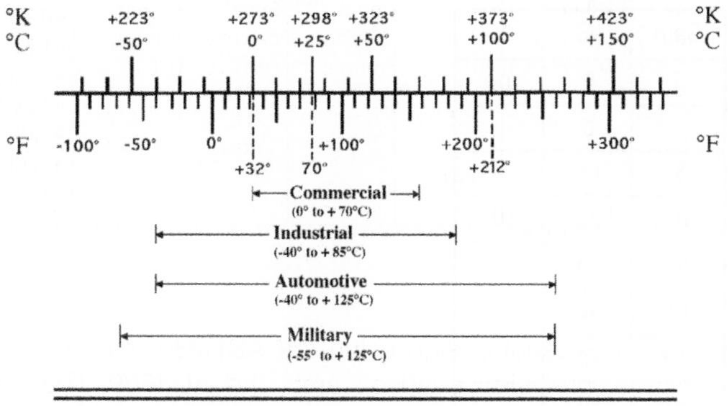

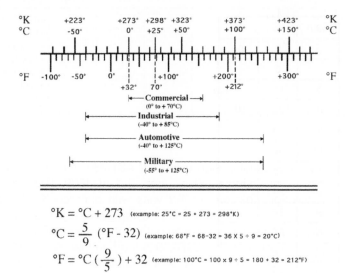

$$°K = °C + 273$$ (example: 25°C = 25 + 273 = 298°K)

$$°C = \frac{5}{9}(°F - 32)$$ (example: 68°F = 68-32 = 36 X 5 ÷ 9 = 20°C)

$$°F = °C\left(\frac{9}{5}\right) + 32$$ (example: 100°C = 100 x 9 ÷ 5 = 180 + 32 = 212°F)

Figure A.1. Temperature Scale Conversion

Typical Characteristics:	Metal-foil	Wirewound	Thin-Film	Thick-Film	Metal-film
Classification	Ultra-high precision	Ultra-high precision	Very-high precision	Semi-precision	Semi-precision / High precision
Maximum Range (Ω)	250KΩ	1MΩ	25MΩ	2GΩ	10MΩ
Accuracy (%)	0.01 to 0.001%	0.01 to 0.005%	0.01 to 2%	2 to 5%	0.1 to 5%
TCR (±ppm/°C)	0.5 to 10 ±ppm/°C	1 to 20 ±ppm/°C	10 to 100 ±ppm/°C	25 to 300 ±ppm/°C	15 to 100 ±ppm/°C
Tracking TCR (±ppm/°C)	0.5 to 3 ±ppm/°C	Not applicable	1 to 5 ±ppm/°C	2 to 50 ±ppm/°C	Not applicable
Long-term stability (±ppm/year)	5 ±ppm/year	20 ±ppm/year	50 to 100 ±ppm/year	30 to 1000 ±ppm/year	1000 ±ppm/year
Cost	High	Medium	Medium	Low	Low
Disadvantages:	Cost	Low to medium frequency; size; inductive effects	Often fragile; low power	Poor TCR, and long-term stability	Differences between mfrs.; long-term stability
Principal U.S. manufacturers:	* Imperial Astronics * Isotek * Wilbrecht Elect. * Vishay	* Isotek * Ohmite * Precision Resistor Co. * Prime Tech./ General Resistance * Process Instr./ Julie Research * Vishay	* AVX Corp. * Caddock * Intl. Mfg. Services * IRC * Kamaya Corp. * Mini-Systems * Thin Film Tech * Vishay	* AVX Corp. * Intl. Mfg. Services * IRC * Kamaya Corp. * Mini-Systems * Ohmite * Vishay	* Brel Intl. Corp. * IRC * Kamaya Corp. * Thin Film Tech * Vishay

NOTES:
1. Contact manufacturers for more information.
2. See Contact Info in Appendix D.
3. Percentage to ppm conversion:
% = ppm x 0.0001
Thus 0.01% = 100-ppm
4. Parameters vary between manufacturers.

Figure A.2. Comparison of various types of precision resistors.

Gain	dB	%
1	0	100
2	6	
5	14	
10	20	10
20	26	
30	30	
40	32	
50	34	
60	36	
70	37	
80	38	
90	39	
100	40	1
250	48	
500	54	
1K	60	0.1
5K	74	
10K	80	0.01
20K	87	
50K	94	
100K	100	0.001
500K	114	
1M	120	0.0001

Figure A.3. *Gain (Av) vs dB conversion table.*

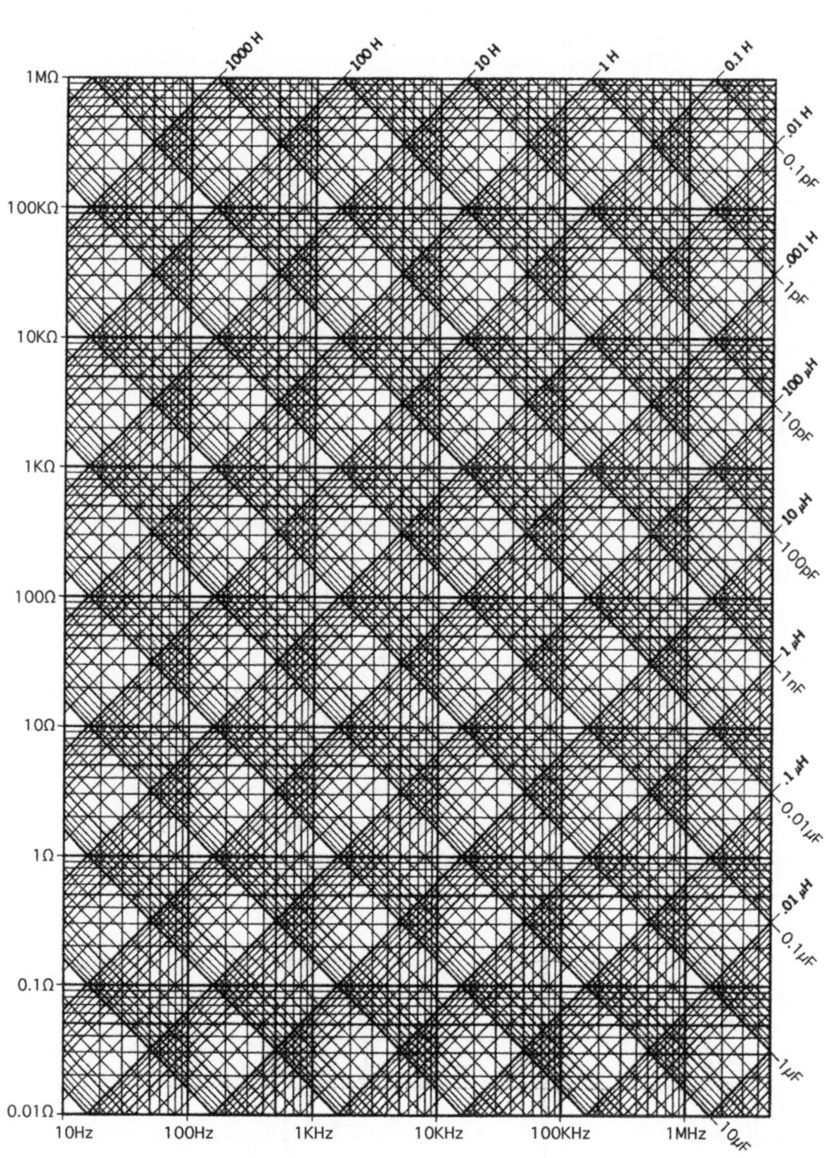

Figure A.4. *Reactance chart for filter design.*

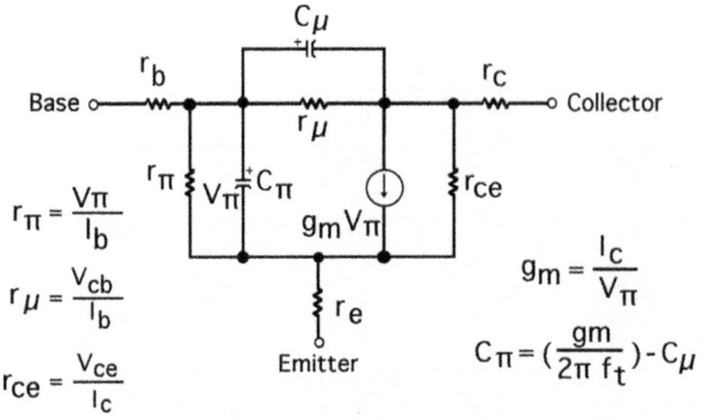

Figure A.5. The small-signal model for the BJT.

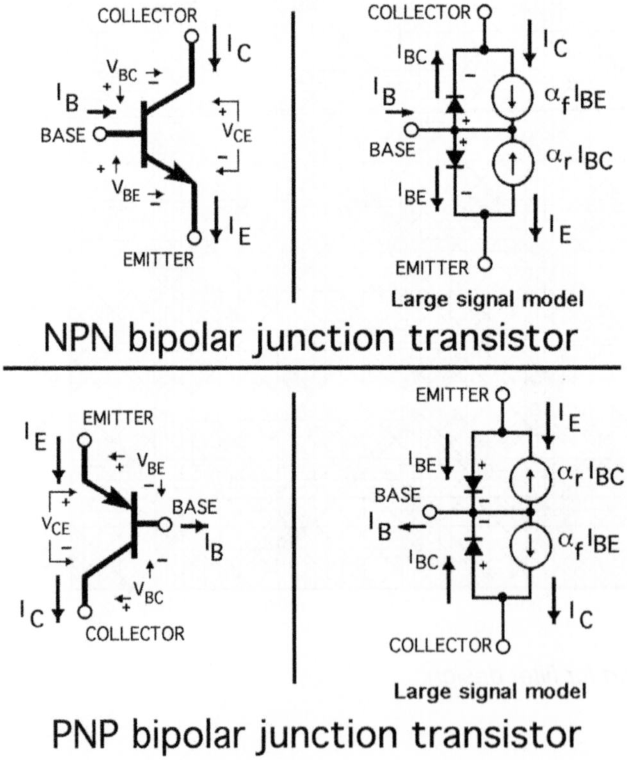

Figure A.6. Simplified BJT models.

Transistor operating modes			
Characteristic:	Common Base (CB)	Common Emitter (CE)	Common Collector (CC)
Current Gain:	No (< x1)	Yes	Yes
Voltage Gain:	Yes	Yes	No (< x1)
Power Gain:	Yes	Yes	Yes
Z_{in} (Ω) typical:	Low (< 100Ω)	Low (1KΩ)	High (>250KΩ)
Z_{out} (Ω) typical:	High (1MΩ)	High (50KΩ)	Low (< 500Ω)
Gain Formula:	$h_{fb} = \frac{I_C}{I_E}$	$h_{fe} = \frac{I_C}{I_B}$	$h_{fc} = \frac{I_E}{I_B}$

Figure A.7. *Operating modes.*

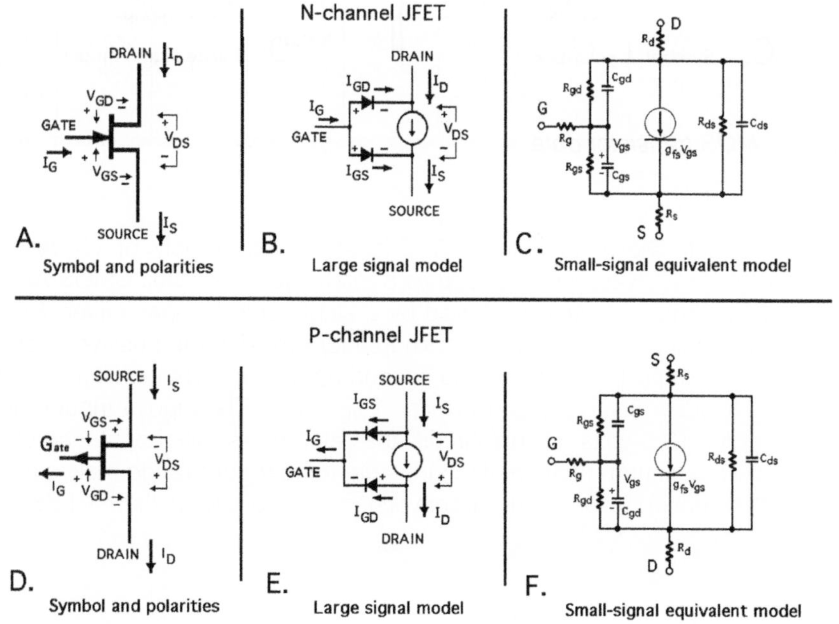

Figure A.8. *JFET models.*

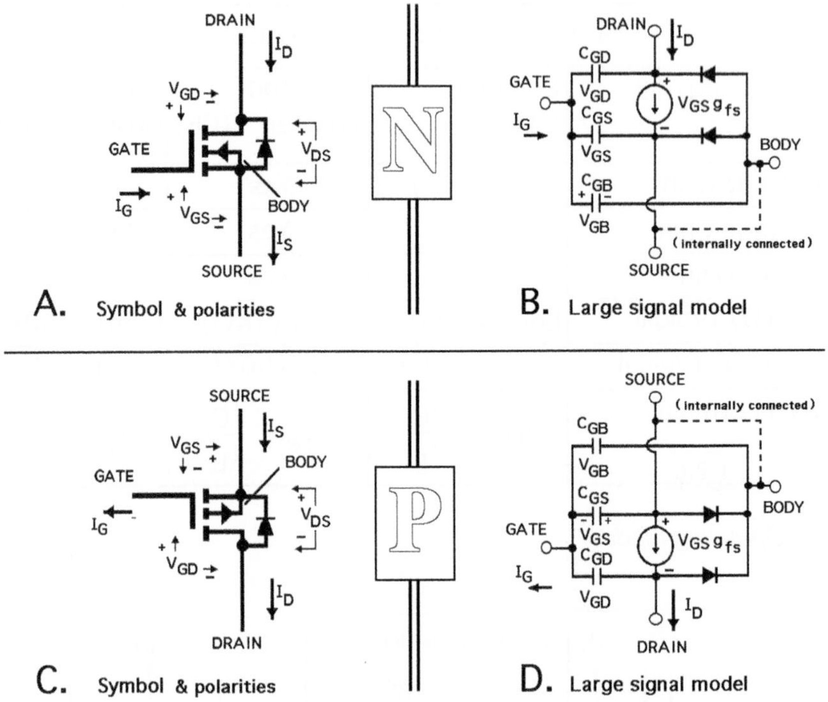

Figure A.9. *MOS transistor data.*

Reference Volts								ppm	%
1.22V	2.048V	2.5V	3.0V	3.3V	4.096V	5.0V	10.0V	1M	100
0.12V	0.204V	0.25V	0.30V	0.33V	0.41V	0.5V	1.0V	100K	10
12mV	20mV	25mV	30mV	33mV	41mV	50mV	0.10V	10K	1
1.2mV	2.0 mV	2.5mV	3mV	3.3mV	4.1mV	5mV	10mV	1K	0.1
120μV	204μV	250μV	300μV	330μV	410μV	500μV	1.0mV	100	0.01
12μV	20μV	25μV	30μV	33μV	41μV	50μV	100μV	10	0.001
1.2μV	2.0μV	2.5μV	3μV	3.3μV	4.1μV	5μV	10μV	1	0.0001
122nV	204nV	250nV	300nV	330nV	410nV	500nV	1.0μV	0.1	0.00001
12nV	20nV	25nV	30nV	33nV	41nV	50nV	100nV	0.01	0.000001
1.2nV	2.0nV	2.5nV	3nV	3.3nV	4.1nV	5nV	10nV	0.001	0.0000001

Figure A.10. PPM to % converter for some popular references.

Bits	Full-scale ranges @ 1/2 LSB resolution, and 100KHz bandwidth Spectral noise density in nV/$\sqrt{Hz}$								
	1.024V Ref	1.24V Ref	2.048V Ref	2.500V Ref	3.000V Ref	3.300V Ref	4.096V Ref	5.000V Ref	10V Ref
8	1040 nV	1280nV	2048 nV	2576 nV	3088 nV	3392	4208	5152 nV	10304 nV
10	260	320	512	644	772	848	1052	1288	2576
12	65	80	128	160	193	212	263	322	643
14	16	20	32	40	48	53	65	80	161
16	4	5	8	10	12	13.25	16.4	20	40
18	1	1.25	2	2.5	3	3.3	4.1	5	10
20	0.25	0.312	0.5	0.625	0.75	0.828	1.02	1.25	2.5
22	0.062	0.078	0.125	0.156	0.187	0.207	0.256	0.314	0.628
24	0.016	0.0195	0.031	0.039	0.046	0.051	0.064	0.078	0.157
NOTES: 1. Computed for 1/2-LSB resolution.									

Figure A.11. Allowable noise levels for n-bit systems, and for 1.024V to 10V ranges.

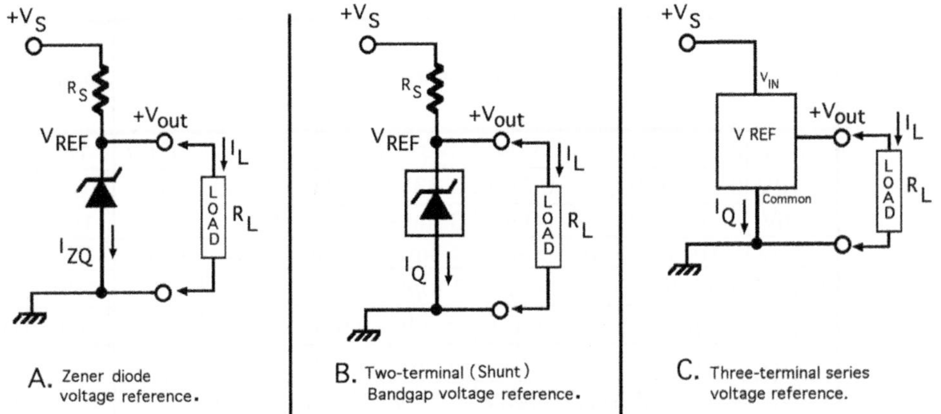

$$I_{TOTAL} = I_{ZQ} + I_L$$

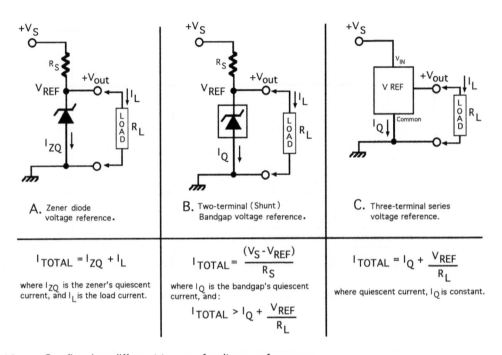

where I_{ZQ} is the zener's quiescent current, and I_L is the load current.

$$I_{TOTAL} = \frac{(V_S - V_{REF})}{R_S}$$

where I_Q is the bandgap's quiescent current, and:

$$I_{TOTAL} > I_Q + \frac{V_{REF}}{R_L}$$

$$I_{TOTAL} = I_Q + \frac{V_{REF}}{R_L}$$

where quiescent current, I_Q is constant.

Figure A.12. Configuring different types of voltage references.

Bits	Required drift (ppm/°C)	Full-scale ranges 1/2 LSB resolution (mV)			
		1.22V Ref	2.5V Ref	5V Ref	10V Ref
4	312.32	39.04 mV	78.08 mV	156.16 mV	312.32 mV
5	156.16	19.52	39.04	78.08	156.16
6	78.08	9.76	19.52	39.04	78.08
7	39.04	4.88	9.76	19.52	39.04
8	19.52	2.44	4.88	9.76	19.52
9	9.76	1.22	2.44	4.88	9.76
10	4.88	0.61	1.22	2.44	4.88
11	2.44	0.305	0.61	1.22	2.44
12	1.22	0.152	0.305	0.61	1.22
13	0.61	0.076	0.152	0.305	0.61
14	0.305	0.038	0.076	0.152	0.305
15	0.152	0.019	0.038	0.076	0.152
16	0.076	0.009	0.019	0.038	0.076
17	0.038	0.0047	0.009	0.019	0.038
18	0.019	0.00238	0.0047	0.009	0.019
19	0.009	0.0012	0.0023	0.0047	0.009
20	0.0047	0.0006	0.0012	0.0023	0.0047
21	0.0023	0.0003	0.0006	0.0012	0.0023
22	0.0012	0.00015	0.0003	0.0006	0.0012
23	0.0006	0.000075	0.00015	0.0003	0.0006
24	0.0003	0.00004	0.000074	0.00015	0.0003

NOTES: 1. Computed for 100°C temperature span.
2. A tempco of 0.3ppm/°C is needed to maintain 1/2 LSB at 14-bits, over an operating temperature range of 100°. Shorter temperature spans can support a higher tempco.

Figure A.13. Allowable tempco drift requirements for n-bit systems.

Appendix B

Glossary

Absolute zero: Refers to the coldest theoretical temperature (–273.15°C or –459.67°F), at which the energy of molecular motion is zero.

Active: One of the BJT's three possible states. This one is a predetermined state facilitated by biasing it halfway (the Q-point), between saturation and cutoff, as used as an AC amplifier.

A/D or ADC: Abbreviation of analog-to-digital converter, a device that converts an analog level to a digital equivalent.

Alpha (α): A figure of merit, and a measure of the current gain between the emitter and collector of a BJT, in the common-base configuration.

Ambient temperature: The current temperature of the surrounding environment, such as room temperature (68°F), or the normal temperature within a circuit's enclosure.

Amplified ΔV_{BE}: Applies to a particular voltage within a bandgap reference circuit.

Anode: The positive terminal of a diode, made of P-type material.

Avalanche breakdown: Refers to a reverse-biased semiconductor junction, where a sudden change from a high electrical resistance to very low resistance can cause damage to the device, unless it is current-limited. Relates to one of two breakdown mechanisms within a zener diode. Above approximately 8 volts, breakdown is caused by this effect.

Axial leads: Usually refers to a metal lead coming from each end of a component, either semiconductor (e.g., a diode) or passive (e.g., capacitor, resistor, inductor), along its axis.

Bandgap: The bandgap is one of the key properties of any semiconductor and is a measure of the energy difference between the valence and conduction bands. The value of this energy gap depends on the particular semiconductor material involved (e.g., Si, GaAs, Ge).

Base: In a bipolar transistor, it is the region that controls current flow between the emitter and the collector.

Base-emitter saturation voltage $V_{BE(sat)}$: In a BJT, it is the voltage required to correctly forward-bias (NPN) or reverse-bias (PNP) this junction.

Beta (β): With a bipolar transistor, this refers to its forward current gain in the common-emitter configuration, specified at a particular collector-emitter voltage, collector current, base current, and temperature. Beta is simply expressed as a number (e.g., 100).

Bias: Refers to either a voltage or current that is purposely applied to a circuit or device, which establishes certain operating conditions. Also applies to a diode, whether it is forward-biased or reverse-biased.

BiCMOS: A device such as an op amp or DAC, which contains both bipolar and CMOS transistors.

BiFET: A device such as an op amp, which contains both bipolar and field-effect transistors. These may be bipolar and MOS, or more usually bipolar and JFETs at the inputs.

Bipolar: With a BJT, both electrons and holes are involved in the current flow.

BJT: Abbreviation of bipolar junction transistor.

Body: In a MOSFET the substrate terminal, which if available separately should be connected to the most negative point in the circuit. Most devices have the body internally connected.

Box method: Applies to monolithic voltage references where manufacturers define, test, and graph the output voltage drift over the operating temperature range. The resulting graph looks like a box shape. It is the most commonly used method. The boundaries are formed by the manufacturer's guaranteed minimum and maximum limits for the output voltage (vertical, top, and bottom), over the defined temperature range (horizontal, left, and right). The maximum temperature coefficient is shown by the box's diagonal corners.

Breakdown voltage: This is the particular voltage at which breakdown occurs and the normally high electrical resistance suddenly changes to a very low resistance. Beyond this point the device is close to being destroyed.

Brokaw cell: A type of bandgap voltage reference invented by Paul Brokaw of Analog Devices.

Bulk metal-foil: A type of ultra-high-precision metal-foil resistor.

Buried-zener reference: A type of very-high-precision voltage reference IC, which relies on an architecture of the same name. Invented by Robert Dobkin in the early 1970s.

Butterfly method: Applies to monolithic voltage references where manufacturers define, test, and graph the output voltage drift over the operating temperature range. It takes its name from the shape of these lines created on the graph (it looks like a butterfly). It is a more detailed set of data than the other methods and uses a central reference point of +25°C. It shows the minimum and maximum slope lines passing through the central reference point, along with various data points along each line. Often used with very-high-precision voltage references.

Calibration: A process of comparing a tested value or condition with a known standard, and then adjusting the circuit being tested.

Cascode: Refers to a combination of two transistors (e.g., BJTs, JFETs, or MOSFETs), which improves high-frequency operation and impedance.

Channel-diodes: With the JFET, the formation of a diode when either forward-biasing or reverse-biasing its junctions.

Cathode: The negative terminal of a diode, made of N-type material.

Celsius temperature scale (Centigrade, °C): A decimal temperature scale used by most world countries and for most technical measurements. The value 0°C is assigned to the freezing point of water, while 100°C relates to its boiling point. The scale was named after the Swedish astronomer Anders Celsius in the early 1700s.

Channel: A path that conducts current between the source and the drain of either a JFET or MOS field-effect transistor.

CMOS: Abbreviation of complementary metal oxide semiconductor. A MOS circuit containing both N-channel and P-channel transistors on the same substrate. Available as either digital (logic and microprocessor ICs) or analog ICs (e.g., op amps, DACs, ADCs).

Collector: One of the three distinct regions of a bipolar transistor, which functions as one of the main current carrying terminals. It is reverse-biased with respect to the base terminal.

Compensation: A corrective technique that increases or decreases a voltage or current level in response to an external stimulus, such as an increase in temperature.

Complementary-bipolar (CB): A process enabling the formation of both NPN and PNP BJTs on the same substrate.

Conductance (g): The ability to conduct, and the reciprocal of resistance.

Constant current: A current that does not vary with respect to input or output conditions.

CRD: Abbreviation of current regulator diode. Equivalent to a single JFET current source with a resistor inserted between its gate and source, but fully monolithic. Its geometry is specially designed and enhanced for precise current regulation.

Current-density ratio: Refers to a bandgap voltage reference where the current through one transistor's collector is deliberately made several times higher than that through a second transistor. Current-density ratios are commonly 8:1 or 10:1.

Current mirror: Also known as a current reflector, it can connect to either rail and usually provides multiple current sources/sinks that either mirror (1:1 match) or are arranged in preset current ratios (2:1, 3:1).

Current scaling: Refers to current sources that generate input-to-output currents that generate integer multiples or fractions on a reference current.

Current sink: Usually comprised of N-type devices, which connect between the load and a negative supply, or ground.

Current source: Usually comprised of P-type devices, which connect between a (more) positive supply rail and a grounded or negative load.

Cutoff: One of the BJT's three possible states, with this being the off state.

Czochralski process: Refers to the most common chemical crystal-growing process used in the manufacture of most semiconductors.

D/A or DAC: Abbreviation of digital-to-analog converter, a device that converts a digital input code to an equivalent analog output level.

Delta (δ): The difference between two values (e.g., the delta of 5 volts and 3 volts is 2 volts).

Depletion layer: This is the region extending on both sides of a reverse-biased semiconductor junction, in which free carriers are removed from the vicinity of the junction. It is also called the space-charge region.

Depletion mode: Applies to all JFETs and some MOSFETs, where devices are normally in the on condition with no applied gate voltage. When the gate-to-source voltage is changed from zero to some other value, this changes the value of the drain current to a lower value.

Diffusion: A process of adding impurities to a semiconductor material, so that they slowly permeate the bulk material or selected areas of it, and so change its electrical characteristics.

Digital-compensation: With voltage references, it refers to the ability of compensating for temperature effects, voltage drift, and so on digitally with the use of a ROM lookup table.

DIL: Abbreviation of dual-inline (IC package), which may be made of either plastic or ceramic, and which has two rows of leads that are inserted into a prearranged area on a printed circuit board and then soldered permanently in place.

DMOS: Refers to double-diffused MOS. A process developed in Japan and introduced in the United States and elsewhere in the early1970s.

Dopant: An element (e.g., aluminum, boron, phosphorus) usually added in tiny amounts during the manufacture of semiconductor material. It can either be a donor or an acceptor, depending on which type of semiconductor material is being created.

Drain: One of the three terminals of any FET transistor, whether JFET or MOSFET, which receives current from the source terminal.

Drain saturation current (I_{DSS}): The maximum limiting current that can flow between the drain and source of an FET.

Drift: A gradual change in value from the original nominal value.

Electron-volt (eV): The standard unit of measurement for an electron's energy. One eV is defined as the energy acquired by an electron moving through a potential of one volt in a vacuum.

Emitter: One of the three distinct regions of a bipolar transistor, which functions as one of its main current-carrying terminals. The emitter current is the largest current in the transistor, comprising both the collector and the base currents.

Emitter area ratio: Applies to the creation of bandgap cells at the semiconductor design level, where two otherwise identical transistors have their emitter areas deliberately made a different size from one another (e.g., 10:1).

Enhancement mode: Applies to how some MOSFET transistors work. Operation is such that no current flows between the drain and source with zero voltage applied to the gate. When voltage of the correct polarity and magnitude is applied to the gate, the transistor turns on and current flows. Most power MOSFETs are examples of this type.

Epitaxial: An added layer of semiconductor material that takes on the same crystalline orientation as the substrate semiconductor material. The two materials can be the same or different, but the crystal lattice of the base material controls the orientation of the atoms in the added layer.

Extrinsic semiconductor material: Semiconductor material whose electrical properties have been altered by the addition of dopants.

Fahrenheit (°F): A temperature scale in which the melting point of ice is specified as 32°F and the boiling point of water at 212°F. It is widely used in the United States and the United Kingdom as a reference for ambient or local weather temperatures but is not usually applied in technical or scientific work. This scale, which is the oldest in use, was named after Gabriel Daniel Fahrenheit, who first introduced it during the early 1700s.

Feed-forward technique: A design technique invented in 1969 by Robert Dobkin, who was then working at National Semiconductor, which enables op amps to achieve a higher slew rate and bandwidth.

FET: Abbreviation of field-effect transistor, in which a voltage applied at the gate controls conduction between the source and the drain and involves only one type of charge carrier.

FGA™: Abbreviation of floating gate analog, a new voltage reference topology invented in 2003 at Xicor Corp. (now a part of Intersil Corp.). Relies on storing an analog voltage on the gate of a MOSFET.

Forward-bias: An electrical condition where a positive voltage is applied to the P-type material (i.e., the anode of a diode) and a negative or less positive voltage is applied to the N-type material (i.e., the cathode). Under such conditions, the diode will support conduction.

Gain: The amplification factor between the input and output of a device such as a transistor or an IC amplifier. It can be expressed as current gain, voltage gain, or power gain as a number or in decibels.

Gate: One of the three terminals of any FET transistor, whether JFET or MOSFET. Because it is located between the source and drain terminals it acts as the control terminal, and usually has a very high input impedance.

Gate charge: A characteristic directly related to a MOSFET's capacitance and switching, used to determine the gate drive requirements.

Gate-source threshold voltage $V_{GS(th)}$: Applies only to enhancement-mode MOS devices and is the voltage necessary to just turn the MOSFET on, at a very low current level. Some MOSFETs have a low $V_{GS(th)}$ value, so that their inputs are logic-level compatible.

G_m: Refers to the small signal forward transconductance characteristic of the BJT. G_m measures the effect of a change in collector current (I_C), for a specific change in base-emitter voltage (V_{BE}), referenced to common-emitter mode.

Heatsink: A rigid metal structure used to absorb and conduct heat away from (heat-sensitive) semiconductor devices. Heatsinks are usually thermally enhanced by using special metal alloys. Typically, a power transistor is mechanically fastened to the heat-sink.

I_C: Refers to a BJT's collector current.

I_{DSS}: The drain saturation current or maximum limiting current that can flow between the drain and source of a JFET or depletion-mode DMOS FET.

I_{GSS}: The very small gate-to-source reverse leakage current of JFETs and MOSFETs, usually measured in pico-amps or nano-amps, that applies at a specified gate-to-source voltage. The last letter indicates that the drain is shorted to the source. This leakage current approximately doubles for every 10°C rise in temperature.

Initial accuracy (aka initial error): With a voltage reference, this applies to the initial output voltage tolerance or error, following an initial application of power to the circuit. Usually specified with a fixed input voltage, at room temperature, and with no load.

Intrinsic semiconductor: The pure basic semiconductor material with neither donors nor acceptors added.

Ion-implantation: A very precise technique used in making certain semiconductors where high-energy dopant ions are blasted at the surface and so create specific P or N regions.

I_{PTAT}: Abbreviation of current proportional to absolute temperature, such as when a transistor's base-emitter voltage (V_{BE}) is correctly forward-biased, it can be used as a predictable reference.

JFET: Abbreviation of junction field-effect transistor. A voltage-controlled transistor.

Junction: The abrupt interface between two semiconductor regions composed of P-type and N-type materials. Applies to most semiconductors, particularly the BJT and the JFET.

Kelvin scale (°K): Refers to an absolute temperature scale, where 0°K is absolute zero, the coldest theoretical temperature (–273.15°C or –459.67°F). The boiling point of water is 373.15°K (+100°C or +212°F). The Kelvin scale is named after the British physicist Lord Kelvin, who first proposed it.

Knee: The operating point for most P-N junctions, where the reverse current changes from a very low value to heavy conductance.

Leakage current: A small undesirable current flow, often related to temperature, voltage levels, or bias conditions.

Line regulation: When applied to a voltage reference or voltage regulator, it is a measure of the device's ability to keep the output voltage at a constant level, irrespective of changes in input voltage (ΔVout/ΔVin). May be expressed in μV/V or ppm/V.

Linearity error: Refers to one of the small sources of error inherent in any A/D or D/A converter.

Load regulation: When applied to a voltage reference or voltage regulator, it is a measure of the device's ability to keep the output voltage at a constant level, irrespective of changes in output load current (ΔVout/ΔIout). May be expressed in μV/mA or ppm/mA.

MOSFET: Abbreviation of metal-oxide semiconductor field-effect transistor. A voltage-controlled transistor.

NPN Transistor: A BJT having an N-type emitter and collector, and a P-type base.

On-state resistance ($R_{DS(on)}$): An FET's resistance when conducting.

Passivation: An important factor in creating rugged diodes, which are achieved by forming a small planar junction under the insulating silicon dioxide layer. Helps reduce capacitance and creates a more stable, reliable device, with a low reverse leakage current.

Peak operating voltage (POV): Refers to the maximum voltage that can be applied to the CRD, at 25°C room temperature.

Periodic table of elements: This is the commonly used classification and table of chemical elements, in the order of their atomic numbers.

Photolithography: A technique borrowed from the printing industry in the production of printed material. With semiconductor manufacturing, it refers to a technique using a light source together with selective areas of masking, to transfer a particular pattern of circuit interconnects onto the surface of the semiconductor wafer material.

Planar: A process invented by Dr. Jean Hoerni at Fairchild Semiconductor in 1958. With a BJT, it has its base, collector, and emitter all on one plane.

PNP transistor: A BJT having a P-type emitter and collector and an N-type base.

Polycrystalline: A crystalline material, with random orientations.

PPM: Abbreviation of parts per million.

Q-point: A particular operating point for an amplifier, biased at a certain voltage and current.

Rectification: The ability of a diode to convert an AC waveform into a DC level.

Reference voltage: A particular voltage level that functions as a reference within a circuit.

Repeatability: The ability to produce again identically.

Safe operating area (SOA): Relates to BJT and MOSFET power transistors. An area limited by breakdown voltage, current rating, on-resistance, power dissipation, and maximum junction temperature. Exceeding any one of these could be fatal to the device.

Saturated: One of the BJT's three possible states, with this being the full-on state.

Series reference: A three-pin (or higher) voltage reference IC, used in series with the load.

Shunt reference: A two-pin voltage reference IC or zener, used in parallel with the load.

Silicon-gate: Applies to MOS transistors or MOS-based ICs, where their gate terminals are specially processed using silicon. It leads to faster, smaller, denser, lower-power ICs.

Source: One of the three terminals of any FET transistor, whether JFET or MOSFET, which sends current to the drain terminal.

Stability: A state of constant equilibrium; resistance to change or deterioration.

Super-beta transistor: A type of ultra-high-gain transistor first developed by David Talbert, the silicon processing specialist for Bob Widlar. They are used in many of today's linear ICs and have betas typically between 1K and 5K, but also very-low-voltage breakdowns.

Surface zener: A typical zener diode junction where breakdown occurs at its surface, as opposed to beneath it (as with the IC buried-zener type).

Temperature coefficient (TC or tempco): Refers to an average of the small changes in a reference's output voltage, or a current source's output current, caused by changes in temperature. Usually expressed in ppm/°C or in %/°C. Ideally, the output voltage will not shift by any measurable amount over the temperature range (e.g., 0 ppm/°C). An increase in output voltage/current with temperature has a positive tempco (e.g., +100 ppm/°C), whereas a decrease has a negative tempco (e.g., –100 ppm/°C).

Thermal gradient: Refers to slight differences in temperature across an IC package, components, and circuit boards, which tend to degrade the accuracy and stability of a reference or other precision circuitry. An undesirable feature.

V_{BE}: This refers to the base-emitter voltage of a BJT, typically around 0.7 volt.

$V_{GS(th)}$: Applies only to enhancement-mode MOSFETs and is the gate-source threshold voltage necessary to just turn the MOSFET on, so that a very low forward current flows.

Voltage reference: Typically a monolithic IC containing precision circuitry that provides a constant reference voltage, irrespective of changes in input voltage, output load current, or temperature.

V_{PTAT}: Abbreviation of voltage proportional to absolute temperature. When a transistor's base-emitter voltage (V_{BE}) is correctly forward-biased, it can be used as a predictable voltage reference and is usually measured as x-mV/°C. Some voltage reference ICs have a dedicated V_{PTAT} output terminal, as do all monolithic temperature sensors.

Zener diode: A specially made diode deliberately intended to operate in the reverse voltage mode. Acts as a voltage reference or a basic clamp.

Zener effect: Relates to one of two breakdown mechanisms within a zener diode. Below approximately 5 volts, breakdown is caused by this effect.

Appendix C

Bibliography

C.1 PART I—Current Sources

Chapter 1

"Electronics on the threshold of a new millenium," *Electronic Engineering Times*, CMP Publications, Twenty-fifth Anniversary Issue, 978, October 1997.

Lundquist, Eric (Ed.), "The 30th Anniversary of the Integrated Circuit," *Electronic Buyers News*, CMP Publications, September 1988.

Mattera, Lucinda (Ed.), "Celebrating 50 years of technology," *Electronic Design*, Penton Publications, Vol. 50, No. 22, October 2002.

Scrupski, Stephen (Ed.), "Forty years," *Electronic Design*, Penton Publications, Vol. 40, No. 24, November 1992.

Chapter 2

Caddock Electronics. "High-performance film resistors," General catalog, 2002.

"Elegant architectures yield precision resistors," *EDN* magazine, July 1992.

Grant, Doug, and Scott Wurcer. "Avoiding passive-component pitfalls," Analog Devices.

Linear Technology. App. note 42.

National Semiconductor data sheet: LM134 family.

Texas Instruments/Burr-Brown data sheet: REF200.

Vishay Intertechnology. Product guide, "A comprehensive guide to trimmers."

Chapter 3

Baliga, B. Jayant. *Modern Power Devices.* New York: John Wiley & Sons, 1987.

Central Semiconductor. "Leaded Semiconductor Selection Guide," Discretes/BJT 2003.

Dascalu, D. *Transit-Time Effects in Unipolar Solid State Devices.* Tunbridge Wells, England: Abacus Press, 1974

Fairchild Semiconductor CD-ROM. Product Specifications, 2002.

General Electric. *Transistor Manual*, 7th ed., GE Semiconductor Products Division, 1964.

Ghandi, Sorab K. *Semiconductor Power Devices*. New York: John Wiley & Sons, 1977.

Kocsis, Miklos, and Adam Hilger. "High-speed silicon planar-epitaxial switching diodes." Bristol, England: Institute of Physics, 1976.

Motorola Semiconductors. "Small-signal transistors, FETs, and diodes device," Data book DL126/D, 1993.

Motorola Semiconductors. "Rectifier Applications Handbook," 3rd ed., HB214/D, 1993.

National Semiconductor. "Discrete diode, BJT, and JFET products," Data book, 1996.

Neudeck, Gerold W. *The PN Junction Diode*, 2nd ed., in *Modular Series on Solid State Devices*, Vol. 2. Reading, MA: Addison-Wesley, 1989.

Streetman, Ben G. *Solid State Electronic Devices*, 3rd ed. Englewood Cliffs, NJ: Prentice Hall, 1990.

Sze, S. M. *Semiconductor Devices: Pioneering Papers*. World Scientific Pub Co., 1991.

Sze, S. M. *Physics of Semiconductor Devices*, 2nd ed., New York: John Wiley & Sons, 1981.

Van der Ziel, A. *Solid State Physical Electronics*. Englewood Cliffs, NJ: Prentice-Hall, 1976.

Chapter 4

Central Semiconductor. "Leaded Semiconductor Selection Guide," Discretes/BJT 2003.

Fairchild Semiconductor. CD-ROM, Product Specifications, 2002.

Feldman, J. *The Physics and Circuit Properties of Transistors*. New York: John Wiley & Sons, 1972.

General Electric. *Transistor Manual*, 7th ed. GE Semiconductor Products Division, 1964.

Gorton, W. S. "The Genesis of the Transistor, Memorandum for Record," in S. Millman (Ed.), *A History of Engineering and Science in the Bell System: Physical Sciences 1925–1980*. AT&T Bell Laboratories.

Motorola Semiconductors. "Small-signal transistors, FETs, and diodes device," Data book DL126/D, 1993.

National Semiconductor. "Discrete diode, BJT, and JFET products," Data book, 1996.

National Semiconductor. LB 41, "Precision reference uses only 10-microAmps," 1978.

Nosov, Y. R. *Switching in Semiconductor Devices.* New York: Plenum Press, 1969.

Riordan, Michael, and Lillian Hoddeson. *Crystal Fire: The Birth of the Information Age,* Sloan Technology Series. New York: W.W. Norton & Co., 1997.

Shockley, William. "How We Built the Transistor," *New Scientist,* December 1972.

Sze, S. M. *Physics of Semiconductor Devices,* 2nd ed., New York: John Wiley & Sons, 1981.

Warner, R. M., and B. L. Grung. *Transistors: Fundamentals for the Integrated Circuit Engineer.* New York: John Wiley & Sons, 1983. (Reprinted by Kreiger, Malabar, FL.)

Chapter 5

Analog Devices. Data sheets: MAT02, MAT03, MAT04, SSM2210, SSM2220.

Linear Integrated Systems. Data sheets: LSIT120A, LSIT130A, LS301-03, LS310-313, LS318, LS351, LS352, LS358, LS3250A/B, LS3550, IT124.

National Semiconductor. Data sheet: LM194/394, LM3046.

National Semiconductor. AN-222, "Super matched bipolar transistor pair sets new standard for low drift and noise," 1979.

ON Semiconductor. Data sheets: BC846, BC856.

Chapter 6

Central Semiconductor. "Leaded Semiconductor Selection Guide," Discretes/JFET 2003.

Crystalonics. Data book.

Evans, Arthur D. (Ed.). *Designing with Field-effect Transistors.* New York: McGraw-Hill, 1981.

Fairchild. Semiconductor CD-ROM, Product Specifications, 2002.

InterFET Semiconductor. Data book, 1996

Motorola Semiconductors. "Small-signal transistors, FETs, and diodes device," Data book, DL126/D, 1993.

National Semiconductor. Data book, "Discrete diode, BJT, and JFET products," 1996.

On Semiconductor. Selector guide, JFETs.

Siliconix (Vishay). Data book, Low power discretes, 1994.

Siliconix (Vishay). App. notes: AN101, AN102, AN103, AN106.

National Semiconductor. AN-32, "FET circuit applications," 1970.

Chapter 7

Siliconix (Vishay). Data book, Low power discretes, 1994

Siliconix (Vishay). App. note AN901.

Siliconix (Vishay). Power products, Data book.

SuperTex. Data book, 1996.

SuperTex. App. note, AN-D1, AN-D2, AN-D11, AN-D12, AN-D15.

Chapter 8

Fairchild Semiconductor. App. note, AN9010, "MOSFET Basics," 2000.

Fairchild Semiconductor. App. note, AN7500 "Understanding Power MOSFETs," 1999.

Goodenough, Frank (Ed.). "Dense MOSFET enables portable power control," *Electronic Design*, Penton Publications, April 1997.

International Rectifier Corp. App. note, "Power MOSFET Basics."

International Rectifier Corp. IRactive Design Tool, CD-ROM.

IXYS Corp. App. note, "Application examples with switchable regulator."

IXYS Corp. Data sheet: IXC and IXCP series of current regulators.

Motorola Semiconductors. TMOS Power MOSFET Transistor Device, Data book , DL135, 1996.

Severns, Rudy (Ed.). "MOSPOWER applications" book, Siliconix (Vishay), 1984.

SuperTex. Data book, 1996.

Chapter 9

Advanced Linear Devices. Data sheets: ALD1101, ALD1102, ALD1103, ALD1108xx, ALD1109xx, ALD1110E, ALD110800/A, ALD110802, ALD110900/A, ALD110902, ALD1148xx, ALD1149xx, and ALD114904A.

Advanced Linear Devices. Linear Products and Analog/Digital ASIC Data book.

National Semiconductor. AN-88, "CMOS linear applications," 1973.

RCA. Solid State CMOS Integrated Circuits Data book, SSD-250C.

Chapter 10

Burr-Brown. Applications Handbook, App. note AB-165, "Implementation and applications of current sources and current receivers," Mark Stitt, 1994.

National Semiconductor. Data sheet: LM134 family.

National Semiconductor. LB 41, "Precision reference uses only 10-microAmps," 1978.

Texas Instruments/Burr-Brown. Data sheet: REF200.

Chapter 11

Analog Devices. Designer's CD-ROM Reference Manual, 1996.

Burr-Brown. IC Data book, Linear Products, 1997.

Linear Technology. Linear Data book, 1990.

Linear Technology. Linear Data book, 1992.

Linear Technology. LinearView CD-ROM, 1996.

Maxim Integrated Products. Full-line Data catalog CD-ROM, 2001.

National Semiconductor. AN-3, "Drift compensation techniques for integrated DC Amplifiers," R. J. Widlar, 1967.

National Semiconductor. AN-4, "Monolithic Op Amp: The universal linear component," R. J. Widlar, 1968.

National Semiconductor. AN-29, "IC Op Amp beats FETs on input current," R. J. Widlar, 1969.

National Semiconductor. AN-241, "Working with high impedance op amps," R. J. Widlar, 1980.

National Semiconductor. Linear Applications Handbook, 1994.

National Semiconductor. Linear/Mixed Signal Designer's Guide, 2000.

National Semiconductor. Operational Amplifiers Data book, 1995.

Precision Monolithics (ADI). Analog Integrated Circuits Data book, Vol. 10, 1990.

C.2 PART II—Voltage References

Chapter 12

Analog Devices. Data Converter Reference Manual, Vol. II, 1992.

Bryant, James. "Voltage references," Ask the Applications Engineer #11, Analog Devices.

Goodenough, Frank (Ed.). "Sub 5-V Voltage reference mimics the performance of buried zeners,," *Electronic Design*, Penton Publications, October 1997.

Jung, Walt. "The Ins and outs of green regulators / references." *Electronic Design,* Analog Applications Issue, June 1994.

Jung, Walt, Walt Kester, and James Bryant. "Voltage references and low dropout linear regulators," Analog Devices.

Knapp, Ron. "Back-to-basics approach yields stable references," *EDN,* June 1988.

Knapp, Ron. "Selection criteria assist in choice of optimum reference," *EDN,* February 1988.

Lee, Mitchell. "Understanding and applying voltage references," Application Note, AN#82, Linear Technology, 1999.

Maxim International Products. "Selecting voltage references," *Engineering Journal,* No. 14.

Miller, Perry, and Doug Moore. "Precision voltage references," *Analog Applications Journal,* Texas Instruments, November 1999.

Nasraty, Roya. "XFET references," Analog Dialog Vol. 32, No. 1, Analog Devices, 1998.

National Semiconductor. *Linear/Mixed Signal Designer's Guide*, 2000.

"Voltage references enable precision conversions," *EDN* (Supplement), June 1999.

Williams, Jim. "A Standards Lab Grade 20-bit DAC with 0.1ppm/°C Drift," Application Note, AN#86, Linear Technology.

Chapter 13

Central Semiconductor. "Leaded Semiconductor Selection Guide," Discretes/Zener, 2003.

Crystalonics. Data sheet: CM697.

Fairchild Semiconductor. CD-ROM, Product Specifications, 2002.

Fairchild Semiconductor. Data sheet: 2N2222A.

Motorola Semiconductor. Device data book, "TVS and Zener diodes," DL150, 1992.

National Semiconductor. Data book, "Discrete diode, BJT, and JFET products," 1996.

National Semiconductor. Data sheet: LMC6061.

SuperTex. Data sheets: DN3525 and VN0104.

Todd, Carl David. *Zener and Avalanche Diodes*. New York: John Wiley & Sons, 1970.

Chapter 14

Analog Devices. Data sheets: AD580, AD588, AD1580, ADR290, ADR420, and ADR430.

Brokaw, Paul A. "More about the AD580 monolithic voltage regulator," Analog Devices.

Intersil. x60008 data sheet.

Linear Technology. LT6650 data sheet.

Maxim. MAX6018 data sheet.

National Semiconductor. AN-56, "1.2V Reference," 1971.

National Semiconductor. AN-161, "IC voltage reference has 1-ppm per degree drift," 1976.

National Semiconductor. AN-211, "New op amp ideas," R. J. Widlar, 1978.

National Semiconductor. Data sheets: LM10, LM199, and LM4041.

Semtech. SC4436 data sheet.

SuperTex. LP0701 data sheet.

Thaler. VRE4100 data sheet.

Chapter 15

Advanced Linear Devices. Data sheets: ALD1102 and ALD110902.

Analog Devices. Data sheets: ADR01, ADR280, ADR390, ADR431A, ADR510, ADR52x, ADR525, AD586, AD688, AD780, AD820, AD5337, AD7701, AD7533, AD8609, OP1177, and REF02.

Intersil. Data sheets: CA3420, EL8186, ISL60002, ISL60007, x9119, x60250, and x60008.

Linear Technology. Data sheets: LT1013, LT1021, LT1218, LT1389, LT1461, LT1634, LT6650, LTC1285, LTC1448, and LTZ1000.

Maxim. Data sheets: MAX400, MAX5143, MAX6006-9, MAX6037, MAX6126, MAX6129, MAX6138, MAX6325, and MAX9030.

National Semiconductor. Data sheets: LM4041, LM4051/ADJ, LM4130, LM4140/A, and LM6144A.

Texas Instruments. Data sheets: REF102 and TLE2021.

Appendix D

Contact Information

D.1 SEMICONDUCTOR MANUFACTURERS

Advanced Linear Devices, Inc.
415 Tasman Dr.
Sunnyvale, CA 94089
(408) 747-1155
www.aldinc.com

Advanced Power Technology, Inc.
405 SW Columbia St.
Bend, OR 97702
(800) 522-0809
www.advancedpower.com

Analog Devices, Inc.
804 Woburn St.
Wilmington, MA 01887
(978) 935-5565
www.analog.com

Calogic LLC
237 Whitney Pl.
Fremont, CA 94539
(510) 656-2900
www.calogic.com

Central Semiconductor Corp.
145 Adams Ave.
Hauppauge, NY 11788
(631) 435-1110
www.centralsemi.com

Crystalonics Inc.
2805 Veterans Hwy., Unit #14
Ronkonkoma, NY 11779
(631) 981-6140
www.crystalonics.com

Diodes, Inc.
3050 E. Hillcrest Dr., #200
Westlake Village, CA 91362
(805) 446-4800
www.diodes.com

EIC Semiconductor, Inc.
12698 Schabarum Ave.
Irwindale, CA 91706
(800) 342-7763
www.eicsemi.com

Exar Corp.
48720 Kato Rd.
Fremont, CA 94538
(510) 668-7000
www.exar.com

Fairchild Semiconductor Corp.
82 Running Hill Rd.
S. Portland, ME 04106
(408) 822-2152
www.fairchildsemi.com

General Semiconductor (Vishay Semiconductors)
10 Melville Park Rd.
Melville, NY 11747
(631) 847-3000
www.gensemi.com

Infineon Technologies Corp.
1730 N. First St.
San Jose, CA 95112
(888) INFINEON (463-4636)
www.infineon.com

InterFET Corp.
715 N. Glenville Dr., Ste. #400
Richardson, TX 75081
(972) 238-1287
www.interfet.com

International Rectifier Corp.
233 Kansas St.
El Segundo, CA 90245
(310) 322-3331
www.irf.com

Intersil Corp.
675 Trade Zone Blvd.
Milpitas, CA 95035
(408) 935-4300
(888) 352-6832
www.intersil.com

IXYS Corp.
3540 Bassett St.
Santa Clara, CA 95054
(408) 982-0700
www.ixys.com

Linear Integrated Systems, Inc.
4042 Clipper Ct.
Fremont, CA 94538
(800) 359-4023
www.linearsystems.com

Linear Technology Corp.
1630 McCarthy Blvd.
Milpitas, CA 95035
(408) 432-1900
www.linear.com

Maxim Integrated Products, Inc.
120 San Gabriel Dr.
Sunnyvale, CA 94086
(408) 737-7600
www.maxim-ic.com

Micrel, Inc.
1849 Fortune Dr.
San Jose, CA 95131
(408) 944-0800
www.micrel.com

Microchip Tech., Inc.
2355 W. Chandler Blvd.
Chandler, AZ 85224
(800) 437-2767
www.microchip.com

Microsemi Corp.
8700 E. Thomas Rd.
Scottsdale, AZ 85252
(480) 941-6300
www.microsemi.com

National Semiconductor Corp.
2900 Semiconductor Dr.
Santa Clara, CA 95052
(408) 721-5000
www.national.com

ON Semiconductor, Inc.
5005 E. McDowell Rd.
Phoenix, AZ 85008
(602) 244-3569
www.onsemi.com

Philips Semiconductors, Inc.
1109 McKay Dr.
San Jose, CA 95131
(800) 835-4555
www.semiconductors.philips.com

Powerex, Inc.
200 E. Hillis St.
Youngwood, PA 15697
(800) 451-1415
www.pwrx.com

Semtech Corp.
652 Mitchell Rd.
Newbury Park, CA 91320
(805) 498-2111
www.semtech.com

Siliconix (Vishay Semiconductors)
2201 Laurelwood Rd.
Santa Clara, CA 95056
(800) 554-5565
www.vishay.com
www.siliconix.com

Solitron Devices, Inc.
3301 Electronics Way
W. Palm Beach, FL 33407
(561) 848-4311
www.solitrondevices.com

ST Microelectronics, Inc.
1310 Electronics Dr.
Carrollton, TX 75006
(972) 466-6000
www.st.com

Supertex Inc.
1235 Bordeaux Dr.
Sunnyvale, CA 94089
(408) 744-0100
www.supertex.com

Teccor Electronics, Inc.
1801 Hurd Dr.
Irving, TX 75038
(972) 580-7777
www.teccor.com

Texas Instruments, Inc.
12500 TI Blvd.
Dallas, TX 75243
(800) 842-2737
www.ti.com

Thaler Corp.
2015 Forbes N.
Tucson, AZ 85745
(800) 827-6006
www.thaler.com

Vishay Semiconductors, Inc.
2201 Laurelwood Rd.
Santa Clara, CA 95056
(800) 554-5565
www.vishay.com

Xicor Inc. (see Intersil Corp.)

Zetex, Inc.
47 Mall Dr.
Commack, NY 11725
(631) 543-7100
www.zetex.com

D.2 ELECTRONICS DISTRIBUTORS

Arrow Electronics Co.
(800) 777-2776
www.arrow.com

Avnet Cilicon
(800) 332-8638
www.em.avnet.com

Digi-Key Corp.
(800) 344-4539
www.digikey.com

Future Electronics Co.
www.futureelectronics.com

Garrett Electronics Corp.
(800) 767-0081
www.garrettelec.com

Jameco Electronics Co.
(800) 831-4242
www.jameco.com

Master Distributors, Inc.
(310) 452-1229
www.onlinecomponents.com

Memec Unique
(408) 952-7160
www.memec.com

Mouser Electronics, Inc.
(800) 346-6873
www.mouser.com

Newark InOne
(800) 4-NEWARK (463-9275)
www.newarkinone.com

NU-Horizons Electronics Corp.
(800) 747-6846
www.nuhorizons.com

Solid State Inc.
(800) 631-2075
www.solidstateinc.com

TTI, Inc.
(800) CALL-TTI (225-5884)
www.ttinc.com

Waldom Electronics, Inc.
(800) 435-2931
www.waldom.com

D.3 PRECISION PASSIVES MANUFACTURERS

Alpha Electronics Corp. of America
13805 1st Ave. N, Ste. #500
Minneapolis, MN 55441
(763) 258-8550
www.alpha-amer.com

AVX Corp.
801 17th Ave. S
Myrtle Beach, SC 29578
(843) 448-9411
www.avxcorp.com

Bourns, Inc.
1200 Columbia Ave.
Riverside, CA 92507
(909) 781-5500
(877) 4-BOURNS (426-8767)
www.bourns.com

Brel Intl. Components Co.
1621 W. University Pkwy.
Sarasota, FL 34243
(800) 237-4564
www.brelintl.com

Caddock Electronics, Inc.
17271 N. Umpqua Hwy.
Roseburg, OR 97470
(541) 496-0700
www.caddock.com

Cal-Chip Electronics, Inc.
59 Steamwhistle Dr.
Ivyland, PA 18974
(800) 915-9576
www.calchip.com

Huntington Electric, Inc.
550 Condit St.
Huntington, IN 46750
(219) 356-0778
www.heiresistors.com

Illinois Capacitor, Inc.
3757 W. Touhy Ave.
Lincolnwood, IL 60712
(847) 675-1760
www.illcap.com

Imperial Astronics
300 Cypress Ave.
Alhambra, CA 91801
(626) 284-9917

IRC, Inc.
4222 S. Staples St.
Corpus Christi, TX 78411
(888) 472-3282
(828) 264-8867
www.irctt.com

Intl. Mfrg. Services, Inc.
50 Schoolhouse Lane
Portsmouth, RI 02871
(401) 683-9700
www.ims-resistors.com

Isotek Corp.
435 Wilbur Ave.
Swansea, MA 02777
(508) 673-2900
(800) LOW-OHMS (569-6467)
www.isotekcorp.com

Julie Research Co.
(see Process Instruments, Inc.)

Kamaya, Inc.
4000 Transportation Dr.
Fort Wayne, IN 46818
(219)489-1533
www.kamaya.com

Kemet Electronics Corp.
P.O. Box 5928
Greenville, SC 29606
(864) 963-6300
www.kemet.com

Mini-Systems, Inc.
20 David Rd.
P.O. Box 69
N. Attleboro, MA 02761
(508) 695-0203
www.mini-systemsinc.com

MWS Wire Industries, Inc.
31200 Cedar Valley Dr.
Westlake Village, CA 91362
(818) 991-8553
www.mwswire.com

Ohmite Manufacturing Co.
3601 Howard St.
Skokie, IL 60076
(847) 675-2600
www.ohmite.com

Precision Resistor Co.
10601 75th St. N.
Largo, FL 33777
(727) 541-5771
www.precisionresistor.com

Prime Technology Co.
Twin Lakes Rd.
N. Branford, CT 06471
(203) 481-5721
www.primetechnology.com

Process Instruments, Inc.
615 E. Carson St.
Pittsburgh, PA 15203
(412) 431-4600
www.procinst.com

Sierra-KD Components Co
5200 Sigstrom Dr.
Carson City, NV 89706
(775) 887-5700
www.sierrakd.com

Thin-Film Technology Corp.
1980 Commerce Dr.
N. Mankato, MN 56003
(507) 625-8445
www.thin-Film.com

Vishay Intertechnology, Inc.
63 Lincoln Hwy.
Malvern, PA 19355
(610) 644-1300
www.vishay.com

Wilbrecht Electronics
1400 Energy Park Dr., #18
St. Paul, MN 55108
(651) 659-0919
www.wilbrecht.com

D.4 INSTRUMENTATION MANUFACTURERS

Agilent Technologies, Inc.
5301 Stevens Ck Blvd.
Santa Clara, CA 95051
(800) 452-4844
www.agilent.com

B &K Precision Corp.
22820 Savi Ranch Pkwy.
Yorba Linda, CA 92887
(714) 921-9095
www.bkprecision.com

Environment Associates, Inc.
9604 Variel Ave.
Chatsworth, CA 91311
(800) 354-1522
www.eatest.com

Espec Corp.
425 Gordon Indl. Ct. SW
Grand Rapids, MI 49509
(800) 537-7320
www.espec.com

Fluke Corp.
6920 Seaway Blvd.
Everett, WA 98204
(800) 443-5853
www.fluke.com

Keithley Instruments Co.
28775 Aurora Rd.
Cleveland, OH 44139
(440) 248-0400
(800) 552-1115
www.keithley.com

Kepco, Inc
131-38 Sanford Ave.
Flushing, NY 11352
(718) 461-7000
www.kepcopower.com

Le Croy Corp.
700 Chestnut Ridge Rd.
Chestnut Ridge, NY 10977
(845) 425-2000
www.lecroy.com

National Instruments, Inc.
11500 N. Mopac Expwy, Bldg. B
Austin, TX 78759
(888) 280-7645
www.ni.com

Omega Engineering, Inc.
One Omega Dr.
Stamford, CT 06907
(888) TC-OMEGA (826-6342)
www.omega.com

Rohde & Schwarz, Inc.
8661-A Robert Fulton Dr.
Columbia, MD 21046
(888) 837-8772
www.rohde-schwarz.com

Stanford Research Systems, Inc.
1290-D Reamwood Rd.
Sunnyvale, CA 94089
(408) 744-9040
www.thinksrs.com

Tektronix Inc.
Howard Vollum Ind. Pk.
Beaverton, OR 97077
(800) 835-9433
www.tek.com

Test Equipment Connection Corp.
30 Skyline Dr.
Lake Mary, FL 32746
(800) 615-8378
www.testequipmentconnection.com

D.5 Magazines & Periodicals

Compliance Engineering
www.ce-mag.com

ECN Magazine
www.ecnmag.com

EDN Magazine
www.edn.com

EE Times
www.eet.com

Electronic Design
www.elecdesign.com

Electronic News
www.electronicnews.com

Electronic Products
www.electronicproducts.com

Evaluation Engineering
www.evaluationengineering.com

NASA Tech Briefs
www.techbriefs.com

Power Electronics Technology
www.powerelectronics.com

RF Design Magazine
www.rfdesign.com

Semiconductor International
www.semiconductor.net

Sensors Magazine
www.sensorsmag.com

Test & Measurement World
www.tmworld.com

Index

D

E

F

M

N

O

P

T

W

X

Y

Z

Printed and bound by CPI Group (UK) Ltd, Croydon, CR0 4YY
03/10/2024
01040336-0011